Modern Applications of Engineering Physics

Modern Applications of Engineering Physics

Editor: Kate Fellows

NY RESEARCH PRESS

New York

Published by NY Research Press
118-35 Queens Blvd., Suite 400,
Forest Hills, NY 11375, USA
www.nyresearchpress.com

Modern Applications of Engineering Physics
Edited by Kate Fellows

International Standard Book Number: 978-1-63238-605-2 (Hardback)

Cataloging-in-Publication Data

Modern applications of engineering physics / edited by Kate Fellows.
    p. cm.
Includes bibliographical references and index.
ISBN 978-1-63238-605-2
1. Physics. 2. Engineering. I. Fellows, Kate.
QC23.2 .M63 2018
530--dc23

Contents

Preface

The purpose of the book is to provide a glimpse into the dynamics and to present opinions and studies of some of the scientists engaged in the development of new ideas in the field from very different standpoints. This book will prove useful to students and researchers owing to its high content quality.

Engineering physics is a combination of physics, mathematics and the principles of engineering. The subject uses the classical and modern concepts of physics for improved technological developments. Some of the branches of engineering physics are biomechanics, cryogenics, digital electronics, nuclear engineering, systems engineering, solid-state physics, energy engineering, etc. The topics included in this book on engineering physics are of utmost significance and are bound to provide incredible insights to readers. With state-of-the-art inputs by acclaimed experts of this field, this book targets students and professionals alike.

At the end, I would like to appreciate all the efforts made by the authors in completing their chapters professionally. I express my deepest gratitude to all of them for contributing to this book by sharing their valuable works. A special thanks to my family and friends for their constant support in this journey.

Editor

Surface-enabled propulsion and control of colloidal microwheels

T.O. Tasci[1], P.S. Herson[2,3], K.B. Neeves[1,4] & D.W.M. Marr[1]

Propulsion at the microscale requires unique strategies such as the undulating or rotating filaments that microorganisms have evolved to swim. These features however can be difficult to artificially replicate and control, limiting the ability to actuate and direct engineered microdevices to targeted locations within practical timeframes. An alternative propulsion strategy to swimming is rolling. Here we report that low-strength magnetic fields can reversibly assemble wheel-shaped devices *in situ* from individual colloidal building blocks and also drive, rotate and direct them along surfaces at velocities faster than most other microscale propulsion schemes. By varying spin frequency and angle relative to the surface, we demonstrate that microwheels can be directed rapidly and precisely along user-defined paths. Such *in situ* assembly of readily modified colloidal devices capable of targeted movements provides a practical transport and delivery tool for microscale applications, especially those in complex or tortuous geometries.

[1] Chemical and Biological Engineering Department, Colorado School of Mines, Golden, Colorado 80401, USA. [2] Department of Anesthesiology, University of Colorado, Denver, Colorado 80045, USA. [3] Department of Pharmacology, University of Colorado, Denver, Colorado 80045, USA. [4] Department of Pediatrics, University of Colorado, Denver, Colorado 80045, USA. Correspondence and requests for materials should be addressed to D.M. (email: dmarr@mines.edu).

At the microscale, fluid dynamics are unique because viscous forces dominate over inertial forces, a condition typically characterized by Reynolds numbers (Re) less than unity. Because propulsion schemes that rely on inertial forces cannot be used, translation requires approaches adapted to overcome the inherent reversibility of low-Re flows by breaking symmetry[1–3]. For example, microorganisms use undulating or rotating flagella and cilia for motility. In the push for technological devices small enough to move through microscale channels (10–100 μm) over macroscale distances (>1 cm), such as those found in human vasculature, there is appreciable effort in developing equivalent artificial approaches. Propulsion schemes based on catalytic methods[4–6] or on cellular machinery analogues[7–9] have shown good progress; however, significant challenges remain. Though catalytic swimmers can reach high speeds[10], they require available solute fuel for propulsion and concentration gradients for direction[11]. While top–down fabrication of flagella has led to speeds comparable to some microorganisms, they cannot be reconfigured for applications in dynamic or varying environments.

Colloidal assembly is a promising alternative. These methods provide bottom-up fabrication using simple colloidal building blocks as components of microstructures that are rapidly and reversibly assembled into a variety of sizes and shapes. Fabrication is initiated via specific[12,13] or non-specific[14] interactions or supplemented with applied electrical[15], optical[16] or magnetic fields[17] to enable switching and direct control of size, structure and function[18]. Used with superparamagnetic colloids, magnetic fields are well-suited to assemble structures *in situ* that are easily manipulated and rapidly disassembled after use. Fields of only a few milli-Tesla (mT) create sufficient dipole strength to induce colloidal assembly. Static applied fields align particles into chains, while rotating fields create net isotropic interactions that can lead to compact aggregates[19]. In our studies, we use superparamagnetic colloids and balance magnetic and viscous forces with appropriate field strengths and rotational frequencies to create reversible close-packed assemblies that subsequently spin due to their net dipole interacting with the dynamic applied field[20,21]. While rotating magnetic fields can construct microwheels and create a driving torque, the reversible nature of low-Re flows dictates that spinning symmetric objects suspended in fluid do not translate. For net movement to occur, symmetry, either in the device or in the surrounding geometry, must be broken. In an approach particularly appropriate for microenvironments where surface to volume ratios are high and surfaces are plentiful, one way to break the symmetry is with a nearby wall.

Here, we show that rotating magnetic fields can be used to assemble and spin microwheels that, when canted relative to the surface, roll smoothly and quickly with a high degree of directional control. We demonstrate this propulsion mechanism with microwheels composed of 1, 2, 3, 7 and even 19 paramagnetic colloidal particles with translation speeds >100 μm s[−1]. These results demonstrate a rapid and reversible microdevice assembly and powering method that overcomes many of the limitations inherent in biomimetic artificial micropopulsion strategies.

Superparamagnetic beads assemble into microwheels by isotropic interactions induced by the in-plane rotating magnetic field (Fig. 1a,b) with wheel size controlled by local bead density. Spinning microwheels lying flat on a surface have no net motion. For translation to occur they must be inclined relative to the surface; therefore, to propel microwheels, we introduce a normal component to the magnetic field to orient the field rotation axis towards the surface plane (Supplementary Movies 1 and 2). With addition of a field in the z direction, $B_z = B_{z0} \cos(\omega_f t - \phi_z)$, both symmetric and asymmetric microwheels reorient off the surface to a defined camber angle, θ_c (Fig. 1c,d), and begin to translate. Apparent in this approach is the similarity of microwheels to rolling tires where friction with the road, combined with tire rotation, propels wheels forward. One difference between microwheels and tires is that the camber angle can vary from lying flat, $\theta_c = 90°$, and spinning without translation to fully

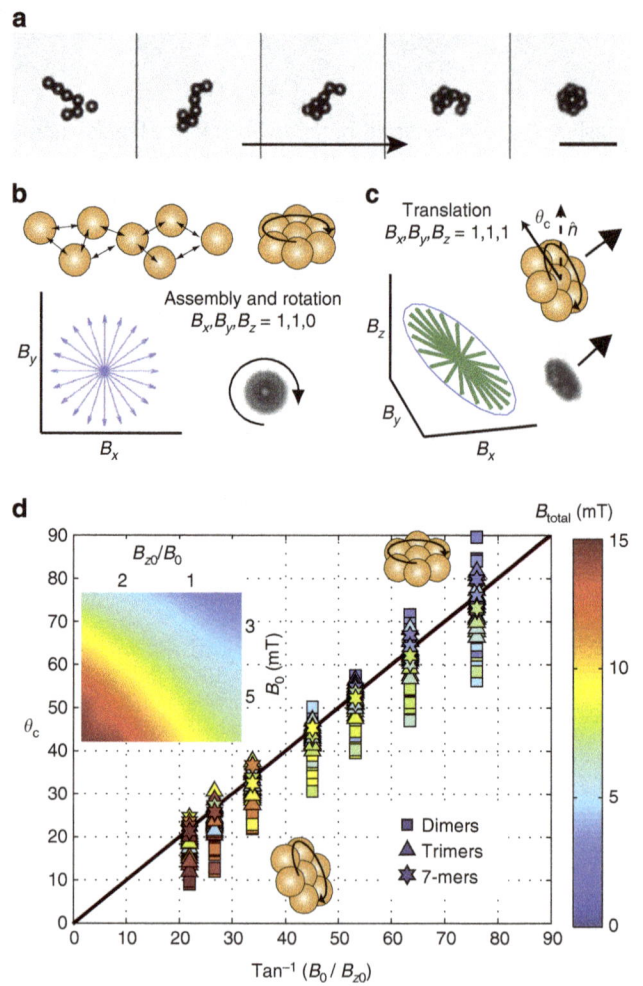

Figure 1 | Field-induced assembly and rotation. (a,b) With application of the rotating magnetic field $B_x + B_y$ in the surface plane, colloids assemble via isotropic interactions and 'sit and spin' (scale bar, 20 μm). **(c)** With addition of a normal variable-phase component (B_z), the field rotation axis is oriented towards the surface plane, wheels 'stand up' at a camber angle, θ_c, and roll along the surface. **(d)** θ_c measured during wheel translation as a function of the applied field rotation axis set via $\tan^{-1}(B_0/B_{z0})$. Data points ($n = 368$) are coloured by the magnetic field magnitude $B_{total} = (B_0^2 + B_{z0}^2)^{1/2}$ from low (blue) to red (high). The black line corresponds to perfect alignment between microwheel and field rotation axis. The inset identifies total field magnitudes as field ratios are varied.

Results

Microwheel assembly and translation mechanism. We begin by setting the in-plane field magnitudes equal as $B_{x0} = B_{y0} = B_0$ varied sinusoidally to create a circularly rotating magnetic field in the xy plane: $B_x = B_0 \cos(\omega_f t - \phi_x)$ and $B_y = B_0 \cos(\omega_f t - \phi_y)$ where $\omega_f = 2\pi f$ is the field angular frequency, f the field frequency, and ϕ_x and ϕ_y are phase angles with $\phi_x - \phi_y = \pi/2$.

upright, $\theta_c = 0°$, and rolling (Fig. 1b,c). Important parameters influencing the rolling velocity V include the number of particles comprising the microwheel n, its angular frequency ω, and, as wheels rotate, the outer circumferential velocity $V_\omega > \omega R$.

For tires, the rolling velocity, V, and circumferential velocity, V_ω, are almost equal, a fact used in an automobile speedometer to gauge speed and distance travelled. Microwheels however spin much faster than they roll, suggesting a significant fluid layer between the wheel and wall that warrants an approach based on wet friction rather than dry friction. To clarify the mechanism of rolling velocity as a function of angular frequency, we use a force balance in the normal and rotational directions (Fig. 2). First, a sum of forces in the normal, z direction:

$$\sum F_z = L - N = L - g \cdot M \cos(\theta_c) = 0, \quad (1)$$

where F_z is the force in the z direction, N is the normal force, g is the gravitational constant, M is the mass, and the L, is the load. Solving for the load gives

$$L = g \cdot M \cos(\theta_c) = g \cdot n \cdot m \cos(\theta_c), \quad (2)$$

with n the number of particles in a wheel and m the buoyant mass of a single particle. For the rolling, x direction, we have

$$\sum F_x = F_f - F_d = 0, \quad (3)$$

where F_f is the friction force and F_d is the drag force. Since circumferential velocity is greater than translation velocity, $V_\omega > V$, the drag and friction forces oppose one another. We use

$$F_f = \mu_k L, \quad (4)$$

with μ_k as the wet friction coefficient, which is referred to variously as the fluid friction, viscous friction, or the hydrodynamic lubrication region in the context of load and friction in bearings. Petroff first examined this situation and found that the wet friction coefficient is proportional to velocity after balancing frictional torque with that torque required to shear the intervening fluid layer[22]. Following his approach, we begin by approximating the fluid shear stress as

$$\tau = \eta(\partial v / \partial z) \sim \eta(V^*/h), \quad (5)$$

where η is the dynamic viscosity of the fluid, v is the velocity vector, V^* is the wheel edge velocity and h is the gap between wheel and surface. This shear stress corresponds to a torque, T,

$$T = \tau A \cdot R, \quad (6)$$

with A as the contact area and R as the wheel radius. Recognizing that the frictional torque, T_f, is

$$T_f = F_f \cdot R = \mu_k L \cdot R, \quad (7)$$

and equating the two torques leads to

$$\mu_k = (\eta A / hL) V^* = (\eta / hP) V^* = \mu^* V^*, \quad (8)$$

with pressure $P = L/A$ and defining $\mu^* \equiv \eta/(hP)$. With V^* the fluid velocity between the wheel and wall $= V_\omega - V$ and approximating $V_\omega \gg V$, we calculate the friction force as

$$F_f = \mu_k L = \mu^* V_\omega L = \mu^* \omega R \cdot L = \mu^* \omega R \cdot n \cdot mg \cos(\theta_c) \quad (9)$$

To approximate drag force during translation, we neglect the presence of the wall and employ low Re results for the edge-wise drag on a disk[23]

$$F_d = 32\eta R V / 3. \quad (10)$$

Using this, we equate the drag force and frictional force and solve for the rolling velocity

$$V = (3mg/32\eta)\mu^* \cdot \omega \cdot n \cdot \cos(\theta_c). \quad (11)$$

We define a weighted angular frequency, ω^*, as

$$\omega^* \equiv \omega \cdot n \cdot \cos(\theta_c), \quad (12)$$

and then can write the rolling velocity as

$$V = (3mg/32\eta)\mu^* \cdot \omega^*. \quad (13)$$

Microwheel rolling. Figure 3a shows the rolling velocity, V, as a function of a weighted angular frequency, ω^*, that accounts for microwheel size and angular frequency. The scaling argument presented above predicts a slope of $(3mg/32\eta)\mu^*$ on plots of V versus ω^*. The data show good agreement with this scaling for weighted angular frequencies below 200 rad s^{-1}. Above this the scaling deviates from the data, likely because the assumption that the wheel velocity is much greater than the fluid velocity is no longer valid. Wheels roll along the surface at speeds of up to $90 \mu\text{m s}^{-1}$ with applied field frequencies up to 50 Hz over the range $0 < \theta_c < 90°$. Even single particles roll as long as the surface-parallel component of the rotational axis is non-zero. Wheels composed of 2, 3, 7 and even 19 particles, though not strictly round, exhibit smooth motion as they rapidly spin and translate across flat surfaces for values of $B_{z0}/B_0 < 2.5$ (Fig. 3b; Supplementary Movies 3–6); at higher values motion becomes unstable. Velocities $> 120 \mu\text{m s}^{-1}$ were achieved with dimers and 19-mers at higher amplitude fields, but observation times were limited due to the size of the field-of-view.

As the data condense to a single line at low to moderate ω in Fig. 3, we can conclude that friction is not stick-slip and wheel speeds increase with wheel size for a given angular frequency due both to the increased load and increased fluid velocities near the wall. A useful feature of this approach is that different sized microwheels, and thus different speeds for a given field rotation frequency, can be assembled from the same building blocks by changing the bulk colloid concentration. In fact, wheels composed of particle numbers other than those studied in Fig. 3 do roll as well; however, structural isomers in these systems make quantification difficult.

Directional control of microwheels. Targeting applications require not only microwheel propulsion but also the ability to direct them to desired locations. Unlike tires, microwheels can be oriented at very high θ_c and, as a result, can experience significant lateral forces and side slip. Defining the side slip angle, θ_s, as the difference between the rolling direction (heading) and the wheel rotation plane (pointing) directions we observe that lateral forces push wheels towards the wheel rotational axis (Fig. 4). The influence on side slip on heading is best illustrated by an example; Supplementary Movie 7 shows two wheels programmed to make a circular motion, where one wheel has a camber angle that causes it to lean inward towards the centre of the circle and another has a equal and opposite camber angle that causes it to

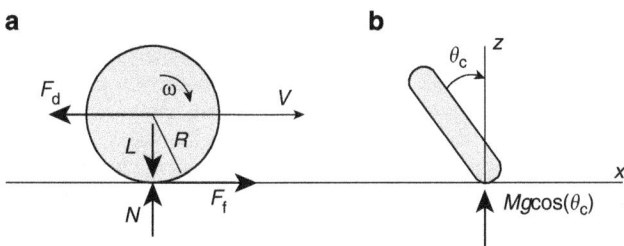

Figure 2 | Scaling analysis. (a) Side view and **(b)** front view of translating microwheel modelled as a disk. The important parameters include F_d, drag force; F_f, friction force; L, load; N, normal force from wall; M, mass of the wheel; θ_c, camber angle; ω, angular frequency; g, gravitational constant; and R, radius.

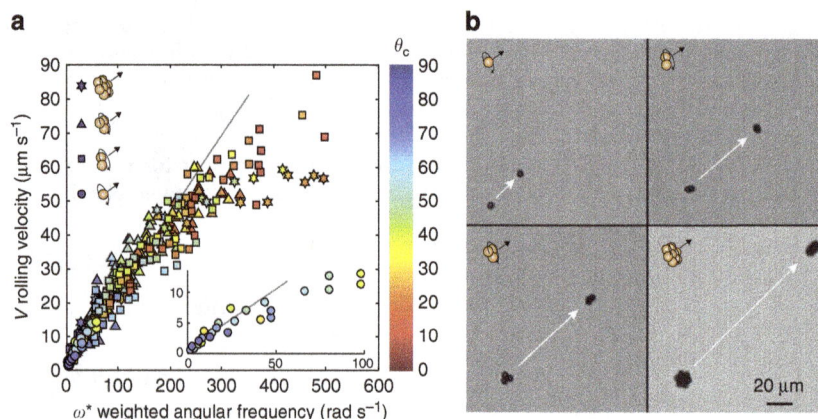

Figure 3 | Wheel rolling. (a) Rolling velocity for microwheels created from 1 ($n=25$), 2 ($n=168$), 3 ($n=140$) and 7 ($n=35$) colloidal particles as a function of weighted angular frequency, $\omega^* \equiv \omega \cdot n \cdot \cos(\theta_c)$. Data points coloured according to orientation with upright (red) to lying nearly flat (blue). Line indicates the slope ($3mg/32\eta$) based on wet friction scaling arguments. To compare results for spherical monomers with those for disk-like wheels, we use the drag force for spheres at low Re, $F_d = 6\pi\eta RV$ resulting in $V = (mg/6\pi\eta)\mu^* \cdot \omega \cdot n \cdot \cos(\theta_c)$. We therefore scale the monomer results by $32/(3 \cdot 6\pi) = 16/9\pi$ inset shows unscaled monomer data. (b) Three seconds of translation under identical field conditions demonstrate that larger wheels roll faster (Supplementary Movie 3). Supplementary Movies 4–6 show that increased angular frequency and lower camber angles lead to increased speeds as well.

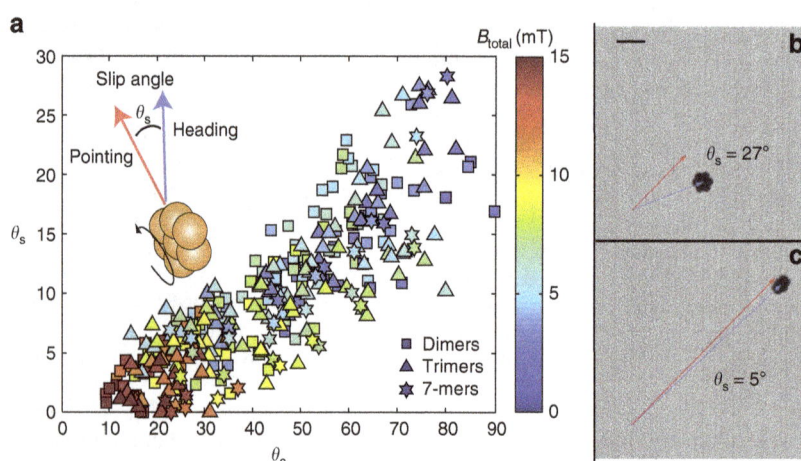

Figure 4 | Side slip. (a) As θ_c increases, heading and pointing directions separate as characterized by increasing wheel side slip angle θ_s ($n=343$). (b) $\theta_c = 76°$, $V = 14\,\mu m\,s^{-1}$ (scale bar, 20 μm), (c) $\theta_c = 28°$, $V = 39\,\mu m\,s^{-1}$.

lean outward. The wheel that leans inward gives a tighter circle compared to the wheel that leans outward.

The microwheel heading angle θ_d (the angle between the heading vector and the positive x axis) is related to the phase angle of the fields and the camber angle by $\theta_d = (\phi_y - \phi_x) + (\phi_z - \phi_x) + \theta_s = \phi_z - \phi_x + \theta_s - \pi/2$. We can use the heading to redirect microwheels with a simple phase shift in the z-field. As a result, speed and heading changes can be actuated immediately in preprogrammed patterns (Fig. 5a; Supplementary Movie 8). Alternatively, the direction of microwheels can be manually manipulated by keyboard controls (Fig. 5b; Supplementary Movie 9). This manual control was used here to assemble microwheels of different sizes, for example the 7-mer in Fig. 5c (Supplementary Movie 10). Note that, during assembly of the 7-mer, asymmetric wheels also translate.

Discussion

With velocity and directional control, microwheels have potential application for cargo transport in complex or tortuous networks of channels, as in the vasculature and within microfluidic devices where surfaces are ubiquitous. Reversible assembly of

microwheels *in situ* has significant advantage over pre-fabricated devices as individual particles are injectable and small enough to pass through blood capillaries or microfluidic channels. Once the applied magnetic field is turned off, wheels disassemble into micron-sized components removable via the body's natural defence mechanisms or filtered out in microfluidic platforms. Here and with a bulk field for assembly and powering, microwheels can be fabricated with mass parallelization and function on walls of high curvature because of their ability to roll at high camber angles. The enhanced mobility is particularly useful in medicine where similar superparamagnetic colloids are used for enhanced imaging or modified with surface or embedded moieties for applications such as drug delivery[24,25] or targeted hyperthermia[26]. Magnetic bead-based methods are also used in microfluidic platforms for immunoassays, DNA hybridization, surface patterning and magnetic mixing, sorting and separations[27]. The magnitude of the magnetic fields we employ is low (typically 2–15 mT), making the approach feasible for use *in vivo* or within distributed microdevices significant distances away from field magnets. Equivalent translation rates by magnetophoresis alone would require extremely high field

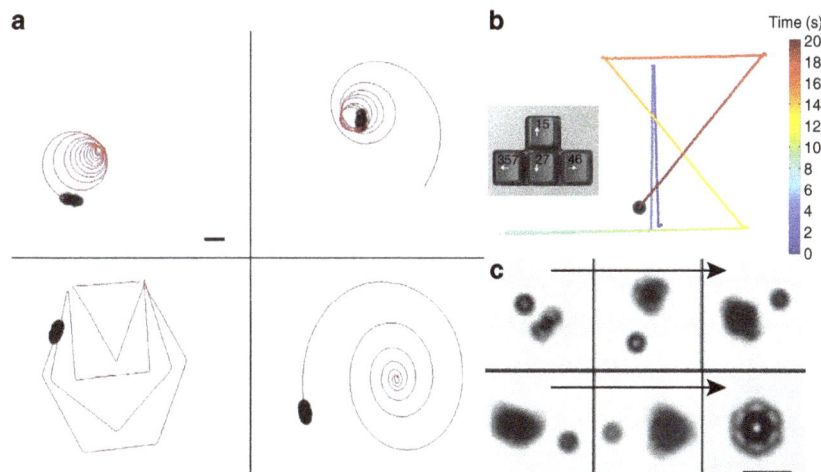

Figure 5 | Directional control. Supplementary movies demonstrate (**a**) automated patterns (Supplementary Movie 8; scale bars, 10 μm), (**b**) manual control (Supplementary Movie 9) and (**c**) stepwise microwheel assembly (Supplementary Movie 10).

gradients[28] and field magnitudes hundreds of times higher than required for rolling.

We note that other approaches use varying magnetic fields to induce translation of small clusters along surfaces[29–31]. For example, asymmetric colloidal dimers have been shown to translate when subjected to a rotating field; however, because of the nature of the field which requires both a.c. and d.c. components and need for cluster asymmetry, velocities are relatively slow ($\sim 10\,\mu m\,s^{-1}$). In these experiments, the underlying translation mechanism relies on asymmetric dissipation with the wall resulting in periodic but irregular motion. Importantly, symmetric systems such as monomers do not translate. Here, we exploit a translation mechanism that relies on wet friction using *in situ*-assembled symmetric and asymmetric colloidal wheels to provide smooth motion along surfaces with user-defined directionality and instantaneous reorientation. These features are a direct result of the unique wheel-like structures that provide not only high speeds and directional control but can be immediately dis- and re-assembled unlike non-reconfigurable microstructures propelled by magnetic fields[29,30].

We show that angular frequency, camber angle, and wheel size all drive wheel motion with upper limits to rolling velocity bounded by ωR, parameters whose values can be increased with larger field magnitudes. While we present data of velocities approaching $100\,\mu m\,s^{-1}$, higher fields, which can be maintained only briefly in our current experimental setup, do lead to velocities $>100\,\mu m\,s^{-1}$, faster than microfabricated flagella and related techniques[8,32,33]. While also higher than most microorganisms including *Escherichia coli* at $20–45\,\mu m\,s^{-1}$ (refs 34,35), a number of bacteria have top speeds an order-of-magnitude greater[35] showing both the effectiveness with which natural flagella can function and how far microfabricated approaches currently lag behind Nature. While there clearly remains work to challenge the quickest microorganisms, our wheel-based rolling may prove a faster, simpler and better controlled microdevice propulsion strategy than swimming in microscale environments where abundant surfaces are available.

Methods

Systems studied. We use superparamagnetic Dynabeads M-450 Epoxy (Life Technologies Corporation, Carlsbad, CA, USA) of diameter 4.5 μm, density $= 1.6\,g\,cm^{-3}$ suspended in 1% sodium dodecyl sulfate (Sigma-Aldrich) solution. To create a sample chamber, double-sided tape with an 8 mm diameter hole was placed on a glass slide and, after sample was added, sealed with a

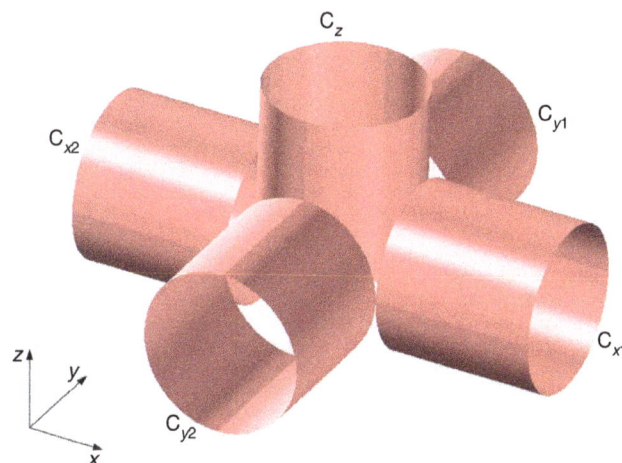

Figure 6 | Experimental setup. Magnetic field system consisting of five air-cored solenoid coils.

coverslip. The chamber was then positioned at the centre of the magnetic field system and wheel motion recorded with a microscope (Zeiss Axioplan 2 Imaging) and high-speed camera (Epix SV 643M) operated at 400 frames per s.

Magnetic field generation. External magnetic fields were generated with five identical air-cored copper solenoid coils of 50 mm inner diameter, 51 mm length, 400 turns and current capacity 3.5 A (Fig. 6). The field generated at the centre of the experimental setup has three components $B_x = B_{x0}\cos(\omega_f t)$; $B_y = B_{y0}\cos(\omega_f t - \pi/2)$ and $B_z = B_{z0}\cos(\omega_f t - \phi_z)$. B_x was generated by coils C_{x1} and C_{x2}, B_y was created by C_{y1} and C_{y2}; B_z was generated by coil C_z.

Sinusoidal voltage waveforms were generated using Matlab (Mathworks, Inc., Natick, MA, USA) and an analog-output card (National Instruments, NI-9263) and then amplified (Behringer EP2000) before being applied to individual solenoids. To monitor coil currents, an analog input data acquisition card was used (National Instruments, NI-USB-6009). The resulting magnetic fields were estimated using a custom Matlab code solving the fields of the solenoids for a given current. Predictions of the code were validated by exciting the coils with constant currents and measuring the field with a gaussmeter (VGM Gaussmeter, Alphalab Inc.).

Data reduction and analysis. A total of 368 experiments were performed on microwheels composed of monomers, dimers, trimers and 7-mers. In the experiments, different conditions were tested by varying the magnetic flux densities from 2.1 to 15 mT and field frequencies from 10 to 80 Hz. Image processing code was written in Matlab to detect the microwheel assemblies and provide rolling velocity V, rotation frequency f and camber angle θ_c (Supplementary Movie 11). Rolling velocity was determined by dividing travelled distance by travel duration. Rotation frequencies were determined by recording pixel sum fluctuations caused

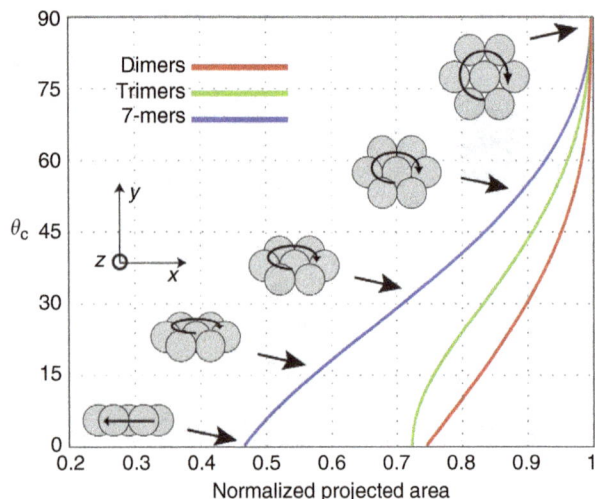

Figure 7 | Camber angle standard curves. Camber angle as a function of wheel projected area (viewed from above).

by microwheel rotation and taking the Fourier transform. Standard camber angle curves (Fig. 7) were analytically calculated for each microwheel assembly size (dimers, trimers and 7-mers) and used to determine θ_c for each experiment by measuring the mean projected area occupied by the translating microwheel.

Directional control of the microwheels was performed with custom Matlab code to vary the sinusoidal voltage applied to the C_z coil, create a phase shift of the B_z field $B_z = B_{z0} \cos(\omega_z t - \phi_z)$, and a net field rotation axis reorientation. This shift could be manually controlled with keyboard arrow keys, allowing one to easily direct microwheels to desired locations. With the same code, the path of the microwheels was programmed as shown in Fig. 5a.

Reynolds numbers. For our systems either a wheel or particle Re can be employed to demonstrate the dominance of viscous over inertial forces. Here, $Re = \rho VD/\mu$, where ρ and μ are fluid density and viscosity, V and D are the rolling velocity and diameter of the wheel. One can also consider a particle Reynolds number (Re_p) where we set V to the circumferential velocity V_ω and D to the single particle diameter instead. Typical conditions for the experiments presented here have $Re \sim 10^{-2}$ and $Re_p \sim 10^{-1}$; however, at the fastest speeds and largest wheels, Re can approach 0.20.

References

1. Taylor, G. Analysis of the swimming of microscopic organisms. *Proc. R. Soc. A* **209**, 447–461 (1951).
2. Purcell, E. M. Life at low Reynolds number. *Am. J. Phys.* **45**, 3–11 (1977).
3. Lauga, E. & Powers, T. R. The hydrodynamics of swimming microorganisms. *Rep. Prog. Phys.* **72**, 096601–096601 (2009).
4. Wang, Y. *et al.* Bipolar electrochemical mechanism for the propulsion of catalytic nanomotors in hydrogen peroxide solutions. *Langmuir* **22**, 10451–10456 (2006).
5. Ebbens, S. J. & Howse, J. R. In pursuit of propulsion at the nanoscale. *Soft Matter* **6**, 726–726 (2010).
6. Baraban, L. *et al.* Catalytic Janus motors on microfluidic chip: deterministic motion for targeted cargo delivery. *ACS Nano* **6**, 3383–3389 (2012).
7. Dreyfus, R. *et al.* Microscopic artificial swimmers. *Nature* **437**, 862–865 (2005).
8. Ghosh, A. & Fischer, P. Controlled propulsion of artificial magnetic nanostructured propellers. *Nano Lett.* **9**, 2243–2245 (2009).
9. Zhang, L. *et al.* Artificial bacterial flagella: fabrication and magnetic control. *Appl. Phys. Lett.* **94**, 064107 (2009).
10. Solovev, A. A., Mei, Y., Bermúdez Ureña, E., Huang, G. & Schmidt, O. G. Catalytic microtubular jet engines self-propelled by accumulated gas bubbles. *Small* **5**, 1688–1692 (2009).
11. Howse, J. R. *et al.* Self-motile colloidal particles: from directed propulsion to random walk. *Phys. Rev. Lett.* **99**, 048102 (2007).
12. Glotzer, S. C. & Solomon, M. J. Anisotropy of building blocks and their assembly into complex structures. *Nat. Mater.* **6**, 557–562 (2007).
13. Sacanna, S., Irvine, W. T. M., Chaikin, P. M. & Pine, D. J. Lock and key colloids. *Nature* **464**, 575–578 (2010).
14. Lin, M. Y. *et al.* Universality in colloid aggregation. *Nature* **339**, 360–362 (1989).
15. Prieve, D. C., Sides, P. J. & Wirth, C. L. 2-D assembly of colloidal particles on a planar electrode. *Curr. Opin. Colloid Int. Sci.* **15**, 160–174 (2010).
16. Terray, A., Oakey, J. & Marr, D. W. M. Microfluidic control using colloidal devices. *Science* **296**, 1841–1844 (2002).
17. Sawetzki, T., Rahmouni, S., Bechinger, C. & Marr, D. W. M. *In situ* assembly of linked geometrically coupled microdevices. *Proc. Natl Acad. Sci. USA* **105**, 20141 (2008).
18. Bharti, B. & Velev, O. D. Assembly of reconfigurable colloidal structures by multidirectional field-induced interactions. *Langmuir* **31**, 7897–7908 (2015).
19. Fermigier, M. & Gast, A. P. Structure evolution in a paramagnetic latex suspension. *J. Magn. Magn. Mater.* **122**, 46–50 (1993).
20. Biswal, S. & Gast, A. Rotational dynamics of semiflexible paramagnetic particle chains. *Phys. Rev. E* **69**, 041406 (2004).
21. Melle, S., Calderón, O. G., Rubio, M. A. & Fuller, G. G. Microstructure evolution in magnetorheological suspensions governed by mason number. *Phys. Rev. E* **68**, 11 (2003).
22. Szeri, A. Z. *Fluid Film Lubrication* (Cambridge Univ Press, 2011).
23. Tanzosh, J. P. & Stone, H. A. Transverse motion of a disk through a rotating viscous fluid. *J. Fluid Mech.* **301**, 295–324 (1995).
24. Barratt, G. Colloidal drug carriers: achievements and perspectives. *Cell. Mol. Life Sci.* **60**, 21–37 (2002).
25. Pouponneau, P., Leroux, J.-C., Soulez, G., Gaboury, L. & Martel, S. Co-encapsulation of magnetic nanoparticles and doxorubicin into biodegradable microcarriers for deep tissue targeting by vascular MRI navigation. *Biomaterials* **32**, 3481–3486 (2011).
26. Jordan, A., Scholz, R., Wust, P., Fähling, H. & Felix, R. Magnetic fluid hyperthermia (MFH): cancer treatment with AC magnetic field induced excitation of biocompatible superparamagnetic nanoparticles. *J. Magn. Magn. Mater.* **201**, 413–419 (1999).
27. Häfeli, U., Schütt, W., Teller, J. & Zborowski, M. *Scientific and Clinical Applications of Magnetic Carriers* (Plenum Press, 1997).
28. Abbott, J. J. *et al.* How should microrobots swim? *Int. J. Robot. Res.* **28**, 1434–1447 (2009).
29. Tierno, P., Golestanian, R., Pagonabarraga, I. & Sagués, F. Magnetically actuated colloidal microswimmers. *J. Phys. Chem. B* **112**, 16525–16528 (2008).
30. Tierno, P., Golestanian, R., Pagonabarraga, I. & Sagués, F. Controlled swimming in confined fluids of magnetically actuated colloidal rotors. *Phys. Rev. Lett.* **101**, 4 (2008).
31. Tierno, P., Güell, O., Sagués, F., Golestanian, R. & Pagonabarraga, I. Controlled propulsion in viscous fluids of magnetically actuated colloidal doublets. *Phys. Rev. E* **81**, 011402 (2010).
32. Pak, O. S., Gao, W., Wang, J. & Lauga, E. High-speed propulsion of flexible nanowire motors: theory and experiments. *Soft Matter* **7**, 8169–8181 (2011).
33. Servant, A., Qiu, F., Mazza, M., Kostarelos, K. & Nelson, B. J. Controlled *in vivo* swimming of a swarm of bacteria-like microrobotic flagella. *Adv. Mater.* **27**, 2981–2988 (2015).
34. Berg, H. C. & Brown, D. A. Chemotaxis in *Escherichia coli* analysed by three-dimensional tracking. *Nature* **239**, 500–504 (1972).
35. Herzog, B. & Wirth, R. Swimming behavior of selected species of Archaea. *Appl. Environ. Microb.* **78**, 1670–1674 (2012).

Acknowledgements

We acknowledge support from the NSF (CAREER, CBET-1351672, KN) and NIH (R21NS082933) and thank N. Wu for helpful discussions.

Author contributions

T.O.T. built, performed and designed the experiments and wrote the custom codes. K.B.N., D.W.M.M. and P.S.H. conceived the project. T.O.T., K.B.N. and D.W.M.M. analysed the experimental results and wrote the manuscript.

Additional information

2

Lasing in silicon–organic hybrid waveguides

Dietmar Korn[1,*], Matthias Lauermann[1,*], Sebastian Koeber[1,2], Patrick Appel[1], Luca Alloatti[1], Robert Palmer[1], Pieter Dumon[3], Wolfgang Freude[1,2], Juerg Leuthold[1,2,†] & Christian Koos[1,2]

Silicon photonics enables large-scale photonic–electronic integration by leveraging highly developed fabrication processes from the microelectronics industry. However, while a rich portfolio of devices has already been demonstrated on the silicon platform, on-chip light sources still remain a key challenge since the indirect bandgap of the material inhibits efficient photon emission and thus impedes lasing. Here we demonstrate a class of infrared lasers that can be fabricated on the silicon-on-insulator (SOI) integration platform. The lasers are based on the silicon–organic hybrid (SOH) integration concept and combine nanophotonic SOI waveguides with dye-doped organic cladding materials that provide optical gain. We demonstrate pulsed room-temperature lasing with on-chip peak output powers of up to 1.1 W at a wavelength of 1,310 nm. The SOH approach enables efficient mass-production of silicon photonic light sources emitting in the near infrared and offers the possibility of tuning the emission wavelength over a wide range by proper choice of dye materials and resonator geometry.

[1] Institute of Photonics and Quantum Electronics (IPQ), Karlsruhe Institute of Technology (KIT), 76131 Karlsruhe, Germany. [2] Institute of Microstructure Technology (IMT), Karlsruhe Institute of Technology (KIT), 76344 Eggenstein-Leopoldshafen, Germany. [3] Department of Information Technology, IMEC, 9000 Gent, Belgium. * These authors contributed equally to this work. † Present address: Laboratory for Electromagnetic Fields and Microwave Electronics (IFH), Swiss Federal Institute of Technology (ETH), Zürich 8092, Switzerland. Correspondence and requests for materials should be addressed to C.K. (email: christian.koos@kit.edu).

Silicon photonics allows fabrication of nanophotonic devices using commercial CMOS facilities and is therefore a highly attractive platform for large-scale photonic integration[1,2]. However, while a wide variety of silicon-based optical and electro-optical devices has been demonstrated over the last years[3], efficient on-chip light sources still represent a challenge[4] due to the indirect bandgap of silicon. Previously reported all-silicon light sources rely on stimulated Raman scattering as a gain mechanism[5–8]. One drawback of these schemes is that they require coupling of external pump lasers to single-mode on-chip waveguides. In early demonstrations, Raman laser cavities were formed by incorporating nanophotonic silicon waveguides into fibre-based off-chip laser cavities to enable synchronization of the cavity round-trip time with a pulsed pump[5,6]. These devices cannot be miniaturized. In contrast to that, silicon photonic Raman lasers with on-chip cavities are compact and enable continuous-wave (CW) lasing[7,8], but require either strong pump lasers in combination with reverse-biased p-i-n-junctions or provide only limited output power in the microwatt range. Hybrid approaches, in which silicon is combined with direct-bandgap III–V compound semiconductors, allow for electrically pumped amplifiers[9] and lasers[10,11], but fabrication requires sophisticated and technologically challenging die-to-wafer bonding processes or advanced technology for the direct growth of III–V quantum dots[12] on silicon. Regarding monolithic integration of light sources on silicon, an electrically pumped CW germanium-on-silicon laser has been demonstrated by using a combination of tensile strain and n-doping of the germanium to enable direct-bandgap transitions in thin germanium layers that are grown on silicon substrates[13]. More recently, lasing has been shown without introducing mechanical strain by using a germanium–tin alloy[14] on silicon. However, fabrication of such devices requires advanced crystal growth techniques and technologically challenging fabrication processes. As an alternative, combinations of erbium-doped active cladding materials and SOI waveguides have been proposed[15,16] and experimentally investigated[17,18]. However, erbium features a rather small emission cross section and hence small gain. As a consequence, lasing in integrated erbium-clad devices has so far only been demonstrated for low-loss silicon nitride waveguides[19], but not for high-index-contrast SOI waveguides. Regarding

peak output power, even the most outstanding on-chip silicon-based lasers are currently limited to ~100 mW or less[12,20,21].

In this work, we demonstrate that lasing can be achieved by combining standard silicon-on-insulator (SOI) waveguides with dye-doped organic cladding materials. This concept of silicon–organic hybrid (SOH) integration is particularly well-suited for flexible and low-cost mass-production of silicon photonic light sources emitting in the near infrared. In a proof-of-principle experiment, we demonstrate pulsed lasing at room temperature with peak output powers of up to 1.1 W at a wavelength of 1,310 nm. Gain is provided by a near-infrared dye that was previously demonstrated to enable lasing in plastic waveguides[22]. More general, exploiting the virtually unlimited variety of organic optical cladding materials, SOH integration allows to complement silicon photonics with novel functionalities while still preserving the strengths of highly standardized CMOS processing[23]. Our proof-of-principle demonstration of SOH light sources complements recent work on SOH integration, comprising high-speed all-optical signal processing[24], broadband electro-optic modulators[25,26] and highly efficient low-power phase shifters[27].

Results

Concept and fabrication of SOH lasers. The basic idea of an SOH laser is illustrated in Fig. 1a. The devices consist of SOI waveguides, which are terminated at both ends with Bragg reflectors[28] and which are covered by a fluorescent organic cladding material suitable for stimulated emission when optically pumped. For efficient light emission, the interaction of the guided optical mode with the active cladding must be maximized. This can be accomplished by using a narrow silicon strip waveguide, for which a large fraction of the guided mode reaches into the cladding, Fig. 1b. Alternatively, a slot waveguide can be used, which consists of two closely spaced silicon rails, Fig. 1c. In both cases, the dominant horizontal electric field component (E_x) of the optical quasi-TE mode experiences strong field discontinuities at the high-index-contrast sidewalls. For the slot waveguide, this leads to an especially pronounced field enhancement within the slot[29], and hence to a strong interaction with the active cladding.

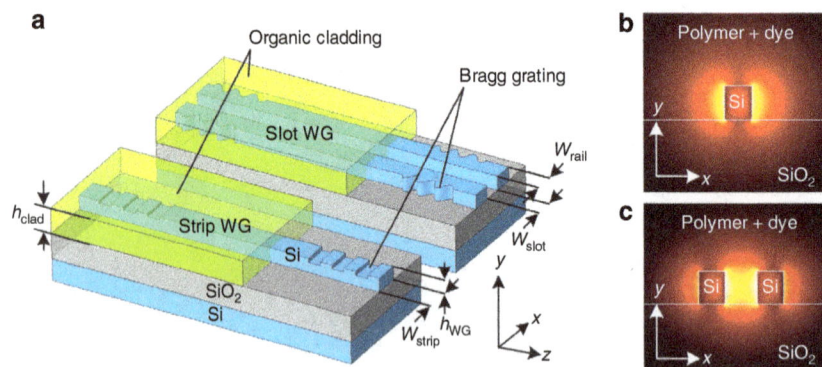

Figure 1 | SOH laser concept. (a) Light is guided by SOI strip or slot waveguides consisting of thin silicon nanowires (width $w_{strip} \approx 150\text{-}500$ nm, height $h_{WG} \approx 200\text{-}350$ nm) that are optically isolated from the silicon substrate by a thick oxide ($h_{SiO2} \approx 2$ µm). Optical gain is provided by a fluorescent organic cladding material (thickness $h_{clad} \approx 500 \pm 50$ nm), which entirely covers the strip or fills the slot ($w_{rail} \approx 100\text{-}200$ nm, $w_{slot} \approx 50\text{-}200$ nm). The optical pump is either launched from above or injected into the waveguide at one of the facets. Bragg reflectors can be used to provide wavelength-selective optical feedback. Interaction of the guided light with the active cladding is maximized by the design of the waveguides. **(b)** Dominant electric field component (E_x) of the fundamental quasi-TE mode for a narrow strip waveguide (colour coding: lighter colours for higher magnitude). A large fraction of the guided mode propagates in the cladding. **(c)** Dominant electric field component (E_x) of the fundamental quasi-TE mode for a slot waveguide consisting of two tightly spaced silicon rails. Discontinuities of the dominant horizontal electric field component lead to a strong field enhancement within the slot region and hence to a strong interaction with the active cladding.

We prove the viability of the concept by investigating a simple test structure. To this end, strip and slot waveguides of 4.8-mm length were fabricated using a state-of-the-art SOI CMOS-based process[30]. The waveguides are embedded into a solid active cladding consisting of a poly(methyl methacrylate; PMMA) matrix doped with 1 wt% of the commercially available dye IR26 (refs 22,31) having a maximum fluorescence at 1,150 nm, see Supplementary Fig. 1. The cladding is deposited in a single post-processing step using standard spin-coating techniques. Scanning electron microscope (SEM) images of coated and uncoated samples can be found in Supplementary Fig. 2, showing that the PMMA cladding fills the slot completely without forming any voids. To enable laser operation in a wide wavelength range, we omit the wavelength-selective Bragg reflectors shown in Fig. 1 and exploit spurious back-reflection from cleaved waveguide facets and from on-chip grating coupler (GC) structures, see Fig. 2a,b. For the cleaved facets, power reflection factors between 4 and 8% are estimated. Light emission from the cleaved facets is coupled to lensed standard single-mode fibres (SMF). For coarse alignment of the fibres, we use 1,550-nm light coupled to the SOH waveguide via the GC. The GC is optimized for operation at a wavelength of 1,550 nm and exhibit spurious back-reflection when operated at the laser emission wavelength of 1,310 nm. This reflection amounts to a few per cent and is comparable to that of the cleaved facet. A more detailed description of device fabrication can be found in the section 'Fabrication of SOH lasers' of the Methods.

Lasing could be demonstrated despite the comparatively low quality of the Fabry–Perot laser resonator, underlining the high potential of using dye-doped active claddings as gain media. In the experiment, the devices are pumped from above by a free-space line-focus beam using a pulsed laser with a wavelength of 1,064 nm, a pump pulse duration of 0.9 ns (full width at half maximum, FWHM), and a pulse energy of up to 1.2 mJ at a repetition rate of 13.7 Hz. The duration of the emitted laser pulses amounts to ~ 0.6 ns. Note that this is much longer than the cavity round-trip time of the laser, and pulsed operation is caused solely by the fact that the pump source is switched on only for certain time intervals. The effective lifetime of the excited state amounts to ~ 10 ps and is much shorter than the durations of the pump and the emission pulses. We may hence assume that the lasing process is close to its steady-state. The experimental setup is explained in more detail in the section 'Experimental demonstration of laser emission' of the Methods, which is followed by an estimation of the pump and emission power levels.

Characterization and experimental proof of lasing. We measured the laser output power in the SMF as a function of the pump power for both the strip and the slot waveguide, Fig. 2c,d. In both cases, a clear threshold can be observed at a launched average pump power of approximately 2.3 mW for the strip, and approximately 1.3 mW for the slot waveguide. The absorbed peak power at threshold in the vicinity of the waveguide can be roughly estimated to be 38 W for the strip, and 24 W for the slot waveguide, taking into account specific parameters of the individual waveguides, see the sections 'Experimental demonstration of laser emission' and 'Estimation of emission power levels' of the Methods and Supplementary Note 1 for more details.

The existence of the threshold indicates laser emission. The measured threshold level is in reasonable agreement with theory, see the section 'Consistence of resonator characteristics and threshold pump power' of the Methods for a more detailed discussion. To rule out any laser look-alikes, we investigate

further criteria formulated by Samuel et al.[32] Below threshold, only amplified spontaneous emission is to be seen, which increases exponentially with the pump power, see insets of Fig. 2c,d. Above threshold, the output power increases linearly with the pump power. For very high pump powers, the laser power saturates. The saturation is attributed to absorption bleaching at the pump wavelength and to pump-induced free-carrier absorption (FCA) in the SOI waveguide, see the section 'Optically induced losses and dynamical behaviour' of the Methods for a more detailed discussion.

Moreover, we investigate the emission spectra from the strip and slot waveguides below and above threshold, see Fig. 2e,f. Broadband amplified spontaneous emission can be observed for operation below threshold, see insets of Fig. 2e,f (logarithmic scale). When pumped above threshold, the emission spectrum narrows considerably. In Fig. 2e and f, the observed linewidth appears slightly larger than the resolution bandwidth of the spectrometer (RBW = 5 nm). We attribute this to a multitude of different longitudinal cavity modes which oscillate simultaneously at every pump pulse, see Supplementary Note 2 and Supplementary Fig. 3 for a more detailed description.

Above threshold, the optical output of slot waveguides and of narrow strip waveguides is laterally single-mode, which can be inferred from the observation that there is a single well-defined optimum spot when coupling to a lensed SMF. For strip waveguides, lasing in higher-order lateral modes can be observed for waveguide widths of ~ 300 nm or more as reported in more detail in the next section 'Influence of waveguide geometry'. For the devices shown in Fig. 2, the emitted light is predominantly polarized in the horizontal direction as is expected for lasing of the quasi-TE mode. The polarization extinction ratio (ER) is about 8 dB for both devices in Fig. 2. To confirm that the dye is indeed responsible for lasing, we prepared reference samples without dye in the PMMA cladding. These samples do not show noticeable light emission. Moreover, without the silicon waveguide but with dye in the cladding, only spontaneous emission is observed. These findings exclude any laser look-alikes and confirm the working principle of the SOH laser concept.

Influence of waveguide geometry. Regarding the influence of waveguide geometry on the performance of the SOH lasers, we find that lasing with high output powers can be achieved with a wide range of waveguide dimensions and that the output power is clearly related to the overlap of the guided mode with the active cladding. The geometry-dependent output power levels of different waveguide geometries are shown in Fig. 3. For the strip waveguide, we vary the width, Fig. 3a, whereas for the slot waveguide, the rail width is fixed to 170 nm and the slot width is varied, Fig. 3b. The length of the active section amounts to 4 mm for all devices. As before, the resonator is formed by back-reflection from a cleaved waveguide facet and from a GC operated far from its design wavelength of 1,550 nm. The experimental setup is the same as before and described in the section 'Experimental demonstration of laser emission' of the Methods. The average pump power is fixed to 5 mW. The coloured areas of each bar in Fig. 3 represent the respective contributions of quasi-TE (blue) and quasi-TM polarization (green) to the total output power. For the strip waveguide, the laser power is largest when the strip width is smallest, that is, when the mode fields extend far into the active cladding. The second maximum at $w_{strip} = 375$ nm is due to lasing not only of the fundamental mode, but also of the next higher-order quasi-TE_{10} mode, which also strongly interacts with the cladding. The polarization ER reaches a maximum of $(18 \pm 2$ dB) for the narrowest strip waveguides we investigated.

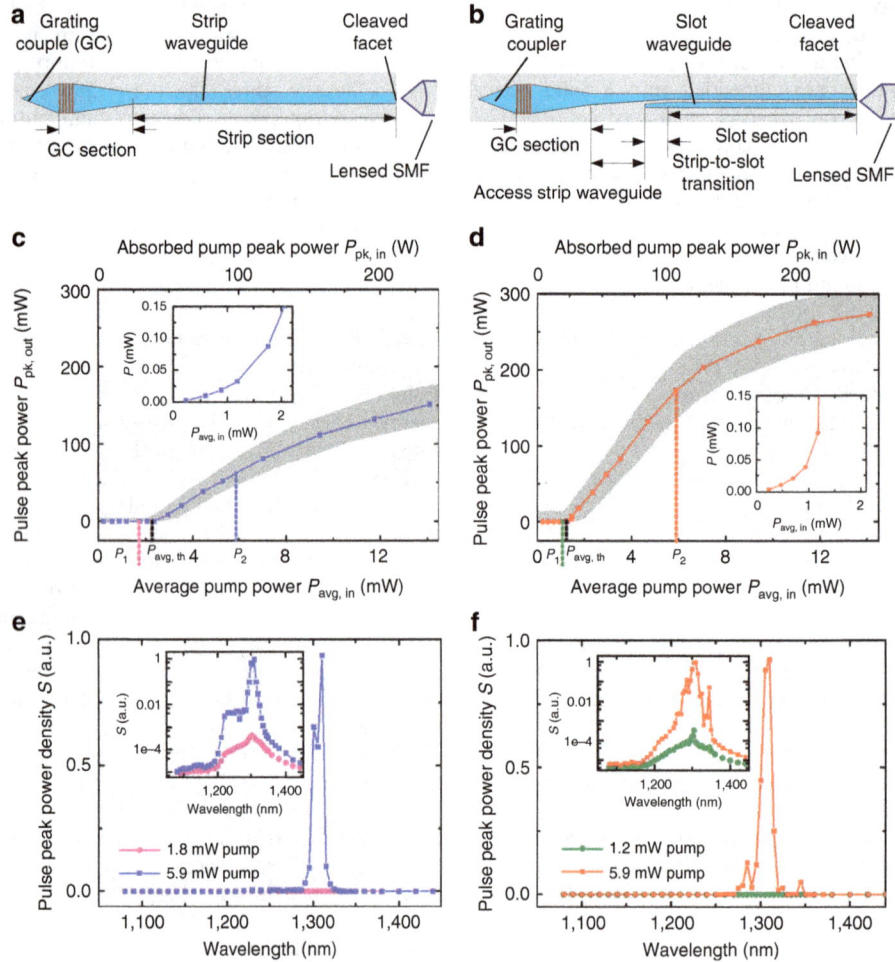

Figure 2 | Experimental proof of lasing in SOH strip and slot waveguides. The cladding consists of the commercially available dye IR26 (ref. 22) dispersed in a PMMA matrix. Cavity mirrors are formed by one cleaved waveguide facet and a GC. The GC is designed for coupling 1,550 nm light from an optical fibre to the strip and exhibits substantial back-reflection at the laser emission wavelength of 1,310 nm. For both the strip and the slot waveguide, the cavities are ~4.8 mm long. The laser output power is measured in a lensed SMF that collects light from the waveguide facet. (**a**) Strip waveguide consisting of a 450 μm long GC section and a 4.3 mm long strip section (waveguide height $h_{WG} \approx 220$ nm, width $w_{strip} \approx 210$ nm). (**b**) Slot waveguide comprising a 450 μm long GC section, a 235 μm long access strip waveguide, a 300 μm long strip-to-slot transition, and a 3.8 mm long slot waveguide section (rail width $w_{rail} \approx 180$ nm, slot width $w_{slot} \approx 215$ nm). (**c**) Peak output power $P_{pk, out}$ (all polarizations) in lensed SMF versus illuminating average pump power $P_{avg, in}$ for the strip-waveguide cavity. A clear pump power threshold of $P_{avg, th} = 2.3$ mW can be observed. The measured incident average pump power (bottom scale) $P_{avg, in}$ is used to calculate the absorbed pulse peak power (top scale) $P_{pk,in}$, taking into account the specific parameters of the waveguide, see the section 'Estimation of pump power levels' of the Methods and Supplementary Note 1. The grey-shaded area indicates an estimate of the accuracy of the measurement. For the uncertainty of the pump power, we use a relative standard error of ±14%, see the section 'Estimation of emission power levels' of the Methods for a more detailed explanation. Regarding the uncertainty of the emitted power, we estimate a relative standard error of ±10% for all pump powers of $P_{p1} = 5$ mW or more. Below pump powers of $P_{p1} = 5$ mW, we assume a constant absolute error which corresponds to the ±10% relative standard error at $P_{p1} = 5$ mW. Note that the grey ranges correspond to a coarse, but conservative estimate of the measurement uncertainties. P_1 and P_2 denote the pump powers for which the spectra in **e** are recorded. (**d**) Peak output power $P_{pk, out}$ in SMF versus incident average pump power $P_{avg, in}$ for the slot-waveguide cavity. The grey-shaded areas indicate again an estimate of the accuracy of the measurements, see the section 'Estimation of emission power levels' of the Methods. A threshold pump power $P_{avg, th} = 1.3$ mW is found, which corresponds to 60% of the threshold for the strip waveguide. P_1 and P_2 denote the pump powers for which the spectra in **f** are recorded. (**e**) Emission spectra below (ASE, magenta) and above threshold (blue) for the strip-waveguide cavity. The spectrum is given in arbitrary units of the spectral pulse peak power density S recorded with a resolution bandwidth of 5 nm (inset with logarithmic scale). Above threshold, the emission spectrum narrows considerably. (**f**) Emission spectra below (ASE, green) and above threshold (red) for the slot-waveguide cavity (inset with logarithmic scale). Also here, the emission spectrum narrows considerably above threshold. For all spectra, the resolution bandwidth amounts to 5 nm to allow for detection of weak ASE and strong laser emission with the same measurement system. High-resolution spectra above lasing threshold have been taken at smaller RBW of 0.2 and 0.05 nm, see Supplementary Note 2 and Supplementary Fig. 3.

In Fig. 3b, we consider slot waveguides and vary the slot width while keeping the rails widths at a constant value of 170 nm. The TE mode dominates laser emission, since interaction with the cladding is enhanced by the electric field discontinuities at the high-index-contrast sidewalls of the slot, as can be seen by comparing the field interaction factors[33] $\Gamma_{clad, TE}$ and $\Gamma_{clad, TM}$

for the two polarizations, see Supplementary Table 1 and Supplementary Note 1. Moreover, the emitted laser power increases with slot width. This is to be expected since larger slot widths lead to both larger field interaction factors of the guided mode with the active cladding and to a larger volume in which dye molecules can interact with the guided mode. For

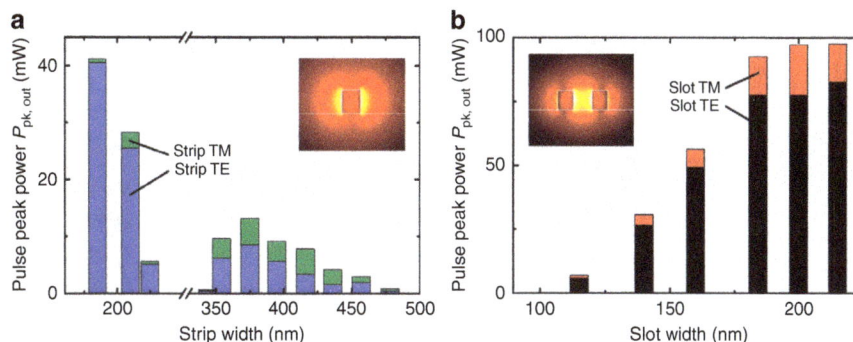

Figure 3 | Geometry-dependent peak output power $P_{pk, out}$ coupled into a lensed SMF for strip and slot waveguides. The resonator relies on back-reflection from one cleaved waveguide facet and from a GC operated far from its design wavelength of 1,550 nm. Quasi-TE and quasi-TM polarizations are measured separately. The average incident pump power is 5 mW for all samples. In the bar diagram, the differently coloured areas represent the contributions of the quasi-TE and the quasi-TM polarization to the total output power; the total bar height corresponds to the total emission. Insets: Dominant electric field magnitudes of the fundamental quasi-TE modes. (**a**) Strip-waveguide cavity. The laser power is largest when the strip width is smallest such that the guided light extends far into the cladding. The secondary maximum at 375 nm is due to lasing of the next higher-order mode (quasi-TE_{10}), which also has a strong overlap with the active cladding, but is not guided for smaller strip widths. (**b**) Slot-waveguide cavity. An increase of the slot width leads to an increase of the field confinement in the cladding and to an expansion of the region in which the active dye interacts with the optical mode. As a consequence, the lasing power increases with slot width. For large slot widths, the fundamental mode is only weakly guided, and the laser power does not increase further. The rail width has only a minor influence (not shown) and is fixed at 170 nm.

very large slot widths, the slot mode is only weakly guided and leaks into the high-index silicon substrate. As a consequence, the output power does not increase further. The polarization ER remains nearly constant and reaches a maximum of 8 ± 2 dB for a slot width of 140 nm. For wider and narrower slots, the ER is slightly smaller.

The optimum choice of the waveguide geometry depends on the desired balance between output power and polarization ER: High power output and a moderate ER when using slot waveguides have to be compared with about half the output power and a high polarization ER obtained from narrow strip waveguides. Using state-of-the-art CMOS fabrication, waveguide dimensions can be reproduced with tolerances of significantly less than 10 nm, which does not influence output power or polarization ER of the SOH lasers to a significant degree. SOH device performance can hence be expected to be resilient against fabrication inaccuracies.

Dynamic emission behaviour. The achievable peak output power of the SOH lasers is remarkable: For an SOH slot waveguide with cleaved facets on both sides, we measured peak output powers of up to 365 mW in the attached SMF, see Fig. 4a,b. The fibre–chip coupling losses are estimated to be (5 ± 1) dB, which leads to peak powers of (30.3 ± 1.0) dBm at the output facet, that is, 1.1 W that could be coupled to an on-chip nanophotonic SOI waveguide. This is the highest peak power emitted from a silicon-based laser with on-chip cavity so far. A more detailed discussion can be found in the section 'Estimation of emission power levels' of the Methods.

The time-dependent emission of the slot waveguide laser is depicted in Fig. 4c for both polarizations, recorded at an average pump power of 5 mW. We observe laser emission into both the quasi-TE and quasi-TM mode, which we attribute to local gain depletion: for large slot widths, the TE and TM modes occupy different cross-sectional domains of the active cladding, see insets in Fig. 4b, and lasing may therefore occur simultaneously in both polarizations. Since the overlap of the quasi-TE slot mode with the active cladding is larger than that of the TM mode, the TE mode experiences higher gain and hence dominates lasing with a polarization ER of 9 dB. The TE and TM emission spectra are

similar—see Supplementary Fig. 3 and Supplementary Note 2 for a more detailed discussion.

Regarding the pulse shapes, we find that the mean FWHM duration of emission amounts to 0.6 ns, which is shorter than the pump pulse FWHM of 0.9 ns. Moreover, the emission pulse features an asymmetric shape and is delayed with respect to the pump pulse. The delay is attributed to the fact that laser emission can only set in once the pump intensity exceeds the threshold level. Note that the relative timing of pump pulse and emission pulses is subject to uncertainties of approximately ± 100 ps due to different propagation delays in the fibre-based measurement setup, see the section 'Optically induced losses and dynamical behaviour' of the Methods for more details. The instantaneous pump power at the onset of laser emission can therefore not be directly associated with the threshold pump power level identified in Fig. 4b. The asymmetric shape of the emission pulse might be caused by nonlinear absorption and subsequent relaxation processes in the active cladding. This aspect requires further investigation.

Discussion
SOH lasers have the potential to cover a broad range of different emission wavelengths between 1.1 and 1.6 μm by using suitable dye materials[34,35]. Due to the high output power, the devices may even be used for exploiting nonlinear optic effects in nanophotonic waveguides. The SOH lasers are remarkably robust: during our experiments, we did not observe significant degradation of the devices, even though they were tested repeatedly over several weeks without taking any specific efforts with respect to encapsulation. This first indication of high stability of the SOH lasers is in good agreement with previous observations, which have shown that photo bleaching of IR26 can be neglected at our pump wavelength of 1,064 nm (ref. 22). A detailed investigation of the stability of SOH lasers is subject of further research.

The devices presented in this paper are first-generation prototypes with considerable room for improvement. In particular, lasing threshold and linewidth of optical emission can be reduced by using optimized Bragg reflectors or ring resonators for optical feedback. Moreover, according to our

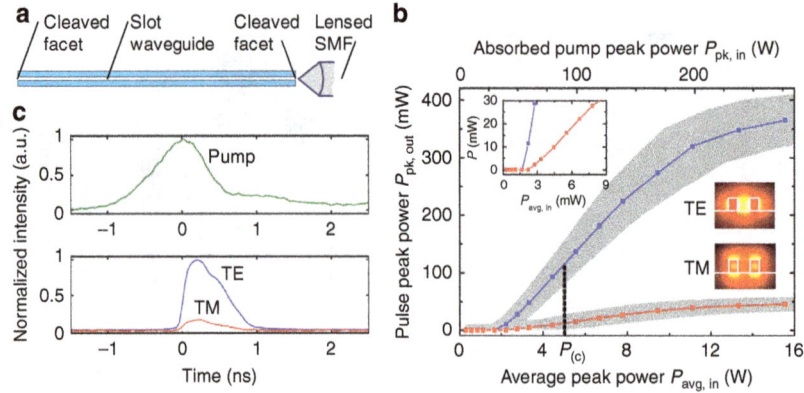

Figure 4 | Lasing in a SOH slot waveguide. In this experiment, cavity mirrors are formed by cleaved waveguide facets on both ends. The cavity length is 3.8 mm, the waveguide height amounts to 220 nm, and for the rail and the slot width, values of $w_{rail} = (160 \pm 15)$ nm and $w_{slot} = (180 \pm 15)$ nm were extracted from scanning electron microscope (SEM) images. (**a**) Schematic top view of the slot waveguide. (**b**) Peak output power in the lensed SMF for quasi-TE and quasi-TM mode versus incident average pump power. The absorbed pump peak power is estimated from the measured incident average pump power, see the section 'Estimation of pump power levels' of the Methods. The grey-shaded areas indicate an estimate of the accuracy of the measurement. For the uncertainty of the pump power, we assume a ±14% relative standard error. Regarding the uncertainty of the emitted power, we estimate a relative standard error of ±10% for all pump powers of $P_{p1} = 5$ mW or more. Below pump powers of $P_{p1} = 5$ mW, we assume a constant absolute error which corresponds to the ±10% relative standard error at $P_{p1} = 5$ mW. More details can be found in the section 'Estimation of emission power levels' of the Methods. Note that the grey ranges correspond to a coarse, but conservative estimate of the measurement uncertainties. Inset: zoom-in of pulse peak power at low pump powers, demonstrating sharp thresholds for both TE and TM mode. (**c**) Temporal shape of the pump pulse at an average power of 5 mW (green) and of the corresponding emission pulses (TE, blue; TM, red). The shape of the pump pulse was measured by averaging over 16 pulses and normalizing to a peak value of 1. Likewise, the emission pulses were measured in the SMF and averaged over 16 pulses. In the plot, the peak of the TE emission has been normalized to 1, and the TM emission is plotted at the same scale. The exact delay between pump and emission cannot be exactly determined due to modal and chromatic dispersion in the standard SMF. The peak pump power was determined with a relative standard error of ±14%; for the peak power of the emitted pulse the relative standard error is ±10%, see the sections 'Estimation of emission power levels' and 'Estimation of pump power levels' of the Methods for a more detailed discussion.

study of the laser dynamics, we expect that better efficiency and lower threshold can be achieved by avoiding FCA as an important loss mechanism of the cavity. To this end, one might consider dyes that allow for pump wavelengths above the absorption edge of silicon[35]. Moreover, the pump efficiency can be improved considerably by guiding the pump light along the SOI waveguide to concentrate it in the active zone. This could be achieved by using an additional polymer waveguide around the SOI waveguide. High duty cycles or CW emission are in general difficult to achieve in dye lasers due to triplet-state excitation and subsequent photo-induced degeneration. This deficiency could be overcome by doping the matrix material with triplet-state quenching or triplet-trapping species of molecules[36], by using optofluidic concepts[37] or by choosing other gain materials such as lanthanide ions or colloidal quantum dots[38,39] that might even be suited for direct electrical pumping[40].

Nevertheless, even without CW operation, SOH lasers enable greatly simplified one-step fabrication processes for realizing thousands of light sources directly integrated into silicon photonic circuitry. Such light sources lend themselves to a wide range of applications such as biosensing[41], where pulsed operation with low-duty cycles is sufficient, where cost-efficient mass fabrication is essential to enable disposable chips for one-time use, and where pump efficiency is secondary. Moreover, the high peak power of the SOH lasers might open interesting opportunities in nonlinear infrared spectroscopy. Further investigation of the dynamics, optimization of the active cladding, and the use of better resonators should help enlarging the application range. We therefore believe that the present approach will be the basis for a novel class of silicon photonic on-chip sources that stand out due to their high peak output power and ease of fabrication.

Methods

Fabrication of SOH lasers. Waveguides were fabricated on SOI wafers from SOITEC using a CMOS pilot line based on 193-nm deep-ultra-violet lithography[30]. All waveguides have a height of $h_{WG} = 220$ nm and are optically isolated from the silicon substrate by a buried oxide (SiO$_2$) layer of thickness $h_{SiO2} = 2\,\mu$m.

The gain medium is deposited on the silicon waveguides in a single post-processing step by spin-coating. The active organic cladding consists of a PMMA matrix which is doped with 1 wt% of the commercially available dye IR26 (ref. 22). The final thickness of the cladding amounts to $h_{clad} \approx (500 \pm 50)$ nm. The measured absorption and fluorescence spectra of a liquid dye solution are depicted[31] in Supplementary Fig. 1, exhibiting a fluorescence emission peak at 1,130 nm. When using the dyes in an extended waveguide structure, the emission peak of IR26 shifts to ~1,300 nm due to self-absorption along the waveguide in the overlap region of the emission and the absorption spectra[42]. This is in good agreement with the laser emission wavelength observed in ref. 22.

Experimental demonstration of laser emission. The experimental setup is depicted in Fig. 5. The SOH devices are pumped from top by a pulsed laser at a wavelength of 1,064 nm with a duty cycle of approximately $p_t = 1.23 \times 10^{-8}$. The FWHM of the pump pulse amounts to 0.9 ns, the repetition frequency is 13.7 Hz. The incident pump power is controlled by adjusting the angle of a half-wave plate in front of a polarizing beam splitter. The pump light is polarized in a direction perpendicular to the waveguide axis and focused on the waveguide under test using a cylindrical lens, see Fig. 5a.

To measure emission from the SOH device, a lensed SMF is placed near the facet, denoted as 'Fibre 2' in Fig. 5a. The fibre collects the emitted light with an estimated coupling loss of ~5 ± 1 dB. By coupling an auxiliary light beam at 1,550 nm through the on-chip GC to the SOH waveguide using a second fibre (Fibre 1), we can facilitate the alignment of the lensed Fibre 2 with respect to the waveguide facet. Polarization-maintaining fibres are used throughout the setup, and Fibre 2 is aligned such that the quasi-TE and quasi-TM emission of the SOH laser is coupled to the slow and the fast axis of the PM fibre, respectively. To characterize the laser emission, we use two different detection paths in our setup: A 'high-sensitivity detection' path, corresponding to the upper part in Fig. 5b, and a 'fast-detection' path, represented by the lower part in Fig. 5b.

The high-sensitivity path allows to measure input–output power characteristics and spectral properties of the laser emission. To this end, we use a monochromator and a highly sensitive photodetector with a large dynamic range, followed by an electrical low-pass filter for noise reduction and a standard oscilloscope, see Fig. 5b.

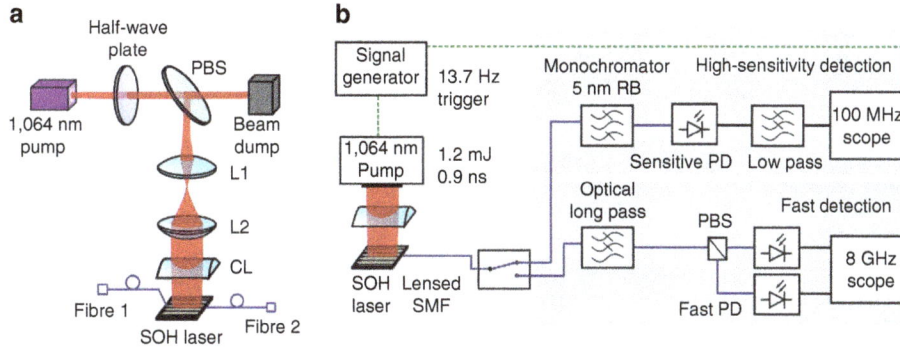

Figure 5 | Measurement setup. (**a**) Pump light at 1,064 nm is focused on the SOH waveguide using a cylindrical lens. Pump power is adjusted by sending the linearly polarized light from the pump laser through a half-wave plate and a polarizing beam splitter. Fibre 1 (cleaved SMF illuminating a GC) is used only to facilitate coarse alignment of Fibre 2 (lensed SMF) by using 1,550 nm light. (**b**) Emission from the SOH laser is collected by the lensed fibre (Fibre 2), which is connected to different detector setups by an optical switch. The upper path is used for high-sensitivity detection. It contains a monochromator and a slow but highly sensitive photodetector to record weak fluorescence. The sensitive PD has a low bandwidth, and a consecutive electrical low-pass filter is used to further suppress noise. The lower 'fast-detection' path is used for time- and polarization-resolved measurements. It is equipped with fast PDs. Residual pump power is blocked by an optical long-pass filter.

The oscilloscope is triggered by the emission of the pump laser and averages over 16 subsequent pulses. Due to the electrical low-pass filter and the bandwidth limitations of both the photodetector and the oscilloscope, the recorded electrical pulse is strongly widened compared with its optical counterpart. However, the peak of the recorded electrical pulse still remains proportional to the received optical power. This setup allows measuring the wavelength-resolved emission spectrum. For high output powers, an attenuator (not shown) was inserted in front of the photodiode.

Time-resolved measurements are made with the fast-detection path. An optical long-pass filter blocks spurious pump light that might be scattered into the lensed fibre, and a polarization beam splitter is used to separate the two polarization states for individual detection. Light pulses with a duration in the (sub-)ns-range are detected with fast photodiodes (NewFocus 25-GHz model 1434, NewFocus 45 GHz model 1014). A high-speed oscilloscope (Tektronix DPO 70804B, 8 GHz bandwidth, 25 GSa s^{-1}) is used to record time-resolved traces. The traces displayed in Fig. 4c have been obtained by averaging over 16 subsequent pulses. We find an average pump pulse duration of 0.9 ns FWHM with a standard error of 0.13 ns (15%). The durations of the emitted SOH laser pulses are shorter than that of the pump pulse. For quasi-TE polarization, the mean FWHM duration amounts to 0.6 ns with a standard error of ± 0.06 ns (10%).

Estimation of emission power levels. For high output powers above the lasing threshold, the peak power levels in the output fibre were measured using the fast-detection path of the setup depicted in Fig. 5, taking into account the responsivity of the fast photodiode and the optical and electrical losses of the various components. To obtain a lower boundary for the on-chip power levels, we assume that the total fibre–chip coupling losses are as low as 5 dB (factor 3.2). This value was estimated from reference measurements at 1,550 nm; the actual losses at 1,310 nm may be slightly higher. The coupling factor also includes losses of 6% due to reflection from the waveguide facet. A measured SOH laser peak power of 365 mW in the SMF hence corresponds to a laser peak power of at least 365 mW × 3.2 × 0.94 = 1.1 W which is coupled out from the waveguide facet and which could be used in an on-chip device that is connected to the SOH laser. To estimate the stochastic variations of the measured emission power, the high-speed detection path depicted in Fig. 5 is used. We record subsequent emission pulses from an SOH slot waveguide similar to the one depicted in Fig. 4a, pumped at powers of $P_{p1} = 5$ mW and $P_{p2} = 15$ mW, both of which are well above the threshold pump power of $P_{p,th} = 2$ mW. At $P_{p1} = 5$ mW, we find relative standard errors of approximately ± 5%, and at $P_{p2} = 15$ mW, the relative standard error amounts to ± 10%. For a conservative estimate, we assume that the relative standard error of the emitted power is ± 10% for all pump powers of $P_{p1} = 5$ mW or more. Below pump powers of $P_{p1} = 5$ mW, we further assume a constant absolute error which corresponds to the ± 10% relative standard error at $P_{p1} = 5$ mW. The range of negative powers is discarded, leading to the grey-shaded areas in Fig. 4b The grey-shaded areas in Fig. 2c,d were constructed in a similar way: For pump powers above $P_{p1} = 5$ mW, we assume a ± 10% relative error, whereas for pump powers below 5 mW, we use a constant absolute error which corresponds to the ± 10% relative standard error at $P_{p1} = 5$ mW. These uncertainty ranges are a coarse, but conservative estimate, which can only give a rough impression of the uncertainties of the measurement data.

For spectrally resolved measurements or for small power levels below the laser threshold, we use the high-sensitivity detection path of our setup. The peak power levels of the deformed pulses in the high-sensitivity path are calibrated by comparison with the corresponding peaks of the true pulse shapes in the

fast-detection path using medium power levels that can reliably be detected in both paths.

Estimation of pump power levels. While the total average pump power is directly accessible by measurement, the absorbed peak pump power needs to be estimated based on further assumptions. The elliptical Gaussian pump spot features a major axis of 8 mm and a minor axis of 0.3 mm, both defined by the FWHM of the intensity on the chip surface. This is much larger than the active area of the SOH waveguide, defined by the region in which pumped dye molecules interact with the lasing waveguide mode. Considering the example of the device depicted in Fig. 4, the length $l_{act, region} = 3.8$ mm of the active region is defined by the length of the slot waveguide section, and the width is estimated to the TE mode field diameter $MFD_x = 0.77$ µm in the lateral direction. The fraction of light that overlaps with the active zone is estimated by integrating the two-dimensional Gaussian distribution over the rectangle of MFD_x and waveguide length in the (x, z)-plane. This integral amounts to $p_{xz} = 0.0027$. To estimate the fraction p_y of pump light absorbed in the active cladding, we need to determine the corresponding absorption coefficient. From a direct transmission measurement using a 1.1-µm-thick IR26 dye-doped polymer layer on glass with the same dye concentration as the cladding material, the absorption cross section of the dispersed dye molecules is found to be $\sigma_p = 1.7 \times 10^{-16}$ cm^2. This is in fair agreement with the value $\sigma_p = 5 \times 10^{-16}$ cm^2 measured in a solution of the dye in 1, 2-dichloroethane[43]. The thickness of the cladding $h_{clad} = (500 \pm 50)$ nm has been measured using a profilometer. Using $\sigma_p = 1.7 \times 10^{-16}$ cm^2 and a dye molecule concentration of $n = 10^{19}$ cm^3, a value of $p_y = 1 - \exp(-\sigma_p N h_{clad}) = 0.08$ is found. The dye molecule number density N is derived from the measured mass ratio before mixing the PMMA matrix with the IR26 dyes. The total percentage of pump light absorbed in the active region is therefore $p_{xyz} = p_{xz} \times p_y = 0.022\%$. Using the measured pump pulse shape and the duty cycle, we find a ratio of average pump power to peak pump power of $p_{avg/peak} = 1.23 \times 10^{-8}$, which leads to a ratio of average incident pump power to 'absorbed' peak pump power of $p = p_{avg/peak}/p_{xyz} = 5.6 \times 10^{-5}$. This ratio is used to relate the top and the bottom power scales in Fig. 4b. Consequently, the average incident threshold pump power of 1.8 mW leads to an estimate of the absorbed peak pump power of 32 W. The same method was used to relate the top and bottom power scales in Fig. 2c,d; the corresponding ratios of average incident power to absorbed peak power are listed in Supplementary Table 1. To estimate the variation of the measured pump power, a fraction of the pump pulse is coupled to a fibre and fed to a high-speed photodiode. From the measurements we find that the standard error of the peak pump power is ∼ 14%.

Consistence of resonator characteristics and threshold pump power. The measured threshold pump powers of the SOH lasers are in reasonable agreement with the losses of the cavities. This is demonstrated by analysing the round-trip losses of a Fabry–Perot resonator with two cleaved facets as used in Fig. 4, and by relating them to the material gain of the active cladding.

The resonator round-trip losses are estimated by measuring the Fabry–Perot fringes in the transmission spectrum of the resonator and by evaluating the fringe contrast, see Supplementary Note 3 for a more detailed discussion. For TE polarization, we find a contrast ratio C of ∼ 0.5 dB between the transmission maxima and the adjacent minima, see Supplementary Fig. 4. According to Supplementary Note 3, this corresponds to a total round-trip loss of $10 \log_{10}(a^2 R^2) = 30.8$ dB, where R denotes the power reflection factor at each facet and where a is the single-pass power transmission factor in the 3.8-mm long waveguide. This result is in good agreement with a bottom-up consideration: we use a finite-element solver[44] to calculate the back-reflection R from the cleaved facet of an SOH waveguide, leading to a value of 6% (-12.2 dB), see Supplementary Table 1. Given the resonator length of $l = 3.8$ mm and the total round-trip loss of 30.8 dB, we hence estimate a propagation loss of ~ 0.9 dB mm^{-1} for the slot waveguide. This is in accordance with typically measured propagation losses of slot waveguides[45] which are of the order of 1 dB mm^{-1}.

At threshold, the round-trip losses of the resonator must be compensated by the round-trip amplification. For TE polarization, this requires a waveguide gain $\Gamma_{\mathrm{clad,TE}} \, g = -\log(aR)/l$ corresponding to 4.1 dB mm^{-1}, where $\Gamma_{\mathrm{clad,TE}} = 0.78$ denotes the field interaction factor of the guided mode with the active cladding, see Supplementary Note 1 for more details. Laser emission in the dye cladding is governed by a transition that has a radiative lifetime[43] of the order of 14 ns and a fluorescence quantum efficiency ϕ ranging from 0.02 to 0.1%, see (refs 43,46). The effective lifetime of the excited state hence amounts to $\phi\tau \approx 3$–14 ps—much shorter than the durations of the pump and the emission pulses. For estimating the pump intensity I_{thresh} at threshold, we may hence use steady-state approximations of the rate equations as described in detail in ref. 47 and Supplementary Note 3. This results in the relation

$$I_{\mathrm{thres}} = \frac{hc}{\lambda_{\mathrm{p}} \sigma_{\mathrm{p}} \tau \phi} \left(\frac{\Gamma_{\mathrm{clad,TE}} g}{\Gamma_{\mathrm{clad,TE}} N \sigma_{\mathrm{e}} - \Gamma_{\mathrm{clad,TE}} g} \right), \qquad (1)$$

where $\lambda_{\mathrm{p}} = 1{,}064$ nm is the pump wavelength, $\sigma_{\mathrm{p}} = 1.7 \times 10^{-16}$ cm^2 denotes the measured absorption cross section at this wavelength, N denotes the volume density of dye molecules, $\tau = 14.4$ ns is the radiative lifetime, $\sigma_{\mathrm{e}} = 0.5 \times 10^{-16}$ cm^2 is the emission cross-section[43], and ϕ is the fluorescence quantum efficiency with typical values ranging from 0.02 to 0.1%, see (refs 43,46), as specified for a liquid solution of the dye molecules.

When applied to the TE emission of the device depicted in Fig. 4, equation (1) leads to theoretically estimated threshold peak pump intensities ranging from 1.9 to 9.5 mW cm^{-2}. This is in reasonable with agreement our experimental estimation of the threshold peak pump intensity of 13.7 mW cm^{-2}. This estimation is based on the launched average threshold pump power of ~ 1.8 mW, the overlap $p_{xz} = 0.0027$ of the active area with the Gaussian pump spot in the x, z-plane, the pump duty cycle of approximately $p_t = 1.23 \times 10^{-8}$, and the area of the active zone having a length of $l = 3.8$ mm and a width of MFD$_x = 0.77$ μm.

The deviations between the measured and the predicted peak pump intensity is attributed to large uncertainties of the quantum efficiency ϕ. Previously published figures range from 0.02 to 0.1% and were measured in liquid dye solutions, see refs 43,46, whereas we use the dyes in a solid polymer matrix. The measured value of 13.7 mW cm^{-2} for the peak pump intensity can be reproduced by equation (1) when assuming a quantum efficiency of $\phi = 0.014\%$—which is comparable to the values obtained for liquid dye solutions. In addition, it turns out that FCA may additionally increase the cavity losses, see the section 'Optically induced losses and dynamical behaviour' of the Methods for more details. This would explain the fact that the experimentally measured threshold is slightly larger than the theoretically predicted value and lead to quantum efficiencies that are even closer to previously published values.

For TM polarization, the measured contrast of the Fabry–Perot fringes is comparable to that for TE polarization. Both polarizations hence experience similar cavity losses. Figure 4b shows a slightly increased threshold pump power of the TM compared with the TE mode—this is attributed to a reduced field interaction factor of $\Gamma_{\mathrm{clad,TM}} = 0.42$ in the cladding compared with $\Gamma_{\mathrm{clad,TE}} = 0.78$. Moreover, the TM mode experiences higher FCA than the TE mode due to a stronger field interaction with the silicon waveguide core, see the section 'Optically induced losses and dynamical behaviour' below.

Optically induced losses and dynamical behaviour. The dynamical behaviour of the laser emission is depicted in Fig. 4c. In this figure, the relative timing of the pump pulse and the emission pulses is subject to uncertainties: The various traces for the pump pulse, the TE emission and the TM emission were measured by an oscilloscope and a photodetector connected to the chip by standard SMF (G.652). For measuring the TE and the TM emission pulse, light was collected from the same fibre facet, and we may assume that both pulse trains experience the same propagation delay in the fibre. This is different for the pump—for measuring the pump pulse trace, we first had to remove the long-pass filter that was used to suppress residual pump light before it reaches the detector. We then moved the lensed fibre (Fibre 2 in Fig. 5) laterally to collect a small portion of 1,064 nm pump light scattered from the surface of the chip. The group delay of the pump pulses from the fibre tip to the detector is slightly different than that of the emission pulses since the optical setup had to be changed slightly and since the optical fibre is operated below its single-mode cutoff wavelength of 1,260 nm. This leads to higher-order mode propagation and hence to further uncertainties of the group delay. The overall uncertainty in relative timing between the pump and the emission pulses is estimated to be ± 100 ps.

We also investigated the dynamics of intra-cavity losses at the emission wavelength of 1,310 nm. The influence of two-photon absorption (TPA) of the emitted light and TPA-induced FCA can be neglected, see Supplementary Note 4. As the only relevant loss mechanism, we identify FCA induced by direct absorption of 1,064 nm pump light in the silicon waveguide core: during the pump pulse, free

carriers accumulate within the core of the silicon waveguide, thereby leading to absorption and considerably increasing the optical losses of the resonator also at the emission wavelength. For a rough quantitative estimate, we assume a linear absorption coefficient of 10 cm^{-1} for the 1,064 nm pump light in the silicon waveguide core[48]. During pumping, photons absorbed in the waveguide create pairs of free carriers with an effective lifetime[45] of the order of 1 ns. Similarly to the considerations made for the active region of the SOH laser, the fraction of pump light that overlaps with the silicon waveguide is estimated to be $p_{xz, \, \mathrm{Si}} = 0.0011$, and the fraction of pump light absorbed in the 220 nm high silicon waveguide core is estimated to $p_{y, \, \mathrm{Si}} = 0.00022$. Using these values, the free-carrier density would reach 6.6×10^{17} cm^{-3} for an average pump power of 1.8 mW, corresponding to the threshold of the laser depicted in Fig. 4. For this carrier density, an empirical model[48] allows us to roughly estimate an upper limit of the FCA-related propagation loss of ~ 5 dB mm^{-1} in the silicon core at the end of the pump pulse. Additional losses of this magnitude may significantly reduce the quality of the optical resonator during pumping and lead to an increased threshold. This is consistent with the observation that the experimentally measured threshold is slightly larger than the theoretically predicted value. We expect that in future devices, FCA can be mitigated by pumping at infrared wavelengths, which are not absorbed in the SOI waveguide core, or by using reverse-biased p-i-n structures that remove free carriers from the silicon core of the waveguides[7]. That would allow to considerably reduce threshold pump powers and to increase the slope efficiencies of the devices.

Summary of resonator and laser emission characteristics. For the quantitative estimations in this paper, various waveguide and resonator parameters are used. These parameters are summarized in Supplementary Table 1 along with threshold and emission power levels of the respective devices. The values are obtained either from experiments or from numerical simulations, for example, for the case of the field interaction factor, effective area[49] and mode field diameter. The underlying mathematical relations are given in Supplementary Note 1.

References

1. Jalali, B. & Fathpour, S. Silicon photonics. *J. Lightw. Technol.* **24**, 4600–4615 (2006).
2. Hochberg, M. & Baehr-Jones, T. Towards fabless silicon photonics. *Nat. Photon.* **4**, 492–494 (2010).
3. Liang, D. & Bowers, J. E. Recent progress in lasers on silicon. *Nat. Photon.* **4**, 511–517 (2010).
4. Bowers, J. E. *et al.* Hybrid silicon lasers: the final frontier to integrated computing. *Opt. Photon. News* **21**, 28–33 (2010).
5. Boyraz, O. & Jalali, B. Demonstration of a silicon Raman laser. *Opt. Express* **12**, 5269 (2004).
6. Boyraz, O. & Jalali, B. Demonstration of directly modulated silicon Raman laser. *Opt. Express* **13**, 796 (2005).
7. Rong, H. *et al.* Low-threshold continuous-wave Raman silicon laser. *Nat. Photon.* **1**, 232–237 (2007).
8. Takahashi, Y. *et al.* A micrometre-scale Raman silicon laser with a microwatt threshold. *Nature* **498**, 470–474 (2013).
9. Park, H. *et al.* A hybrid AlGaInAs-silicon evanescent amplifier. *IEEE Photon. Technol. Lett.* **19**, 230–232 (2007).
10. Fang, A. W. *et al.* Electrically pumped hybrid AlGaInAs-silicon evanescent laser. *Opt. Express* **14**, 9203–9210 (2006).
11. Liang, D. *et al.* Hybrid silicon evanescent approach to optical interconnects. *Appl. Phys. A Mater. Sci. Process* **95**, 1045–1057 (2009).
12. Chen, S. M. *et al.* 1.3 um InAs/GaAs quantum-dot laser monolithically grown on Si substrates operating over 100°C. *Electron. Lett.* **50**, 1467–1468 (2014).
13. Camacho-Aguilera, R. E. *et al.* An electrically pumped germanium laser. *Opt. Express* **20**, 11316–11320 (2012).
14. Wirths, S. *et al.* Lasing in direct-bandgap GeSn alloy grown on Si. *Nat. Photon.* **9**, 88–92 (2015).
15. Barrios, C. A. & Lipson, M. Electrically driven silicon resonant light emitting device based on slot-waveguide. *Opt. Express* **13**, 10092–10101 (2005).
16. Pintus, P., Faralli, S. & Pasquale, F. D. Low-threshold pump power and high integration in Al2O3:Er slot waveguide lasers on SOI. *IEEE Photon. Technol. Lett.* **22**, 1428–1430 (2010).
17. Tengattini, A. *et al.* Toward a 1.54 um electrically driven Erbium-doped silicon slot waveguide and optical amplifier. *J. Lightw. Technol.* **31**, 391–397 (2013).
18. Isshiki, H., Jing, F., Sato, T., Nakajima, T. & Kimura, T. Rare earth silicates as gain media for silicon photonics. *Photonics Research* **2**, A45 (2014).
19. Hosseini, E. S. *et al.* CMOS-compatible 75 mW erbium-doped distributed feedback laser. *Opt. Lett.* **39**, 3106 (2014).
20. Sun, X. *et al.* Electrically pumped hybrid evanescentSi/InGaAsP lasers. *Opt. Lett.* **34**, 1345 (2009).
21. Tanaka, S. *et al.* High-output-power, single-wavelength silicon hybrid laser using precise flip-chip bonding technology. *Opt. Express* **20**, 28057 (2012).

22. Morishita, T., Yamashita, K., Yanagi, H. & Oe, K. 1.3μm solid-state plastic laser in dye-doped fluorinated-polyimide waveguide. *Appl. Phys. Express* **3**, 092202 (2010).

23. Koos, C. *et al.* Silicon-organic hybrid (SOH) and plasmonic-organic hybrid (POH) integration. *J. Lightw. Technol.* doi:10.1109/JLT.2015.2499763 (2015).

24. Koos, C. *et al.* All-optical high-speed signal processing with silicon–organic hybrid slot waveguides. *Nat. Photon.* **3**, 216–219 (2009).

25. Lauermann, M. *et al.* Low-power silicon-organic hybrid (SOH) modulators for advanced modulation formats. *Opt. Express* **22**, 29927 (2014).

26. Koeber, S. *et al.* Femtojoule electro-optic modulation using a silicon–organic hybrid device. *Light Sci. Appl.* **4**, e255 (2015).

27. Pfeifle, J., Alloatti, L., Freude, W., Leuthold, J. & Koos, C. Silicon-organic hybrid phase shifter based on a slot waveguide with a liquid-crystal cladding. *Opt. Express* **20**, 15359–15376 (2012).

28. Wang, X., Grist, S., Flueckiger, J., Jaeger, N. A. F. & Chrostowski, L. Silicon photonic slot waveguide Bragg gratings and resonators. *Opt. Express* **21**, 19029 (2013).

29. Xu, Q., Almeida, V. R., Panepucci, R. R. & Lipson, M. Experimental demonstration of guiding and confining light in nanometer-sizelow-refractive-index material. *Opt. Lett.* **29**, 1626–1628 (2004).

30. Selvaraja, S. K., Bogaerts, W., Dumon, P., Van Thourhout, D. & Baets, R. Subnanometer linewidth uniformity in silicon nanophotonic waveguide devices using CMOS fabrication technology. *IEEE J. Sel. Topics Quantum Electron* **16**, 316–324 (2010).

31. Kranitzky, W., Kopainsky, B., Kaiser, W., Drexhage, K. H. & Reynolds, G. A. A new infrared laser dye of superior photostability tunable to 1.24 μm with picosecond excitation. *Opt. Commun.* **36**, 149–152 (1981).

32. Samuel, I. D. W., Namdas, E. B. & Turnbull, G. A. How to recognize lasing. *Nat. Photon.* **3**, 546–549 (2009).

33. Brosi, J.-M. *et al.* High-speed low-voltage electro-optic modulator with a polymer-infiltrated silicon photonic crystal waveguide. *Opt. Express* **16**, 4177–4191 (2008).

34. Elsaesser, T. & Kaiser, W. in *Dye Lasers: 25 Years.* (ed. Stuke, M.) **70**, 95–109 (Springer, 1992).

35. Zhang, J. & Zhu, Z. Novel heptamethine thiapyrylium infrared laser dyes of superior photostability tunable from 1.35 to 1.65 μm. *Opt. Commun.* **113**, 61–64 (1994).

36. Zhang, Y. & Forrest, S. R. Existence of continuous-wave threshold for organic semiconductor lasers. *Phys. Rev. B* **84**, 241301 (2011).

37. Schmidt, H. & Hawkins, A. R. The photonic integration of non-solid media using optofluidics. *Nat. Photon.* **5**, 598–604 (2011).

38. Schaller, R. D., Petruska, M. A. & Klimov, V. I. Tunable near-infrared optical gain and amplified spontaneous emission using PbSe nanocrystals. *J. Phys. Chem. B* **107**, 13765–13768 (2003).

39. Rogach, A. L., Eychmüller, A., Hickey, S. G. & Kershaw, S. V. Infrared-emitting colloidal nanocrystals: synthesis, assembly, spectroscopy, and applications. *Small* **3**, 536–557 (2007).

40. Heo, J., Jiang, Z., Xu, J. & Bhattacharya, P. Coherent and directional emission at 1.55 μm from PbSe colloidal quantum dot electroluminescent device on silicon. *Opt. Express* **19**, 26394 (2011).

41. Iqbal, M. *et al.* Label-free biosensor arrays based on silicon ring resonators and high-speed optical scanning instrumentation. *IEEE J. Sel. Top. Quantum Electron.* **16**, 654–661 (2010).

42. Casalboni, M. *et al.* 1.3 μm light amplification in dye-doped hybrid sol-gel channel waveguides. *Appl. Phys. Lett.* **83**, 416 (2003).

43. Benfey, D. P., Brown, D. C., Davis, S. J., Piper, L. G. & Foutter, R. F. Diode-pumped dye laser analysis and design. *Appl. Opt.* **31**, 7034–7041 (1992).

44. CST - Computer Simulation Technology. 3d electromagnetic simulation software. Available at https://www.cst.com/.

45. Vallaitis, T. *et al.* Optical properties of highly nonlinear silicon-organic hybrid (SOH) waveguide geometries. *Opt. Express* **17**, 17357–17368 (2009).

46. Semonin, O. E. *et al.* Absolute photoluminescence quantum yields of IR-26 Dye, PbS, and PbSe quantum dots. *J. Phys. Chem. Lett.* **1**, 2445–2450 (2010).

47. Shank, C. V. Physics of dye lasers. *Rev. Mod. Phys.* **47**, 649–657 (1975).

48. Vardanyan, R. R., Dallakyan, V. K., Kerst, U. & Boit, C. Modeling free carrier absorption in silicon. *J. Contemp. Phys.* **47**, 73–79 (2012).

49. Koos, C., Jacome, L., Poulton, C., Leuthold, J. & Freude, W. Nonlinear silicon-on-insulator waveguides for all-optical signal processing. *Opt. Express* **15**, 5976–5990 (2007).

Acknowledgements

This work was supported by the European Research Council (ERC Starting Grant 'EnTeraPIC', number 280145), the Alfried Krupp von Bohlen und Halbach Foundation, the EU-FP7 projects SOFI (grant 248609) and PhoxTrot, the Center for Functional Nanostructures (CFN) of the Deutsche Forschungsgemeinschaft (DFG), the Karlsruhe Nano-Micro Facility (KNMF), the Karlsruhe School of Optics and Photonics (KSOP), the Initiative and Networking Fund of the Helmholtz Association, and the Helmholtz International Research School for Teratronics (HIRST). We acknowledge support by Deutsche Forschungsgemeinschaft and Open Access Publishing Fund of Karlsruhe Institute of Technology. We are grateful for technological support by the Light Technology Institute (LTI) at Karlsruhe Institute of Technology, and by ePIXfab (silicon photonics platform).

Author contributions

D.K. and M.L. designed and performed the experiments, fabricated SOH devices, analysed the data and wrote the manuscript. S.K. and P.A. supported experiments and device fabrication and analysed the data. L.A. designed the SOI waveguides. R.P. and D.K. numerically determined reflection factors. P.D. coordinated chip tapeout and fabrication. J.L. supported analysis of the data. C.K. and W.F. designed the experiments, supported analysis of the data and wrote the manuscript. The manuscript was reviewed by all authors.

Additional information

3

Single-ion adsorption and switching in carbon nanotubes

Adam W. Bushmaker[1], Vanessa Oklejas[1], Don Walker[1], Alan R. Hopkins[1], Jihan Chen[2] & Stephen B. Cronin[2]

Single-ion detection has, for many years, been the domain of large devices such as the Geiger counter, and studies on interactions of ionized gasses with materials have been limited to large systems. To date, there have been no reports on single gaseous ion interaction with microelectronic devices, and single neutral atom detection techniques have shown only small, barely detectable responses. Here we report the observation of single gaseous ion adsorption on individual carbon nanotubes (CNTs), which, because of the severely restricted one-dimensional current path, experience discrete, quantized resistance increases of over two orders of magnitude. Only positive ions cause changes, by the mechanism of ion potential-induced carrier depletion, which is supported by density functional and Landauer transport theory. Our observations reveal a new single-ion/CNT heterostructure with novel electronic properties, and demonstrate that as electronics are ultimately scaled towards the one-dimensional limit, atomic-scale effects become increasingly important.

[1] Physical Sciences Laboratories, The Aerospace Corporation, 355 S. Douglas Street, El Segundo, California 90245, USA. [2] Department of Electrical Engineering, The University of Southern California, 3601 Watt Way, Los Angeles, California 90089, USA. Correspondence and requests for materials should be addressed to A.W.B. (email: adam.bushmaker@aero.org).

Electronic conduction in one-dimensional (1D) channels has been the focus of many research efforts over the past 15 years, and during that time carbon nanotubes (CNTs) have often served as the prototypical system for studying 1D conduction. The reason for the strong interest is both because of the interesting fundamental physics that occur in such systems[1-3] and because of the enhanced functional performance that has been predicted to occur in CNT-based field effect transistors (FETs) such as high linearity and high operation frequency[4,5]. In addition, CNT devices have been developed for new applications using the unique properties of this new material, such as sensors[6], radio frequency (RF) nanoelectromechanical (NEM) devices[7] and flexible electronics and displays[8]. CNT FETs are now being considered as a promising new technology for next-generation micro- and nano-electronic devices and circuitry.

Owing to their small physical size, the conductivity of CNTs and other molecular systems is highly susceptible to changes in charge states of defects nearby or in contact with the material[9], or even by direct chemical activity on the surface of the CNT[10,11]. Similar switching behaviour has also been observed in molecular electronic systems[12], and also as random telegraph noise in scaled deep-submicron silicon devices[13]. As dimensions in nanoelectronic devices are scaled further and further towards the single-atom, 1D channel limit, effects such as those listed above will become increasingly important, even to the point where they dominate the device operation. Single neutral atom detection techniques have shown only small, barely detectable responses[14-17], and there is relatively little information in the literature on ion adsorption on the surfaces of solids from gas. This is despite the various industrial techniques that make use of ions from corona discharge for surface treatment, such as a surface treatment of polymers to improve adhesion and other surface properties[18], and surface charging of insulators as a contactless electrode for both processing and characterization in the semiconductor microelectronics industry[19,20]. Liquid-phase ion–surface interaction is also an important topic, for applications such as ionic gating of graphene[21].

In this study, we report the effect of single gaseous ion adsorption on the conductance of isolated, suspended, single-walled CNT FETs, which were electrically characterized *in situ* during exposure to gaseous ions created by ionizing radiation and high-voltage corona discharge. Positive ions are found to increase the resistance of the CNT by several orders of magnitude, while negative ions have no effect on electrical conduction. Multiple simultaneous ion adsorption events are detected, leading to quantized increase in total device resistance. The gate voltage dependence of electrical conduction during ion adsorption is measured and characterized, and *in situ* electrical heating of the suspended CNTs is used to drive off adsorbed ions. These experimental results are modelled computationally using density functional theory (DFT) and Landauer electrical transport theory.

Results

Electrical current during exposure to gaseous ions. Figure 1a shows a scanning electron image of one of the CNT FET devices used in this study. The CNT is suspended over the trench, making electrical contact on either side, with a gate electrode in the bottom of the trench. Figure 1b shows the drain current plotted versus time for the CNT, showing large decreases in drain current observed during exposure to ionized nitrogen gas. The observed transients are characterized by sudden, discrete reductions in current, which had durations ranging from milliseconds to minutes, followed by an equally sudden recovery back to the pristine state. The transient events were observed during exposure to ionized air, Ar, N_2, He and O_2, and ceased occurring when the

surrounding gas was removed using a vacuum, or when the source of ionization was removed.

Multiple ion interactions. In Fig. 2a,b, the device current and resistance during exposure to higher gas ionization rates are plotted, showing multiple, simultaneous switching events, each adding a quantized resistance of R_{ion} to the total device resistance. The current was measured every 4 ms using a two-point method, and the resistance was calculated using Ohms law with the drain-source voltage of 0.1 V. To reduce noise, the displayed resistance is a running average of 50 data points. After collection of each data set, there is a dead period during which data are transferred to the computer. Equally spaced dashed lines separated by 3.45 GΩ were added to Fig. 2b as a guide to the eye, highlighting the quantized steps in resistance.

Figure 3a shows a close up of exemplary data from the first frame in Fig. 2b. To analyse the statistics of the multi-level switching data, a histogram of resistance levels was calculated for the entire period of exposure to ions, with bin widths of 400 MΩ. This analysis encompassed 50 data sets such as those shown in Fig. 3a, or 1,000 s of time series data. Grouping can clearly be observed in the resistance data, with one dominant peak at the axis (actual resistance is ~0.003 GΩ), two prominent peaks at ~3 and 6 GΩ and two smaller peaks at 10 and 14 GΩ, corresponding to the quantized steps in resistance observed in Fig. 2. The average peak spacing for all peaks observed was found to be $R_{ion} = 3.45$ GΩ. These five peaks are attributed to the pristine CNT resistance, and the resistance of the CNT with one, two, three and four ions, respectively, adsorbed to the CNT surface at different locations. If the arrival of individual ions on the CNT surface is uncorrelated, as one would expect, then the probability of observing a given number of adsorbed ions in any given measurement should form a Poisson's distribution. To test this hypothesis, the measured resistances were tabulated into bins separated by R_{ion} (dashed lines denote bin edges), the distribution normalized and the result plotted in Fig. 3c (solid blue line with open circles). A Poisson's distribution was also fit to the data, given by

$$f(k; \lambda) = \frac{\lambda^k e^{-\lambda}}{k!}, \qquad (1)$$

where f is the probability of occupation, k is the number adsorbed ions and λ is the expected value for individual ion adsorption, and also the only fitting parameter. The resulting Poisson's distribution was plotted in Fig. 3c (dashed red line with solid circles), along with the fit value $\lambda = 0.26$. As can be seen, the data fit a Poisson's distribution well, with the exception of the anomalous peak of four ions, which were observed as a single long switching event in one data set. To evaluate the goodness of fit, the χ^2-value was calculated for a fit including all resistances shown ($k = 0$ through 5) and was found to be 1.87. However, when we ignore the outlier data set mentioned above by only considering data below 12 GΩ ($k = 0$ through 3), the calculated value for χ^2 is only 0.028, indicating a better fit, and that these data points are well described by a Poisson's distribution. A cartoon model illustrating multiple ion defects along the length of the CNT is shown in Fig. 3d, and explains this observed behaviour. Individual ions make their way through diffusive transport to the surface of the CNT and adsorb independently, each adding a resistance of several GΩ in series to the total resistance of the CNT.

Switching event duration. To investigate the statistics of desorption time, the duration of large number of switching events was measured during exposure to gaseous ions at a lower

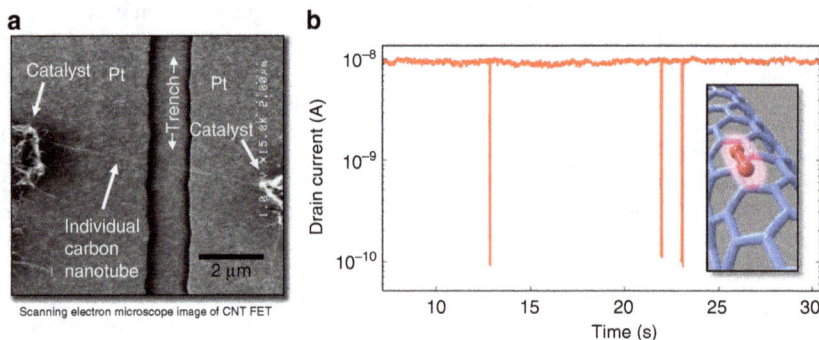

Figure 1 | Device layout and switching transients caused by single-ion adsorption. (**a**) Scanning electron microscope image of CNT FET device and (**b**) plot of drain current versus time showing switching transients observed during ionized gas exposure. The inset shows a cartoon image of a gas molecule adsorbed on the surface of a carbon nanotube.

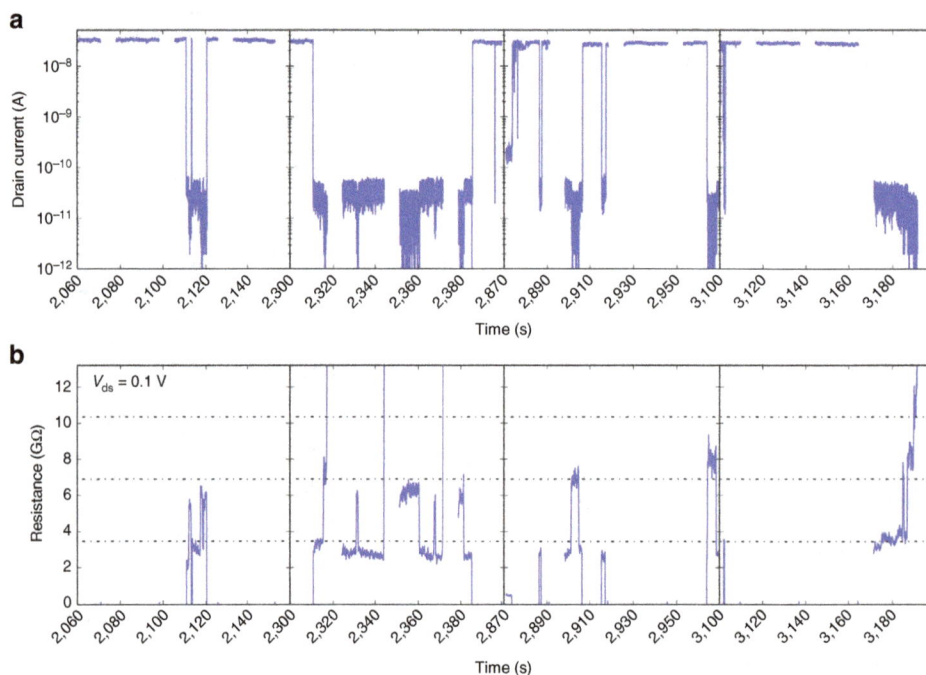

Figure 2 | Drain current and calculated resistance plotted versus time. (**a**) The drain current shows numerous sharp reductions in magnitude, and (**b**) calculated resistance shows quantized equally spaced resistance states. The resistance was calculated using Ohms law with the drain-source voltage of 0.1 V. To reduce noise, the displayed resistance is a running average of 50 data points.

ionization rate, and plotted in Fig. 4. The drain current is plotted versus time for an exemplary data set in Fig. 4a. For lower ionization rates and shorter adsorbed ion residency times, multiple steps are not observed because of the decreased probability for multiple simultaneous interactions. A total of 475 adsorption events were recorded over a time period of 3 h and 10 min. The events were automatically characterized by a computer algorithm, which calculated the total time the current for each event spent below a detection threshold (noted in Fig. 4a). The event duration varied from event to event, and the distribution of event durations calculated for this data set is shown in Fig. 4b. The number of observations for a given event duration decays exponentially, with an average duration of 200 ms. Similar switching behaviour has been observed before in CNT FETs[9,22], as random telegraph signal (RTS). RTS behaviour is somewhat commonly observed in both CNT FET and conventional scaled metal oxide semiconductor MOSFET devices, and is attributed to charge fluctuations in defects at the semiconductor/oxide interface. The associated capture and

emission times for typical RTS systems take an exponential probability distribution (as do the adsorbed ion residency durations here), and are determined by the temperature and state energy[13]. In this context, the average ion residency duration can be interpreted as the emission time for a random telegraph-like system, where the emission time corresponds to the gaseous ion desorbing from the CNT surface.

Temperature effects. To investigate the effect of temperature on the adsorbed ion residency, electrical current was passed through a CNT FET device, increasing the temperature of the suspended CNT *in situ* through Joule heating. Figure 5 shows data from this experiment, in which the device was exposed to a high flux of argon ions generated by corona discharge. After ion exposure, the device resistance increased substantially, remaining in its high resistance state for several minutes. After this period of persistent degradation, a larger drain-source voltage of 0.9 V was applied to the device for a period of 80 s, inducing 600 nA of electrical

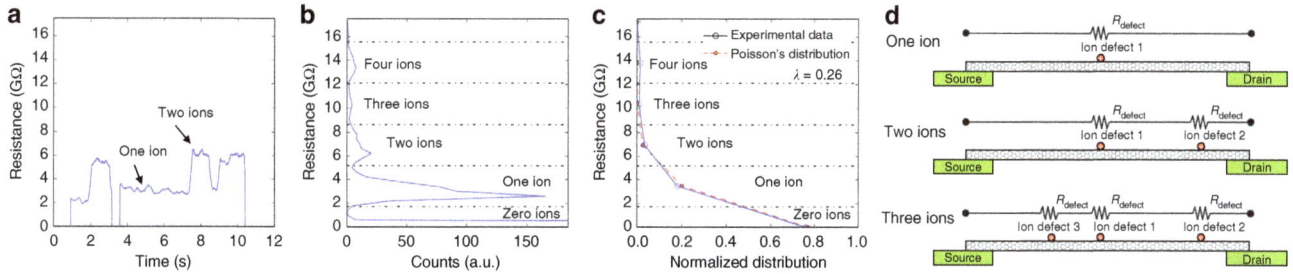

Figure 3 | Statistical analysis of multi-level switching and Poissonian model for ion adsorption. (a) Resistance plotted versus time for an exemplary data set of 5,000 data points showing multi-level switching. (b) Resistance measurement histogram for an entire ion exposure period including 50 total data sets similar to those shown in **a**. (**c**) Normalized distribution of data (solid blue line with open circles) in binned resistance states corresponding to observed levels in **b**, plotted along with a fit Poissonian distribution (red dashed line, solid circles), and (**d**) Poissonian model for multiple, independent, simultaneously adsorbed ions. The ion defect resistance for this data set was found to be 3.45 GΩ.

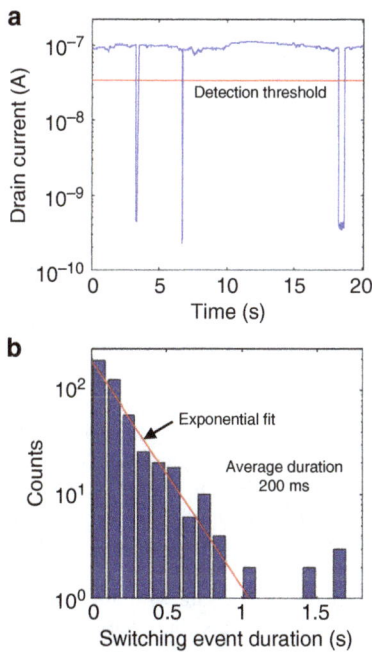

Figure 4 | Statistical analysis of ion residency time. (**a**) Drain current plotted versus time during gaseous ion exposure, with the threshold for event detection denoted with a red line, and (**b**) histogram plotting duration of a large number of switching events, with exponential fit. The average switching event duration for this experiment was found to be 200 ms.

current to flow through the device, heating the CNT in the process. After this treatment, the device resistance was restored to its pristine state. This switching and recovery cycle was repeated four times. Eventually, the device did not recover, possibly because of irreversible chemical or structural changes in the device during prolonged exposure to elevated temperatures. From previous measurements on similar CNT devices using Raman spectroscopy as an *in situ* temperature probe[23,24], the electrical power dissipated in the device during the heating steps shown in Fig. 5 (540 nW) is estimated to increase the temperature of the CNT by ∼45 °C above ambient room temperature. Under such bias conditions, the majority of electrical power dissipation occurs in the CNT itself, as opposed to the contacts because of strong optical phonon scattering[25]. It is possible that ion desorption is assisted by nonequilibrium electron and phonon populations in the CNT, which have been shown to occur in CNT devices at high-voltage biases[26].

Ion polarity effects. To determine the polarity of the ionic species interacting with the CNT, an ionization drift chamber was constructed, and electric fields were applied across the chamber containing the CNT and the surrounding ionized gas. This allowed us to drive either positive or negative ions towards the CNT, depending on the polarity of the electric field. Only positive ions were found to cause the transient switching events to occur, while electrons and negative ions did not (see Supplementary Fig. 1). The electric field applied was 17.3 V cm^{-1}. Ions were not accelerated significantly above room temperature thermal energy, and the electric field is not expected to influence adsorption. Further details on ion polarity measurements are described in Supplementary Note 1.

Contact effects. All data collected in the near-threshold regime of device operation (gate voltage data near $V_G = 0$ V and all time series data) displayed a pristine device resistance of several MΩ. The contact resistance in CNT FET devices with Pt Schottky contacts (such as those used in this study) is in the range of several tens to hundreds of kΩ (refs 27,28), which is relatively small compared with the pristine channel resistance (to say nothing of the several GΩ device resistance during exposure to ions). Thus, the contacts are not expected to play a major role in the transport phenomenon observed here. It is possible that the ions interact with the contacts, increasing the contact resistance of the CNT devices. This is unlikely, however, because of the fact that up to four resistance quanta are observed, and there are only two contact resistances to be modified (source and drain). In addition, it is unlikely that a single ion would have a strong effect on the Pt metal in the contact because of charge screening from the large free electron density in the bulk metal. Finally, the observed change in subthreshold swing requires some interaction with the gate. Ions on the surface of the metal contacts would be screened from the gate by the large free electron density in the metal nearby.

DFT modelling. To model the effects of adsorbed ionic species on the electrical properties of the CNT, DFT modelling was performed using a 1.4-nm-long section of an (8,0) semiconducting CNT with and without adsorbed N_2^+. First, calculations were performed to find the equilibrium CNT–N_2^+ configuration, which gave an equilibrium ion–surface separation of 3.0 Å (ion distance to the CNT axis was 6 Å). Results indicate that the binding energy for N_2^+ on the surface of the CNT is -11.95 ± 0.02 eV, which is considerably more favourable than the binding energy for neutral N_2 (-0.16 ± 0.02 eV). Lowdin charge analysis shows that there is also significant charge transfer to the ion of 0.8 electrons from the CNT to the ion. The densities

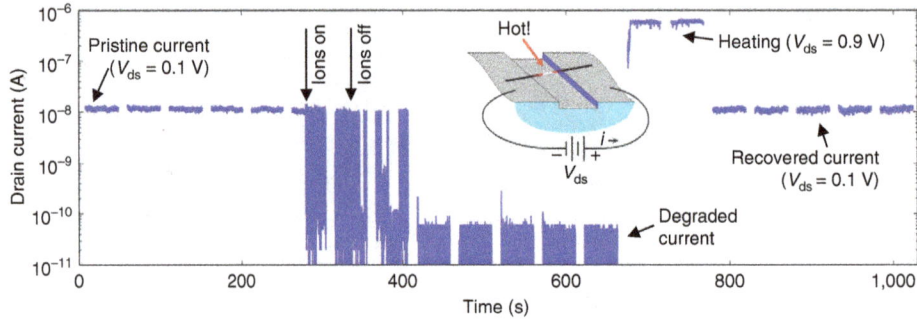

Figure 5 | Ion adsorption and *in situ* thermally induced desorption. Drain current data plotted versus time before, during and after ion exposure and local electrical heating. After ion exposure, the CNT remains in a high resistance state until application of high-voltage bias, which heats the suspended CNT. The estimated temperature increase during the electrical heating phase is 45 °C.

of states for the pristine CNT and CNT with adsorbed N_2^+ are plotted in Fig. 6a, in which the CNT–N_2^+ band states are shifted down by ~ 450 meV. It is hypothesized that, despite the presence of free electrons on the CNT and significant charge transfer to the adsorbed ion, the ion does not recombine with an electron to form a free neutral molecule because the total system energy is lower for the adsorbed ion state, as outlined in Fig. 6b. For more information on DFT calculations, see Supplementary Note 2 and Supplementary Fig. 2.

Gate voltage dependence and electrical transport modelling. Data showing the gate voltage dependence of the CNT FET drain current in both the pristine state (without ions) and transient switched state (during ionized gas exposure) are plotted in Fig. 7b. Data taken while the CNT was in the pristine state are represented as red circles; data taken while the CNT was exposed to ions are represented by cyan circles; and the dark cyan and dark red lines represent a theoretical transport model, discussed below. The pristine CNT FET shows p-type behaviour, with the device being ON for negative gate voltages and an ON/OFF ratio of $\sim 1 \times 10^4$. The devices are not strained significantly by the application of gate voltage, as determined by Raman spectroscopy[29]. The subthreshold swing of the drain current in the pristine state is ~ 110 mV per decade. The gate voltage dependence of the drain current from the defected CNT state is similar to that of the pristine state, although the threshold voltage is shifted more negative, and the subthreshold swing is increased substantially to 400 mV per decade. Both data sets are fit by a theoretical transport model, as outlined below. The negative threshold voltage shift is caused by the presence of positive charge near the 1D channel, and the increase in subthreshold swing is characteristic of the lowered effective gate efficiency for the defect. The subthreshold swing for a FET, which is the reciprocal value of the subthreshold slope, is described by

$$S = \frac{1}{\alpha}\ln(10)k_B T, \qquad (2)$$

where $\alpha = C_G/C_{Total}$ is the gate efficiency, with gate capacitance C_G and total capacitance C_{Total}.

To model the effect of the added ion potential on device electrical characteristics with greater fidelity, numerical calculations were performed to solve the discretized quantum mechanical Hamiltonian for the electron eigenstates and quantum mechanical carrier transmission coefficients[30] for the new system consisting of CNT with adsorbed ion potential U_{Ion}. On the basis of these results, a Landauer transport model was used to calculate the gate voltage dependence of the drain current in a CNT with an adsorbed ion, where the potential on the CNT

is calculated by solving Poisson's equation[31]:

$$U_{CNT} = \frac{eQ_{CNT}}{C_{\Sigma}} + e\alpha V_G + U_{Ion}, \qquad (3)$$

where Q_{CNT} is the charge on the CNT, V_G is the gate voltage and U_{Ion} is the potential associated with the adsorbed ion, which is given by

$$U_{Ion} = \frac{q_{Ion}}{4\pi\varepsilon_0\sqrt{z^2 + h^2}}, \qquad (4)$$

where h is the ion distance from the CNT centre axis, z is the distance along the CNT axis relative to the adsorption site and q_{Ion} is the charge of the ion. The band structure modified by U_{Ion} is shown in Fig. 7a, and it forms a potential barrier in the CNT valence band, reducing current, increasing the valence band–Fermi energy difference and shifting the threshold gate voltage towards more negative values. An additional result of this study is that the ion potential forms a single bound electron state near the conduction band edge. The state energy is too high to contribute to device electrical behaviour in the p-type ON state; however, the result is nonetheless interesting and may have important implications for electrical behaviour in n-type devices. The nanotube experimental $I–V_G$ data with and without ion exposure were fit with model parameters (Supplementary Note 3), and the resulting curves plotted in Fig. 7b along with the experimental data. When tunnelling effects are taken into account (see Supplementary Note 4 and Supplementary Figs 3 and 4), the estimated value for q_{Ion} based on the fit was $+0.18$ elementary charges, which agrees with the DFT results showing significant charge transfer to the adsorbed ion. The value for h was found to be 7 Å, which also corresponds well with DFT results for the equilibrium N_2^+–CNT axis separation of 6 Å. The fit values for q_{Ion} and h are independent of the fit value for α_{ion}, as the first two change the threshold voltage, while the later changes the subthreshold swing.

Decreased gate efficiency. The defect potential gate efficiency α_{ion} was found to be 14%, which is ~ 3.5 times lower than the gate efficiency for the device as a whole (50%). Similar changes in the subthreshold swing are often observed in silicon FET devices when mid-gap energy defect states are created at the Si/SiO$_2$ interface. These traps add an effective capacitance term $C_{it} = q^2 D_{it}$ associated with the defect density of states into the calculation for subthreshold swing[32]. This mechanism is not sufficient, however, to explain the increase in the subthreshold swing observed in the experimental $I–V_G$ data, as these devices are clean, as-grown, suspended CNTs, with very low defect concentrations. In addition, there is insufficient evidence from theoretical modelling of the ion–CNT system for a large enough

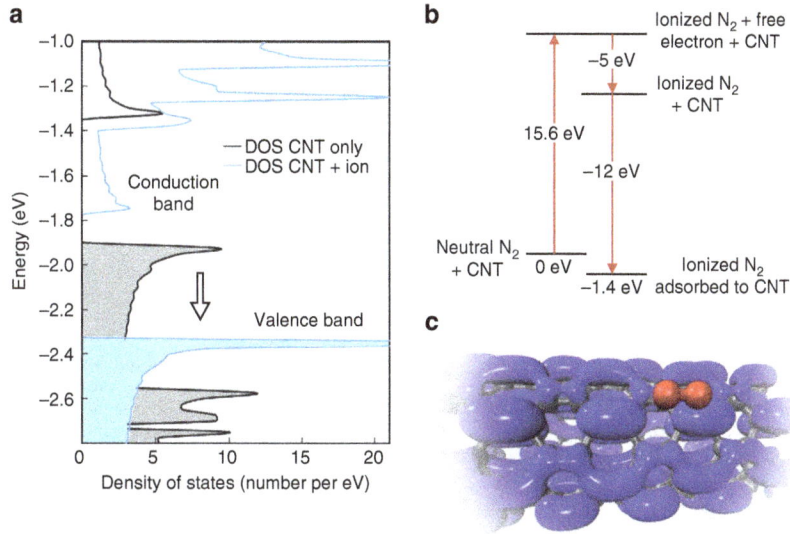

Figure 6 | Density functional theory results. (**a**) Calculated density of states with (cyan) and without (black) adsorbed ion plotted versus energy, (**b**) ionization/adsorption energy diagram for N_2^+ adsorbed to the surface of an (8,0) CNT and (**c**) electron density plot for the gamma point conduction band wavefunction on the CNT with adsorbed N_2^+ ion. The equilibrium adsorbed ion–CNT separation was found to be ~3 Å, with significant charge transfer of 0.8 electrons to the adsorbed ion. The large binding energies found for ion adsorption on CNTs indicate quasi-stable ion residency, which explains the long (milliseconds to minutes) observed lifetimes.

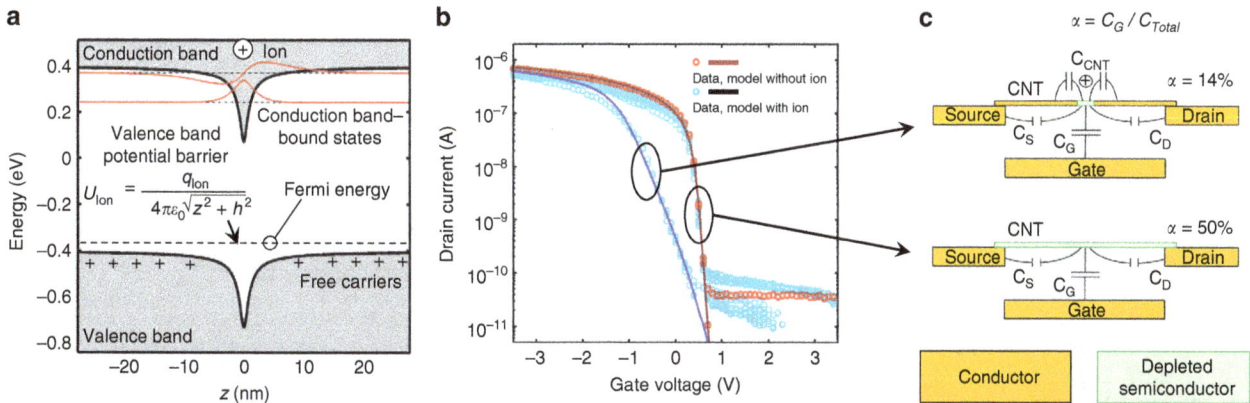

Figure 7 | Ion potential and gate voltage dependence. (**a**) CNT potential plotted versus distance along the CNT axis with adsorbed ion. (**b**) Drain current plotted versus gate voltage showing experimental data from CNT in both pristine (red circles) and defected (cyan circles) states, along with corresponding results from a Landauer transport model (dark red and dark cyan lines) of the CNT with the adsorbed ion potential shown in **a**. (**c**) Lumped capacitance model for pristine and ion-adsorbed subthreshold CNT FETs. The ion charge q_{ion} and separation distance h used to calculate the results plotted in **a,b** were 0.18 elementary charges and 7 Å, respectively, in concordance with DFT results.

density of mid-gap states to warrant this mechanism as the sole explanation for the observed increase in the subthreshold swing.

An alternative mechanism is proposed to explain the decrease in gate efficiency and increase in subthreshold swing. When the adsorbed ion modifies the potential along the length of the CNT as depicted in Fig. 7, there is an additional stray capacitance contribution from the nearby conducting portion of the CNT (C_{CNT}), which is added to the total system capacitance $C_{Total,ion} = C_G + C_{S,D} + C_{CNT}$, as shown in Fig. 7c. When this additional source of stray capacitance becomes available, the gate efficiency is decreased, leading to an increased subthreshold swing, according to equation 2. This additional capacitance can also be thought of as capacitive coupling to the partially filled density of states in the valence band of nearby sections of the CNT. Owing to the ion potential, these states are mid-gap in the defect region, as can be seen in Fig. 7a. Hence, this coupling is

analogous to the effective capacitance of interface traps in silicon FET devices. Given the gate capacitance of $8\,pF\,m^{-1}$, this reduction in alpha corresponds to an additional stray capacitance of $C_{CNT} = 40\,pF\,m^{-1}$. For comparison, the quantum capacitance of an individual CNT is $368\,pF\,m^{-1}$. These values are compatible with the proposed model of capacitive coupling to the partially filled density of states of nearby CNT; however, a more thorough, quantitative analysis of the electrostatics involved is needed for a complete understanding of the changes observed in the subthreshold swing. The Landauer transport model with ion potential included fits the data reasonably well, except for in the case of the OFF state, which is dominated by the noise floor and leakage currents, and above-threshold, where the measured current is lower than that predicted by the model. This discrepancy might be accounted for by defect potential-induced charge-carrier scattering.

Discussion

The transient switching events observed during exposure to ionized gas are attributed to the ionized gas molecules from the surrounding atmosphere adsorbing on the CNT and behaving as defects that modify the electrical properties of the CNT channel. This hypothesis is well supported by the data in Figs 2 and 3, which show discrete quantized resistance increases in the CNT resistance when exposed to ionized gas, and also by the fact that the effect is only observed during exposure to positive ions.

It may seem surprising that a potential well caused by such a small charge could reduce the current passing through the CNT by over two orders of magnitude. On further inspection, however, this seems reasonable. The potential barrier created by the above ion configuration has a barrier height of ~ 0.37 eV and a full-width half-maximum of 2.5 nm. This alone would pose a significant impediment to charge transport, but perhaps not to the degree seen in the data. This figure for barrier width, however, ignores the energy dependence of charge carrier transport. For the time series data, the gate bias voltage was held at 0 V, which corresponds approximately to the threshold voltage of this device. In this state, the Fermi energy is well inside the forbidden band gap, and thus only the charge carriers closest to the valence band edge contribute to charge transport. A weighted average of the barrier width based on the Fermi distribution of carrier population yields a much larger *effective* barrier width of 22 nm. Thus, when the device is biased into this subthreshold state, it is very sensitive to external potential variations. However, when the device is fully turned on (for large negative gate voltages), the potential perturbation has a much smaller effect, as is seen in Fig. 7. The extreme sensitivity to electrically charged defects observed here can also be understood by contrasting the CNT with two- and three-dimensional systems. In these systems, charge carriers can bypass localized potential barriers by going around them. This is not possible in the severely restricted current path of the 1D CNT.

In summary, we report on the extreme sensitivity of nano-electronic devices to single gaseous ion adsorption, which is attributed to single-ion-induced charge depletion in the 1D channel. Ion adsorption on the CNT is modelled with DFT, which predicts large binding energies, an ion–CNT surface separation of 3 Å and a significant charge transfer. The gate voltage dependence of the CNT current during ion exposure is modelled with a numerical solution of the discretized quantum mechanical Hamiltonian combined with a Landauer electrical transport model, with good agreement between theoretical modelling and experimental results. These experiments have profound impact on our current understanding of the interaction of single ions in 1D nano-electronic materials, and demonstrate the basis for a powerful system to study ion adsorption dynamics[33] at the single-ion level, and also charge transport in CNTs as well as other nano-electronic systems such as single-atom transistors[34], nanowires[35] and quantum dots[36]. Equally important is the demonstration of a new single-ion/CNT-based 1D heterostructure, with novel electronic properties such as resonant tunnelling and single-electron bound states. In addition, the response of nano-electronic devices to ionized gasses has implications for microelectronic devices operating while being exposed to ionizing radiation, as, for instance, in the space radiation environment[37,38]. The effects observed here raise the prospect of sensitive gas detectors with short exposure times and miniscule detection limits operating at room temperature with almost complete noise immunity. Several important questions remain, such as the cause for the large differences observed in ion–surface adsorption lifetime, and also the nature of interactions between the CNT and ions of different chemical compositions.

Methods

CNT device fabrication. The isolated, suspended CNTs were grown by using chemical vapour deposition at temperatures between 800 and 900 °C, argon-bubbled through ethanol as the carbon source and lithographically defined catalyst islands consisting of a Fe–Mo mixture on an oxide support[27]. After chemical vapour deposition, nearly defect-free, isolated, suspended CNTs are produced, as evidenced by Raman spectroscopy. CNTs span 500-nm-deep trench structures with widths from 500 nm to 2 μm, formed in a silicon substrate with 1 μm SiO_2 per 100-nm-low stress SiN_x (Fig. 1a).

Electrical characterization and ion generation methods. Electrical measurements were performed with an Agilent 4,155 semiconductor parameter analyser with $V_{ds} = 0.1$ V, and all measurements were performed at room temperature. Gaseous ions were generated using ionizing radiation exposure with 50-MeV protons from the 88″ cyclotron facility at the Lawrence Berkeley National Laboratory, and also with gamma radiation (1.17/1.33 MeV) from a Co-60 gamma ray source at The Aerospace Corporation. Gaseous ions were also generated using high-voltage corona discharge from commercially available benchtop air ionizers. The resulting ions generated by these methods are in thermal equilibrium with the surrounding neutral gas molecules.

Theoretical modelling and simulation. DFT calculations were performed using the plane-wave pseudopotential method implemented using the Quantum ESPRESSO DFT package (v. 5.1)[39], using the spin-polarized Perdew-Burke-Ernzerhof (PBE) gradient-corrected functional[40]. Further details on DFT implementation and results are discussed in Supplementary Note 2. Device-level numerical simulations were performed with Matlab on an engineering workstation. Eigen functions of Schrödinger's equation were obtained by solving a discretized Hamiltonian matrix, and energy-dependent transmission coefficients through the ion potential barrier were calculated using the propagation matrix method[30]. The details on implementation of the Landauer transport model are reported elsewhere[28,31].

References

1. Yao, Z., Kane, C. L. & Dekker, C. High-field electrical transport in single-wall carbon nanotubes. *Phys. Rev. Lett.* **84,** 2941 (2000).
2. Kong, J. *et al.* Quantum interference and ballistic transmission in nanotube electron waveguides. *Phys. Rev. Lett.* **87,** 106801 (2001).
3. Deshpande, V. V. & Bockrath, M. The one-dimensional Wigner crystal in carbon nanotubes. *Nat. Phys.* **4,** 314–318 (2008).
4. Durkop, T., Getty, S. A., Cobas, E. & Fuhrer, M. S. Extraordinary mobility in semiconducting carbon nanotubes. *Nano Lett.* **4,** 35–39 (2004).
5. Baumgardner, J. *et al.* Inherent linearity in carbon nanotube field-effect transistors. *Appl. Phys. Lett.* **91,** 052107 (2007).
6. Kong, J. *et al.* Nanotube molecular wires as chemical sensors. *Science* **287,** 622–625 (2000).
7. Sazonova, V. *et al.* A tunable carbon nanotube electromechanical oscillator. *Nature* **431,** 284 (2004).
8. Sun, D.-m. *et al.* Flexible high-performance carbon nanotube integrated circuits. *Nat. Nano* **6,** 156–161 (2011).
9. Liu, F., Wang, K. L., Li, C. & Zhou, C. Study of random telegraph signals in single-walled carbon nanotube field effect transistors. *IEEE Trans. Nanotechnol.* **5,** 441–445 (2006).
10. Goldsmith, B. R., Coroneus, J. G., Kane, A. A., Weiss, G. A. & Collins, P. G. Monitoring single-molecule reactivity on a carbon nanotube. *Nano Lett.* **8,** 189–194 (2007).
11. Sorgenfrei, S. *et al.* Label-free single-molecule detection of DNA-hybridization kinetics with a carbon nanotube field-effect transistor. *Nat. Nano* **6,** 126–132 (2011).
12. Donhauser, Z. J. *et al.* Conductance switching in single molecules through conformational changes. *Science* **292,** 2303–2307 (2001).
13. Ralls, K. S. *et al.* Discrete resistance switching in submicrometer silicon inversion layers: individual interface traps and low-frequency 1/f noise. *Phys. Rev. Lett.* **52,** 228–231 (1984).
14. Jensen, K., Kim, K. & Zettl, A. An atomic-resolution nanomechanical mass sensor. *Nat. Nano* **3,** 533–537 (2008).
15. Schedin, F. *et al.* Detection of individual gas molecules adsorbed on graphene. *Nat. Mater.* **6,** 652–655 (2007).
16. Kim, J.-H. *et al.* Single-molecule detection of H_2O_2 mediating angiogenic redox signaling on fluorescent single-walled carbon nanotube array. *ACS Nano* **5,** 7848–7857 (2011).
17. Choi, Y. *et al.* Single molecule sensing with carbon nanotube devices. *Proc. SPIE* **8814,** doi:10.1117/12.2025661 (2013).

18. Gerenser, L. J., Elman, J. F., Mason, M. G. & Pochan, J. M. E.s.c.a. studies of corona-discharge-treated polyethylene surfaces by use of gas-phase derivatization. *Polymer* **26**, 1162–1166 (1985).

19. Comizzoli, R. B. Uses of corona discharges in the semiconductor industry. *J. Electrochem. Soc.* **134**, 424–429 (1987).

20. Schroder, D. K. Contactless surface charge semiconductor characterization. *Mater. Sci. Eng. B* **91–92**, 196–210 (2002).

21. Ang, P. K., Chen, W., Wee, A. T. S. & Loh, K. P. Solution-gated epitaxial graphene as pH sensor. *J. Am. Chem. Soc.* **130**, 14392–14393 (2008).

22. Wang, N.-P., Heinze, S. & Tersoff, J. Random-telegraph-signal noise and device variability in ballistic nanotube transistors. *Nano Lett.* **7**, 910–913 (2007).

23. Deshpande, V. V., Hsieh, S., Bushmaker, A. W., Bockrath, M. & Cronin, S. B. Spatially resolved temperature measurements of electrically heated carbon nanotubes. *Phys. Rev. Lett.* **102**, 105501–105504 (2009).

24. Bushmaker, A. W., Deshpande, V. V., Hsieh, S., Bockrath, M. W. & Cronin, S. B. Gate voltage controllable non-equilibrium and non-ohmic behavior in suspended carbon nanotubes. *Nano Lett.* **9**, 2862–2866 (2009).

25. Pop, E. *et al.* Negative differential conductance and hot phonons in suspended nanotube molecular wires. *Phys. Rev. Lett.* **95**, 155505–155508 (2005).

26. Bushmaker, A. W., Deshpande, V. V., Bockrath, M. W. & Cronin, S. B. Direct observation of mode selective electron – phonon coupling in suspended carbon nanotubes. *Nano Lett.* **7**, 3618–3622 (2007).

27. Cao, J., Wang, Q. & Dai, H. Electron transport in very clean, as-grown suspended carbon nanotubes. *Nat. Mater.* **4**, 745–749 (2005).

28. Bushmaker, A. W., Amer, M. R. & Cronin, S. B. Electrical transport and channel length modulation in semiconducting carbon nanotube field effect transistors. *IEEE Trans. Nanotechnol.* **13**, 176–181 (2014).

29. Chang, C.-C. *et al.* A new lower limit for the ultimate breaking strain of carbon nanotubes. *ACS Nano* **4**, 5095–5100 (2010).

30. Levi, A. F. J. *Applied Quantum Mechanics* (Cambridge University Press, 2006).

31. Rahman, A., Jing, G., Datta, S. & Lundstrom, M. S. Theory of ballistic nanotransistors. *IEEE Trans. Electron Devices* **50**, 1853–1864 (2003).

32. Sze, S. M. & Ng, K. K. *Physics of Semiconductor Devices* (John Wiley & Sons, 2006).

33. Horinek, D. & Netz, R. R. Specific ion adsorption at hydrophobic solid surfaces. *Phys. Rev. Lett.* **99**, 226104 (2007).

34. Park, J. *et al.* Coulomb blockade and the Kondo effect in single-atom transistors. *Nature* **417**, 722–725 (2002).

35. Singh, N. *et al.* High-performance fully depleted silicon nanowire (diameter ≤5nm) gate-all-around CMOS devices. *IEEE Electron Device Lett.* **27**, 383–386 (2006).

36. Michler, P. *et al.* A quantum dot single-photon turnstile device. *Science* **290**, 2282–2285 (2000).

37. Estrup, P. J. Surface effects of gaseous ions and electrons on semiconductor devices. *IEEE Trans. Nucl. Sci.* **12**, 431–436 (1965).

38. Bushmaker, A. W. *et al.* Single event effects in carbon nanotube-based field effect transistors under energetic particle radiation. *IEEE Trans. Nucl. Sci.* **61**, 2839 (2014).

39. Giannozzi, P. *et al.* QUANTUM ESPRESSO: a modular and open-source software project for quantum simulations of materials. *J. Phys. Condens. Matter* **21**, 395502 (2009).

40. Perdew, J. P., Burke, K. & Ernzerhof, M. Generalized gradient approximation made simple. *Phys. Rev. Lett.* **77**, 3865–3868 (1996).

Acknowledgements

We acknowledge Rocky Koga, Jeffrey George and Steve Bielat of The Aerospace Corporation for assistance with operation of the LBNL 88″ cyclotron, and Steve Moss, Ron Lacoe and Robert Nelson of The Aerospace Corporation and Cory Cress of NRL for discussions regarding the interpretation of the data. The portion of the work performed at The Aerospace Corporation was funded by the Independent Research and Development Program at The Aerospace Corporation. DFT calculations were performed on a SGI Silicon Graphics Inc. integrated cluster environment (SGI ICE X) system, with wall time provided by the DoD HPC Modernization Office. A portion of this work was performed in the UCSB nanofabrication facility, part of the NSF funded NNIN network. Sample fabrication was supported by the US Department of Energy, Office of Basic energy Sciences, Division of Materials sciences and Engineering under Award No. DE-FG02-07ER46376.

Author contributions

J.C. and S.B.C. fabricated CNT FET samples; A.W.B., D.W. and A.R.H. designed the study, procured experimental hardware and collected the data; A.W.B., V.O. and D.W. analysed the data; V.O. performed DFT theory calculations; and A.W.B. performed device-level modelling and wrote the paper. All authors discussed results and commented on the manuscript.

Additional information

4

In situ stress observation in oxide films and how tensile stress influences oxygen ion conduction

Aline Fluri[1], Daniele Pergolesi[1], Vladimir Roddatis[2], Alexander Wokaun[1] & Thomas Lippert[1,3]

Many properties of materials can be changed by varying the interatomic distances in the crystal lattice by applying stress. Ideal model systems for investigations are heteroepitaxial thin films where lattice distortions can be induced by the crystallographic mismatch with the substrate. Here we describe an *in situ* simultaneous diagnostic of growth mode and stress during pulsed laser deposition of oxide thin films. The stress state and evolution up to the relaxation onset are monitored during the growth of oxygen ion conducting $Ce_{0.85}Sm_{0.15}O_{2-\delta}$ thin films *via* optical wafer curvature measurements. Increasing tensile stress lowers the activation energy for charge transport and a thorough characterization of stress and morphology allows quantifying this effect using samples with the conductive properties of single crystals. The combined *in situ* application of optical deflectometry and electron diffraction provides an invaluable tool for strain engineering in Materials Science to fabricate novel devices with intriguing functionalities.

[1] Department for Energy and Environment, Paul Scherrer Institut, 5232 Villigen-PSI, Switzerland. [2] Institut für Materialphysik, Universität Göttingen, Friedrich-Hund-Platz 1, Göttingen 37077, Germany. [3] Department of Chemistry and Applied Biosciences, Laboratory of Inorganic Chemistry, Vladimir-Prelog-Weg 1-5/10, ETH Zürich, Zürich CH-8093, Switzerland. Correspondence and requests for materials should be addressed to D.P. (email: daniele.pergolesi@psi.ch) or to T.L. (email: thomas.lippert@psi.ch).

Stress-induced lattice distortions, that is, a tensile or compressive strain of the crystal structure, can significantly influence the physicochemical characteristics of materials by enhancing/inhibiting specific properties or by enabling new functionalities not allowed in the unperturbed structure. Strain engineering offers indeed a new route to tune the characteristics of a material for example by modifying the electronic bandgap[1], multiferroic[2] or catalytic[3] properties, thermal conductivity[4] and charge transport[5–8].

To investigate the effects of strain, epitaxial thin films are typically used where the strain is induced by a lattice mismatch at the interface with the substrate. For highly ordered epitaxial films, established models exist, which assume elastic distortions and coherent growth, that is, a 1:1 matching between all lattice planes[9–11]. The upper limit of the strain is in this case the lattice mismatch between film and substrate. In general, however, we often deal with semi-coherent heterointerfaces where it is energetically favourable to reduce the elastic energy by introducing dislocations. Already a mismatch exceeding ~1% is typically not expected to be fully accommodated[9–11]. The growth mode (layer-by-layer or island-like) also influences the effective strain through the surface morphology and relaxation behaviour[12–14].

Monitoring the stress and the growth mode in situ (during the growth) allows investigating fundamental mechanisms of stress generation and evolution from the initial nucleation to the onset of relaxation. Literature concerning the in situ diagnostic of stress/strain generation and evolution during the epitaxial growth of oxide films is scarce. Here, we address this topic measuring simultaneously in situ the stress with a multi-beam optical stress sensor (MOSS) and the growth mode by reflection high-energy electron diffraction (RHEED) during pulsed laser deposition (PLD).

When a stressed film grows, the substrate bends minimizing the elastic energy. Stoney's equation[10] correlates the substrate curvature $1/\rho$ with the film stress σ as

$$\frac{1}{\rho} = 6\frac{(1-v)}{Y\tau_s^2}\tau\sigma \qquad (1)$$

where ρ being the ray of curvature, τ and τ_s the film and substrate thicknesses, v and Y the Poisson ratio and the Young's modulus of the substrate and σ the average film stress. As sketched in Fig. 1, the MOSS uses the deflection of laser beams to measure the change in curvature, and thus to monitor in situ and in-plane the stress of the growing film[10]. In contrast, conventional X-ray diffraction analysis yields the strain, ex situ and out-of-plane.

The in- and out-of-plane strain of a film, ε_{xx} and ε_{zz}, are related through the Poisson ratio[9] v as

$$\varepsilon_{zz} = \frac{2v}{v-1}\varepsilon_{xx} \qquad (2)$$

For most materials $0 < v < 0.5$, implying that an in-plane tensile (compressive) strain of the unit cell corresponds to an out-of-plane compressive (tensile) strain.

Different in situ strain diagnostics have been applied for thin films. The strain evolution of $Ba_{0.5}Sr_{0.5}TiO_3$ thin films grown on MgO was investigated using real-time X-ray diffraction[15]. This is undoubtedly a very powerful approach but it requires a synchrotron light source[16] precluding its application as a routine measurement technique. RHEED was also employed to monitor the in-plane strain evolution, for example, during the growth of $BaTiO_3$ on $SrTiO_3$ (refs 9,14). This specific application of RHEED is very rare, even considering the widespread use of high-pressure RHEED. In fact, to minimize electron scattering, it can only be attempted at low pressure, a growth condition unsuitable for many oxides. A wafer curvature measurement similar to MOSS

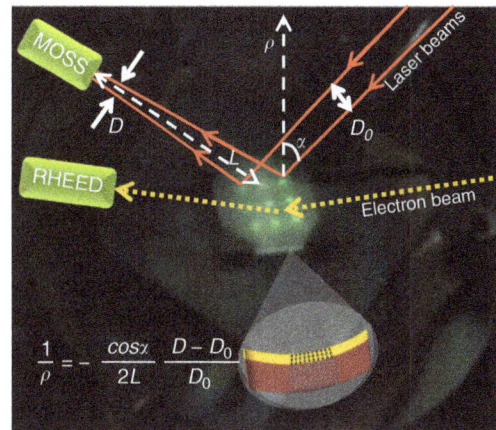

$$\frac{1}{\rho} = -\frac{\cos\gamma}{2L}\frac{D-D_0}{D_0}$$

Figure 1 | Working principle of the multi-beam optical stress sensor (MOSS). $10 \times 10\,mm^2$ MgO substrate on the sample holder of the PLD system equipped with MOSS and RHEED. A 3×3 array of parallel laser beams (visible as bright spots on the substrate surface) is reflected by the substrate towards a CCD camera that records the relative distance between the laser spots. The paths of two laser beams of the MOSS and of the electron beam of the RHEED are sketched. The growth of a strained layer induces a change of curvature $(1/\rho)$ of the substrate and a change of the direction of the reflected laser beams. The effect of a stress-induced curvature of the substrate is illustrated in cross-section for the case of an in-plane tensile strained film $(\rho > 0)$. The relative curvature change is obtained by measuring the change in the relative distance $(D-D_0)/D_0$ between the laser beams; D_0 being the distance at the beginning of the growth, L the optical path length and α the incident angle. From the CCD image, the MOSS software calculates the average of $(D-D_0)/D_0$ using all nine beams.

was used during PLD of $BaTiO_3$ and $SrTiO_3$ ultra-thin films on a Pt cantilever[17]. The main advantages of MOSS over similar deflection techniques are that one can use any substrate, no position sensitive detector is needed and that the noise is reduced through multiple beams[10].

Several examples of MOSS (or similar techniques) applications for studying the stress evolution during the growth of semiconductor or metal thin films exist[18–24], whereas reports on MOSS diagnostic with oxide materials are very rare[25]. To our knowledge, this technique has never been applied simultaneously with RHEED during PLD, the most widespread method for the growth of oxide films. In particular, by coupling the MOSS with RHEED, the stress state of the growing film can be correlated with the growth mode from the initial nucleation to stress relaxation, thus providing a unique diagnostic tool.

15 at.% Sm-doped CeO_2 (SDC) is the material selected for this fundamental study. SDC is an oxygen-ion conductor used as electrolyte in solid oxide fuel cells (SOFCs), environmentally sustainable electrochemical energy converters[26]. The oxygen ion conductivity σ_{ion} is described with the Arrhenius equation

$$\sigma_{ion} = \frac{\sigma_0}{T}e^{-\frac{E_A}{k_B T}} \qquad (3)$$

T being the temperature, σ_0 the pre-exponential factor, E_A the activation energy and k_B the Boltzmann constant. Currently, the most accepted hypothesis is that tensile strain lowers the activation energy for the oxygen-ion hopping by increasing the migration volume[7,8,27]. Consequently, larger ionic conductivities could be achieved at lower temperatures, with important consequences for technological applications.

Here, as a case study for the first MOSS plus RHEED application during PLD of oxide materials, the fundamental question of how strongly homogeneous strain can affect the ionic conductivity in a single crystal is addressed. This fundamental question is of interest as controversial results have been reported on the magnitude of the effect of strain on the ionic conductivity[8,27,28]. The increase in conductivity ascribed to strain effects spreads over several orders of magnitude showing the need to evaluate the effect using samples where any potential side effect is minimized and where the strain/stress state is unambiguously identified.

To achieve this goal, thin films of high crystalline quality—as close to a single crystal as experimentally possible—but with different strain values are employed as model systems. The MOSS is particularly suited for this investigation as the stress state can be probed in-plane, precisely in the direction of the ion migration during conductivity measurements, and *in situ* in a temperature range similar to the operating temperature of SDC for technological applications. This study confirms that tensile strain indeed leads to a decrease in activation energy showing that the effect is relatively small, roughly a factor of 2 at around 350 °C. But more importantly, we show here the potentials of MOSS diagnostic as an invaluable tool for strain engineering in Materials Science.

Results

Strain generation and evolution on different substrates. To study the strain generation and evolution in SDC films, MgO, $NdGaO_3$ (NGO) and $LaAlO_3$ (LAO) single-crystal substrates were used; Table 1 reports their crystallographic properties, the expected epitaxial relations and the in-plane lattice misfit with 15 at.% SDC (cubic fluorite structure with a lattice parameter of 5.43 Å). NGO and LAO provide a lattice mismatch in a suitable range to expect epitaxial films of SDC in tensile and compressive in-plane strain, respectively. On the contrary, the large lattice mismatch between SDC and MgO may prevent any crystallographic matching leading to a fully relaxed growth from the start. Figure 2 compares the MOSS *in situ* stress characterizations with the X-ray diffraction *ex situ* strain measurements of (001)-oriented SDC films on the different substrates. The MOSS measurements are reported as stress-thickness product (directly proportional to the curvature, from the Stoney equation (1)) in function of thickness, a plot commonly used in the literature for wafer curvature measurements[19–25].

As expected, during the growth on MgO, the stress-thickness product, that is, the substrate curvature, remains constant indicating that the film grows without exerting any force on the substrate, that is, stress free. Correspondingly, the X-ray diffraction analysis shows a fully relaxed crystalline structure. The large lattice misfit is accommodated by a high density of interfacial misfit dislocations that fully release the stress as revealed by transmission electron microscopy, discussed below. For the SDC film grown on NGO, the positive curvature indicates that the film exerts a compressive force on the substrate surface and bends it: the film is under in-plane tensile stress, which originates at the very beginning of the growth. The constant slope of the stress-thickness product (Fig. 2a) shows that the stress is

constant to a film thickness of 33 nm. An average stress of ~ 3.3 GPa can be calculated. The tensile in-plane stress is in agreement with the compressive out-of-plane strain of -0.52% measured *ex situ* by XRD. The XRD analysis of the 33-nm-thick SDC film on LAO reveals an average out-of-plane tensile strain of 0.37%. Comparing this value with the theoretical lattice mismatch of 0.49% between the two materials, we conclude that $\sim 25\%$ of the theoretical lattice mismatch is lost, most probably through interfacial misfit dislocations. Accordingly, the *in situ* MOSS measurement (Fig. 2a) shows in this case a constant compressive in-plane stress. In accordance to the slightly smaller strain value measured out-of-plane on LAO compared with NGO, the MOSS shows a smaller value (~ 3 GPa) of average residual and effective stress for the same total film thickness.

When a strained epitaxial film grows, elastic energy is accumulated until it becomes energetically favourable for dislocations lines to move through the crystal, relieving the strain[9–11]. Experimentally, mismatch values up to around 1% can be expected to be fully accommodated[11]. With the materials and growth condition selected here for SDC, 0.4–0.5% strain was induced starting from a theoretical lattice mismatch of ~ 0.5–0.6%. One possible explanation for the stress relaxation could be the growth mechanism. During the SDC growth on all the three different substrates, RHEED did not show a two-dimensional layer-by-layer growth. Instead, RHEED revealed a three-dimensional growth mode, also known as island or layer-plus-island growth mode. Such a growth mechanism along the (001) crystallographic orientation of ceria films has been reported and explained[29]. This growth mode facilitates the nucleation of dislocations[12,13], which assist the strain relaxation.

Tensile strained SDC films for electrical characterization. To allow a reliable in-plane electrical characterization of thin ion conducting films, the growth platform must add a negligible contribution to the total conductance of the sample. Among the substrates used in this study, only MgO fulfils this condition. As an example, at around 600 °C, the resistance of an NGO substrate is similar to that of a 10-nm-thick SDC film with the same area[30]. However, although MgO is an ideal substrate for conductivity measurements, it only allows the growth of fully relaxed SDC films.

In previous studies, $SrTiO_3$ (STO) was used as a thin epitaxial buffer layer on MgO to favour a highly ordered growth of ceria films on an insulating substrate[31,32]. More recently, it was found that an additional thin inter-layer of $BaZrO_3$ (BZO) with a thickness of 5 nm grown between MgO and STO improves the crystalline quality of the STO layer[33]. The same sample design described in ref. 33 was used here. The 5-nm-thin STO seed layer does not significantly affect the conductance of the sample while providing a lattice mismatch of $\sim 1.6\%$ (for the fully relaxed structure) with respect to SDC, which will induce an in-plane tensile strain in epitaxially grown SDC.

Figure 3a shows the XRD analysis of three epitaxial SDC films, 13, 20 and 83 nm-thick, grown on the (001)-oriented

---

Table 1 | Characteristics of the single crystal substrates used for this study.

Substrate	Crystalline structure		Expected epitaxial relation	Lattice mismatch (%)
NdGaO$_3$ (110)	Orthorhombic perovskite $a = 5.43$ Å $c = 5.50$ Å	In-plane Out-of-plane	SDC(1$\bar{1}$0)\|\|NGO(1$\bar{1}$0) SDC(110)||NGO(001) SDC(001)||NGO(110)	0.64, 0.39
LaAlO$_3$ (001)	Pseudocubic perovskite $a = 3.82$ Å	In-plane Out-of-plane	SDC(110)||LAO(100) SDC(1$\bar{1}$0)||LAO(010) SDC(001)||LAO(001)	-0.49, -0.49
MgO (001)	face-centered cubic $a = 4.21$ Å		None	≈ -29

Expected epitaxial relations and lattice mismatch with 15 at.% Sm-doped CeO$_2$ are displayed. The lattice mismatch, given in percent, is calculated as $(a_{substrate} - a_{SDC})/a_{substrate}$.

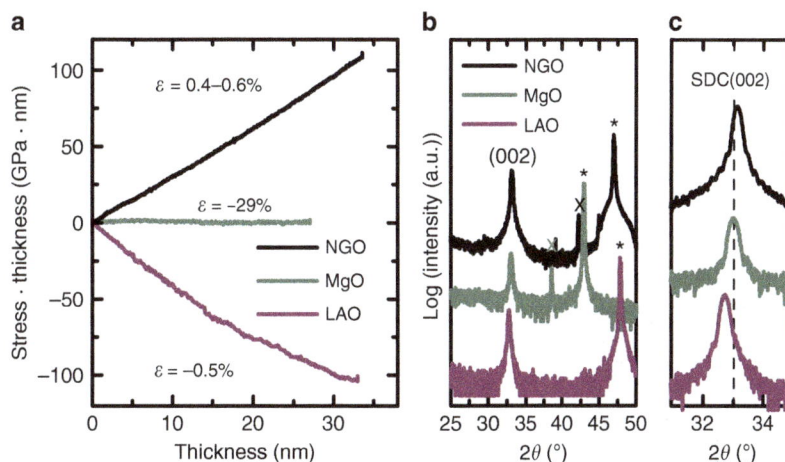

Figure 2 | MOSS and XRD analysis of SDC films grown on NGO and MgO and LAO. (a) Stress-thickness product, evaluated using the Stoney equation and the elastic properties of the substrates[57-59], as a function of thickness. A positive (negative) stress value indicates an in-plane tensile (compressive) stress for the film. ε is the lattice mismatch (Table 1). **(b)** $\omega/2\theta$ scans. *indicates the substrate peak and X is an instrumental artefact (K_β reflection of substrate). **(c)** Magnification of the region around the (002) diffraction peak of SDC. The dashed line indicates the angular position for the fully relaxed SDC structure. The peak shifts are in accordance with the sign of the stress-thickness product.

MgO + BZO + STO template platform (hereafter we will refer to this template platform as MgO-BS). In spite of the small thickness of the layer, the $\omega/2\theta$ scans allow a rough estimation of the out-of-plane compressive strain of the STO layer between 0.1 and 0.5%, thus the misfit between the growth platform and SDC is expected to be in the range of 1.7–2.1% in-plane.

The BZO + STO layers were deposited at around 765 °C. The 20- and 83-nm-thick SDC films on MgO-BS were deposited at 650 °C, the same temperature used for the films on MgO, NGO and LAO. To probe the effect of temperature on the developed stress, the 13-nm-thick film was grown at the same temperature used for BZO and STO.

The shift of the diffraction peak of these three films towards larger 2θ angles indicates different out-of-plane compressive strain values (Fig. 3b). To quantify the strain in-plane, reciprocal space maps (RSMs) were recorded. As an example, the RSM acquired along the (204) asymmetric diffraction of the 20-nm-thick SDC film grown on MgO-BS is reported in Fig. 3c. The 20-nm-thick film shows the largest in-plane tensile strain, that is, 0.35%.

As expected, the crystalline structure relaxes with increasing thickness: for the 83-nm film, an in-plane strain of 0.24% is found. The 13-nm-thick film grown at the higher temperature shows the smallest in-plane tensile strain value of $\sim 0.1\%$. This can be explained by considering that for a film to relax dislocations have to nucleate or existing dislocation lines have to be mobile. The mobility of a dislocation can be described by a periodic energy barrier that can be overcome by accumulating elastic energy or by providing thermal energy[9], which facilitates the relaxation process.

Figure 3d, e shows the results of the *in situ* MOSS plus RHEED measurements performed during the growth of the SDC films on the MgO-BS template platform. Figure 3d reports an example of the wafer curvature measurements showing the capability of MOSS diagnostic to probe *in situ* the in-plane stress of the different layers during the growth of a complex heterostructures.

Owing to the very good lattice matching between MgO and BZO, no significant wafer curvature (stress) can be observed during the growth of the 5-nm-thick BZO layer. On the contrary, the lattice mismatch between BZO and STO (as large as 7%) induces an in-plane tensile stress to the STO film, as revealed by the positive wafer curvature measured by MOSS. It is worth

highlighting that for such thin layers the MOSS allows the stress state of the film to be unambiguously identified *in situ* much more effectively than by evaluating *ex situ* the angular position of the broad peaks in the $\omega/2\theta$ scans. Within this experiment, the MOSS diagnostic allows resolving the change of the substrate curvature induced by the deposition of films with thicknesses in the ≥ 1 nm range.

Finally, the substrate continues to bend in the same direction during the growth of the SDC film on top of STO indicating an in-plane tensile strain of the oxygen ion conductor.

The simultaneous RHEED measurements are shown in Fig. 3e. The RHEED patterns of BZO and STO indicate an almost two-dimensional growth. But, as previously discussed, the RHEED pattern for SDC clearly shows a three-dimensional growth mode from the very beginning of the film nucleation.

Figure 3f shows the *in situ* MOSS measurements of the 20- and 83-nm-thick SDC films grown on MgO-BS, as well as those of the SDC films grown on NGO and MgO for comparison. The data for the 20- and 83-nm-film grown with the same deposition parameters on MgO-BS are in good agreement. At the thickness of 20 nm, the MOSS allows estimating an average in-plane stress of ~ 1.5 GPa for both films.

The extent of the average tensile stress retained in the SDC layer on MgO-BS was smaller than on NGO (and on LAO in compressive stress). The much larger lattice misfit and the presence of a larger density of crystalline defects in the double buffer layer as compared with a single-crystal substrate can explain the larger stress relaxation.

The stress-thickness curve shows a constant slope up to 65 nm for increasing film thickness until the curve becomes flat between 70 and 75 nm. A constant curvature indicates that the total elastic energy in the film remains constant even though the thickness increases. This can be related to theoretical models for epitaxial thin film relaxation[9-11] according to which above a critical thickness enough elastic energy is accumulated that it becomes favourable for dislocation lines to move. The average stress is then gradually reduced and the total elastic energy remains constant while the film continues to grow. In other words, MOSS measurements allow identifying *in situ* the onset of stress relaxation in the film.

The measured ratio of stress (MOSS) over strain (RSM), equal to $Y/(1-v)$, is the same for both the 20- and 83-nm-thick films and

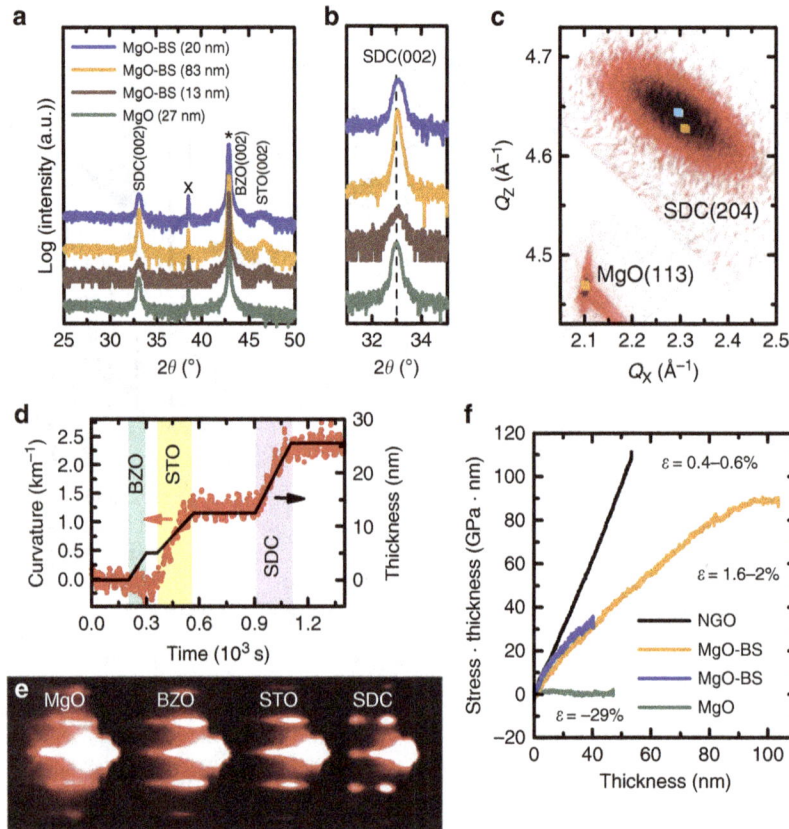

Figure 3 | XRD, MOSS and RHEED analysis of SDC on MgO + BZO + STO. (**a**) $\omega/2\theta$ scans. * and X indicate the substrate peak and an instrumental artefact (K_β reflection of substrate), respectively. (**b**) Magnification of the region around the (002) diffraction peak of SDC. The stress-free film on MgO is included for comparison. (**c**) RSM of the (204) reflection of the 20-nm SDC film. The (113) asymmetric diffraction of the MgO substrate was used for alignment. Orange square indicates the unstrained literature values and blue square the centre of the SDC(204) reflection. (**d**) representative MOSS measurement of wafer curvature during the growth of a complete heterostructure and (**e**) simultaneous RHEED measurement. For the RHEED diagnostic, the electron beam is parallel to the (110) direction of the substrate, thus parallel to (110) of BZO and STO, which show a cube-on-cube growth, but parallel to (100)—equivalently (010)—direction of SDC due to the in-plane 45° rotation of the unit cell of the ionic conductor with respect to the template platform. (**f**) MOSS measurements during the growth of the SDC films with different stress states.

in agreement with the literature[34] showing that the MOSS indeed reflects directly the average stress of the film.

The morphology and microstructure can both influence the evolution of the strain and the activation energy for oxygen ion conduction[35]. The local microstructure of the samples selected for the electrical characterization was investigated using transmission electron microscopy (TEM). Figure 4a, b shows representative examples of SDC films grown on MgO-BS and MgO, respectively.

Figure 4c shows a high-resolution bright-field (BF) STEM image of the cross-section of a SDC film grown on MgO-BS. Selected area electron diffraction and fast Fourier transform analysis reveal very good crystallographic quality showing that the [001] zone axis of the SDC film slightly precesses within only $\sim 1°$ on the scales of the order of 100 nm. The SDC film grown directly on the MgO (Fig. 4d) demonstrates a cube-on-cube growth in spite of the very large lattice misfit as already reported in the literature[36,37]. Further, the film shows the presence of local isolated domains rotated by 45° around the substrate surface normal. These isolated defects are not present in the film on MgO-BS. The very large lattice mismatch between SDC and MgO results in a lower crystallographic quality of the SDC film, which shows an average grain size of 40 nm and the SDC/MgO interface characterized by an almost continuous line of misfit dislocations that release the stress still preserving the epitaxial relation. The same kind of interface was observed between Gd-doped ceria and MgO (ref. 38). As reported in ref. 33, this difference in

morphology does not influence the ionic conduction. These measurements support the XRD and MOSS results that showed a fully relaxed (001) oriented SDC structure for the thin film grown on MgO and the presence of a measurable in-plane tensile strain for the films grown on MgO-BS.

Influence of tensile strain on ionic conductivity. AC impedance spectroscopy was applied for electrical characterization using patterned Pt electrodes on the film surface. Figure 5a shows the grain interior (bulk) ionic conductivity of three different Sm-doped ceria samples from the literature[39,40] compared with the measured conductivity of the MgO-BS template platform, which is more than two orders of magnitudes smaller, that is, negligible. Figure 5b shows the electrical characterization of SDC films with different strain values. Activation energy and conductivity values are in very good agreement with reference data for SDC grain interior[39–41], showing that the grain interior contribution dominates. The complex impedance plane plots (Supplementary Discussion and Supplementary Fig. 1) show a clear polarization of the Pt electrodes at low frequencies, indicating that the dominant charge carriers are indeed oxygen ions, as is well established for SDC under the selected experimental conditions. This leaves no doubt about the nature of the charge carriers.

The activation energy, calculated by fitting the linearized Arrhenius equation (3) to the data, shows a clear trend with respect to the film strain, that is, it decreases with increasing

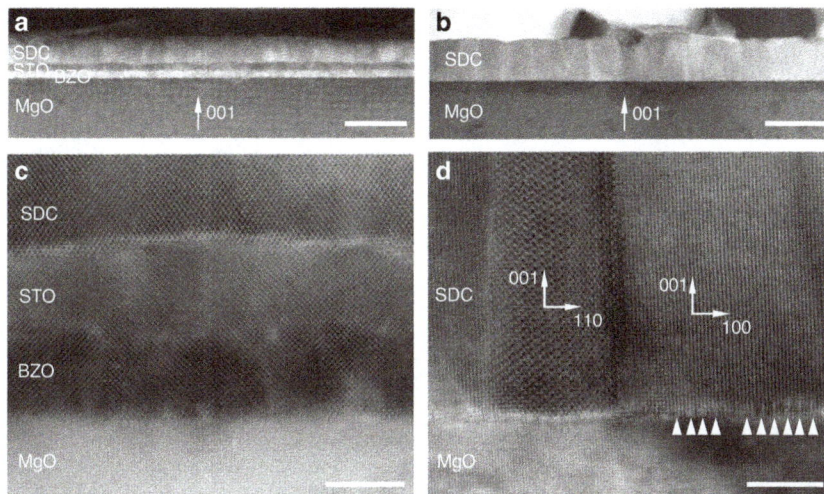

Figure 4 | TEM analysis of SDC on MgO-BS (left) and on MgO (right). Low-magnification cross-sectional HAADF-STEM image of (**a**) ~21-nm-thick SDC film on MgO-BS and (**b**) ~32-nm-thick SDC film on MgO. Scale bars correspond to 50 nm (**a**), 50 nm (**b**), 5 nm (**c**), 5 nm (**d**). Bright-field high-resolution STEM image of the SDC film on MgO-BS (**c**) and on MgO (**d**). One of the local defects (tilted grains) is shown in **d** together with the positions, marked with arrowheads, of some interfacial misfit dislocation at the SDC/MgO interface. The TEM micrographs were acquired along [010] direction of MgO.

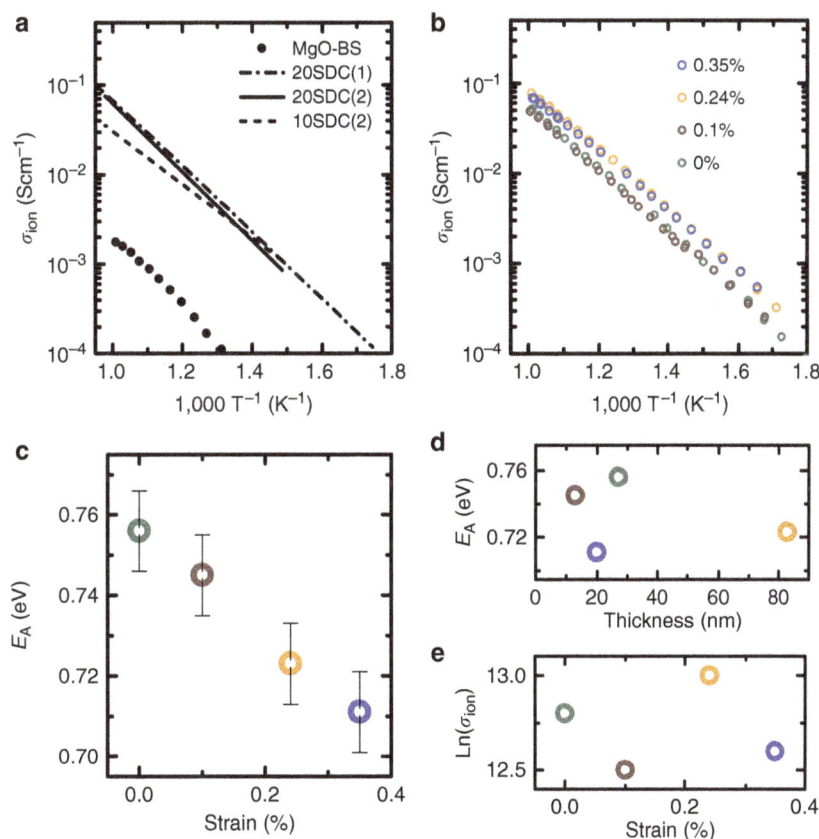

Figure 5 | Impedance spectroscopy measurements of SDC films with different strain. (**a**) Conductivity measurement of the MgO + BZO + STO (MgO-BS) template platform compared with literature references of unstrained Sm-doped ceria bulk: 20 at.% Sm (20SDC(1) (ref. 39)) and 10 and 20 at.% Sm (10SDC(2) and 20SDC(2) (ref. 40)). (**b**) Comparison of conductivity measurements of the SDC films fabricated for this study with different in-plane tensile strain values. (**c**) Activation energy for ion migration as a function of tensile strain. The error bar is determined form the standard deviation of the fit and empirically by comparison of independent measurements. (**d**) Activation energy versus SDC thickness and (**e**) Arrhenius pre-exponential factor of the SDC films with different strain values.

tensile strain (Fig. 5c). Interface effects (for example, dislocations[42]) could also influence the activation energy, resulting in a thickness dependence of the activation energy. This is not the case here (Fig. 5d) suggesting that it is indeed the strain that influences the activation energy. Further, the pre-exponential factor σ_0 of the Arrhenius equation varies little (Fig. 5e), showing that the charge carrier density is not affected by the strain.

The change of conductivity is negligible in the high-temperature range, whereas for example at 350 °C (a temperature relevant for miniaturized SOFCs), the bulk ionic conductivity is a factor of two larger compared with the unperturbed crystalline structure as a result of 0.35% tensile lattice strain, which lowers the activation energy by ~ 0.05 eV.

The observed influence of the strain values on the activation energy is very similar to that reported in a theoretical study on undoped ceria[43]. From the experimental point of view, literature reports on the effect of strain on conductivity for epitaxial doped ceria films are scarce. In ref. 44, for example, a very large increase in conductivity with increasing tensile strain was reported. Here a surprisingly large value of strain of 2% was estimated for 10 at.% Gd-doped ceria films as thick as 250 nm. However, in this study, larger strain results in higher ionic conductivity, but also in an increased defect density, yielding a much higher activation energy. The same material coupled with Er_2O_3 was used for the fabrication of multi-layered microdots (111)-oriented on sapphire[45]. The reported activation energy of the single doped-ceria layer is in good agreement with the typical values reported for this material, even though the total conductivity is surprisingly smaller than the typical bulk conductivity of Gd-doped ceria[39,41] suggesting that conduction pathways other than through the bulk dominate. However, the authors reported that a change of the compressive in-plane strain of the ceria layers from about 0.1 to 1.1% results in an increase of the activation energy by 0.2 eV. Assuming an almost linear trend of activation energy versus strain for strain values within $\pm 1\%$, the reported effect in ref. 45 is in line with our measurements.

Comparing the effect on SDC to a different oxygen ion conductor, that is, yttria-stabilized zirconia (YSZ), a similar increase in conductivity can be extrapolated for this particular strain value (0.35%) from a theoretical study on the influence of strain in YSZ[46]. The result reported here is also in agreement with the ~ 3.4-fold enhanced conductivity measured for YSZ/Y_2O_3 0.8% strained multilayers with columnar morphology[47]. Few papers report no effect at all, for example, using CeO_2/YSZ multi-layered heterostructures[32], where the typical YSZ bulk conductivity was measured for all samples. In that work, the non-uniform strain along the layers revealed by high-resolution TEM and RSM could not be quantified and the residual and effective strain could be small. Moreover, only the temperature range above 400 °C was investigated, whereas in the present work, impedance measurements were acquired down to a minimum temperature of 280 °C, and it is precisely at lower temperatures where the effect is more evident.

On the contrary, as already pointed out, many other papers report very large strain effect on the ionic conductivity which are not in agreement with this or the above-mentioned studies. Several examples can be found in ref. 8. Finally, the effect of the local strain arising from buckling in free-standing polycrystalline membranes of doped ceria[48], as eventually relevant for devices, is discussed elsewhere, as well as the effect of strain in polycrystalline samples (sintered pellets or film crystallized by post-annealing), where the origin of strain is the specific local micro-morphology[49], which is completely different to the almost single crystalline samples in our study.

Discussion

Our results on the role of strain on the ionic conductivity in Sm-doped ceria shows that, as reported in many literature contributions, lower activation energies indeed result from tensile lattice distortions in the direction of the migration of the charge carriers. Our study was conducted on samples as closely resembling a uniformly strained single crystal as it was experimentally possible showing the characteristic bulk conductivity of the material under investigation.

It should be highlighted that the in-plane geometry of the electrical characterization dictates stringent limitations to the choice of the substrate. As a consequence, a relatively small value of residual and effective strain could be preserved. However, it was shown that significantly larger strain can be achieved. As an example, extrapolating the observed effect of strain on the activation energy, around 350 °C, a 0.5% tensile strained ion conductor, as achieved for films grown on NGO, would show the same conductivity as the unstrained conductor at around 450 °C.

Beside the specific application to oxygen ion conductors, which supports the thesis of a relatively small effect of strain on ionic conductivity, the most important result of our study is that it has been clearly shown that the MOSS is a valuable tool for investigating *in situ* the mechanism of stress generation and evolution in complex multi-layered oxide heterostructures. The stress states arising for different lattice mismatches as well as the relaxation behaviour in the epitaxial oxide films grown by PLD could be clearly monitored and were found to be in very good agreement with standard *ex situ* XRD analysis. Moreover, the stress state of the film can be clearly identified also for very thin layers in the nanometre range, well below the limit of conventional XRD analysis. Finally, the MOSS diagnostic of oxide materials has been coupled for the first time with RHEED during PLD.

Strain-controllable ionic or mixed ionic/electronic conductivity can boost the development of oxide heterostructures as active components for micro-electrochemical devices[45], such as micro-SOFCs[50], electrochemical sensors, resistive switching memories[51], or as active catalytic surfaces. But the potential of the MOSS plus RHEED diagnostics for lattice strain engineering goes far beyond this, providing a tool for tuning material properties for a broad range of applications (multiferroicity, catalysis, nano-device fabrication) in many different disciplines opening unprecedented opportunities in Materials Science.

Methods

Thin film deposition and *in situ* techniques. The thin films were fabricated in a PLD system from Twente Solid State Technology with integrated MOSS and RHEED systems (k-Space Associates, Inc.). The vacuum chamber, equipped with load-lock chamber, has a base pressure of about 10^{-8} mbar. Gas inlet lines allow setting the required background partial pressure during the deposition. A radiant resistive heater sets the desired deposition temperature. The substrate temperature was monitored with a pyrometer. The target carousel can hold up to five different targets. A custom-made substrate holder allows the simultaneous alignment of the substrate surface with the ablation plume, the RHEED and the MOSS systems.

$10 \times 10 \times 0.5$ mm^3 substrates from CrysTec GmbH are used. For MgO, a special substrate preparation is required where the substrates are cut in small wafers after polishing in order to obtain negligible surface curvature. This procedure results in a not atomically flat surface, as revealed by RHEED. By annealing the substrates in O_2 at 1,000 °C for 12 h, two-dimensional surface reconstruction was obtained. The substrate holder supports the substrate only in the corners without mechanical constrain to allow it to bend freely. As the MgO substrates have a transmission wavelength range of 0.2–8 μm, to ensure an accurate temperature reading and an efficient substrate heating, the unpolished surface of the substrates is coated with Pt (sputtered with 40 W at 3×10^{-2} mbar for 4 min, ~ 400 nm) used as heat absorber. Such a thin layer does not affect the overall elastic properties of the substrate; neither has it influenced the MOSS measurements of the relative changes in the substrate curvature.

Ceramic sintered pellets of BZO, STO and 15 at.% SDC, prepared in our laboratories, were ablated using a 248-nm KrF excimer laser (Coherent Lambda Physics GmbH) with pulse lenght of 25 ns. A spot size of the laser on the target of 1.2 mm^2, a fluence of 1.2 J cm^{-2}, a frequency of 4 Hz and a target to substrate distance of 5 cm were used. The oxygen back ground pressure was set to 5×10^{-2} mbar. BZO and STO were deposited at ~ 740 °C and SDC, unless otherwise indicated, at 650 °C. The deposition rates have been accurately calibrated by X-ray reflectometry to be 0.14 Å pulse^{-1} for BZO, 0.09 Å pulse^{-1} for STO and 0.11 Å pulse^{-1} for SDC. The samples were well reproducible, for example, concerning crystallinity, stress state, conductive properties and so on.

For the *in situ* wafer curvature measurement with the MOSS, a 3×3 array of laser beams is used. Typically, the measured wafer curvature is converted into the stress-thickness product via Stoney's equation and plotted as a function of thickness (see Supplementary Methods for more details; Supplementary Fig. 2). The main advantages of MOSS over similar techniques, for example, using a cantilever as substrate, are that any substrate can be used, no position-sensitive detector is needed, and the noise is reduced through multiple beams.

***Ex situ* characterization.** XRD characterizations were performed with a Siemens D500 diffractometer. The RSMs were recorded for the MgO(113) and the SDC(204) diffraction peak separately with the integration time for the film peak being 20 times higher. The RSMs consist of rocking curves, recorded for a range of 2θ values. The centre of the diffraction is determined by fitting individual rocking curves in the software Diffract.EVA.

A difference in the thermal expansion coefficients could lead to different *in* and *ex situ* strain values. However, the linear expansion coefficients for the materials used in this work are all around $10^{-5} K^{-1}$ (NGO (ref. 52), MgO (ref. 53), LAO (ref. 54), SDC (ref. 55), STO (ref. 54), BZO (ref. 56)) and the expansion mismatch is negligible (for example, 0.003% for SDC grown at 700 °C on MgO).

Cross-sectional TEM specimens were prepared by mechanical polishing and low-voltage Ar^+ ion milling for the final thinning. TEM and STEM investigations were performed on a Titan 80-300 (FEI) Environmental Transmission Electron Microscope equipped with an imaging-side aberration corrector. The experiments were conducted at an acceleration voltage of 300 kV. Atomic-resolution Z contrast images were obtained by HAADF and BF STEM imaging. The inner and outer collection angles of the detector were 70 and 200 mrad, respectively.

The impedance spectroscopy measurements were carried out under a flow of O_2 in a tube furnace. Rectangular parallel Pt electrodes were used, defined by stainless steel masks except for the characterization of the growth platform, where interdigitated electrodes were patterned by ultraviolet photolithography (Supplementary Methods and Supplementary Fig. 3). The electrodes were deposited by magnetron sputtering at room temperature. Pt was sputtered with 40 W at 3×10^{-2} mbar for 2 min (100–200 nm) on a Ti sticking layer sputtered with 20 W at 7×10^{-2} mbar for 1.5 min (~5 nm). The target to substrate distance was 4 cm. The electrodes were wired to the read-out electronics using Ag paste and Au wires. The impedance was measured using a Solartron 1260 impedance/gain-phase analyser with a bias voltage of 1 V in the frequency range between 1 Hz and 1 MHz, varying the temperature between 300 and 720 °C. Data analysis was performed with EC-Lab (V10.31), where the data were fit to the response of a RC parallel circuit.

References

1. Healy, N. *et al.* Extreme electronic bandgap modification in laser-crystallized silicon optical fibres. *Nat. Mater.* **13,** 1122–1127 (2014).
2. Ramesh, R. & Spaldin, N. A. Multiferroics: progress and prospects in thin films. *Nat. Mater.* **6,** 21–29 (2007).
3. Strasser, P. *et al.* Lattice-strain control of the activity in dealloyed core-shell fuel cell catalysts. *Nat. Chem.* **2,** 454–460 (2010).
4. Li, S. *et al.* Strain-controlled thermal conductivity in ferroic twinned films. *Sci. Rep.* **4,** 6375 (2014).
5. Llordés, A. *et al.* Nanoscale strain-induced pair suppression as a vortex-pinning mechanism in high-temperature superconductors. *Nat. Mater.* **11,** 329–336 (2012).
6. Giri, G. *et al.* Tuning charge transport in solution-sheared organic semiconductors using lattice strain. *Nature* **480,** 504–508 (2011).
7. Kilner, J. A. Ionic conductors: feel the strain. *Nat. Mater.* **7,** 838–839 (2008).
8. Yildiz, B. "Stretching" the energy landscape of oxides - effects on electrocatalysis and diffusion. *MRS Bull.* **39,** 147–156 (2014).
9. Hanbücken, M. *Stress And Strain In Epitaxy: Theoretical Concepts, Measurements and Applications* (Elsevier, 2001).
10. Suresh, S. & Freund, L. B. *Thin Film Materials: Stress, Defect Formation And Surface Evolution* (Cambridge Univ., 2006).
11. Ayers, J. E. *Heteroepitaxy Of Semiconductors: Theory, Growth, and Characterization* (CRC Press Taylor & Francis Group, 2007).
12. Jesson, D. E., Pennycook, S. J., Baribeau, J.-M. & Houghton, D. C. Surface stress, morphological development, and dislocation nucleation during strained-layer epitaxy. *MRS Online Proc. Library* **317,** 31–37 (1993).
13. Yang, W. H. & Srolovitz, D. J. Cracklike surface instabilities in stressed solids. *Phys. Rev. Lett.* **71,** 1593–1596 (1993).
14. Zhu, J., Li, Y. R., Zhang, Y., Liu, X. Z. & Tao, B. W. Effects of compressive and tensile stress on the growth mode of epitaxial oxide films. *Ceram. Int.* **34,** 967–970 (2008).
15. Bauer, S. *et al.* The power of in situ pulsed laser deposition synchrotron characterization for the detection of domain formation during growth of $Ba_{0.5}Sr_{0.5}TiO_3$ on MgO. *J. Synchrotron Radiat.* **21,** 386–394 (2014).
16. Brown, A. S. & Losurdo, M. in *Handbook of Crystal Growth* 2nd Edn (ed. Thomas, F. Kuech) 1169–1209 (Elsevier, 2015).
17. Premper, J., Sander, D. & Kirschner, J. *In situ* stress measurements during pulsed laser deposition of $BaTiO_3$ and $SrTiO_3$ atomic layers on Pt(0 0 1). *Appl. Surf. Sci.* **335,** 44–49 (2015).
18. Jacobs, R. N. *et al.* Dynamic curvature and stress studies for MBE CdTe on Si and GaAs substrates. *J. Electron. Mater.* **44,** 3076–3081 (2015).
19. Chason, E. *et al.* Growth of patterned island arrays to identify origins of thin film stress. *J. Appl. Phys.* **115,** 123519 (2014).
20. Scharf, T., Faupel, J., Sturm, K. & Krebs, H.-U. Intrinsic stress evolution in laser deposited thin films. *J. Appl. Phys.* **94,** 4273–4278 (2003).
21. Floro, J. A., Chason, E., Cammarata, R. C. & Srolovitz, D. J. Physical origins of intrinsic stresses in Volmer-Weber thin films. *MRS Bull.* **27,** 19–25 (2002).
22. Abadias, G., Fillon, A., Colin, J. J., Michel, A. & Jaouen, C. Real-time stress evolution during early growth stages of sputter-deposited metal films: influence of adatom mobility. *Vacuum* **100,** 36–40 (2014).
23. Chason, E., Shin, J. W., Hearne, S. J. & Freund, L. B. Kinetic model for dependence of thin film stress on growth rate, temperature, and microstructure. *J. Appl. Phys.* **111,** 083520 (2012).
24. Floro, J. A. & Chason, E. Measuring Ge segregation by real-time stress monitoring during $Si_{1-x}Ge_x$ molecular beam epitaxy. *Appl. Phys. Lett.* **69,** 3830–3832 (1996).
25. Michotte, S. & Proost, J. *In situ* measurement of the internal stress evolution during sputter deposition of ZnO:Al. *Sol. Energy Mater. Sol. Cells* **98,** 253–259 (2012).
26. Wachsman, E. D. & Lee, K. T. Lowering the temperature of solid oxide fuel cells. *Science* **334,** 935–939 (2011).
27. Jiang, J. & Hertz, J. L. On the variability of reported ionic conductivity in nanoscale YSZ thin films. *J. Electroceram.* **32,** 37–46 (2014).
28. Korte, C. *et al.* Coherency strain and its effect on ionic conductivity and diffusion in solid electrolytes - an improved model for nanocrystalline thin films and a review of experimental data. *Phys. Chem. Chem. Phys.* **16,** 24575–24591 (2014).
29. Pergolesi, D., Fronzi, M., Fabbri, E., Tebano, A. & Traversa, E. Growth mechanisms of ceria- and zirconia-based epitaxial thin films and heterostructures grown by pulsed laser deposition. *Mater. Renew. Sustain. Energy* **2,** 1–9 (2012).
30. Petric, A. & Huang, P. Oxygen conductivity of $Nd(SrCa)Ga(Mg)O_{3-\delta}$ perovskites. *Solid State Ionics* **92,** 113–117 (1996).
31. Sanna, S. *et al.* Fabrication and electrochemical properties of epitaxial samarium-doped ceria films on $SrTiO_3$-buffered MgO substrates. *Adv. Funct. Mater.* **19,** 1713–1719 (2009).
32. Pergolesi, D. *et al.* Tensile lattice distortion does not affect oxygen transport in Yttria-stabilized zirconia-CeO_2 heterointerfaces. *ACS Nano* **6,** 10524–10534 (2012).
33. Pergolesi, D. *et al.* Probing the bulk ionic conductivity by thin film heteroepitaxial engineering. *Sci. Tech. Adv.Mater* **16,** 015001 (2015).
34. Wachtel, E. & Lubomirsky, I. The elastic modulus of pure and doped ceria. *Scripta Mater.* **65,** 112–117 (2011).
35. Göbel, M. C., Gregori, G., Guo, X. & Maier, J. Boundary effects on the electrical conductivity of pure and doped cerium oxide thin films. *Phys. Chem. Chem. Phys.* **12,** 14351–14361 (2010).
36. Pérez Casero, R. *et al.* Epitaxial growth of CeO_2 on MgO by pulsed laser deposition. *Appl.Surf. Sci.* **109-110,** 341–344 (1997).
37. Chen, L. *et al.* Electrical properties of a highly oriented, textured thin film of the ionic conductor Gd:$CeO_{2-\delta}$ on (001) MgO. *Appl. Phys. Lett.* **83,** 4737–4739 (2003).
38. Sanna, S. *et al.* Enhancement of the chemical stability in confined δ-Bi_2O_3. *Nat. Mater.* **14,** 500–504 (2015).
39. Sameshima, S. *et al.* Electrical conductivity and diffusion of oxygen ions in rare-earth-doped ceria. *Nippon Seramikkusu Kyokai Gakujutsu Ronbunshi/J. Ceramic Soc. Jpn* **108,** 1060–1066 (2000).
40. Giannici, F. *et al.* Structure and oxide ion conductivity: local order, defect interactions and grain boundary effects in acceptor-doped ceria. *Chem. Mater.* **26,** 5994–6006 (2014).
41. Balazs, G. B. & Glass, R. S ac impedance studies of rare earth oxide doped ceria. *Solid State Ionics* **76,** 155–162 (1995).
42. Sun, L., Marrocchelli, D. & Yildiz, B. Edge dislocation slows down oxide ion diffusion in doped CeO_2 by segregation of charged defects. *Nat. Commun.* **6,** 1–10 (2015).
43. De Souza, R. A., Ramadan, A. & Horner, S. Modifying the barriers for oxygen-vacancy migration in fluorite-structured CeO_2 electrolytes through strain: a computer simulation study. *Energy Environ. Sci.* **5,** 5445–5453 (2012).
44. Mohan Kant, K., Esposito, V. & Pryds, N. Strain induced ionic conductivity enhancement in epitaxial $Ce_{0.9}Gd_{0.1}O_{2-\delta}$ thin films. *Appl. Phys. Lett.* **100,** 033105 (2012).
45. Schweiger, S., Kubicek, M., Messerschmitt, F., Murer, C. & Rupp, J. L. M. A microdot multilayer oxide device: Let us tune the strain-ionic transport interaction. *ACS Nano* **8,** 5032–5048 (2014).
46. Kushima, A. & Yildiz, B. Oxygen ion diffusivity in strained yttria stabilized zirconia: Where is the fastest strain? *J. Mater. Chem.* **20,** 4809–4819 (2010).
47. Aydin, H., Korte, C., Rohnke, M. & Janek, J. Oxygen tracer diffusion along interfaces of strained Y_2O_3/YSZ multilayers. *PCCP* **15,** 1944–1955 (2013).

48. Shi, Y., Bork, A. H., Schweiger, S. & Rupp, J. L. M. The effect of mechanical twisting on oxygen ionic transport in solid-state energy conversion membranes. *Nat. Mater.* **14**, 721–727 (2015).

49. Rupp, J. L. M. *et al.* Scalable oxygen-ion transport kinetics in metal-oxide films: impact of thermally induced lattice compaction in acceptor doped ceria films. *Adv. Funct. Mater.* **24**, 1562–1574 (2014).

50. Evans, A., Bieberle-Hütter, A., Rupp, J. L. M. & Gauckler, L. J. Review on microfabricated micro-solid oxide fuel cell membranes. *J. Power Sources* **194**, 119–129 (2009).

51. Messerschmitt, F., Kubicek, M., Schweiger, S. & Rupp, J. L. M. Memristor kinetics and diffusion characteristics for mixed anionic-electronic $SrTiO_{3-\delta}$ bits: The memristor-based cottrell analysis connecting material to device performance. *Adv. Funct. Mater.* **24**, 7448–7460 (2014).

52. Sasaura, M., Miyazawa, S. & Mukaida, M. Thermal expansion coefficients of high-T_c superconductor substrate $NdGaO_3$ single crystal. *J. Appl. Phys.* **68**, 3643–3644 (1990).

53. Tsay, Y.-f., Bendow, B. & Mitra, S. S. Theory of the temperature derivative of the refractive index in transparent crystals. *Phys. Rev. B* **8**, 2688–2696 (1973).

54. Char, K., Antognazza, L. & Geballe, T. H. Study of interface resistances in epitaxial $YBa_2Cu_3O_{7-x}$/barrier/$YBa_2Cu_3O_{7-x}$ junctions. *Appl. Phys. Lett.* **63**, 2420–2422 (1993).

55. Hyodo, J., Ida, S., Kilner, J. A. & Ishihara, T. Electronic and oxide ion conductivity in $Pr_2Ni_{0.71}Cu_{0.24}Ga_{0.05}O_4$/$Ce_{0.8}Sm_{0.2}O_2$ laminated film. *Solid State Ionics* **230**, 16–20 (2013).

56. Zhao, Y. & Weidner, D. J. Thermal expansion of $SrZrO_3$ and $BaZrO_3$ perovskites. *Phys. Chem. Minerals* **18**, 294–301 (1991).

57. Guennou, M., Bouvier, P., Garbarino, G. & Kreisel, J. Structural investigation of $LaAlO_3$ up to 63 GPa. *J. Phys. Condens. Matter.* **23**, 395401 (2011).

58. Krivchikov, A. I. *et al.* Structure, sound velocity, and thermal conductivity of the perovskite $NdGaO_3$. *Low Temp. Phys.* **26**, 370–374 (2000).

59. Ahrens, T. J. *Mineral Physics & Crystallography: A Handbook of Physical Constants* (American Geophysical Union, 1995).

Acknowledgements

The research leading to these results has received funding from the European Community's Seventh Framework Programme (FP7/2007–2013) under grant agreement no. 290605 (COFUND: PSI-FELLOW), as well as from the Swiss National Foundation for Science (SNFS) under grant agreement no. 200021_126783. We thank Johann Michler, Madoka Hasegawa and Vipin Chawla of the Laboratory for Mechanics of Materials and Nanostructures at EMPA, Thun (CH) for ultraviolet photolithography of interdigitated micro-electrodes.

Author contributions

A.F. and D.P. set up the experimental facility, fabricated and characterized the films, and wrote the manuscript. V.R. performed the TEM measurements. T.L. and D.P. designed and supervised the experiment. A.W. helped in the discussion and interpretation of results. All authors contributed in the revision of the manuscript.

Additional information

Competing financial interests: The authors declare no competing financial interests.

5

A series connection architecture for large-area organic photovoltaic modules with a 7.5% module efficiency

Soonil Hong[1,2], Hongkyu Kang[2,3], Geunjin Kim[1,2], Seongyu Lee[1,2], Seok Kim[1,2], Jong-Hoon Lee[1,2], Jinho Lee[2,4], Minjin Yi[3], Junghwan Kim[2,3], Hyungcheol Back[1,2], Jae-Ryoung Kim[3] & Kwanghee Lee[1,2,3,4]

The fabrication of organic photovoltaic modules via printing techniques has been the greatest challenge for their commercial manufacture. Current module architecture, which is based on a monolithic geometry consisting of serially interconnecting stripe-patterned subcells with finite widths, requires highly sophisticated patterning processes that significantly increase the complexity of printing production lines and cause serious reductions in module efficiency due to so-called aperture loss in series connection regions. Herein we demonstrate an innovative module structure that can simultaneously reduce both patterning processes and aperture loss. By using a charge recombination feature that occurs at contacts between electron- and hole-transport layers, we devise a series connection method that facilitates module fabrication without patterning the charge transport layers. With the successive deposition of component layers using slot-die and doctor-blade printing techniques, we achieve a high module efficiency reaching 7.5% with area of 4.15 cm^2.

[1] School of Materials Science and Engineering, Gwangju Institute of Science and Technology, Gwangju 61005, Republic of Korea. [2] Heeger Center for Advanced Materials, Gwangju Institute of Science and Technology, Gwangju 61005, Republic of Korea. [3] Research Institute for Solar and Sustainable Energies, Gwangju Institute of Science and Technology, Gwangju 61005, Republic of Korea. [4] Department of Nanobio Materials and Electronics, Gwangju Institute of Science and Technology, Gwangju 61005, Republic of Korea. Correspondence and requests for materials should be addressed to H.K. (email: gemk@gist.ac.kr) or to K.L. (email: klee@gist.ac.kr).

Bulk heterojunction solar cells, which are built on photoactive nanocomposites of electron-donating and electron-withdrawing organic semiconductors, are good candidates to be a ubiquitous renewable energy source that allows for integration with portable and wearable electronic applications[1,2]. Moreover, these organic solar cells (OSCs) are considered representative of the research field of printed electronics, because the solution processability of organic semiconductors enables cost-efficient, high-volume/throughput printing production with roll-to-roll manufacturing facility[3-8]. To realize this photovoltaic technology, research has focused mainly on enhancing device performance, extending device lifetime and developing up-scaling techniques for transitioning from small-area laboratory-scale devices to large-area industrial-scale modules[3-14]. Although the impressive progress made in the past two decades has led to considerable improvements in both the efficiency and operational stability of OSCs, the fabrication of large-area printed modules still suffers from significantly reduced power conversion efficiencies (PCEs), amounting to less than half the efficiencies of small-sized laboratory cells.

The current module architecture possesses an inherent weakness with regard to area loss, so-called aperture loss, which is well known to be a major contributor to the drastic performance degradation observed in large-area printed OSC modules[6-8]. The module geometry is based on a monolithic structure composed of several serially interconnecting subcells that are patterned into stripes with sufficiently narrow widths ($W \sim 10$ mm) to enable the sheet resistance of transparent electrodes to be neglected. However, such monolithic interconnections inevitably produce area loss to ensure contact areas for series connections between subcells. Furthermore, because of the low (millimetre scale) patterning resolutions of current printing techniques, using these techniques to create regularly spaced stripe-patterned subcells worsens unwanted area loss, resulting in very poor module PCEs with low geometric fill factors (FF, ratios between photoactive and total areas). Despite intense research efforts to reduce aperture loss by using laser ablation or metal-filament patterning techniques, only a few methods for realizing high-efficiency printed modules without additional patterning processing have been reported[15-17].

Here we demonstrate a new module architecture for manufacturing large-area printed OSC modules without the aid of additional and complicated post-patterning processing. By introducing an innovative series connection concept based on the charge recombination characteristic that occurs at the contacts between charge transport layers (CTLs), we design a monolithic interconnection that enables facile and efficient module fabrication without patterning the CTLs and producing the considerable aperture loss. Therefore, through consecutive printing processes using doctor-blade and slot-die machines, we successfully fabricate a large-area module that exhibits a high module PCE of 7.5% with a high geometric FF of 90%.

Results

Module architecture and fabrication. A schematic illustration of our module architecture is shown in Fig. 1a. The module has three inverted-type subcells consisting of three main component layers sandwiched between an indium tin oxide (ITO)

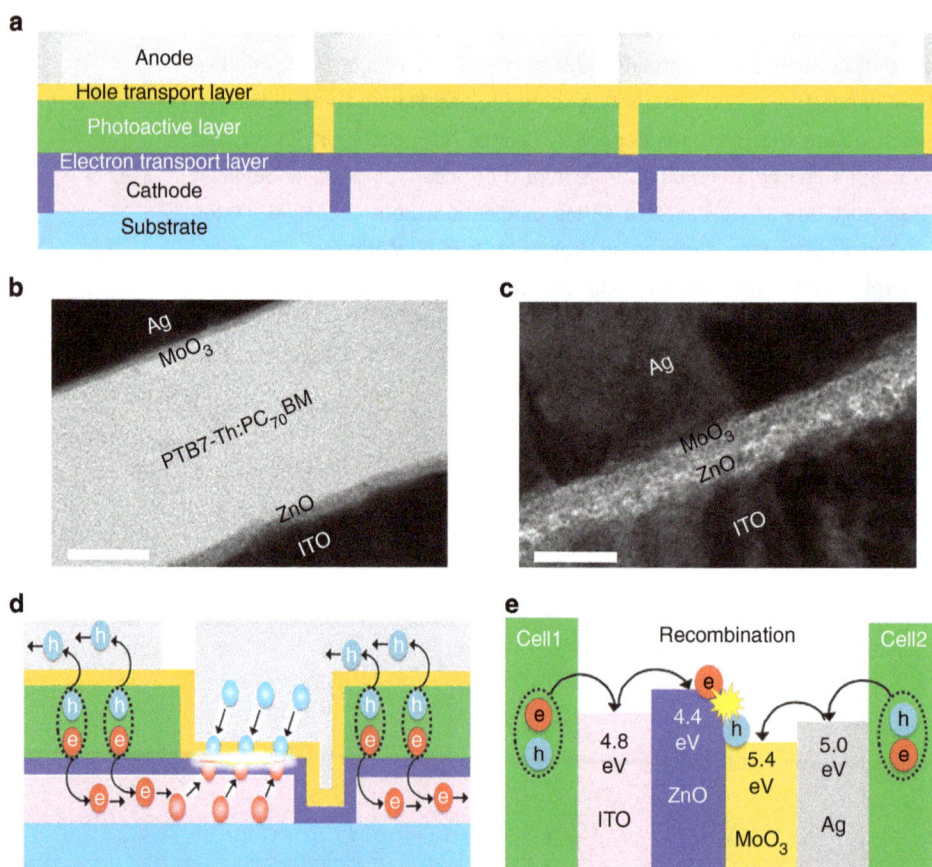

Figure 1 | Schematic illustration of the module. (a) Conceptual module structure consisting of patternless electron-transport and hole-transport layers and one patterned photoactive layer. **(b,c)** Corresponding cross-sectional TEM images of the active area (scale bar, 50 nm) **(b)** and series connection region (scale bar, 25 nm) **(c)**. **(d)** A schematic image of charge recombination as it occurs in our module. **(e)** Energy level diagrams of series connection region components.

cathode and a silver (Ag) anode. For a photoactive layer, we used a bulk heterojunction composite comprising an electron-donating poly(thieno[3,4-b]thiophene-alt-benzodithiophene) derivative (PTB7-Th) and an electron-accepting [6,6]-phenyl-C_{71}-butyric acid methyl ester ($PC_{70}BM$) (ref. 18). To obtain inverted-type subcells, we introduced two CTLs, including sol-gel zinc oxide (ZnO) as an electron-transport layer and molybdenum oxide (MoO_3) as a hole-transport layer (HTL), between the photoactive layers and their respective electrodes[19–21]. In the module fabrication, the photoactive layer was patterned in stripes onto the patterned ITO cathodes ($W = 13.5$ mm) with a slight blank offset (0.5 mm), whereas the two CTLs were deposited on the surfaces of the ITO and photoactive layer in a single-layer form without any stripe patterning. Series connections between adjacent subcells were achieved by forming stripe-patterned Ag anodes ($W = 13.5$ mm) with a subtle blank offset (0.5 mm) relative to the patterned photoactive layers. By printing the component layers with a doctor-blade machine for non-patterned single-layer forms and a slot-die machine for stripe patterning, we succeeded in fabricating a monolithic printed module with a high geometric FF of 90% without the use of any post-patterning processing (Supplementary Fig. 1 and Supplementary Note 1).

Series connection mechanism in the module. The most prominent feature of our module configuration is that the CTLs

were not patterned in stripe, in contrast to the conventional modules fabricated with all stripe-patterned component layers (Supplementary Fig. 2). Cross-sectional images taken with a high-resolution transmission electron microscope (TEM) clearly show not only all-component layers (ITO/ZnO/PTB7-Th:$PC_{70}BM$/MoO_3/Ag) in the active area, but also the CTLs (ZnO/MoO_3) sandwiched between the ITO and Ag electrodes in a series connection region (SCR), in which the counter electrodes of the adjacent subcells vertically overlap (Fig. 1b,c). Because of the CTLs embedded within the SCRs, our module operation will be quite different from that of typical modules in which the counter electrodes are in direct contact. We can expect the series connection mechanism of our monolithic module to be similar to that of existing multi-junction OSCs; the photogenerated charge carriers (that is, holes and electrons) from neighbouring subcells transport along the counter electrodes and are injected into the CTLs, thereby leading to series connections between subcells and voltage gains via charge recombination at the interface between the CTLs in the SCRs (Fig. 1d,e and Supplementary Fig. 3)[22,23].

Characteristics of the SCRs. To investigate the impact of the SCRs on module operation, we partitioned the module into a SCR unit cell and two subcells consisting of ITO/ZnO/MoO_3/Ag and ITO/ZnO/PTB7-TH:$PC_{70}BM$/MoO_3/Ag, respectively. Using current density–voltage (J–V) and electrochemical impedance

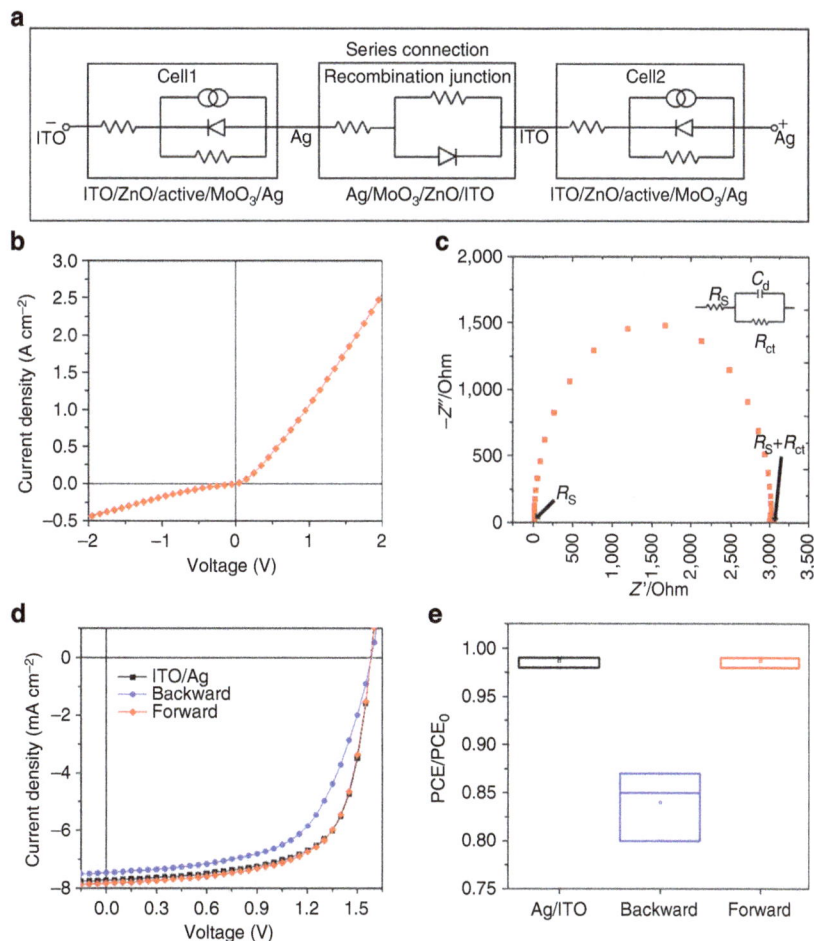

Figure 2 | Equivalent circuit of series-connected OSCs with an SCR unit cell. (**a**) The circuit comprised of two OSC unit cells with an SCR unit cell. (**b**) J-V characteristic of an SCR unit cell (Ag/MoO_3/ZnO/ITO) in the dark. (**c**) The Nyquist plot obtained from the EIS analysis of SCR (Ag/MoO_3/ZnO/ITO). (**d**,**e**) The J-V characteristics (**d**) and performance deviations (**e**) of OSCs with an SCR unit cell under AM 1.5G with 100 mW cm^{-2} (PCE_0 pertains to OSCs without any unit cell, boxes are measured values and rectangular points are average values).

spectroscopy (EIS) measurements, we formulated the equivalent circuit of the SCR unit cell (Fig. 2a). The J–V characteristic of the SCR unit cell shows the rectifying property originating from the electrical junction between the electron-transporting ZnO and hole-transporting MoO_3 layers (Fig. 2b). Meanwhile, the Nyquist plot obtained via EIS analysis exhibits a low series resistance (R_s) of $12\,\Omega$ and a relatively high shunt resistance (R_{sh}) of $3,000\,\Omega$ in the SCR unit cell (Fig. 2c and Supplementary Fig. 4). By combining these results, we can define the SCR unit cell as an electrical component composed of a diode and two resistors. On connecting this SCR unit cell between two subcells, we observed the dependence of device performance on the polarity of the SCR unit cell (Fig. 2d,e). The resulting equivalent circuit reveals that the series connection of the subcells is dominantly affected by the low R_s of the forward connection and the high R_{sh} of the backward connection. Because our module structure connects the forward SCRs with neighbouring subcells, the SCRs are expected to allow loss-free charge recombination of the adjacent subcells in the module operation.

Optimization of printing processes. To fabricate our printed module, we employed two kinds of printing techniques using doctor-blade and slot-die machines (Supplementary Fig. 5). Both printing machines have similar control factors, which depend on the viscosity of the solution, the amount of feeding solution, the coating speed and the substrate temperature. By delicately adjusting these parameters, we achieved high-quality printed films with smooth and uniform film morphologies (Supplementary Fig. 6). In particular, we designed a slot-die coating head with a positive shim mask to create a meniscus guide (Supplementary Fig. 7); the photoactive PTB7-Th:PC_{70}BM solution was ejected through a narrow slot and followed the shim mask pattern, thereby forming a meniscus between the mask and substrate via capillary action (Fig. 3a). To simply control the film thickness (t), we changed the coating speed (S) while fixing other coating parameters. As shown in Fig. 3b, the thicknesses of PTB7-Th:PC_{70}BM films follow power law equation of $t \approx S^{0.62}$, which has been demonstrated in meniscus coating methods using low-viscosity organic solutions[24,25]. After optimizing the film thicknesses, we obtained a high-quality printed photoactive layer with a thickness of $\sim 125\,nm$ that exhibited an optimal PCE of 8.5% (Fig. 3c, Supplementary Fig. 8 and Supplementary Table 1).

Performance of large-area printed OSC modules. Figure 4a displays a photograph of the complete large-area printed module with optimized film thicknesses. The current–voltage (I–V) and current density–voltage (J–V) characteristics of the module, which are measured via a large-scale calibrated solar simulator under standard illumination conditions, are shown in Fig. 4b,c and Table 1. The best performance of the new module ($4.15\,cm^2$) yielded a remarkable PCE of 8.1% with an open-circuit voltage (V_{oc}) of 2.36 V, a short-circuit current density (J_{sc}) of $5.53\,mA\,cm^{-2}$, and a FF of 62%. The V_{oc} (2.36 V) of the module is nearly three times larger than the V_{oc} (0.79 V) of the small-area reference; the J_{sc} ($5.53\,mA\,cm^{-2}$) of the module is exactly one-third of that ($16.6\,mA\,cm^{-2}$) of the reference; and the FF (62%) of the module is almost comparable to that of the reference (66%). Considering its geometric FF of 90%, the module exhibits a high module PCE of 7.3%. One of the modules exhibiting the best PCE was sent to the Korea Institute of Energy Research and was returned to our laboratory with a certificated module PCE of $\sim 7.5\%$, as shown in Fig. 4d. To the best of our knowledge, this PCE is the highest value in printed solar modules to date in scientific literature. These outstanding results indicate that the subcells were perfectly interconnected via effective charge recombination at the interfaces between the CTLs in the SCRs.

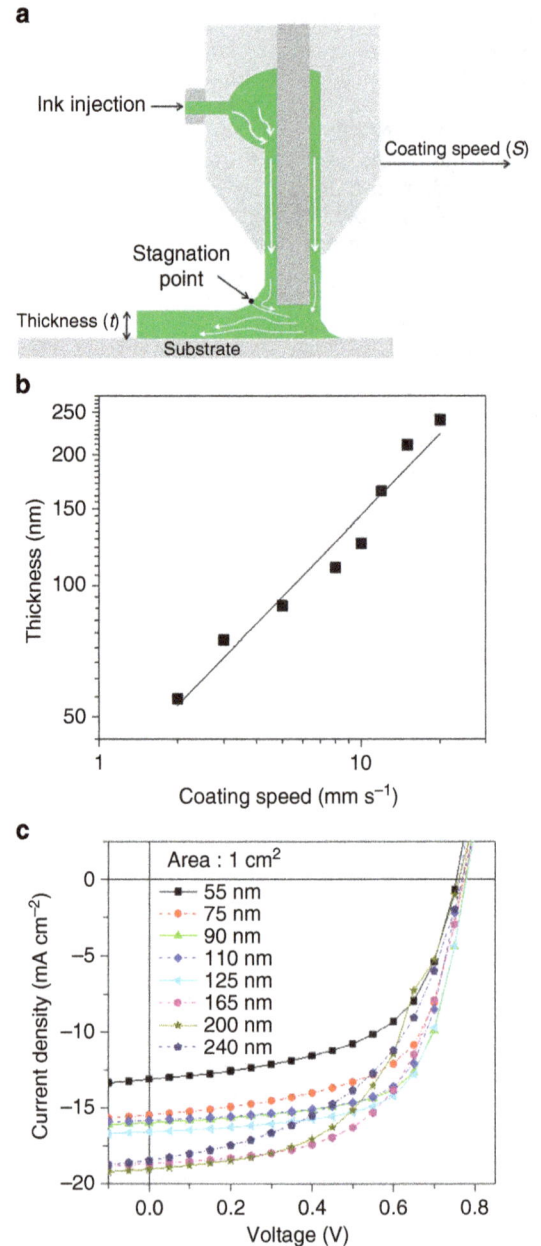

Figure 3 | OSCs fabricated using the printing method. (**a**) Schematic of meniscus formation and the streamlines near the stagnation point in the slot-die coating using a positive-shim style mask. (**b**) Thicknesses of PTB7-Th:PC_{70}BM films coated via the slot-die coating method using various coating speeds from 2 to $30\,mm\,s^{-1}$ (log scale). (**c**) J–V characteristics of OSCs fabricated using the slot-die coating method at various thicknesses.

Although increasing the length (size) of the module causes a slight decrease in the FF and J_{sc} values due to a few concomitant defect sites (for example, pinholes and fine dusts) within the printed films, we can overcome this problem through defect-free printing process in clean rooms, thereby enlarging the module size without suffering serious performance loss[6,26]. In addition, we introduced a solution-processed MoO_3 layer for the module fabrication[27]. By printing the MoO_3 layer, we obtained reasonable average efficiencies of 7.7% for large-area single cells and 6.9% for module, thereby demonstrating that our module can operate well even when we used all-printed CTLs (Fig. 4e,f and Supplementary Table 2).

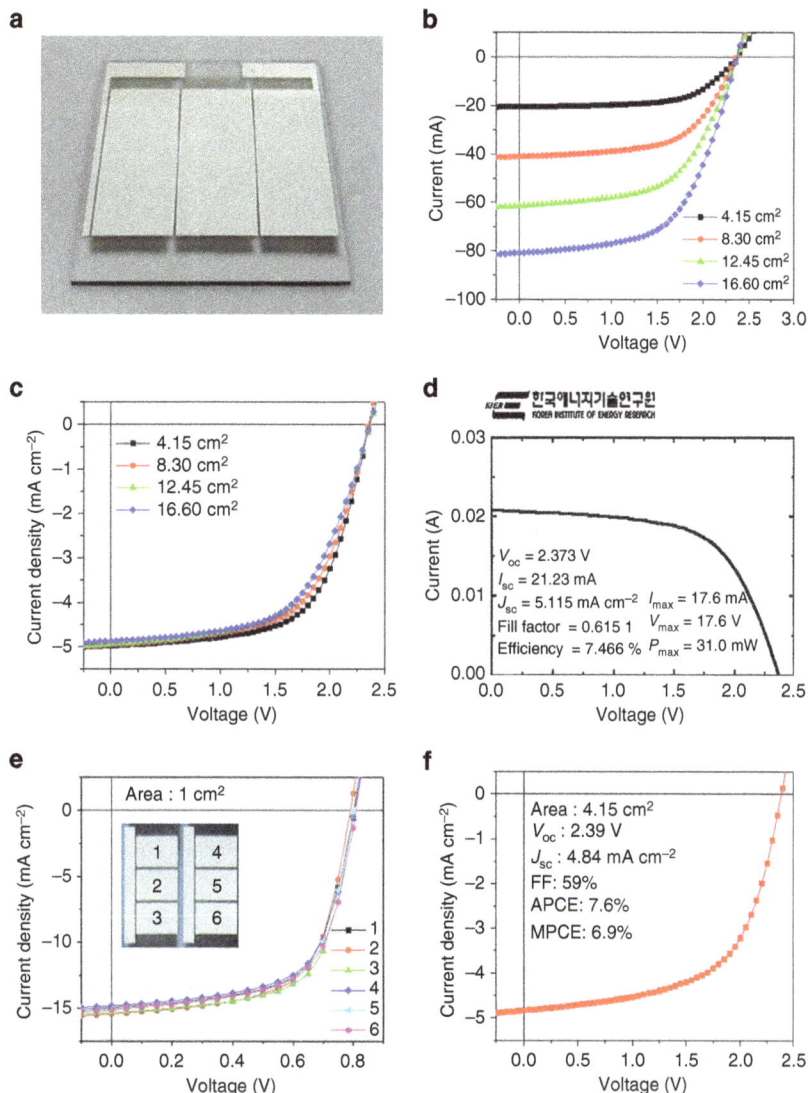

Figure 4 | OSC modules fabricated using the printing method. (a) A photograph image of our module (size of 60×44.5 mm). (b,c) I–V (b) and J–V (c) characteristics of OSC modules of various sizes depending on module lengths from 1 to 4 cm. (d) Korea Institute of Energy Research-certified J–V characteristics of our OSC module. (e,f) J–V characteristics of single OSCs in different positions (inset: photograph image of printed OSCs) (e) and OSC module (f) by using the printed MoO_3 layer. APCE is the PCE of the module in active area and MPCE is the PCE of the module in total area.

Discussion

Our work demonstrates a new scientific perspective, in that we have designed a simple and efficient module structure by developing a novel series connection method and a remarkable technical advance towards manufacturing printed modules for next-generation photovoltaic systems. By using a charge recombination feature arising from the electrical junctions between CTLs, we succeed in fabricating a monolithic module without the use of additional complex patterning processes. This monolithic module has achieved a certificated module PCE of 7.5% with a high geometric FF of 90%. We expect that this new approach presents a simple and useful means for transitioning from small-area laboratory-scale OSCs to large-area industrial-scale OSC modules.

Methods

Material preparation. The ZnO precursor was prepared by dissolving zinc acetate dihydrate ($Zn(CH_3COO)_2 \cdot 2H_2O$, Aldrich, 99.9%, 1 g) and ethanolamine ($NH_2CH_2CH_2OH$, Aldrich, 99.5%, 0.5 g) in isopropyl alcohol ($CH_3OCH_2CH_2OH$, Aldrich, 99.8%, 50 g) via stirring for 24 h. The PTB7-Th:PC_{70}BM solution was prepared by blending PTB7-Th (1-material) and PC_{70}BM (Nano-C) at a ratio of

Table 1 | Performance parameters of the OSC modules with increasing area.

Area (cm²)	V_{OC} (V)	I_{SC} (mA)	J_{SC} (mA cm^{-2})	FF (%)	APCE (%)	MPCE (%)
4.15	2.36	20.7	4.98	62	8.1	7.3
8.30	2.36	41.1	4.95	60	7.7	7.0
12.45	2.38	61.6	4.94	57	7.4	6.7
16.60	2.37	80.9	4.87	58	7.4	6.7

APCE, the PCE of the module in active area; MPCE, the PCE of the module in total area.

1:2 in chlorobenzene solvent with 1,8-diiodooctane additive (3% by volume) to obtain a total concentration of 12 mg ml^{-1}. The MoO_3 solution was prepared by dissolving bis(acetylacetonato) dioxomolybdenum (Sigma Aldrich) in a cosolvent of methanol and 1-butanol.

Single-cell fabrication. ITO/glass substrates were cleaned with detergent, after which they were sequentially washed via ultrasonic treatment in de-ionized water, acetone and IPA. The ZnO solution was coated onto the ITO/glass substrate using a doctor-blade coater (Coatmaster 509 MC, Erichsen in Germany) at 40 °C, then annealed at 150 °C for 20 min in air. The PTB7-Th:PC$_{70}$BM composite solution was coated on top of the ZnO layer in air using a slot-die coating method (Slot-die coater, iPen in South Korea) at room temperature. The pumping rate for coating PTB7-Th:PC$_{70}$BM solution was 0.1 ml min^{-1} when using a 50-μm-thick mask, and the film thickness was controlled by the coating speed of the slot-die header. To complete the single-cell device fabrication, MoO$_3$ (as a HTL) and Ag (as a top electrode) were deposited sequentially by thermal evaporation in a vacuum with a pressure of 10^{-6} torr.

Module fabrication. The ITO/glass preparation and ZnO coating process is the same as that described for single-cell fabrication. Before the module fabrication, the patterned ITO glass substrate (the stripe width of 13.5 mm and the gap of 0.5 mm) was prepared by wet-etching processing using a typical acid etchant. For the patterned PTB7-Th:PC$_{70}$BM coating process, we used a slot-die coating machine with a 50-μm-thick three-stripe mask. The coating speed and pumping rate used to coat the PTB7-Th:PC$_{70}$BM solution are 10 mm s^{-1} and 0.4 ml min^{-1}, respectively, and the optimized thickness of PTB7-Th:PC$_{70}$BM is ∼125 nm. The MoO$_3$ (as a HTL) was deposited onto the patterned photoactive layer with no patterned mask by using thermal evaporation in a vacuum with a pressure of 10^{-6} torr. The solution-processed MoO$_3$ layer was deposited on the photoactive layer by using a doctor-blade coating machine. Module-device fabrication was completed by thermal evaporation of the Ag metal top electrode (200 nm thickness) in a vacuum with a pressure of 10^{-6} torr. In contrast to MoO$_3$ deposition, Ag was deposited via a patterned mask, which is used to obtain a series connection in the module.

Characterization and analysis. The current–voltage (I–V) characteristics were recorded using an Iviumsoft apparatus with simulated AM 1.5 illumination (100 mW cm^{-2}) via a solar simulator (Abet Technologies Sun 3000) in normal air conditions. The thicknesses of the coated films were measured using a thickness profile metre (Surfcorder ET 3000, Kosaka Laboratory, Ltd.). Cross-sectional TEM samples of the printed OSC module were prepared using a dual beam-focused ion beam (Helios NanoLabTM). The TEM images were obtained using field emission TEM (FEI TecnaiTM G2 F30 Super-Twin) operated at 200 kV. The topographies of surface images were characterized using atomic force spectroscopy.

References

1. Zhang, Z. *et al.* A lightweight polymer solar cell textile that functions when illuminated from either side. *Angew. Chem.* **126**, 11755–11758 (2014).
2. Zhang, Z. *et al.* Integrated polymer solar cell and electrochemical supercapacitor in a flexible and stable fiber format. *Adv. Mater.* **26**, 466–470 (2014).
3. Li, G., Zhu, R. & Yang, Y. Polymer solar cells. *Nat. Photon.* **6**, 153–161 (2012).
4. Krebs, F. C., Espinosa, N., Hösel, M., Søndergaard, R. R. & Jørgensen, M. 25th anniversary article: rise to power – OPV-based solar parks. *Adv. Mater.* **26**, 29–39 (2014).
5. Espinosa, N., Hösel, M., Jørgensen, M. & Krebs, F. C. Large scale deployment of polymer solar cells on land, on sea and in the air. *Energy Environ. Sci.* **7**, 855–866 (2014).
6. Krebs, F. C., Tromholt, T. & Jørgensen, M. Upscaling of polymer solar cell fabrication using full roll-to-roll processing. *Nanoscale* **2**, 873–886 (2010).
7. Krebs, F. C. Polymer solar cell modules prepared using roll-to-roll methods: knife-over-edge coating, slot-die coating and screen printing. *Sol. Energ. Mat. Sol. C.* **93**, 394–412 (2009).
8. Søndergaard, R. R., Hösel, M. & Krebs, F. C. Roll-to-roll fabrication of large area functional organic materials. *J. Polym. Sci. B Polym. Phys.* **51**, 16–34 (2013).
9. Chen, J.-D. *et al.* Single-junction polymer solar cells exceeding 10% power conversion efficiency. *Adv. Mater.* **27**, 1035–1041 (2015).
10. He, Z. *et al.* Single-junction polymer solar cells with high efficiency and photovoltage. *Nat. Photon.* **9**, 174–179 (2015).
11. Liu, Y. *et al.* Aggregation and morphology control enables multiple cases of high-efficiency polymer solar cells. *Nat. Commun.* **5**, 5293 (2014).
12. Krebs, F. C. *Stability and Degradation of Organic and Polymer Solar Cells* (Wiley, 2012).
13. Jørgensen, M. *et al.* Stability of polymer solar cells. *Adv. Mater.* **24**, 580–612 (2012).
14. Andersen, T. R. *et al.* Scalable, ambient atmosphere roll-to-roll manufacture of encapsulated large area, flexible organic tandem solar cell modules. *Energy Environ. Sci.* **7**, 2925–2933 (2014).
15. Kang, H., Hong, S., Back, H. & Lee, K. A new architecture for printable photovoltaics overcoming conventional module limits. *Adv. Mater.* **26**, 1602–1606 (2014).
16. Spyropoulos, G. D. *et al.* Flexible organic tandem solar modules with 6% efficiency: combining roll-to-roll compatible processing with high geometric fill factors. *Energy Environ. Sci.* **7**, 3284–3290 (2014).
17. Lee, J. *et al.* Seamless polymer solar cell module architecture built upon self-aligned alternating interfacial layers. *Energy Environ. Sci.* **6**, 1152–1157 (2013).
18. Liao, S.-H., Jhuo, H.-J., Cheng, Y.-S. & Chen, S.-A. Fullerene derivative-doped zinc oxide nanofilm as the cathode of inverted polymer solar cells with low-bandgap polymer (PTB7-Th) for high performance. *Adv. Mater.* **25**, 4766–4771 (2013).
19. Sun, Y., Seo, J. H., Takacs, C. J., Seifter, J. & Heeger, A. J. Inverted polymer solar cells integrated with a low-temperature-annealed sol-gel-derived ZnO film as an electron transport layer. *Adv. Mater.* **23**, 1679–1683 (2011).
20. Sun, Y. *et al.* Efficient, air-stable bulk heterojunction polymer solar cells using MoO$_x$ as the anode interfacial layer. *Adv. Mater.* **23**, 2226–2230 (2011).
21. He, Z. *et al.* Enhanced power-conversion efficiency in polymer solar cells using an inverted device structure. *Nat. Photon.* **6**, 591–595 (2012).
22. Kong, J. *et al.* Building mechanism for a high open-circuit voltage in an all-solution-processed tandem polymer solar cell. *Phys. Chem. Chem. Phys.* **14**, 10547–10555 (2012).
23. Chen, C.-C. *et al.* An efficient triple-junction polymer solar cell having a power conversion efficiency exceeding 11%. *Adv. Mater.* **26**, 5670–5677 (2014).
24. Hong, S., Lee, J., Kang, H. & Lee, K. Slot-die coating parameters of the low-viscosity bulk-heterojunction materials used for polymer solar cells. *Sol. Energ. Mat. Sol. C.* **112**, 27 (2013).
25. Vak, D. *et al.* 3D printer based slot-die coater as a lab-to-fab translation tool for solution-processed solar cells. *Adv. Energy Mater.* **5**, 1401539 (2015).
26. Jeong, W.-I., Lee, J., Park, S.-Y., Kang, J.-W. & Kim, J.-J. Reduction of collection efficiency of charge carriers with increasing cell size in polymer bulk heterojunction solar cells. *Adv. Func. Mater.* **21**, 343–347 (2011).
27. Murase, S. & Yang, Y. Solution processed MoO$_3$ interfacial layer for organic photovoltaics prepared by a facile synthesis method. *Adv. Mater.* **24**, 2459–2462 (2012).

Acknowledgements

We thank the Heeger Center for Advanced Materials (HCAM) at the Gwangju Institute of Science and Technology (GIST) of Korea for help with device fabrication and measurements. This research was supported by a grant from the National Research Foundation of Korea (NRF) funded by the Korean government (MSIP) (NRF-2014R1A2A1A09006137), the Technology Development Program to Solve Climate Changes of the NRF funded by the (MSIP) (NRF-2015M1A2A2057510), the R&D program of MSIP/COMPA (2015K000199), and the 'Basic Research Projects in High-tech Industrial Technology' Project through a grant provided by GIST in 2015. K.L. also acknowledges support provided by the Core Technology Development Program for Next-generation Solar Cells of the Research Institute for Solar and Sustainable Energies, GIST.

Author contributions

H.K. contributed a key idea. S.H., H.K. and K.L. designed the concept and the required experiments. S.H. performed the fabrication and characterization of devices. G.K. helped with the measurement of EIS. S.L. helped with the measurement of atomic force microscopy. S.K. helped with the fabrication of devices. H.B. helped with the preparation of the MoO$_3$ solution. S.H., H.K. and K.L. prepared the manuscript. K.L. guided and directed the research. All authors discussed the results and contributed to the writing of the paper.

Additional information

6

Mobility overestimation due to gated contacts in organic field-effect transistors

Emily G. Bittle[1,2], James I. Basham[1,3], Thomas N. Jackson[3], Oana D. Jurchescu[2] & David J. Gundlach[1]

Parameters used to describe the electrical properties of organic field-effect transistors, such as mobility and threshold voltage, are commonly extracted from measured current–voltage characteristics and interpreted by using the classical metal oxide–semiconductor field-effect transistor model. However, in recent reports of devices with ultra-high mobility ($>40\,\mathrm{cm^2\,V^{-1}\,s^{-1}}$), the device characteristics deviate from this idealized model and show an abrupt turn-on in the drain current when measured as a function of gate voltage. In order to investigate this phenomenon, here we report on single crystal rubrene transistors intentionally fabricated to exhibit an abrupt turn-on. We disentangle the channel properties from the contact resistance by using impedance spectroscopy and show that the current in such devices is governed by a gate bias dependence of the contact resistance. As a result, extracted mobility values from d.c. current–voltage characterization are overestimated by one order of magnitude or more.

[1] National Institute of Standards and Technology, Engineering Physics Division, 100 Bureau Drive, MS 8120, Gaithersburg, Maryland 20899, USA. [2] Department of Physics, Wake Forest University, 1834 Wake Forest Road, Winston-Salem, North Carolina 27109, USA. [3] Department of Electrical Engineering, The Pennsylvania State University, 121 Electrical Engineering East, University Park, State College, Pennsylvania 16802, USA. Correspondence and requests for materials should be addressed to E.G.B. (email: emily.bittle@nist.gov) or to D.J.G. (email: david.gundlach@nist.gov).

rganic semiconductors (OSCs) remain a topic of considerable interest for basic and applied research. As such, accurate electrical characterization and parameterization of physical properties which govern the device operation of light emitting diodes, field-effect transistors and photovoltaic cells is essential for continued device performance improvement and possible commercialization of organic semiconductor-based devices. Charge carrier mobility is one of several commonly cited physical properties, and describes charge motion under applied electric field. The organic field-effect transistor (OFET) is routinely used as a test structure for extracting mobility in addition to being a key element in circuits. The most commonly used method to evaluate OFET parameters such as field-effect mobility, μ, and threshold voltage, V_{th}, is the classical metal-oxide–semiconductor field-effect transistor (MOSFET) model. This model is described for the two extreme modes of operation above threshold, $|V_{GS}| > V_{th}|$, in equations (1) and (2). For linear mode, $|V_{DS}| < |V_{GS} - V_{th}|$,

$$I_D = \mu_{lin} c_{ox} \frac{W}{L} \left[(V_{GS} - V_{th}) V_{DS} - \frac{V_{DS}^2}{2} \right], \quad (1)$$

and for saturation mode $|V_{DS}| > |V_{GS} - V_{th}|$,

$$I_D = \mu_{sat} c_{ox} \frac{W}{2L} (V_{GS} - V_{th})^2, \quad (2)$$

where V_{GS} is the gate voltage, V_{th} is the threshold voltage, I_D is the drain current, V_{DS} is the drain voltage, μ_{lin} and μ_{sat} are the linear and saturation mobility, respectively, W and L are the width and length of the transistor channel and c_{ox} is the oxide capacitance per unit area. In this paper, we illustrate how this model describing the device behaviour for idealized materials and interfaces that adhere to solid-state band theory can lead to severe inaccuracies in extracted parametric values when used to analyse non-ideal transistors.

We focus on a particular non-ideality in OFET transistor characteristics, shown in Fig. 1a, which appears in many high-mobility single-crystal FETs[1–3], polymer[4–10] and small molecule thin film transistors[11,12]. In this non-ideal case, I_D shows an abrupt change in slope as a function of V_{GS}, whereas in the classical model I_D ($I_D^{1/2}$) is linear with V_{GS} in the linear (saturation) regime as defined in equations (1) and (2). This slope is regularly used to calculate mobility and extrapolate the threshold voltage. As reports of transistors with this non-ideal behaviour become more prevalent in the literature, yielding impressive mobility values, a detailed understanding of the source

of the non-ideal behaviour and its impact on extracted figures of merit has become important. Here, we fabricate a transistor to exhibit non-ideal $I_D - V_{GS}$ characteristics and use impedance spectroscopy to disentangle the contact behaviour from the transistor channel behaviour to directly compare these measurements to d.c. I–V measurements made on the same OFET device. This comparison clarifies the impact that a non-ohmic contact can have on transistor behaviour, and by extension, on the mobility extracted from the I–V data. We show that the mobility can be overestimated by up to one order of magnitude in transistors with pronounced non-ideal current–voltage characteristics, an effect which arises from the gate-bias dependence of the contact resistance.

Results

Material and electrical considerations. For this study, we characterized the electrical properties of single-crystal rubrene field-effect transistors fabricated in a bottom contact, bottom gate geometry (Fig. 1b). The bottom-contact geometry simplifies the electric field distribution and the parasitic gate to source/drain overlap capacitance, and eliminates charge transport in the out-of-plane direction in the OSC. Rubrene was chosen because its single crystals have been shown to exhibit nearly ideal transistor behaviour and are useful for fundamental studies[13]. Small molecule single crystals have lower molecular disorder and more straightforward transport pathways than polymer thin films, which involve a convolution of pi-pi and backbone transport. Furthermore, single crystals do not suffer from pronounced grain boundaries that can lead to large potential drops in the channel as in the case of polycrystalline thin films of small molecule semiconductors[14,15]. A micrograph of a completed field-effect transistor device is shown in Fig. 1c.

Current–voltage characterization. The d.c. I_D–V_{GS} characteristics for a device biased in the saturation regime with $V_{DS} = -20$ V are plotted in Fig. 1a. For both saturation ($|V_{DS}| > |V_{GS} - V_{th}|$) and linear ($|V_{DS}| < |V_{GS} - V_{th}|$) regimes, the current–voltage characteristics change from high slope to low slope as the gate bias is increased from 0 V. The kink where the slope change happens, around -6 V in the saturation characteristics, allows us to define two regions: one at small gate bias ($0 > V_{GS} > -6$ V) with high slope and another at large gate bias (-20 V $< V_{GS} < -7$ V) with low slope. This behaviour deviates from the ideal FET behaviour given by equations (1) and (2),

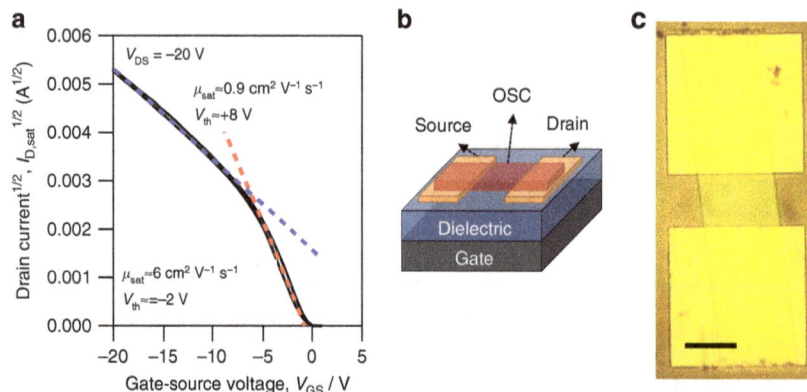

Figure 1 | Current-voltage characteristics for a non-ideal transistor and transistor geometry. (a) Plot of the transfer characteristics in the saturation regime ($V_{DS} = -20$ V) of a rubrene transistor exhibiting non-ideal characteristics. Fit lines in red and blue illustrate the ambiguity associated with characterizing OFETs. **(b)** Bottom gate/bottom contact OFET. **(c)** Image of transistor. Drain and source contact pairs are $250 \times 250 \, \mu m$ squares and the rubrene active area is $140 \, \mu m$ wide $\times 100 \, \mu m$ long. Scale bar is $100 \, \mu m$.

a

b

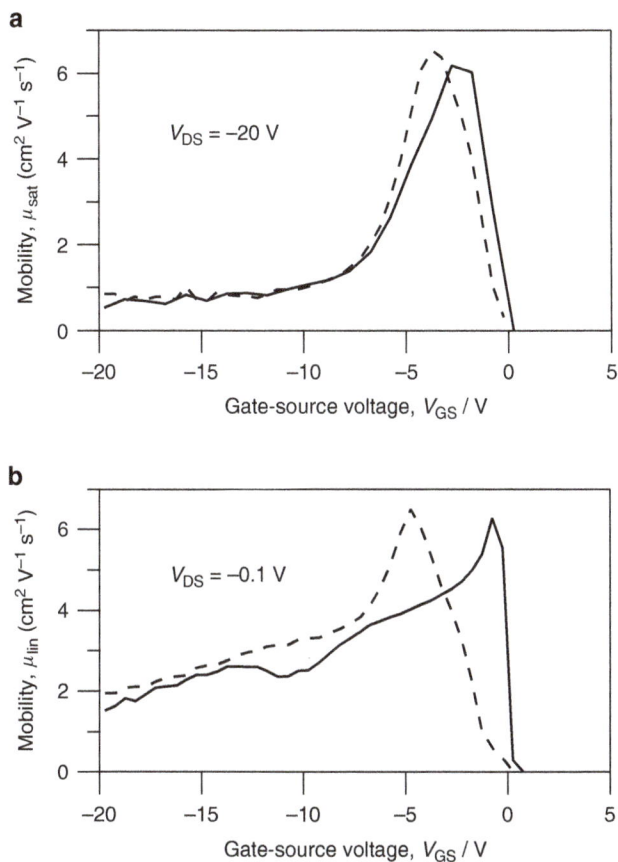

Figure 2 | Overestimation of the mobility. 'Mobility', as defined in equations (1) and (2), are plotted in the saturation (**a**) and linear (**b**) regimes as a function of gate bias, increasing bias direction in solid lines and reverse in dashed lines. Similar results were found for 30 transistors measured during the course of this study.

Figure 3 | Prevalence of overestimation of the mobility. Our estimates of peak mobility, μ_{peak}, at low gate bias and aggregate mobility, μ_{agg}, calculated for higher gate bias using the MOSFET equations applied to hand fits of published data[1,3,5–7,9–11,16]. We also include our data from Fig. 1a and the Supplementary Figs (SI) 1b, 1d and 2. Polymers are given by filled symbols and small molecules are given by open symbols. Selected papers show a change from high to low slope in the transconductance data as gate bias is increased in either the saturation or linear regime for p-type conduction. Lines are guides to the eye, and show the ratio of peak to high gate bias values.

where the slope is constant with V_{GS} ($I_D \propto V_{GS}$ in the linear regime and $I_D^{1/2} \propto V_{GS}$ in the saturation regime.) The slope and intercept of linear fits to the I–V data provide an aggregate mobility and threshold voltage values when evaluated in the MOSFET model. In the saturation regime, these fits yield mobility of $6\,cm^2\,V^{-1}\,s^{-1}$ and threshold voltage of $-2\,V$ at low gate bias, and $0.9\,cm^2\,V^{-1}\,s^{-1}$ and $8\,V$ at high gate bias. The extracted mobility differs by $\sim 6\times$ and V_{th} differs by $10\,V$.

Mobility as a function of gate voltage was extracted using the MOSFET model for the saturation ($V_{DS} = -20\,V$) and linear ($V_{DS} = -0.1\,V$) regimes and is plotted in Fig. 2a,b. The differential mobility values extracted from the MOSFET model are nearly constant for $-20\,V < V_{GS} < -10\,V$, but increase to a peak at $V_{GS} \approx -5\,V$. Hysteresis, likely due to modest charge trapping in the transistor channel, is small. Forward and reverse sweeps show a comparable variation of the mobility with gate bias. The linear and saturation regime mobilities have similar magnitude and variation with gate bias, which creates ambiguity about the intrinsic transistor channel mobility.

In several devices we fabricated using nominally the same processing method the discrepancy in the mobility was as large as $14\times$. We have included two examples in Supplementary Fig. 1. For contrast, we have also included a sample prepared on platinum contacts, which shows nearly ideal behaviour in Supplementary Fig. 2. We have reexamined data from several references[1,3,5–7,9–11,16] with a similar discrepancy between the high and low gate bias regions by applying equations (1) and (2)

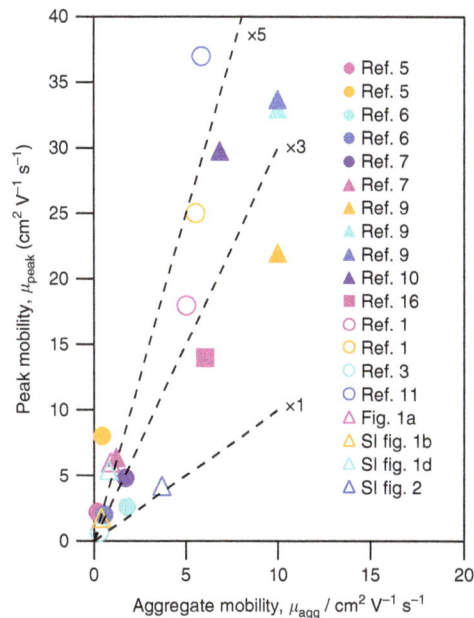

to our estimates from the slopes of published I–V characteristics. The comparison plotted in Fig. 3 show that for most of the data examined, the peak mobility is $\sim 3\times$ to $5\times$ of the aggregate mobility calculated at high gate bias; with largest disparity being $18\times$. It is important to note that the mobility values calculated using the MOSFET model are scalar fit parameters and not intrinsic material parameters. We made an effort in our literature search to include data on polymers, small molecules, as well as semiconductors with different band gaps, but have found no correlation between this behaviour and material type.

Explanations for this non-ideal behaviour vary in the literature. Some have proposed that at low gate bias, the accumulation layer is not tightly confined to the interface and extends into the bulk where there is less effect of the disorder associated with the OSC/dielectric interface. At higher gate bias, the charges become confined to the interface at high gate fields and mobility decreases due to the increased disorder[1], charges become trapped in the gate dielectric at high field[17] or that the high carrier density causes Coulombic interactions between the charges[18]. Other reports have proposed that a high contact resistance restricts current at high gate bias when the channel and contact resistance become comparable[19]. However, this latter interpretation is not consistent with observations by several other groups who report a decrease in contact resistance with increasing gate bias[20,21]. Researchers studying ambipolar operation have speculated that this effect could be due to negative trap filling in low band gap materials[16], which we discuss below.

To further analyse our results in the context of the MOSFET model, we use the aggregate mobility and V_{th} shown in Fig. 1a for the two distinct regions to calculate the I_D versus V_{DS} at various V_{GS} by using equations (1) and (2). The calculated values for I_D

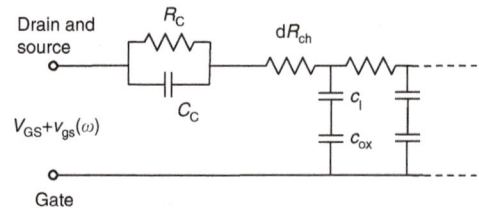

Figure 5 | Model of the transistor impedance. Equivalent circuit used to model a.c. transistor behaviour.

divided into elements of length dx; resistance and capacitance per area,

$$dR_{ch} = \frac{r}{W}dx, \qquad (6)$$

$$c' = \frac{c_{ox}c_I}{c_{ox} + c_I}, \qquad (7)$$

where r is the sheet resistance and c_I is the interfacial capacitance per area. Due to the experimental design, the contact resistance is frequency independent, whereas the distributed channel resistance is frequency dependent, and the two are easily separated. The transmission line model was shown by Hamadani et al.[22] to be successful in the analysis of poly(3-hexylthiophene) transistors when Z_C is included.

The channel and contact resistance were extracted for the reverse (negative to positive V_{GS}) trace by fitting equation (3) to the impedance data and are plotted in Fig. 6a as a function of gate bias. Additional details about the impedance data modelling are included in the Supplementary Figs 3 and 4. The channel resistance varies as $1/V_{GS}$ over the entire bias range. This functional dependence is consistent with linear region MOSFET operation as given by equation (1). The contact resistance exhibits a pronounced gate bias dependence over a small bias range ($0\,\mathrm{V} > V_{GS} > -6\,\mathrm{V}$), where R_C decreases exponentially by a factor of ~5,000 as the amplitude of V_G increases. At high gate bias ($-10 > V_{GS} > -20\,\mathrm{V}$), the contact resistance remains at a constant low value. This functional dependence of contact resistance with gate bias is most consistent with that of a gated Schottky contact, where a relatively abrupt transition from thermionic to thermionic-field emission and finally to field emission (tunnelling) results in an exponential decrease and plateau of the contact resistance. The charge accumulation in the channel provides the necessary conditions for the tunnelling injection process and is analogous to the formation of a highly doped contact region that greatly reduces that depletion region formed at the metal–semiconductor interface and allows for efficient injection.

Figure 6b provides a graphical comparison of $I_D(V_{GS})$ biased in the linear regime at small V_{DS} ($V_{DS} = -0.1\,\mathrm{V}$) to the channel and contact resistances as a fraction of total resistance ($R_T = R_{ch} + R_C$). This comparison best illustrates the influence of R_C (V_{GS}) on I–V characteristics for this device. The large change in contact resistance at low gate bias correlates with high transconductance ($g_m = dI_D/dV_{GS}$) of the transistor I–V characteristics and with the peak in the differential mobility, Fig. 2b. At high gate bias, R_C is low ($\approx 10^3\,\Omega$) relative to the channel resistance ($\approx 10^5\,\Omega$) and nearly constant. In this same bias regime, we observe nearly linear behaviour of $I_D(V_{GS})$ and in the levelling of the differential mobility extracted from the I–V characteristics. We therefore conclude that the mobility peak at $V_{GS} \approx -5\,\mathrm{V}$ is a result of exponentially changing contact resistance relative to the more slowly changing channel resistance.

Electrical contact between metals and organic semiconductors has long been known to have a strong influence on the operation

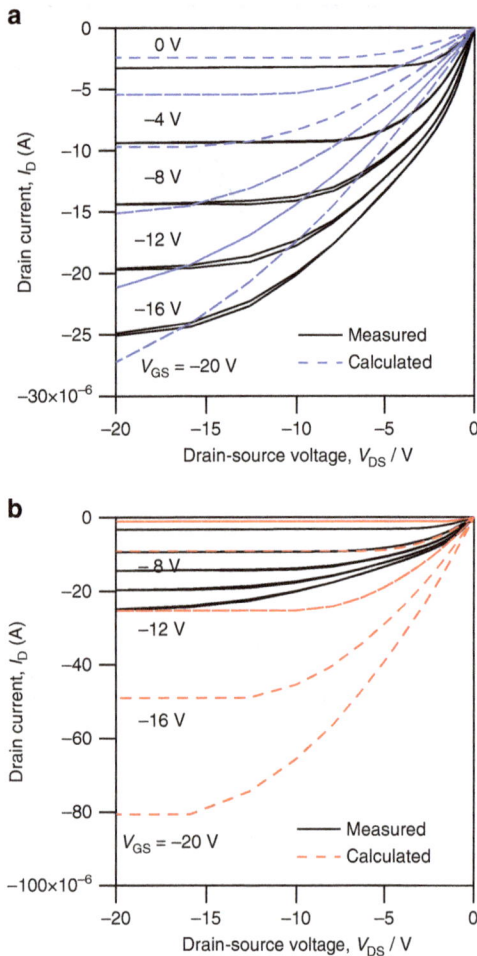

Figure 4 | A comparison of mobility estimation results. Measured I_D versus V_{DS} plots compared with plots calculated from equations (1) and (2) using extracted mobility and V_{th} obtained from the fitting lines to the measured data in Fig. 1a for the device biased in saturation in the two distinct regions; (a) high gate bias and (b) low gate bias.

(V_{DS}, V_{GS}) are plotted in Fig. 4a,b and reveal that the classical MOSFET relationship describing I_D (V_{DS}) provides a poor fit when compared with the measured characteristics. The best agreement to the measured data is obtained by using the aggregate mobility ($\mu_{sat} = 0.9\,\mathrm{cm^2\,V^{-1}\,s^{-1}}$) and corresponding threshold voltage ($V_{th} = +8\,\mathrm{V}$) extracted for high gate bias.

Impedance analysis. We have used impedance spectroscopy to characterize and extricate the components of the transistor for the linear regime (measured at $V_{DS} = 0\,\mathrm{V}$) to clarify what governs device operation in the different gate bias ranges, V_{GS}. The impedance data were analysed by using a combination of a transmission line to model the transistor channel and a parallel RC circuit to model the contacts (R_C and C_C),

$$Z_T = Z_{dist} + Z_C, \qquad (3)$$

$$Z_{dist} = \frac{1}{j\omega WLc'}\lambda\coth\lambda, \quad \lambda = \sqrt{\frac{1}{4}j\omega c'rL^2}, \qquad (4)$$

$$Z_C = \frac{jR_C}{j - \omega R_C C_C}, \qquad (5)$$

shown in Fig. 5 where Z_{dist} is the transistor channel impedance and Z_C is the contact impedance. The transistor channel was

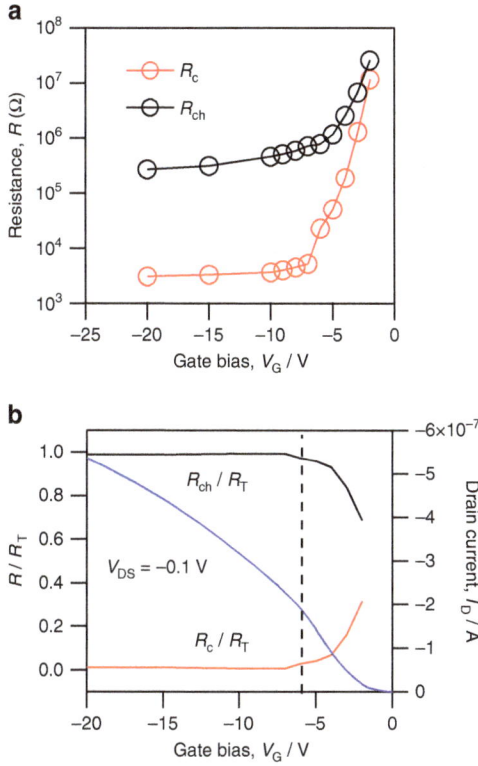

Figure 6 | Comparison of contact and channel resistance. (**a**) R_C and R_{ch} values extracted from fits to impedance data using equation (3) for reverse (negative to positive V_{GS}) sweep in the linear regime. (**b**) Plot of the I-V characteristic (blue) for $V_{DS} = -0.1$ V negative to positive sweep, along with plot of R_C/R_T and R_{ch}/R_T.

and extrinsic performance of organic electronic devices[17,23-28]. OFET measurements of the contact resistance[26,28] and local potential[14,29,30] show that the metal–organic semiconductor interface can be a significant source of potential drop at the injection contact and can be highly influential on I–V characteristics. In particular, transistor behaviour can be significantly impacted by charge injection from the metal electrode into the OFET channel due to the large current density (10^6 times larger than for diodes, such as light-emitting diodes and photovoltaics.) Calculations of the effects of contact resistance on OFETs, modelled by using either a Schottky barrier or low-mobility areas at the contact, show that R_C can be gate-voltage dependent, which significantly impacts the resulting current–voltage (I–V) characterization[31].

The effect of a gated source contact on FET operation has been previously observed and/or induced in devices based on a large number of materials. For example, poly and amorphous silicon and zinc oxide FETs with Schottky contacts (source gated transistors or Schottky source barrier transistors) have been engineered to take advantage of the high transconductance and low output conductance for specific circuit applications[32-35]. Precisely, how the contact affects operations depends on many factors. Carbon nanotube FETs and two-dimensional layered FETs are similar to organic FETs in that conventional doping of the contact region is challenging and Schottky contacts are routinely formed to the semiconductor[36,37]. Injection and transport studies on the former devices[36,37] report injection barriers that are typically less than 0.3 eV and a transition from thermionic to thermionic-field emission to field emission (tunnelling) with applied voltage that is less abrupt than we

report here for rubrene single-crystal FETs. We expect FETs with the pronounced dependence of transconductance on bias to result from a larger injection barrier and in devices where the magnitude of the channel resistance falls within the range of the exponentially decreasing contact resistance. High-mobility organic semiconductors with contacts having large injection barriers would appear to be prone to this specific effect. Similarly, such an effect might be present but not readily observed in low-mobility organic FETs because the transition would likely occur in the subthreshold region. At a minimum, a large injection barrier is expected in devices exhibiting ambipolar operation and confirms that contacts with large injection barriers are less selective than assumed[6,7].

Ambipolar behaviour is often observed in low bandgap OSCs. The poor charge selectivity of contacts that facilitate ambipolar behaviour can, under the appropriate bias stress conditions, result in electron injection and trapping. It has recently been suggested[16] that electron trapping during bias conditions can contribute to non-ideal behaviour and give rise to over-estimation of the field effect mobility. The most likely mechanism being the current–voltage characteristics reflect non-equilibrium measurement conditions where the trapped charge is not neutralized by the injected counter charge. This permits the quasi Fermi level to move more quickly through the band gap and the current to increase more rapidly with increasing gate bias than under equilibrium conditions. Such charge trapping would only further enhance the gated contact-controlled trans-conductance that we report here for wide band gap organic semiconductors because of the resulting electrostatics, which inhibit compression of the depletion region at the Schottky barrier interface and efficient charge injection. The non-ideal behaviour reported for low band gap ambipolar FETs is consistent with both gated-Schottky contacts, further enhanced by electron trapping as well as measurements made under non-equilibrium conditions. A more detailed and careful study is required to disentangle such effects in these systems.

Although the transconductance of our transistor is dominated by the gate-activated R_C for low gate bias, the impedance measurements gave us access to the channel behaviour in this region. We can use this to calculate the true mobility of the device channel above the apparent threshold voltage of the device, which corresponds with the sharp turn-on in channel capacitance seen in Fig. 7a. Channel mobility can be calculated from the accumulated charge in the transistor channel, Q_I, and the sheet resistance, r, in the channel by using equation (8) and the results of impedance modelling[38]. The channel mobility calculated for low gate bias from the channel r and Q_I plotted in Fig. 7b shows that channel mobility increases slowly over this range to a constant value and does not show the pronounced peak as in the differential mobility extracted from device I–V characteristics analysed with the MOSFET model. This impedance-based analysis of the channel mobility further supports our conclusion that the apparent high mobility is due to the effect of the gate voltage dependence of R_C on the total device transconductance and not to a variation of the transistor channel.

$$\mu = \frac{1}{Q_I r}, Q_I = \int_\infty^V c' dV_{GS} \approx \sum_\infty^V c' \delta V_{GS}, \qquad (8)$$

Discussion

The importance of contacts has been widely acknowledged in organic electronic devices[17,19,21,39]. There exists numerous studies using gated four-terminal measurements, gated transfer length measurements and scanned Kelvin probe microscopy, all of which attempt to measure R_C, correct for the reduced V_{DS} and

a

b

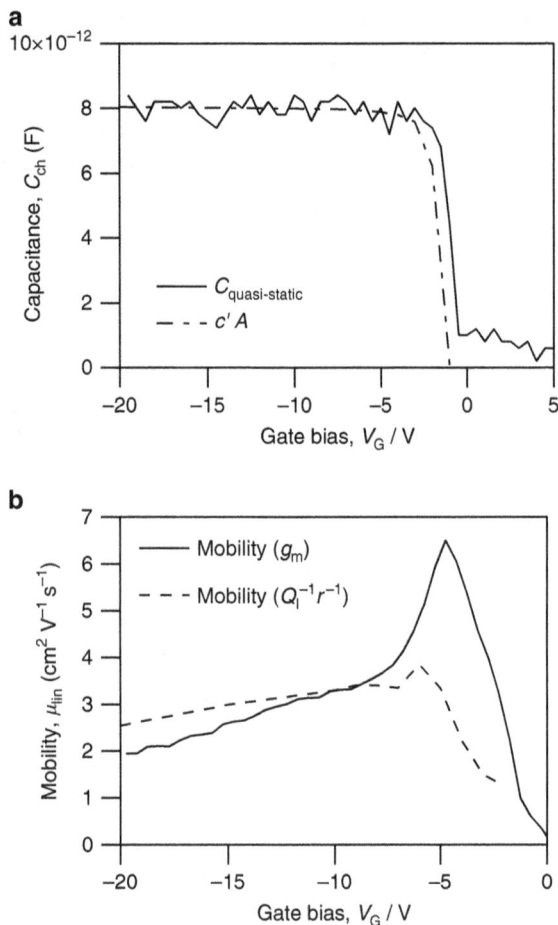

Figure 7 | Determining the mobility. (a) The capacitance in the channel ($c'A$, where A is the area of the transistor). **(b)** Mobility for the reverse (negative to positive V_{GS}) sweep: the differential mobility (solid line) calculated from the linear transistor characteristics using equation (1) and mobility (dashed line) calculated from equation (8) using the channel properties obtained from the fit to impedance data.

V_{GS}, and extract 'intrinsic' channel mobility[1,21,29]. However, these approaches rely on the assumption that the measured dependence of drain current on gate voltage (transconductance) is controlled mainly by the channel properties, the channel potential as a function of position is accurately measured and the channel threshold voltage is given by the intercept of the fitting line. Furthermore, analytical expressions that model the extrinsic transistor d.c. $I-V$ characteristics as a forward-biased diode in series with an ideal transistor do not accurately capture the physics governing device operation or the inherent two-dimensional effects of the drain and gate fields. They often yield fits to data that agree only over a limited bias range. These approaches are best suited to devices with ohmic contacts (linear $I-V$ characteristics) that can be modelled as a resistor in series with the transistor. Efforts have been made to improve the charge injection and extraction at the metal contact to OSCs in OFETs including the use of self-assembled monolayers or contact area doping of the semiconductor[39–43]. When characterizing new OSCs, contact optimization is often not addressed; this can lead to over- or under-estimates of their performance potential.

It is important to note that the method detailed here for device parameterization is not entirely exempt from contact effects. Accurate parameterization near the apparent threshold voltage remains problematic because charge injection still limits channel

charging and channel resistance extraction at the lowest frequency range used in these measurements (20 Hz) and results in the apparent slow increase in channel mobility as shown in Fig. 7b. At longer charging timescales and at sweep rates comparable to $I-V$ measurements, quasi-static capacitance-voltage measurements provide yet another route to characterizing channel accumulation. Plotted in Fig. 7a is the quasi-static capacitance taken at $dV_{GS}/dt = 0.5\,\mathrm{V\,s}^{-1}$ and showing channel accumulation at $V_{GS} \approx -1\,\mathrm{V}$. A small shift in channel accumulation is observed relative to the characteristics extracted by using impedance spectroscopy. Smaller voltage ramp rates are required to reveal larger shifts towards positive threshold voltage but present a significant measurement challenge. This calls into question the accuracy of other important parameters such as threshold voltage and subthreshold slope. These parameters, like mobility, are routinely used in benchmarking performance and to estimate interface trap density with the assumption that both are governed entirely by the channel interface properties.

We show here that by disentangling channel and contact impedance in working transistors, we gain a better understanding of the origin of non-ideal behaviour in the I_D-V_{GS} characteristics of OFETs. Strongly varying contact resistance at low $|V_{GS}|$ results in transistor behaviour that is dominated by charge injection. This leads to an overestimation of the channel mobility by an order of magnitude when extracted using the MOSFET model and ambiguity in the transistor behaviour near the threshold. Analysis of the current–voltage data is not straightforward due to the variety of non-ideal contact and channel effects present in organic field-effect transistors. For accurate measurement of device parameters, such as mobility and subthreshold behaviour, more robust measurements and analysis must be developed along with contact engineering methods to improve charge injection at the metal–organic semiconductor interface.

Methods

Sample preparation. We used pre-fabricated transistor test structures consisting of a heavily doped silicon substrate (gate electrode, n-type $10^{-3}\,\Omega\,\mathrm{cm}$), thermally grown silicon dioxide layer (gate dielectric, $\approx 57\,\mathrm{nm}$) and photolithographically defined metal electrodes (source and drain contacts, 40 nm gold on 5 nm titanium). The completed transistors have channel lengths of 50–100 μm. We used a self-assembled monolayer of octadecyltrichlorosilane (OTS) to improve the semiconductor adhesion and to create a hydrophobic surface on the SiO_2 to eliminate water[3,15,44]. This layer is assembled by immersing the prefabricated substrates overnight in 5 mmol l^{-1} OTS in hexadecane, followed by sonication for 5 min in each of the following solvents: chloroform, isopropyl alcohol and de-ionized water, and then heating the wafer to 150 °C for 10 min.

Rubrene single crystals were grown by physical vapour transport in a tube oven under argon flow and carefully laminated to the surface of the prefabricated substrates. The starting material was 99.99 % rubrene from Sigma-Aldrich and used as received. The transistors that we considered for this study were those where the rubrene crystal occupied the drain-source channel and the contacts pads only; this was done to ensure that the transistor was isolated and to remove parasitic impedance from charged rubrene outside of the transistor area, Fig. 1c.

Electrical characterization. $I-V$ measurements were taken with an Agilent 4155C Semiconductor Parameter Analyzer. Impedance measurements were taken using an Agilent E4980 LCR meter by applying the high potential and current terminals to the gate and low potential and current terminals to the shorted drain and source. A d.c. bias voltage (V_{GS}) is applied to the gate along with a small a.c. signal ($V_{GS}(\omega) = 0.025\,\mathrm{V}$). At each d.c. bias point, the a.c. frequency $f = \omega/2\pi$ is swept from 20 to 2 MHz. Quasi-static capacitance was measured with a Hewlett Packard 4140B pA Meter at $dV_{GS}/dt = 0.5\,\mathrm{V\,s}^{-1}$. $I-V$, impedance and quasi-static capacitance measurements were taken successively in the dark at room temperature in an N_2 gas environment. Computer interfacing was done using Instrument Control[45].

References

1. Takeya, J. *et al.* Very high-mobility organic single-crystal transistors with in-crystal conduction channels. *Appl. Phys. Lett.* **90**, 102120 (2007).
2. Takeya, J. *et al.* In-crystal and surface charge transport of electric-field-induced carriers in organic single-crystal semiconductors. *Phys. Rev. Lett.* **98**, 196804 (2007).

3. Goldmann, C. *et al.* Determination of the interface trap density of rubrene single-crystal field-effect transistors and comparison to the bulk trap density. *J. Appl. Phys.* **99**, 034507 (2006).
4. Li, J. *et al.* A stable solution-processed polymer semiconductor with record high-mobility for printed transistors. *Sci. Rep.* **2**, 754 (2012).
5. Chen, H. *et al.* Highly π-extended copolymers with diketopyrrolopyrrole moieties for high-performance field-effect transistors. *Adv. Mater.* **24**, 4618–4622 (2012).
6. Lee, J. *et al.* Solution-processable ambipolar diketopyrrolopyrrole–selenophene polymer with unprecedentedly high hole and electron mobilities. *J. Am. Chem. Soc.* **134**, 20713–20721 (2012).
7. Lee, J. *et al.* Boosting the ambipolar performance of solution-processable polymer semiconductors via hybrid side-chain engineering. *J. Am. Chem. Soc.* **135**, 9540–9547 (2013).
8. Tseng, H.-R. *et al.* High mobility field effect transistors based on macroscopically oriented regioregular copolymers. *Nano Lett.* **12**, 6353–6357 (2012).
9. Luo, C. *et al.* General strategy for self-assembly of highly oriented nanocrystalline semiconducting polymers with high mobility. *Nano Lett.* **14**, 2764–2771 (2014).
10. Tseng, H.-R. *et al.* High-mobility field-effect transistors fabricated with macroscopic aligned semiconducting polymers. *Adv. Mater.* **26**, 2993–2998 (2014).
11. Yuan, Y. *et al.* Ultra-high mobility transparent organic thin film transistors grown by an off-centre spin-coating method. *Nat. Commun.* **5**, 3005 (2014).
12. Giri, G. *et al.* Tuning charge transport in solution-sheared organic semiconductors using lattice strain. *Nature* **480**, 504–508 (2011).
13. Blülle, B., Häusermann, R. & Batlogg, B. Approaching the trap-free limit in organic single-crystal field-effect transistors. *Phys. Rev. Appl.* **1**, 034006 (2014).
14. Teague, L. C. *et al.* Surface potential imaging of solution processable acene-based thin film transistors. *Adv. Mater.* **20**, 4513–4516 (2008).
15. Jurchescu, O. D. *et al.* Organic single-crystal field-effect transistors of a soluble anthradithiophene. *Chem. Mater.* **20**, 6733–6737 (2008).
16. Phan, H., Wang, M., Bazan, G. C. & Nguyen, T.-Q. Electrical instability induced by electron trapping in low-bandgap donor–acceptor polymer field-effect transistors. *Adv. Mater.* **27**, 7004–7009 (2015).
17. Sirringhaus, H. 25th Anniversary Article: organic field-effect transistors: the path beyond amorphous silicon. *Adv. Mater.* **26**, 1319–1335 (2014).
18. Fratini, S., Xie, H., Hulea, I. N., Ciuchi, S. & Morpurgo, A. F. Current saturation and Coulomb interactions in organic single-crystal transistors. *New J. Phys.* **10**, 033031 (2008).
19. Braga, D. & Horowitz, G. High-performance organic field-effect transistors. *Adv. Mater.* **21**, 1473–1486 (2009).
20. Reyes-Martinez, M. A., Crosby, A. J. & Briseno, A. L. Rubrene crystal field-effect mobility modulation via conducting channel wrinkling. *Nat. Commun.* **6**, 6948 (2015).
21. Hamadani, B. H. & Natelson, D. Temperature-dependent contact resistances in high-quality polymer field-effect transistors. *Appl. Phys. Lett.* **84**, 443–445 (2004).
22. Hamadani, B. H., Richter, C. A., Suehle, J. S. & Gundlach, D. J. Insights into the characterization of polymer-based organic thin-film transistors using capacitance-voltage analysis. *Appl. Phys. Lett.* **92**, 203303 (2008).
23. Kahn, A., Koch, N. & Gao, W. Electronic structure and electrical properties of interfaces between metals and π-conjugated molecular films. *J. Polym. Sci. Part B Polym. Phys.* **41**, 2529–2548 (2003).
24. Scott, J. C. Metal–organic interface and charge injection in organic electronic devices. *J. Vac. Sci. Technol. A* **21**, 521–531 (2003).
25. Miyadera, T., Minari, T., Tsukagoshi, K., Ito, H. & Aoyagi, Y. Frequency response analysis of pentacene thin-film transistors with low impedance contact by interface molecular doping. *Appl. Phys. Lett.* **91**, 013512 (2007).
26. Gundlach, D. J. *et al.* An experimental study of contact effects in organic thin film transistors. *J. Appl. Phys.* **100**, 024509 (2006).
27. Necliudov, P. V., Rumyantsev, S. L., Shur, M. S., Gundlach, D. J. & Jackson, T. N. 1/f noise in pentacene organic thin film transistors. *J. Appl. Phys.* **88**, 5395–5399 (2000).
28. Zimmerling, T. & Batlogg, B. Improving charge injection in high-mobility rubrene crystals: from contact-limited to channel-dominated transistors. *J. Appl. Phys.* **115**, 164511 (2014).
29. Teague, L. C. *et al.* Probing stress effects in single crystal organic transistors by scanning Kelvin probe microscopy. *Appl. Phys. Lett.* **96**, 203305 (2010).
30. Nichols, J. A., Gundlach, D. J. & Jackson, T. N. Potential imaging of pentacene organic thin-film transistors. *Appl. Phys. Lett.* **83**, 2366–2368 (2003).
31. Li, T., Ruden, P. P., Campbell, I. H. & Smith, D. L. Investigation of bottom-contact organic field effect transistors by two-dimensional device modeling. *J. Appl. Phys.* **93**, 4017–4022 (2003).
32. Shannon, J. M. & Gerstner, E. G. Source-gated thin-film transistors. *IEEE Electron Device Lett.* **24**, 405–407 (2003).
33. Sporea, R. A., Guo, X., Shannon, J. M. & Silva, S. R. Source-gated transistors for versatile large area electronic circuit design and fabrication. *ECS Trans.* **37**, 57–63 (2011).
34. Ma, A. M. *et al.* Zinc oxide thin film transistors with Schottky source barriers. *Solid-State Electron.* **76**, 104–108 (2012).
35. Ma, A. M. *et al.* Schottky barrier source-gated ZnO thin film transistors by low temperature atomic layer deposition. *Appl. Phys. Lett.* **103**, 253503 (2013).
36. Appenzeller, J., Radosavljević, M., Knoch, J. & Avouris, P. Tunneling versus thermionic emission in one-dimensional semiconductors. *Phys. Rev. Lett.* **92**, 048301 (2004).
37. Das, S., Chen, H.-Y., Penumatcha, A. V. & Appenzeller, J. High performance multilayer MoS2 transistors with scandium contacts. *Nano Lett.* **13**, 100–105 (2013).
38. Chow, P.-M. D. & Wang, K.-L. A new AC technique for accurate determination of channel charge and mobility in very thin gate MOSFET's. *IEEE Trans. Electron Devices* **33**, 1299–1304 (1986).
39. Wakatsuki, Y., Noda, K., Wada, Y., Toyabe, T. & Matsushige, K. Molecular doping effect in bottom-gate, bottom-contact pentacene thin-film transistors. *J. Appl. Phys.* **110**, 054505 (2011).
40. Gundlach, D. J., Jia, L. L. & Jackson, T. N. Pentacene TFT with improved linear region characteristics using chemically modified source and drain electrodes. *IEEE Electron Device Lett.* **22**, 571–573 (2001).
41. Kymissis, I., Dimitrakopoulos, C. D. & Purushothaman, S. High-performance bottom electrode organic thin-film transistors. *IEEE Trans. Electron Devices* **48**, 1060–1064 (2001).
42. Gundlach, D. J. *et al.* Contact-induced crystallinity for high-performance soluble acene-based transistors and circuits. *Nat. Mater.* **7**, 216–221 (2008).
43. Noda, K., Wada, Y. & Toyabe, T. Intrinsic difference in Schottky barrier effect for device configuration of organic thin-film transistors. *Org. Electron.* **15**, 1571–1578 (2014).
44. Pernstich, K. P. *et al.* Threshold voltage shift in organic field effect transistors by dipole monolayers on the gate insulator. *J. Appl. Phys.* **96**, 6431–6438 (2004).
45. Pernstich, K. P. Instrument Control (iC)—an open-source software to automate test equipment. *J. Res. Natl Inst. Stand. Technol.* **117**, 176 (2012).

Acknowledgements

We thank Oleg Kirillov for design and fabrication of transistor test beds.

Author contributions

E.G.B., J.I.B. and D.J.G. designed the experiments. E.G.B. fabricated and measured OFET devices, and analysed the data. J.I.B. wrote the code used to analyse the impedance data. T.N.J., O.D.J. and D.J.G. supervised the project. E.G.B. and D.J.G. wrote the manuscript; O.D.J. and D.J.G. edited the manuscript. Certain commercial equipment, instruments or materials are identified in this paper to specify the experimental procedure adequately. Such identification is not intended to imply recommendation or endorsement by the National Institute of Standards and Technology, nor is it intended to imply that the materials or equipment identified are necessarily the best available for the purpose.

Additional information

7

Wavefront shaping through emulated curved space in waveguide settings

Chong Sheng[1,*], Rivka Bekenstein[2,*], Hui Liu[1], Shining Zhu[1] & Mordechai Segev[2]

The past decade has witnessed remarkable progress in wavefront shaping, including shaping of beams in free space, of plasmonic wavepackets and of electronic wavefunctions. In all of these, the wavefront shaping was achieved by external means such as masks, gratings and reflection from metasurfaces. Here, we propose wavefront shaping by exploiting general relativity (GR) effects in waveguide settings. We demonstrate beam shaping within dielectric slab samples with predesigned refractive index varying so as to create curved space environment for light. We use this technique to construct very narrow non-diffracting beams and shape-invariant beams accelerating on arbitrary trajectories. Importantly, the beam transformations occur within a mere distance of 40 wavelengths, suggesting that GR can inspire any wavefront shaping in highly tight waveguide settings. In such settings, we demonstrate Einstein's Rings: a phenomenon dating back to 1936.

[1] National Laboratory of Solid State Microstructures & School of Physics, Collaborative Innovation Center of Advanced Microstructures, Nanjing University, Nanjing, Jiangsu 210093, China. [2] Department of Physics and Solid State Institute, Technion, Haifa 32000, Israel. * These authors contributed equally to this work. Correspondence and requests for materials should be addressed to R.B. (email: beken@tx.technion.ac.il) or to H.L. (email: liuhui@nju.edu.cn).

General electromagnetic (EM) beams propagating through linear homogenous media experience diffraction broadening. However, many applications would greatly benefit from having beams that remain very narrow or shape-invariant for large distances. The past two decades have witnessed remarkable progress in wavefront shaping specifically for the purpose of generating non-diffracting beams, such as shape-preserving Bessel beams[1] and accelerating beams in free space[2-5], in plasmonics[6-9] and even in nonlinear materials[10-15]. The concept of shape-invariant wavepackets was extended beyond EM waves, for example to shaping wavefunctions of electrons[16-19] and generating shape-invariant acoustic beams[20,21], and even accelerating surface water gravity waves[22]. All of these shape-invariant wavepackets are not square integrable (they carry infinite power), hence physically they must be truncated, which implies that they stay non-diffracting only for a finite distance[2]. In a similar vein, there are other kind of beams which are a priori designed to stay shape-invariant only for a finite distance, for example, the cosine-Gauss beams[23] and a class of beams that propagate on arbitrary curved trajectories[5,24,25]. Naturally, all of these beams require wavefront shaping: the launch beam must be shaped in a specific structure (amplitude and phase), to stay non-diffracting for the specified distance.

Wavefront shaping for generating non-diffracting optical beams can be achieved by various methods, ranging from annular slits[1], axicon lenses[26], computer generated holograms[24], spatial light modulators[3,28], gratings[7,23,29], metasurfaces[30-32] and diffraction from nanoparticles[4,33]. Importantly, non-diffracting beams can also be generated in inhomogeneous media such as photonic crystal slabs[34-38], photonic crystals[39,40] and photonic lattices[41]. All these too require wavefront shaping, that is typically done externally, outside the medium within which the beam is propagating. However, wavefront shaping can also be done by shaping the EM environment in which the wave is propagating[42,43]. The fact that the propagation of EM waves in static curved space is analogous to that in inhomogeneous media[42-44] is the underlying principle of emulating general relativity (GR) phenomena in transformation optics[42,43,45-52]. In transformation optics, the permittivities and permeabilities are structured to vary according to the curvature of space[53-59], giving rise to unique trajectories[55-57,60] and controlling the diffraction of light[61,62].

Here, we show that using ideas inspired by GR yields efficient beam shaping in waveguide settings. The concept is general, applicable to many cases where wavefront beam shaping in a waveguide platform is required. First, we fabricate the micro-structured optical waveguide with the specific refractive index emulating the curved space environment generated by a massive gravitational object. This dielectric structure yields a very narrow beam that remains non-diffracting for many Rayleigh lengths. Second, with the same experimental system, we demonstrate the Einstein's rings phenomenon, matching Einstein's 80 years old formula. Finally, we present a general formalism to transform Gaussian beams to considerably narrower shape-invariant beams accelerating (bending) along arbitrary trajectories.

Results

Generating non-diffracting beams through gravitational collimation. The first goal is to create a narrow beam that would propagate in a non-diffracting fashion for a considerable distance in a homogeneous medium. We do that by passing a Gaussian beam through a specific refractive index structure, inspired by the

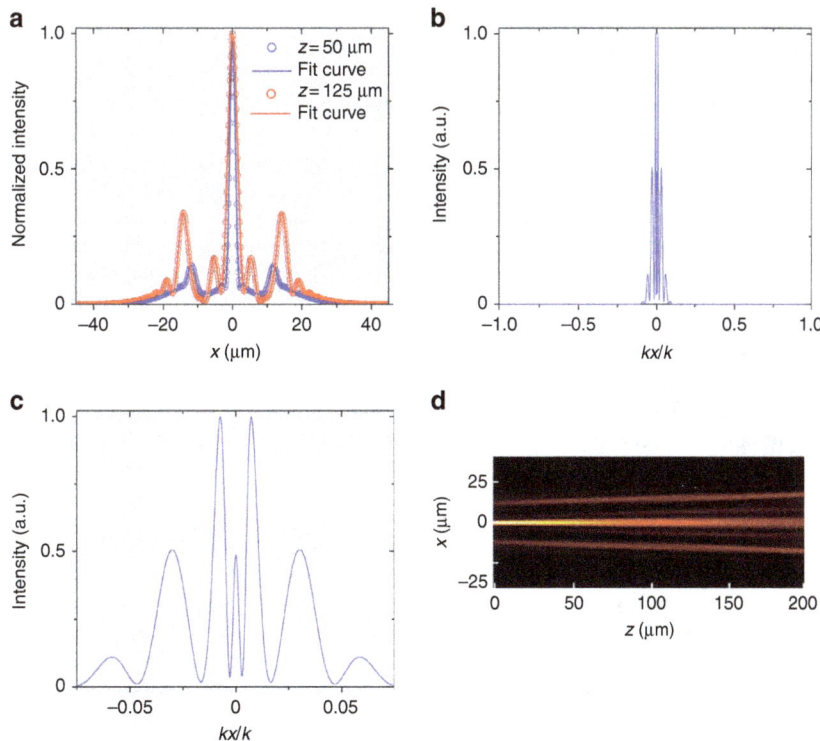

Figure 1 | Calculated propagation of gravitational collimation resulting in a non-diffracting beam. (**a**) The calculated non-diffracting beam fitted to the beam arising from the simulation of the experimental setting. (**b**) Spatial spectrum of the beam displaying two main peaks, as can be seen in **c** showing zoom-in on the central section of the spectrum. The two pronounced peaks correspond to a superposition of non-diffracting cosine and sine distributions, resulting in the narrow non-diffracting beam. (**d**) Simulated propagation of the non-diffracting beam of **a**, for a distance of 200 μm inside a homogenous medium, revealing the non-diffracting property.

gravitational lensing phenomenon occurring around massive stars. We design a specific curvature where the emulated gravitational lensing of the light on the micro-scale can create a very narrow non-diffracting beam. The basic principles of diffraction imply that non-diffracting beams can be constructed when their plane-waves constituents accumulate phase at the same rate. The non-diffracting property of beams depends on the dimensions of the wavepackets, that is, a non-diffracting beam can be a shape-invariant solution to the wave equation in three dimensions (3D) or in two dimensions (2D). In 3D homogeneous media, beams that are structured in both their transverse dimensions exhibit shape-invariant propagation on a straight line in the third dimension include the family of Bessel beams[1]. In 2D, on the other hand, when the beams are structured in a single transverse dimension (for example, when the beam is propagating in a planar waveguide), an ideal non-diffracting beam has a unique shape: two plane waves propagating at opposite symmetric angles with respect to the propagation axis. However, whereas the Bessel beams are localized, that is, they have a main lobe carrying most of the power, the planar case is just an interference grating—which is periodic and cannot be used for applications that require a beam with a single main lobe. Interestingly, providing proper spatial bandwidth to each of the opposite waves in the one-dimensional (1D) case does lead to a localized beam displaying non-diffracting features for some finite distance. More specifically, superimposing two beams whose spectrum in k-space is small compared with the wavenumber, at opposite angles with respect to the propagation axis, gives rise to non-diffracting propagation up to a finite distance, due to the similar rate of phase accumulation of the different modal (plane waves) constituents. Here, we construct such a very narrow non-diffracting beam by drawing on intuition from GR, where it is known that light waves are deflected by the space curvature generated by a massive star[63,64]. We exploit this gravitational lensing effect to construct a field that is a superposition of two beams of a finite spatial bandwidth, propagating at opposite angles with respect to the propagation axis. Such a beam remains non-broadening for a finite distance that can be much larger than the Rayleigh length of its main lobe. An example for such a 1D non-diffracting beam and its spectrum is displayed in Fig. 1a,b, respectively. Figure 1c shows zoom-in on the spectrum, while Fig. 1d presents its simulated propagation—where it is clear that the main lobe remains narrow for a large distance, in spite of the fact that its width is only four wavelength. The two main peaks in the spectrum (Fig. 1c) represent a superposition of cosine/sine distributions, along with a central peak. The width of the spectral peaks is two orders of magnitude smaller than the wavenumber, enabling a non-diffracting property to a finite distance. This structured beam, whose full-width-half-maximum (FWHM) is 2 µm, is approximately shape-preserving for ~200 µm, which corresponds to six Rayleigh lengths (Fig. 1d).

To transform a broad Gaussian beam (FWHM ~30 µm) into this non-diffracting beam in a planar waveguide setting, we fabricate a specific refractive index structure inspired by the concepts of curved space known from GR. Namely, curved space generated by a massive gravitational body leads to gravitational lensing, that can in principle overcome diffraction broadening and cause beam collimation. The planar waveguide has a unique width profile, causing a change in the propagation constant and effectively modifying the refractive index. The structure is shown in Fig. 2a. During the fabrication process, a silver film is deposited on a silica (SiO$_2$) substrate with a thickness of 80 nm, followed by polymethyl methacrylate (PMMA) microsphere powder scattered on the substrate. The microspheres are distributed on the substrate, with a small density and large separation distance between microspheres. The sample processing includes a stage

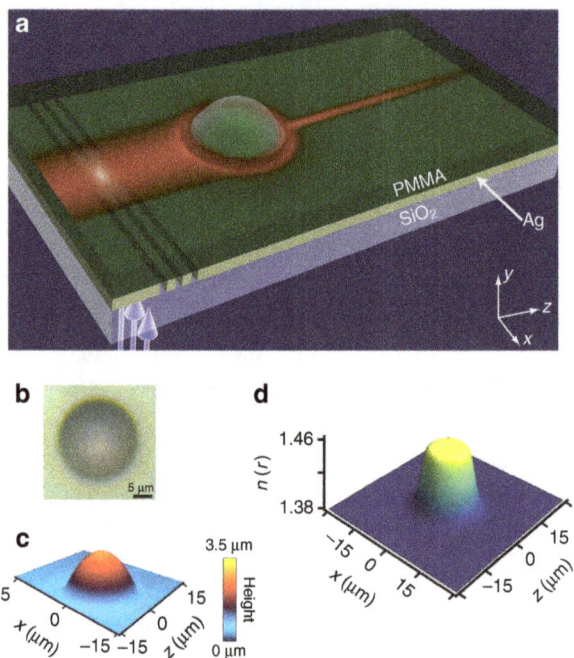

Figure 2 | The sample fabricated for generating a narrow collimated beam. (a) Schematic view of the fabricated waveguide: the inhomogeneous planar waveguide with the specifically designed refractive index structure. The structure is fabricated by depositing a thin silver film on a silica (SiO$_2$) substrate with a thickness of 80 nm, followed by PMMA microsphere powder scattered on the substrate. The blue arrows at the bottom represent the incident 457 nm blue laser light, and the bright spot marks the illumination spot where the light is incident on the grating. (b) Top-view optical microscopy image of the microdroplet. (c) The surface structure of the microdroplet, as mapped by AFM measurements. (d) The effective refractive index structure calculated from c, based on waveguide theory.

where the sample is put on the heating table (300 °C) for 30 s. As the melting temperature of PMMA polymer is ~250 °C, the heating process deforms the PMMA microspheres into domes, just as shown in Fig. 2b,c. In this process, the size of resultant PMMA domes is not uniform, and their diameters can vary greatly, from 1 to 100 µm. For the experiment presented here, we work with one of domes that has an appropriate size, as shown in its optical microscope image in Fig. 2b. The structure is shaped as a dome protruding from the plane of the waveguide (Fig. 2a). This is further confirmed by mapping the surface structure with atomic force microcopy (Asylum Research, MFP-3D-SA, USA), as shown in Fig. 2c. Next, a set of gratings with the period 310 nm are drilled on the sliver film around the microdroplet with focused ion beam (FEI Strata FIB 201, 30 keV, 150 pA). These gratings enable to couple the light into the slab waveguide. Next, we spin-coat the sample with a PMMA photoresist mixed with rare earth (Eu^{3+}) to a thickness of ~1 µm, and subsequently dry the sample in the oven at 70 °C for 2 h. The Eu^{3+} rare earth ions are added to the sample to facilitate fluorescence imaging that will reveal the propagation dynamics of the beam. These Eu^{3+} ions absorb the beam propagating in the slab waveguide, whose wavelength (457 nm) is specifically chosen to excite the rare earth ions, that in turn emit fluorescent light at 615 nm wavelength. We note that, although the 1-µm-thick PMMA layer is not single-mode waveguide for the 457 nm beam, the designed grating allows only one mode to be excited inside the waveguide. Here, only the TM3 mode is excited in our experiment (The grating is designed that only one waveguide mode is excited. Hence, plasmonic modes are not excited in the experiment). The resultant 2D structure of the refractive index is displayed in Fig. 2d, together with a 3D illustration of the entire sample (Fig. 2a). Figure 2c shows the width of the PMMA waveguide as mapped by AFM measurements. From this width, we calculate the refractive index structure displayed in Fig. 2d, which is fitted with the function $n(x,z) = n_0 + a/(1 + (\sqrt{x^2 + z^2}/r_c)^8)$, with $n_0 = 1.37$, $a = 9.22 \times 10^{-2}$, $r_c = 9.69$. Recall that the refractive index of bulk PMMA polymer is 1.49, hence our fabrication process reduces the refractive index according to our design. Specifically, in the region of the dome, the thickness is increased to 3.5 μm, and therefore the effective index of the TM3 waveguide mode is increased from 1.37 to 1.49.

In the experiment, we launch a Gaussian beam of 457 nm wavelength and 11.3 μm FWHM to propagate inside the PMMA layer that acts as a waveguide. The loss in this waveguide is quite small, in spite of the proximity of the thin Ag layer, enabling propagation distances of hundreds of micrometres. The specifically designed refractive index structure focuses the wide beam to a very narrow (2 μm) beam that is subsequently propagating without diffraction for ~200 μm, as expected from the theory. We emphasize that, after passing the 'star', the very narrow beam is propagating in a completely homogeneous medium, hence its non-diffracting property arises solely from the beam structure generated by passing the 'star'. Moreover, whereas most shape-preserving beams are very broad, this beam presents a narrow profile, only 2 μm wide. For comparison, we study the dynamic of a Gaussian beam passing through the same medium numerically and compare it with the experimental results (Fig. 3). We do this by numerically simulating the beam propagation, with the beam propagation method in a medium with the specific refractive index structure conforming to that of the sample used in the experiment (Fig. 2d). In both the experiments and the simulations

the transformation of the wide Gaussian beam to a narrow collimated beam is achieve within a very short propagation distance (~20 μm), allowing the use of this scheme in integrated photonics circuits. In Fig. 3, the diameter of the dome is roughly 25 μm. In the experiment, we can fabricate domes with different diameters, always with circular shape. Naturally, domes of different sizes yield collimation for different propagation distances and with different beam widths.

Experiments emulating the Einstein rings phenomenon. Interestingly, we find that besides producing collimated beams, the same planar 'central potential' index structure can also be used to emulate the phenomenon of Einstein's Rings, which is a famous phenomenon predicted by GR and observed in astronomy[65,66]. The Einstein Ring phenomena occurs when light from a point source is deformed by a mass distribution through gravitation lensing that causes the appearance of a ring around the mass distribution. For this case, the beam approaching the 'star' should emulate the radiation originating from a point source, namely, the wave reaching the 'star' should be a spherical wave. To emulate a point source, we fabricate (with focused ion beam) an arc-shaped grating (period of 310 nm) inside the metal film. This is shown in Fig. 4b, where the radius of the arc is 30 μm. When a plane wave is incident (from below) on the arc grating, the grating transforms it into a spherical wave propagating inside the waveguide layer. The region of incidence on the grating acts as a point source, emitting a spherical wave diverging both to the left and to the right of that point (negative and positive z, respectively). In such a setting, the spherical wavefront produced by the arc-grating emulates the wave radiated outwards from a point source located at the centre of grating arc. When this 1D spherical wavefront is passing

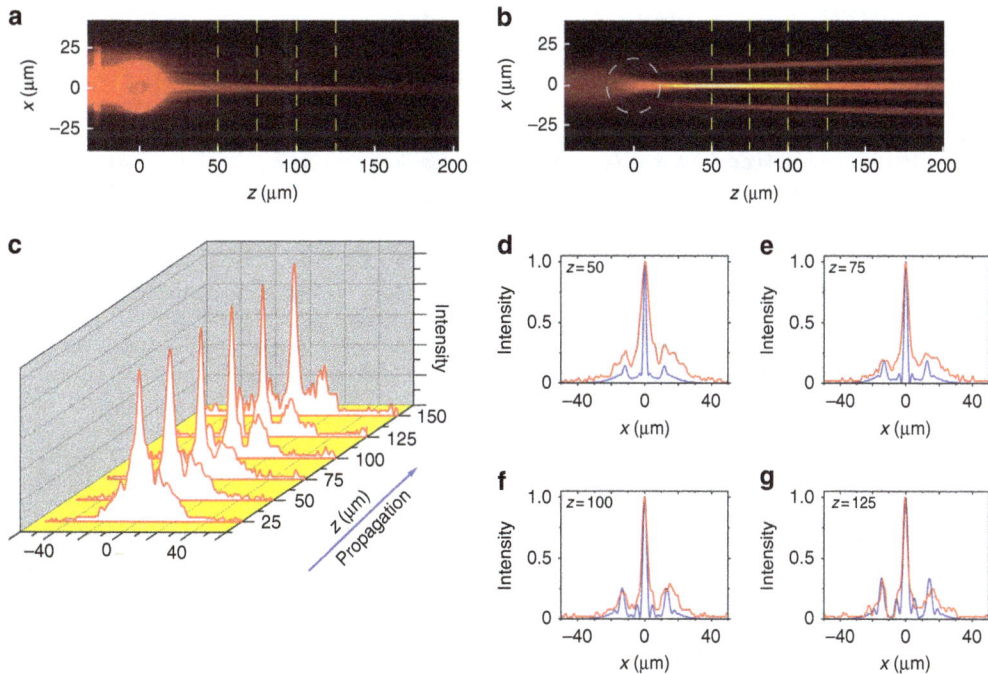

Figure 3 | Experimentally observed propagation dynamics of gravitationally collimated non-diffracting beam. (**a**) Top-view photograph of the experimentally observed results obtained through florescence. A broad Gaussian beam with FWWH 11.3 μm passes through the region of the dome, giving rise to the refractive index profile described in Fig. 2c. The wide Gaussian beam focuses to a narrow collimated beam that is non-diffracting for ~200 μm. The entire beam transformation process occurs within20 μm. (**b**) Simulated results of the same beam showing a similar effect as the experiment. The white dashed circle corresponds to the dome region. (**c**) Normalized intensity profile of the beam for several propagation distances, after passing though the dome region. (**d-g**) Measured (red) and simulated (blue) 1D intensity profiles for $z = 50$ μm, $z = 75$ μm, $z = 100$ μm, $z = 125$ μm, respectively, which correspond to the planes marked by the yellow dashed lines in **a-b**.

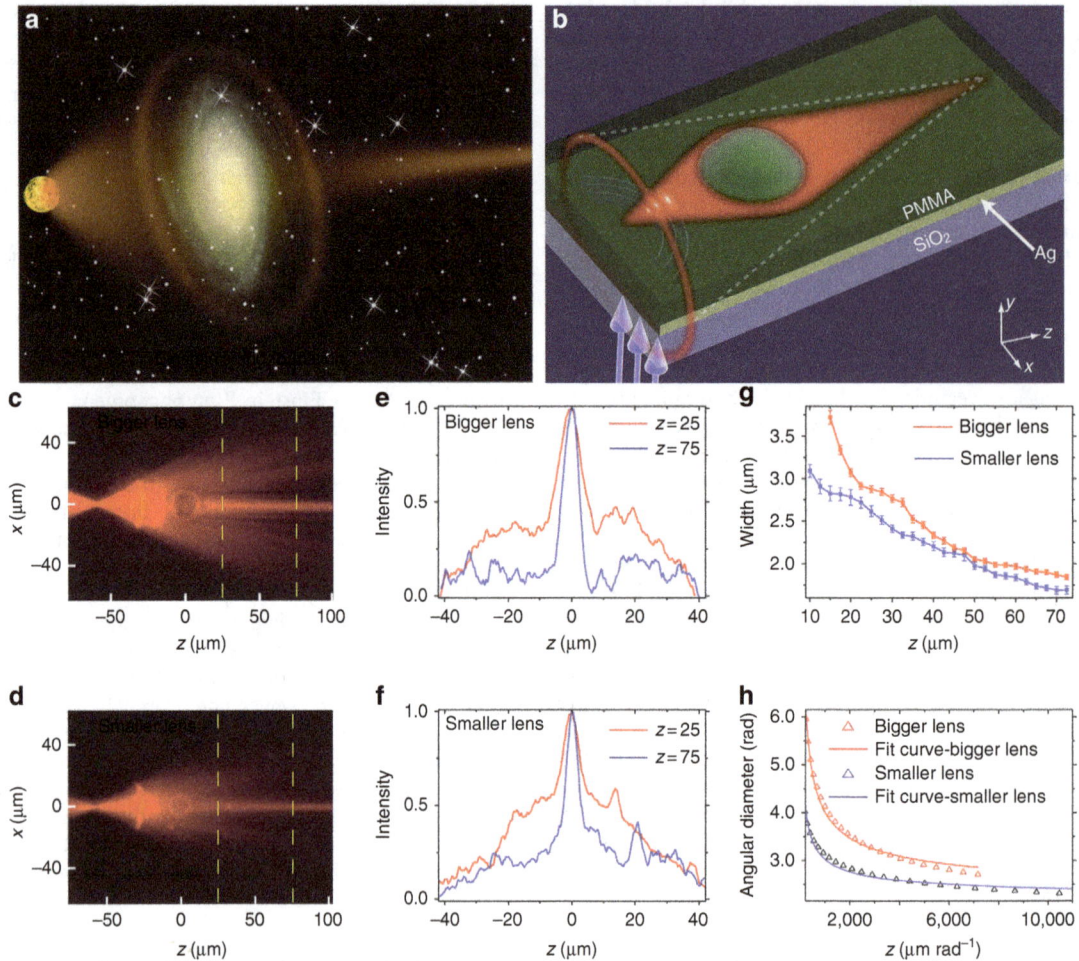

Figure 4 | Experimental emulation of the formation of Einstein's ring. (**a**) Einstein's vision: light from a point source is focused by a gravitational lens, and is subsequently observed as a virtual ring around the mass distribution. (**b**) Schematic view of the fabricated inhomogeneous waveguide. (**c,d**) Experimental results (obtained through florescence) showing a spherical wave passing though the dome region, for two domes of different radii. The inhomogeneous area acts as a gravitational lens on the light. (**e,f**) Measured beam profiles at $z = 25$ (red line), $z = 75$ (blue line), respectively, which correspond to the locations marked by the yellow dashed lines in **c**,**d**. (**g**) Measured beam width as a function of the propagation distance in the homogenous medium after the dome region. (**h**) Fit to Einstein's formula for the angular diameter of the Einstein rings. The calculated angular diameter from the experimental measurement is in a very good agreement with the theoretical formula.

by the star—it is focused and the beam width changes as a function of the propagation distance, as extracted from the experimental data. It is important to emphasize that our optical setting represents Einstein's rings formed by a time-harmonic EM waves, hence the entire dynamics is in space (not in time). Typical results for two different 'stars' (microdroplets with two different radii) are displayed in Fig. 4. As the Radius of the 'star' is larger the convergence of the beam is more extreme, but the final beam is wider (Fig. 4). At this point it is intriguing to compare our emulation results with Einstein's prediction. The Einstein Formula for the angular diameter of the virtual ring[64] is given by $\beta = \sqrt{\alpha_0 R_0 / z}$, that depends on the convergence angle α_0, the radius of the mass distribution R_0 and the distance between the centre of the mass distribution to the observation point. We calculate the angular diameter of the Einstein Ring from the measured convergence angle of the beam, for several different observation points (propagation distances). For a given observation point, the focusing angle of the beam after passing the 'star' gives the slope, from which we calculate the angular diameter of the virtual ring that an observer located at this specific distance (from the 'star') will see. To conform with the

Einstein formula, we calculate the relative angular radius between the two mass distributions (two samples). Namely, instead of calculating the absolute angular radius as a function of z, we calculate the relative angular radius between the results of each sample. We then fit the curve $\beta\left(\frac{z}{\alpha_0}\right) = \sqrt{c / \left(\frac{z}{\alpha_0}\right)}$ with c as a free parameter and compare the relative constant extracted from the experiment with the constant expected from Einstein's formula. In comparing the ratio and not the absolute number, we avoid the factor 2 between the relativistic Einstein formula and our experiment that represents Newtonian dynamics. As Fig. 4h shows, the experiments agree well with theory, although at large z, the experimental values are slightly lower than the model. This minute discrepancy arises from the difference between the fabricated optical potential (refractive index structure) and the $1/r$ gravitational potential of a point source. Consequently, for large values of z (distances), the focusing angle of the light deviates from Einstein's formula, hence the measured focusing angle is somewhat smaller than the theoretical curve.

Shaping beams accelerating on arbitrary trajectories. Finally, we present a general formalism for transforming broad Gaussian beams to accelerating beams that bend along arbitrary (convex) trajectories in a planar waveguide setting. As above, we do that by passing an incident broad Gaussian beam (11.3 μm FWHM) through a miniature refractive index structure that is designed specifically for this task. Accelerating beams are beams with a well-defined peak intensity that propagates along some non-straight trajectory, depending on the phase of the initial beam[4,5,29]. From the point of view of GR, the peak intensity of the beam does not follow geodesics paths[67], which are the shortest paths that light propagated along (by the Fermat principle). This important property of accelerating beams had been exploited for various applications, such as curved plasma channels[65], manipulating microparticles[68,69] and micromachining[70]. We design accelerating beams by utilizing the formalism suggested in ref. 5, for finding the specific 1D phase $\phi(x)$ required for shaping the wavefront of an accelerating beam that will propagate along a specific trajectory. This ID phase can be achieved by a 2D refractive index structure that the beam passes through, and obeys the relation

$$\phi(x) = k_0 \int_{z_i}^{z_f} n(x,z)dz, \qquad (1)$$

under the assumption that the propagation of the beam is in the paraxial regime. Using this method, there is no unique solution for $n(x,z)$. We therefore suggest a simple method that solves equation (1) for one specific refractive index profile to a specified phase, by assuming $n(x,z)$ is constructed from a function that is separable in x, z, namely $n(x,z) = f(x)g(z)$. For simplicity, we take $g(z) = \exp(-z^2/\sigma^2)$, and assume the Gaussian width (in z) is small compared with the propagation distance ($\sigma << z_f - z_i$). This allows setting the boundaries of the integral to infinity which after

integrating over z yields:

$$f(x) = \frac{\phi(x)}{k_0\sqrt{\pi}\sigma}. \qquad (2)$$

It is important to emphasize that the approximation we used for solving the integral of the phase only, causes additional effects. Due to the 2D refractive index distribution the beam is shifted to some different direction of propagation—$z' = ze^{i\theta}$ while propagating through the inhomogeneous area. Consequently, $n(x,z) = \frac{\phi(x)}{k_0\sqrt{\pi}\sigma}\exp(-z^2/\sigma^2)$. To present an example for this method, we find the refractive index profile required to create the phase for an accelerating beam along the trajectory $f(z) = az'^3$. In this specific case, the propagation of the resulting beam can be solved analytically using the method presented in ref. 5. In more complicated cases, a numerical solution for the ordinary differential equation (ODE) is required. We then use equation (1) to calculate the 2D refractive index structure that will provide the beam with the appropriate phase. By simulating the dynamic of a broad Gaussian beam passing through the designed refractive index structure, we find that the main lobe indeed accelerates along the expected trajectory, for a distance of 20 μm as displayed in Fig. 5. In this regime, it is possible to design a beam that will accelerate beam on an arbitrary trajectory. As any accelerating beam, the structure of such a beam involves a main lobe accompanied by oscillations on one side, and exponential decay on the other side. An example is shown in Fig. 5c, where the beam cross-sections at several propagation distances is displayed. This technique for beam shaping inside a slab waveguide is general, and can be used to shape the wavefront of non-diffracting beams accelerating on any convex trajectory, by designing the refractive index structure using equations (1 and 2), which relates the initial phase front (assumed here to be of a broad Gaussian beam) and the desired phase front $\phi(x)$ to the refractive index structure required for such wavefront shaping.

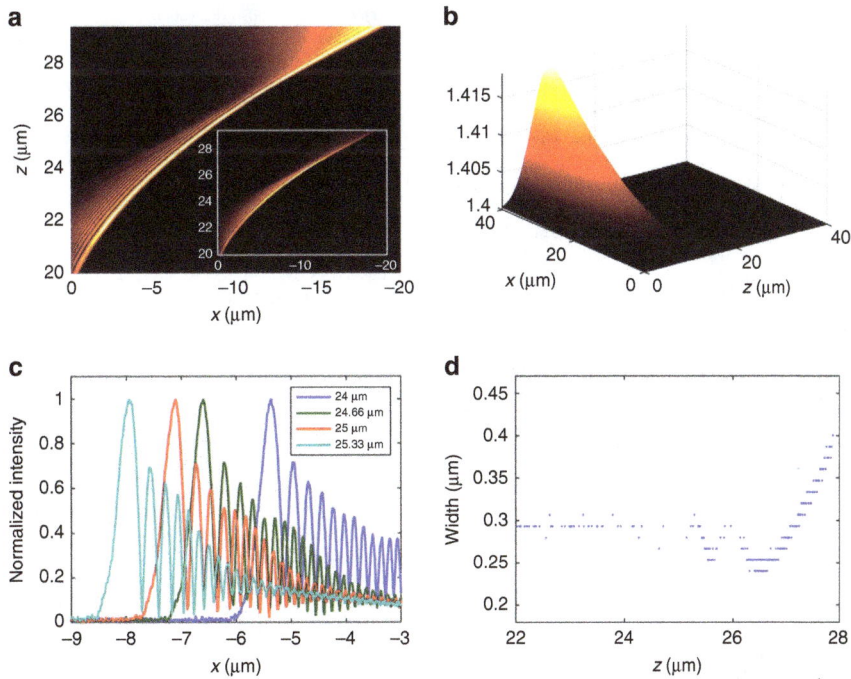

Figure 5 | Accelerating beams propagating along arbitrary trajectories produced by designing the refractive index structure within the initial 10 μm propagation distances in the waveguide layer. (**a**) Simulated evolution of the accelerating beam, where the peak intensity is propagating along the curve $f(z) = az^3$. Inset: the evolution displayed with a non-normalized intensity (**b**) The calculated refractive index structure which transforms a broad Gaussian beam into the narrow non-diffracting accelerating beam of **a**. (**c**) Structure of the accelerating beam for different propagation distance. (**d**) Width of the main lobe as a function of the propagation distance.

Discussion

To conclude, we have presented a method for shaping optical wavefronts in waveguide settings. Our technique is inspired by GR and it provides a platform for emulating the spatial dynamics of EM waves in curved space. This method can be achieved in thin film waveguides and can be implemented in integrated photonics settings. Specifically, we have demonstrated experimentally the construction of a narrow non-diffracting beam, the formation of Einstein's rings, and presented a general method to construct accelerating beans propagating along arbitrary trajectories. This method can be used for shaping any general beam, thereby suggesting a new way of using transformation optics media for beam shaping in waveguide settings with a single dielectric material. In this work, we presented beam shaping in the spatial domain; consequently, our experiments employed only continuous laser beams as our input waves. However, in principle this technique can also be used to shape ultrashort laser pulses with the traditional grating pairs, the lenses and the spatial modulation at the focal plane, all implemented in a slab waveguide geometry with proper design of the planar refractive index structure. This idea will be pursued in our future research.

References

1. Durnin, J., Miceli, J. J. & Eberly, J. H. Diffraction-free beams. *Phys. Rev. Lett.* **58**, 1499–1501 (1987).
2. Siviloglou, G. A. & Christodoulides, D. N. Accelerating finite energy Airy beams. *Opt. Lett.* **32**, 979–981 (2007).
3. Siviloglou, G. A., Broky, J., Dogariu, A. & Christodoulides, D. N. Observation of Accelerating Airy Beams. *Phys. Rev. Lett.* **99**, 213901 (2007).
4. Kaminer, I., Bekenstein, R., Nemirovsky, J. & Segev, M. Nondiffracting accelerating wave packets of Maxwell's equations. *Phys. Rev. Lett.* **108**, 163901 (2012).
5. Greenfield, E., Segev, M., Walasik, W. & Raz, O. Accelerating light beams along arbitrary convex trajectories. *Phys. Rev. Lett.* **106**, 213902 (2011).
6. Salandrino, A. & Christodoulides, D. N. Airy plasmon: a nondiffracting surface wave. *Opt. Lett.* **35**, 2082–2084 (2010).
7. Zhang, P. et al. Plasmonic Airy beams with dynamically controlled trajectories. *Opt. Lett.* **36**, 3191–3193 (2011).
8. Minovich, A. et al. Generation and near-field imaging of airy surface plasmons. *Phys. Rev. Lett.* **107**, 116802 (2011).
9. Epstein, I. & Arie, A. Arbitrary bending plasmonic light waves. *Phys. Rev. Lett.* **112**, 023903 (2014).
10. Wulle, T. & Herminghaus, S. Nonlinear optics of Bessel beams. *Phys. Rev. Lett.* **70**, 1401–1404 (1993).
11. Kaminer, I., Segev, M. & Christodoulides, D. N. Self-accelerating self-trapped optical beams. *Phys. Rev. Lett.* **106**, 213903 (2011).
12. Lotti, A. et al. Stationary nonlinear Airy beams. *Phys. Rev. A* **84**, 021807 (2011).
13. Bekenstein, R. & Segev, M. Self-accelerating optical beams in highly nonlocal nonlinear media. *Opt. Express* **19**, 23706–23715 (2011).
14. Dolev, I., Kaminer, I., Shapira, A., Segev, M. & Arie, A. Experimental observation of self-accelerating beams in quadratic nonlinear media. *Phys. Rev. Lett.* **108**, 113903 (2012).
15. Bekenstein, R., Schley, R., Mutzafi, M., Rotschild, C. & Segev, M. Optical simulations of gravitational effects in the Newton-Schrodinger system. *Nat. Phys.* **11**, 872–878 (2015).
16. Uchida, M. & Tonomura, A. Generation of electron beams carrying orbital angular momentum. *Nature* **464**, 737–739 (2010).
17. Voloch-Bloch, N., Lereah, Y., Lilach, Y., Gover, A. & Arie, A. Generation of electron Airy beams. *Nature* **494**, 331–335 (2013).
18. Grillo, V. et al. Generation of nondiffracting electron bessel beams. *Phys. Rev. X* **4**, 011013 (2014).
19. Kaminer, I., Nemirovsky, J., Rechtsman, M., Bekenstein, R. & Segev, M. Self-accelerating Dirac particles and prolonging the lifetime of relativistic fermions. *Nat. Phys.* **11**, 261–267 (2015).
20. Zhang, P. et al. Generation of acoustic self-bending and bottle beams by phase engineering. *Nat. Commun.* **5**, 4316 (2014).
21. Bar-Ziv, U., Postan, A. & Segev, M. Observation of shape-preserving accelerating underwater acoustic beams. *Phys. Rev. B* **92**, 100301 (2015).
22. Fu, S., Tsur, Y., Zhou, J., Shemer, L. & Arie, A. Propagation dynamics of airy water-wave pulses. *Phys. Rev. Lett.* **115**, 034501 (2015).
23. Lin, J. et al. Cosine-gauss plasmon beam: a localized long-range nondiffracting surface wave. *Phys. Rev. Lett.* **109**, 093904 (2012).
24. Rosen, J. & Yariv, A. Snake beam: a paraxial arbitrary focal line. *Opt. Lett.* **20**, 2042–2044 (1995).
25. Froehly, L. et al. Arbitrary accelerating micron-scale caustic beams in two and three dimensions. *Optics Express* **19**, 16455 (2011).
26. Scott, G. & McArdle, N. Efficient generation of nearly diffraction-free beams using an axicon. *Opt. Eng.* **31**, 2640–2643 (1992).
27. Rosen, J. & Yariv, A. Synthesis of an arbitrary axial field profile by computer-generated holograms. *Opt. Lett.* **19**, 843–845 (1994).
28. Zhang, P. et al. Nonparaxial mathieu and weber accelerating beams. *Phys. Rev. Lett.* **109**, 193901 (2012).
29. Li, L., Li, T., Wang, S. M. & Zhu, S. N. Collimated plasmon beam: nondiffracting versus linearly focused. *Phys. Rev. Lett.* **110**, 046807 (2013).
30. Bomzon, Z., Kleiner, V. & Hasman, E. Formation of radially and azimuthally polarized light using space-variant subwavelength metal stripe gratings. *Appl. Phys. Lett.* **79**, 1587–1589 (2001).
31. Yu, N. et al. Light propagation with phase discontinuities: generalized laws of reflection and refraction. *Science* **334**, 333–337 (2011).
32. Kildishev, A. V., Boltasseva, A. & Shalaev, V. M. Planar photonics with metasurfaces. *Science* **339**, 1232009 (2013).
33. Chen, Z., Taflove, A. & Backman, V. Photonic nanojet enhancement of backscattering of light by nanoparticles: a potential novel visible-light ultramicroscopy technique. *Opt. Express* **12**, 1214–1220 (2004).
34. Yu, X. & Fan, S. Bends and splitters for self-collimated beams in photonic crystals. *Appl. Phys. Lett.* **83**, 3251–3253 (2003).
35. Rakich, P. T. et al. Achieving centimetre-scale supercollimation in a large-area two-dimensional photonic crystal. *Nat. Mater.* **5**, 93–96 (2006).
36. Shih, T.-M. et al. Supercollimation in photonic crystals composed of silicon rods. *Appl. Phys. Lett.* **93**, 131111 (2008).
37. Hamam, R. E., Ibanescu, M., Johnson, S. G., Joannopoulos, J. D. & Soljacic, M. Broadband super-collimation in a hybrid photonic crystal structure. *Opt. Express* **17**, 8109–8118 (2009).
38. Mocella, V. et al. Self-collimation of light over millimeter-scale distance in a quasi-zero-average-index metamaterial. *Phys. Rev. Lett.* **102**, 133902 (2009).
39. Longhi, S. & Janner, D. X-shaped waves in photonic crystals. *Phys. Rev. B* **70**, 235123 (2004).
40. Conti, C. & Trillo, S. Nonspreading wave packets in three dimensions formed by an ultracold bose gas in an optical lattice. *Phys. Rev. Lett.* **92**, 120404 (2004).
41. Manela, O., Segev, M. & Christodoulides, D. N. Nondiffracting beams in periodic media. *Opt. Lett.* **30**, 2611–2613 (2005).
42. Leonhardt, U. Optical conformal mapping. *Science* **312**, 1777–1780 (2006).
43. Pendry, J. B., Schurig, D. & Smith, D. R. Controlling electromagnetic fields. *Science* **312**, 1780–1782 (2006).
44. Laundau, L.D. & Lifshitz, E. M. *The Classical Theory Of Fields* (Butterworth-Heinemann, 1975).
45. Li, J. & Pendry, J. B. Hiding under the carpet: a new strategy for cloaking. *Phys. Rev. Lett.* **101**, 203901 (2008).
46. Alù, A. & Engheta, N. Multifrequency optical invisibility cloak with layered plasmonic shells. *Phys. Rev. Lett.* **100**, 113901 (2008).
47. Smolyaninov, I. I., Smolyaninova, V. N., Kildishev, A. V. & Shalaev, V. M. Anisotropic metamaterials emulated by tapered waveguides: application to optical cloaking. *Phys. Rev. Lett.* **102**, 213901 (2009).
48. Valentine, J., Li, J., Zentgraf, T., Bartal, G. & Zhang, X. An optical cloak made of dielectrics. *Nat. Mater.* **8**, 568–571 (2009).
49. Gabrielli, L. H., Cardenas, J., Poitras, C. B. & Lipson, M. Silicon nanostructure cloak operating at optical frequencies. *Nat. Photon.* **3**, 461–463 (2009).
50. Smolyaninova, V. N., Smolyaninov, I. I., Kildishev, A. V. & Shalaev, V. M. Experimental observation of the trapped rainbow. *Appl. Phys. Lett.* **96**, 211121 (2010).
51. Chen, H., Chan, C. T. & Sheng, P. Transformation optics and metamaterials. *Nat. Mater.* **9**, 387–396 (2010).
52. Zentgraf, T., Liu, Y., Mikkelsen, M. H., Valentine, J. & Zhang, X. Plasmonic luneburg and eaton lenses. *Nat. Nanotechnol.* **6**, 151–155 (2011).
53. Smolyaninov, I. I. Surface plasmon toy model of a rotating black hole. *New J. Phys.* **5**, 147–147 (2003).
54. Leonhardt, U. & Philbin, T. G. General relativity in electrical engineering. *New J. Phys.* **8**, 247 (2006).
55. Genov, D. A., Zhang, S. & Zhang, X. Mimicking celestial mechanics in metamaterials. *Nat. Phys.* **5**, 687–692 (2009).
56. Narimanov, E. E. & Kildishev, A. V. Optical black hole: broadband omnidirectional light absorber. *Appl. Phys. Lett.* **95**, 041106–041106-3 (2009).
57. Cheng, Q., Cui, T. J., Jiang, W. X. & Cai, B. G. An omnidirectional electromagnetic absorber made of metamaterials. *New J. Phys.* **12**, 063006 (2010).
58. Smolyaninov, I. I. & Narimanov, E. E. Metric signature transitions in optical metamaterials. *Phys. Rev. Lett.* **105**, 067402 (2010).
59. Genov, D. A. General relativity: optical black-hole analogues. *Nat. Photon.* **5**, 76–78 (2011).

60. Sheng, C., Liu, H., Wang, Y., Zhu, S. N. & Genov, D. A. Trapping light by mimicking gravitational lensing. *Nat. Photon.* **7,** 902–906 (2013).
61. Batz, S. & Peschel, U. Linear and nonlinear optics in curved space. *Phys. Rev. A* **78,** 043821 (2008).
62. Bekenstein, R., Nemirovsky, J., Kaminer, I. & Segev, M. Shape-preserving accelerating electromagnetic wave packets in curved space. *Phys. Rev. X* **4,** 011038 (2014).
63. Einstein, A. Die Grundlage der allgemeinen relativitätstheorie. *Ann. Phys.* **354,** 769–822 (1916).
64. Einstein, A. Lens-like action of a star by the deviation of light in the gravitational field. *Science* **84,** 506–507 (1936).
65. Hewitt, J. N. *et al.* Unusual radio source MG1131 + 0456: a possible Einstein ring. *Nature* **333,** 537–540 (1988).
66. King, L. J. *et al.* A complete infrared Einstein ring in the gravitational lens system B1938 + 666. *MNRAS* **295,** L41–L44 (1998).
67. Polynkin, P., Kolesik, M., Moloney, J. V., Siviloglou, G. A. & Christodoulides, D. N. Curved plasma channel generation using ultraintense airy beams. *Science* **324,** 229–232 (2009).
68. Baumgartl, J., Mazilu, M. & Dholakia, K. Optically mediated particle clearing using Airy wavepackets. *Nat. Photon.* **2,** 675–678 (2008).
69. Schley, R. *et al.* Loss-proof self-accelerating beams and their use in non-paraxial manipulation of particles' trajectories. *Nat. Commun.* **5,** 5189 (2014).
70. Mathis, A. *et al.* Micromachining along a curve: Femtosecond laser micromachining of curved profiles in diamond and silicon using accelerating beams. *Appl. Phys. Lett.* **101,** 071110–071113 (2012).

Acknowledgements

R.B. gratefully acknowledges the support of the Adams Fellowship Programme of the Israel Academy of Sciences and Humanities. This research was also supported by the ICore Excellence center 'Circle of Light' and a grant from the US Air Force Office for Scientific Research (AFOSR). H.L. gratefully acknowledges the support of the National Natural Science Foundation of China (No's 11321063, 61425018 and 11374151), the National Key Projects for Basic Researches of China (No. 2012CB933501 and 2012CB921500), the Doctoral Programme of Higher Education (20120091140005) and Dengfeng Project B of Nanjing University. C.S. gratefully acknowledge the support of the programme A for Outstanding PhD candidate of Nanjing University.

Author contributions

All authors contributed to all aspects of this work.

Additional information

Competing financial interests: The authors declare no competing financial interests.

8

A three-dimensional actuated origami-inspired transformable metamaterial with multiple degrees of freedom

Johannes T.B. Overvelde[1], Twan A. de Jong[1], Yanina Shevchenko[2], Sergio A. Becerra[1], George M. Whitesides[2,3], James C. Weaver[3], Chuck Hoberman[3,4,5] & Katia Bertoldi[1,6]

Reconfigurable devices, whose shape can be drastically altered, are central to expandable shelters, deployable space structures, reversible encapsulation systems and medical tools and robots. All these applications require structures whose shape can be actively controlled, both for deployment and to conform to the surrounding environment. While most current reconfigurable designs are application specific, here we present a mechanical metamaterial with tunable shape, volume and stiffness. Our approach exploits a simple modular origami-like design consisting of rigid faces and hinges, which are connected to form a periodic structure consisting of extruded cubes. We show both analytically and experimentally that the transformable metamaterial has three degrees of freedom, which can be actively deformed into numerous specific shapes through embedded actuation. The proposed metamaterial can be used to realize transformable structures with arbitrary architectures, highlighting a robust strategy for the design of reconfigurable devices over a wide range of length scales.

[1] John A. Paulson School of Engineering and Applied Sciences, Harvard University, Cambridge, Massachusetts 02138, USA. [2] Department of Chemistry and Chemical Biology, Harvard University, Cambridge, Massachusetts 02138, USA. [3] Wyss Institute for Biologically Inspired Engineering, Harvard University, Cambridge, Massachusetts 02138, USA. [4] Hoberman Associates, New York, New York 10001, USA. [5] Graduate School of Design, Harvard University, Cambridge, Massachusetts 02138, USA. [6] Kavli Institute, Harvard University, Cambridge, Massachusetts 02138, USA. Correspondence and requests for materials should be addressed to K.B. (email: bertoldi@seas.harvard.edu) or to J.C.W, (email: James.Weaver@wyss.harvard.edu) or to C.H. (email: Chuck@hoberman.com).

Metamaterials are rapidly appearing at the frontier of science and engineering due to their exotic and unusual properties obtained from their structure rather than their composition. Using origami, the ancient art of paper folding, programmable metamaterials have been previously created from two-dimensional (2D) sheets through folding along pre-defined creases[1–10]. In particular, origami patterns have proven to be promising in the design of solar panels for space deployment[11], flexible medical stents[12], 3D cell-laden microstructures[13] and flexible electronics[14]. Moreover, it has been shown that cellular metamaterials can be designed by stacking folded layers[4,15,16]. While almost all proposed origami-inspired mechanical metamaterial designs are based on the Miura-ori fold pattern[11], which has a single degree of freedom, there are many other origami-like architectures with multiple degrees of freedom that can be used to design highly flexible and deformable 3D structures.

The current work was initially inspired by snapology, a type of modular unit-based origami invented by Heinz Strobl in which paper ribbons are used to create complex geometric extruded polyhedra (Fig. 1a)[17,18]. While snapology provides the geometric starting point for our research, our focus here is on the foldability of these structures and how this can lead to new designs for transformable metamaterials. Using this design approach, some of the extruded polyhedral geometries, such as the extruded icosahedron shown in Fig. 1b, are stiff and almost rigid, while others have multiple degrees of freedom and can be easily deformed (see Supplementary Movie 1)[19]. On the basis of these observations, here we focus on an extruded cube[20] (Fig. 1c) as a

fundamental building block and demonstrate both analytically and experimentally that such a structure can be used as a unit cell in the design of foldable reprogrammable matter whose shape, volume and stiffness can be markedly altered in a fully predictable fashion. Moreover, we show that the properties of the resulting 3D mechanical metamaterial can be actively tuned and controlled by strategically placing pneumatic actuators on the edges (hinges) of each unit cell.

Results

Characterization of the unit cell. We construct a unit cell by extruding the edges of a cube in the direction normal to each face (Fig. 2a). This results in a 3D structure with 24 faces connected by 36 edges of length L. If, as in the simplified model of rigid origami[21,22], we assume that the faces are rigid and the structure can only fold along the edges, such a unit has three degrees of freedom identified by the angles γ_1, γ_2 and γ_3 (Fig. 2a). Changing these three angles deforms the internal cube into a rhombohedron (a 3D geometry formed from six rhombi) and, more importantly, reconfigures the unit cell into many specific shapes. To describe them, we introduce the vectors $\mathbf{p}_1$, $\mathbf{p}_2$ and $\mathbf{p}_3$, which span the internal rhombohedron (Fig. 2a)

$$\mathbf{p}_1 = [L, \quad 0, \quad 0], \tag{1}$$

$$\mathbf{p}_2 = [L\cos(\gamma_3), \quad L\sin(\gamma_3), \quad 0], \tag{2}$$

$$\mathbf{p}_3 = \left[L\cos(\gamma_2), \quad L\delta, \quad L\sqrt{1 - \cos^2(\gamma_2) - \delta^2}\right], \tag{3}$$

where $\delta = [\cos(\gamma_1) - \cos(\gamma_2)\cos(\gamma_3)]/\sin(\gamma_3)$. It is important to note that, because of contact occurring between faces, not all combinations of γ_1, γ_2 and γ_3 are possible. All possible combinations of the three angles can be found by requiring the third component of $\mathbf{p}_3$ to be real valued,

$$1 - \cos^2(\gamma_2) - \delta^2 = 1 - \cos^2(\gamma_2) - \frac{(\cos(\gamma_1) - \cos(\gamma_2)\cos(\gamma_3))^2}{\sin^2(\gamma_3)}$$

$$\geq 0. \tag{4}$$

We first rewrite equation (4) as

$$1 - \cos^2(\gamma_1) - \cos^2(\gamma_2) - \cos^2(\gamma_3) + 2\cos(\gamma_1)\cos(\gamma_2)\cos(\gamma_3) \geq 0, \tag{5}$$

and then solve it for γ_3 to obtain

$$\gamma_3 \leq \gamma_1 + \gamma_2, \quad \gamma_3 \leq -(\gamma_1 + \gamma_2) + 2\pi,$$
$$\gamma_3 \geq (\gamma_1 - \gamma_2), \quad \gamma_3 \geq -(\gamma_1 - \gamma_2). \tag{6}$$

where we have used $0 \leq \gamma_1, \gamma_2, \gamma_3 \leq \pi$. Note that equation (6) defines a regular tetrahedron with vertices located at $(\gamma_1, \gamma_2, \gamma_3) = (0, 0, 0)$, $(\pi, \pi, 0)$, $(\pi, 0, \pi)$ and $(0, \pi, \pi)$ (Fig. 2b). Therefore, only the combinations of angles contained within this domain are attainable.

As shown in Fig. 2c and Supplementary Movie 2, the unit cell can be transformed into multiple highly distinct shapes by varying γ_1, γ_2 and γ_3. At the centre of the tetrahedron (that is, $(\gamma_1, \gamma_2, \gamma_3) = (\pi/2, \pi/2, \pi/2)$) we find the fully expanded configuration (state #1 in Fig. 2c). For this state and for all other combinations of angles.

Inside the tetrahedron, the faces do not come into contact and therefore all three degrees of freedom can be used to deform the unit cell. Alternatively, for configurations that lie on the surface of the tetrahedron, one degree of freedom is constrained due to contact between the faces. As an example, in Fig. 2c we show the configuration that lies at the centre of each face of the tetrahedron (state #2).

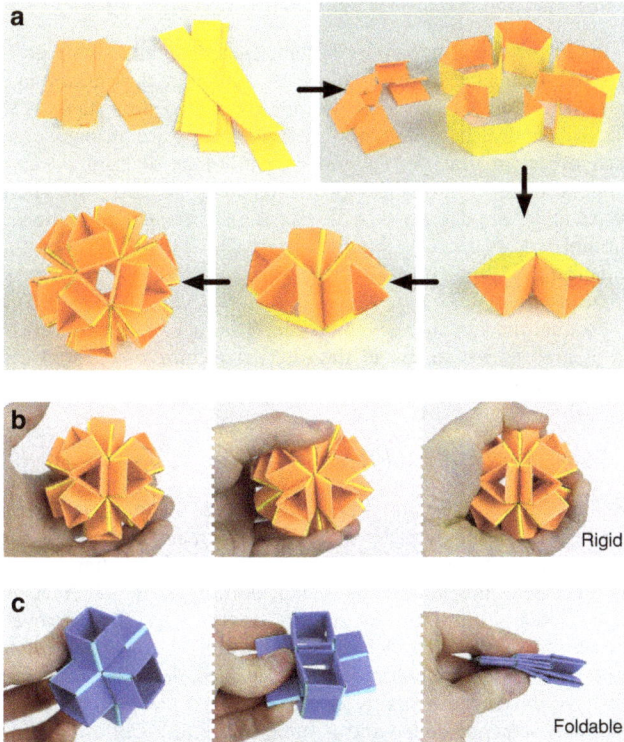

Figure 1 | Our work is inspired by snapology. (a) Snapology is a type of modular, unit-based origami in which paper ribbons are folded, and 'snapped' together to assemble extruded polyhedra, such as the extruded icosahedron shown. (b) Some of the geometries that can be made in this way, including the extruded icosahedron, are almost rigid. (c) In contrast, other geometries, including the extruded cube, have multiple degrees of freedom and can be easily deformed.

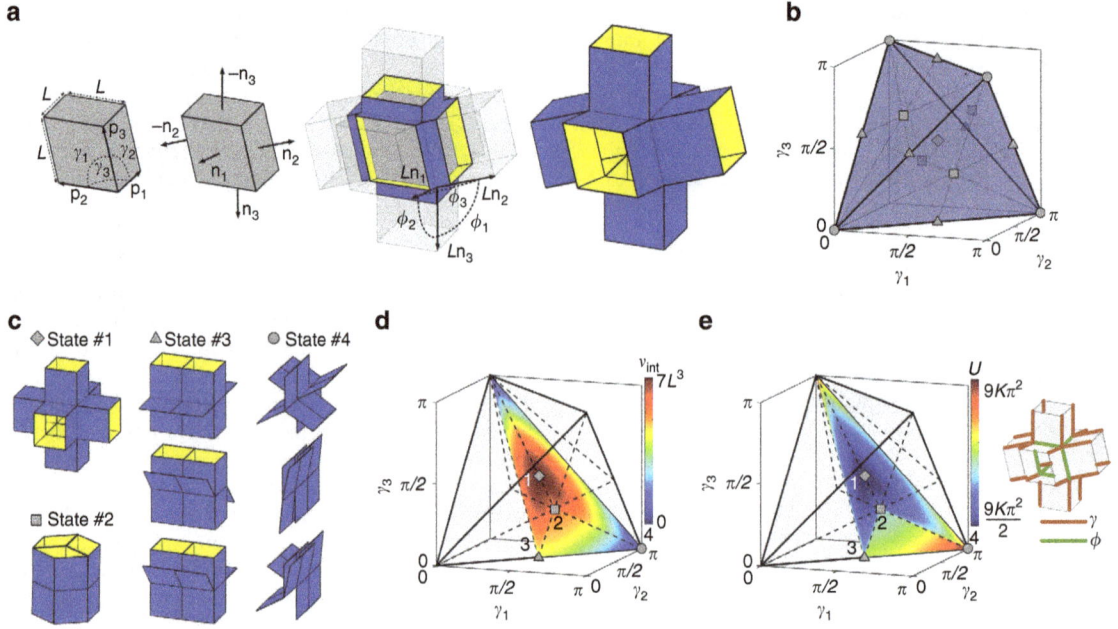

Figure 2 | Analysis of the possible shapes of the extruded cube unit cell. (a) The shape of the unit cell is found by extruding the edges in the direction normal to the faces of a rhombohedron, and can be fully described by the vectors $\mathbf{p}_1$, $\mathbf{p}_2$ and $\mathbf{p}_3$ spanning the internal rhombohedron. **(b)** Regular tetrahedron containing all combinations of angles $(\gamma_1, \gamma_2, \gamma_3)$ that are attainable. **(c)** State #1, #2, #3 and #4 are configurations that lie in the centre of the regular tetrahedron, on the centre of its faces, on the centre of its edges and on its vertices, respectively. **(d)** Contour plot showing the evolution of internal volume (v_{int}) of the unit cell as a function of γ_1, γ_2 and γ_3. **(e)** Contour plot showing the evolution of the strain energy (U) of the unit cell as a function of γ_1, γ_2 and γ_3. Note that the values of v_{int} and U are shown on the boundary of a sub-region of the regular tetrahedron, which, because of the symmetry of the unit cell, contains all possible configurations. Moreover, the orange and green lines on the unit cell indicate edges (hinges) whose energy are specified by the γ and ϕ angles, respectively.

When moving to the edges of the tetrahedron, two of the six extruded rhombi flatten, resulting in shapes similar to that of state #3 (Fig. 2c), which lies in the centre of each edge. These edge states still have three degrees of freedom, one associated with shearing of the four open rhombi and two associated with tilting of the two flattened rhombi (Fig. 2c and Supplementary Fig. 1). However, only one degree of freedom is controlled by the γ angles, as the edge states correspond to singular points in the γ domain. To see this, we focus on the angles ϕ_1, ϕ_2 and ϕ_3 between the vectors normal to the faces of the central rhombohedron (Fig. 2a)

$$\phi_k = \arctan\left(\frac{|\mathbf{n}_i \times \mathbf{n}_j|}{\mathbf{n}_i \cdot \mathbf{n}_j}\right) \qquad (7)$$

for $(i, j, k) \in \{(1, 2, 3), (3, 1, 2), (2, 3, 1)\}$, in which $\mathbf{n}_l$ are the normal vectors (Fig. 2a)

$$\mathbf{n}_l = \mathbf{p}_m \times \mathbf{p}_n, \qquad (8)$$

for $(l, m, n) \in \{(1, 2, 3), (3, 1, 2), (2, 3, 1)\}$.

Since for all states on an edge of the tetrahedron two of the vectors spanning the internal rhombohedron are parallel (that is, $\mathbf{p}_m \times \mathbf{p}_n = 0$), it follows from equation (8) that one of the normal vectors vanishes (that is, $\mathbf{n}_l = \mathbf{0}$). As a result, when we use equation (7) to evaluate ϕ_m and ϕ_n, both the numerator and denominator of the argument of the arctangent function are zero, such that both angles are indefinite (and therefore no longer depend on the γ angles). Furthermore, by inspecting the deformed unit cell, we find that the two indefinite ϕ angles are related to each other (Supplementary Figs 1 and 4, and Supplementary Table 1), so that each flattened rhombi has one additional degree of freedom, $\tilde{\phi}$, leading to a total of three degrees of freedom (Supplementary Fig. 1).

Finally, at the four vertices of the tetrahedron all three degrees of freedom related to the γ angles are constrained, resulting in a state where all six extruded rhombi are folded flat (state #4 in Fig. 2c). For such configuration $\mathbf{p}_1$, $\mathbf{p}_2$ and $\mathbf{p}_3$ are all parallel to each other, so that ϕ_1, ϕ_2 and ϕ_3 cannot be determined as a function of the γ angles (Supplementary Fig. 1). Therefore, these vertex states are described by the six angles between the flattened rhombi, $\tilde{\phi}_a$, $\tilde{\phi}_b$, ..., $\tilde{\phi}_f$ (Supplementary Fig. 1), subjected to the constraint $\sum_{i \in \{a,b,c,d,e,f\}} \tilde{\phi}_i = 2\pi$, leading to a total of five degrees of freedom.

To quantify the changes in geometry induced by variations in γ_1, γ_2 and γ_3, we calculated the internal volume of the unit cell (that is, the volume enclosed by the 24 faces of the extruded unit), v_{int}, which is given by

$$v_{int} = |\mathbf{p}_3 \cdot (\mathbf{p}_1 \times \mathbf{p}_2)| + 2L(\|\mathbf{p}_3 \times \mathbf{p}_1\| + \|\mathbf{p}_1 \times \mathbf{p}_2\| + \|\mathbf{p}_2 \times \mathbf{p}_3\|). \quad (9)$$

In Fig. 2d we show the values of v_{int} on the boundary of a sub-region of the regular tetrahedron, which—because of the symmetry of the unit cell—contains all possible configurations. As expected, the volume is maximum for state #1 (that is, $v_{int} = 7L^3$) and minimum for state #4 (that is, $v_{int} = 0$). Moreover, the results indicate that by varying γ_1, γ_2, and γ_3, any intermediate value for the internal volume can be achieved, demonstrating that we have developed a highly effective, yet simple mechanism to markedly alter the shape and volume of the unit cell.

We can also quantify the amount of energy required to deform the unit cell into a given state. By assuming that each of its 36 active edges acts as a linear rotational spring of stiffness K, the total strain energy, U, can be determined according to

$$U = \frac{K}{2} \sum_{i=1}^{3} \left[4\gamma_i^2 + 4(\pi - \gamma_i)^2 + 2\phi_i^2 + 2(\pi - \phi_i)^2\right], \quad (10)$$

where we assumed that for each edge zero energy is associated to a 180° angle between the two connected faces (so that they form a flat surface) and we have used the fact that 24 edge angles are directly specified by γ_1, γ_2 and γ_3 (orange edges in Fig. 2e) and the remaining 12 edges are defined by equation (7) (green edges in Fig. 2e). While equation (10) can be used to determine the strain energy associated to states lying inside or on the surface of the tetrahedron domain, to characterize the edge states the energy should be modified as

$$U = \frac{K}{2}\sum_{i=1}^{3}\left[4\gamma_i^2 + 4(\pi - \gamma_i)^2\right] + \frac{K}{2}\sum_{j\in\{a,b\}}\left[2\tilde{\phi}_j^2 + 2\left(\pi - \tilde{\phi}_j\right)^2\right] + K\pi^2,$$

$$(11)$$

$\tilde{\phi}_j$ being the additional degrees of freedom related to the tilting of the flattened rhombi (Supplementary Fig. 1). Similarly, for the vertex states we have

$$U = \frac{K}{2}\sum_{j\in\{a,b,c,d,e,f\}}\left[\tilde{\phi}_j^2 + \left(\pi - \tilde{\phi}_j\right)^2\right] + 6K\pi^2, \quad (12)$$

with $\sum_{j\in\{a,b,c,d,e,f\}}\tilde{\phi}_j = 2\pi$.

The results reported in Fig. 2e show that the energy is minimum for state #1 ($U = 9K\pi^2/2$). Since this is the only minimum, we expect it to be the preferred configuration when no external force is applied. Additional energy, ΔU, needs to be applied to deform the system into any other configuration. For example, to change the unit cell from state #1 to state #2, $\Delta U = 11K\pi^2/6$, while $3K\pi^2/2 \leq \Delta U \leq 5K\pi^2/2$ to reach state #3. Note that a range in ΔU exists to deform the unit cell to state #3 and any state on the edges of the tetrahedron, corresponding to different orientations of the flattened extruded rhombi (Fig. 2c and Supplementary Fig. 1). Finally, a fully flattened state (#4) requires an energy of $U = 9K\pi^2$, but also in this case different orientations of the flattened extruded rhombi result in an energy range ($23K\pi^2/3 \leq U \leq 10K\pi^2$). Most importantly, our analysis indicates that the energies required to deform the unit cell lie within a relatively small range ($9K\pi^2/2 \leq U \leq 10K\pi^2$), as the highest energy state is only a factor 20/9 higher then the lowest one. Therefore, we expect that all states can be reached by applying external forces of similar magnitude.

Next, we validated our findings by fabricating and testing a centimetre-scale prototype of the unit cell. To manufacture robust, thin-walled unit cells for which most of the deformation is focused at the edges, we used nearly inextensible polymeric sheets of varying thickness and an efficient stepwise layering and laser-cutting technique (see Fig. 3a,b and the 'Methods' section)[1,5]. In Fig. 3b we show the resulting unit cell; exactly as predicted by our analysis, the structure shapes into the fully expanded state #1 (that is, $\gamma_1 = \gamma_2 = \gamma_3 = \pi/2$) when no force is applied to it. However, the shape and volume of the unit cell can be altered by manually applying a force, and all configurations predicted by our analysis can be easily realized (Fig. 3c).

3D transformable metamaterial. Having identified a highly flexible and deformable 3D unit cell, we next show that it can be used to form a mechanical metamaterial whose shape and volume can be markedly altered. Such metamaterial can be constructed by connecting multiple unit cells through their outer edges. It is important to note that both the number of unit cells and their connections affect the number of degrees of freedom of the assemblies (see Supplementary Fig. 2 for details). In general, we find that large enough connectivity is required for the assembly to be characterized by the same folding modes of the constituent unit cells.

Here, we connected the outer edges of 64 identical unit cells to fabricate a $4 \times 4 \times 4$ cubic crystal (Fig. 3d). The snapshots shown in Fig. 3d (see also Supplementary Movie 3) indicate that the assembly still has three degrees of freedom and deforms in exactly the same manner as each constituent unit cell. Therefore, an external force can trigger a collective behaviour which transforms the mechanical metamaterial into a number of different configurations. In fact, by changing the microstructure (that is, the unit cells) into the various possible configurations described in Fig. 2, the macrostructure of the mechanical metamaterial can be significantly altered and its initial cubic shape deforms either into an extruded hexagon, a rhombohedron, or even a completely flat 2D state (Fig. 3d). Note that for the states inside the regular tetrahedron the total volume occupied by each unit cell, v, can be calculated as

$$v = |\mathbf{p}_3 \cdot (\mathbf{p}_1 \times \mathbf{p}_2)| + 2L(\|\mathbf{p}_3 \times \mathbf{p}_1\| + \|\mathbf{p}_1 \times \mathbf{p}_2\| + \|\mathbf{p}_2 \times \mathbf{p}_3\|)$$
$$+ 8L^3\sqrt{1 + 2\prod_{i=1}^{3}\cos(\phi_i) - \sum_{i=1}^{n}\cos^2(\phi_i)} + 4L^3\sum_{i=1}^{3}\sin(\phi_i) + 24t_f L^2,$$

$$(13)$$

where t_f is the thickness of the extruded faces of the unit cell. As expected, we find that the total volume is maximum for the fully expanded case ($v_{\text{state}\#1} = 27L^3 + 24t_f L^2$) and it is minimum in the fully flat state ($v_{\text{state}\#4} = 24t_f L^2$). Therefore, the maximum volume ratio, $\hat{v}_{\max}$, that can be achieved equals

$$\hat{v}_{\max} = \frac{v_{\text{state}\#1}}{v_{\text{state}\#4}} = \frac{27L}{24t_f} + 1. \quad (14)$$

Equation (14) clearly shows that the maximum change in volume depends on the ratio L/t_f and, therefore, on the material and techniques used to fabricate the extruded geometry. Focusing on the structure fabricated in this study, for which $L = 30$ mm and $t_f = 0.35$ mm, we have $\hat{v}_{\max} = 97.43$. Therefore, approximately a 100-fold volume reduction is achieved when our metamaterial is completely flattened, making the unit cell highly suitable for the design of deployable systems.

Actuation. While so far we have shown that the shape and volume of our metamaterial can be altered by manually applying a force (Fig. 3d), we now explore a possible distributed actuation approach to programme and control its shape. To this end, we position inflatable pockets on three hinges of the unit cells to control the γ angles (Fig. 4a). Upon pressurization, these pockets apply a moment to the hinges, flattening them and forcing the extruded rhombi to change their shape (Fig. 4b and Supplementary Movie 4).

To determine which inflatable pockets should be pressurized to move between different configurations, we first note that the total energy, T, in our system is given by

$$T = U - W, \quad (15)$$

where W indicates the work done to actuate the system and U is the total strain energy defined in equations (10–12). The equilibrium configurations can then be found by minimizing T with respect to all degrees of freedom, yielding

$$\frac{\partial T}{\partial \gamma_i} = \frac{\partial U}{\partial \gamma_i} - \frac{\partial W}{\partial \gamma_i} = 0, \quad \text{for} \quad i = 1, 2, 3, \quad (16)$$

for all the states internal and on the faces of the regular tetrahedron. Since $\partial W/\partial \gamma_i$ is the moment applied by the inflatable pocket to the hinge with angle γ_i, equation (16) indicates that, in order to move between two states, the actuator controlling the γ_i

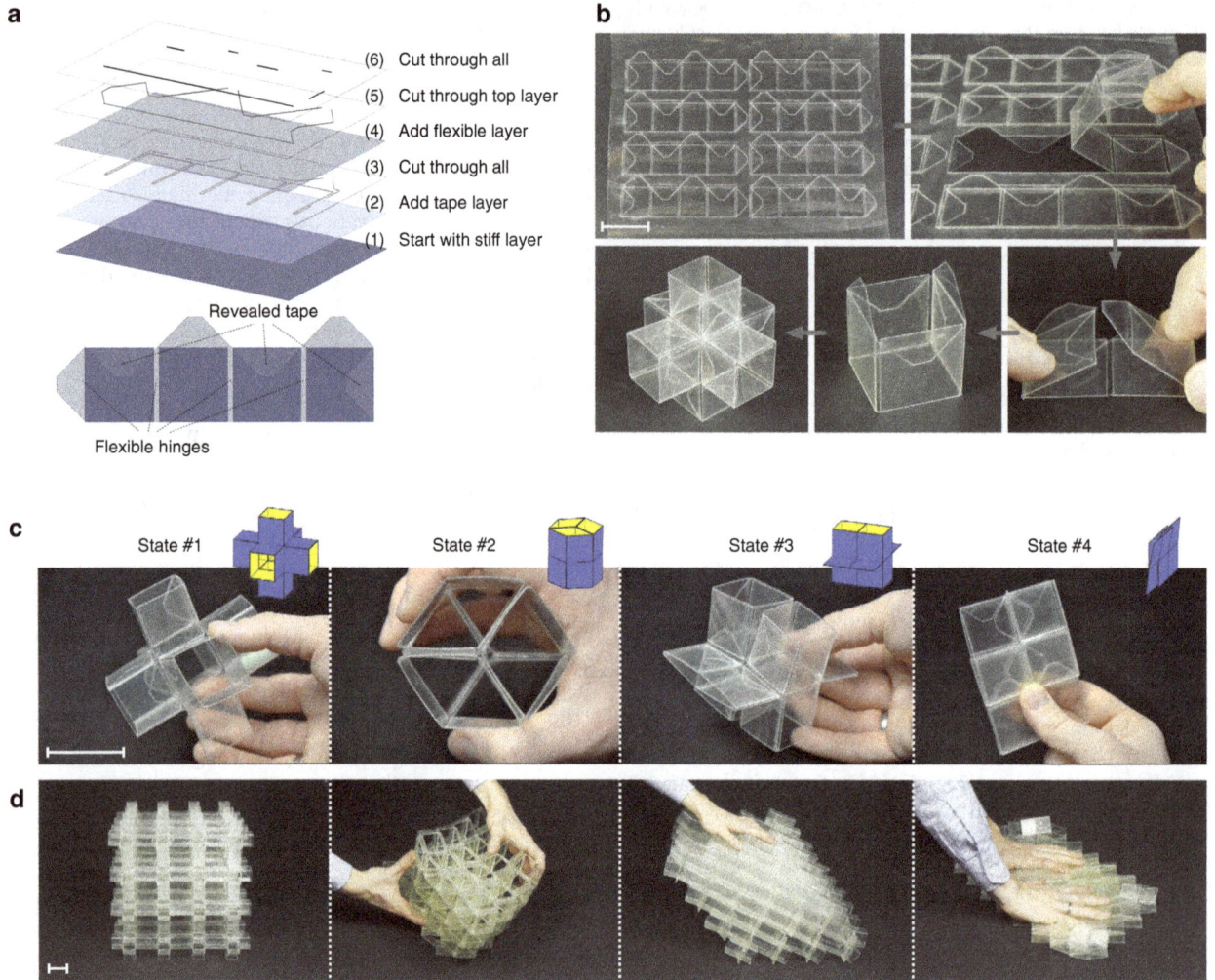

Figure 3 | Fabrication and deformation of a single extruded cube unit cell and the corresponding mechanical metamaterial. (**a**) The unit cells were fabricated using three layers: two outer layers of polyethylene terephthalate (with thickness $t = 0.25$ mm and 0.05 mm) and a layer of double-sided tape ($t = 0.05$ mm) in the middle. The layers were cut in three steps to form flat building blocks with both flexible and rigid regions. (**b**) The extruded rhombi were formed by simply removing the building blocks from the layered sheet, folding them and sticking their ends together using the revealed adhesive tape. To form the unit cell, six cubes were attached together using the double-sided tape incorporated into the layered sheet. (**c**) State #1, #2, #3 and #4 can be realized by simply applying a compressive load. (**d**) A highly flexible mechanical metamaterial with a cubic microstructure was formed by connecting the outer edges of 64 identical unit cells. An external force can trigger a collective behaviour which shapes the cubic crystal into a number of different configurations. Scale bars, 3 cm.

angle need to be pressurized if

$$\frac{\partial U}{\partial \gamma_i} = K \left[8\gamma_i - 4\pi + 2 \sum_{k=1}^{3} (2\phi_k - \pi) \frac{\partial \phi_k}{\partial \gamma_i} \right] \neq 0, \qquad (17)$$
$$\text{for} \quad i = 1, 2, 3.$$

Interestingly, some states inside the regular tetrahedron can be reached without actuation of all three γ angles. In fact, we numerically find that $\partial U/\partial \gamma_i = 0$ defines three surfaces within the regular tetrahedron (Fig. 4c). As a result, when moving between two states connected by a path that remains on one of these three surfaces, no moment needs to be applied to the hinge with angle γ_i.

It follows from our analysis that state #2 and any other state on the faces of the tetrahedron can only be reached if all three inflatable pockets are pressurized to different levels (Fig. 4d). This is because it is not possible to connect state #1 to a state on the face of the tetrahedron through a path that remains on one of the surfaces shown in Fig. 4c. In contrast, any of the edge or vertex

states can be reached from state #1 while remaining on one of these surfaces, indicating that only two pockets need to be inflated (Fig. 4d). The only exception is the highly symmetric state #3, which only requires pressurization of one actuator since two moments are equal to zero when moving in a straight path between state #1 and #3 (Fig. 4c). We also find that, when trying to reach state #4 by inflating two pockets, the six extruded rhombi do not flatten completely (Fig. 4d), suggesting that the force applied by actuators is insufficient to achieve this configurational change. In fact, all extruded rhombi can be completely flattened by placing one of the actuators on a different hinge, and actuating the three actuators simultaneously (Fig. 4e). Note that the unit cell does not fold completely flat, but instead deforms into the state with the lowest strain energy with $\tilde{\phi}_a = \tilde{\phi}_b = \ldots = \tilde{\phi}_f = \pi/3$, as predicted by equation (12). In order to reach a completely flat configuration, additional air pockets should be placed on some of the hinges highlighted in green in Fig. 2e to control the $\tilde{\phi}$ angles.

Having demonstrated that the shape of the unit cell can be controlled by pressurizing embedded inflatable pockets, we now

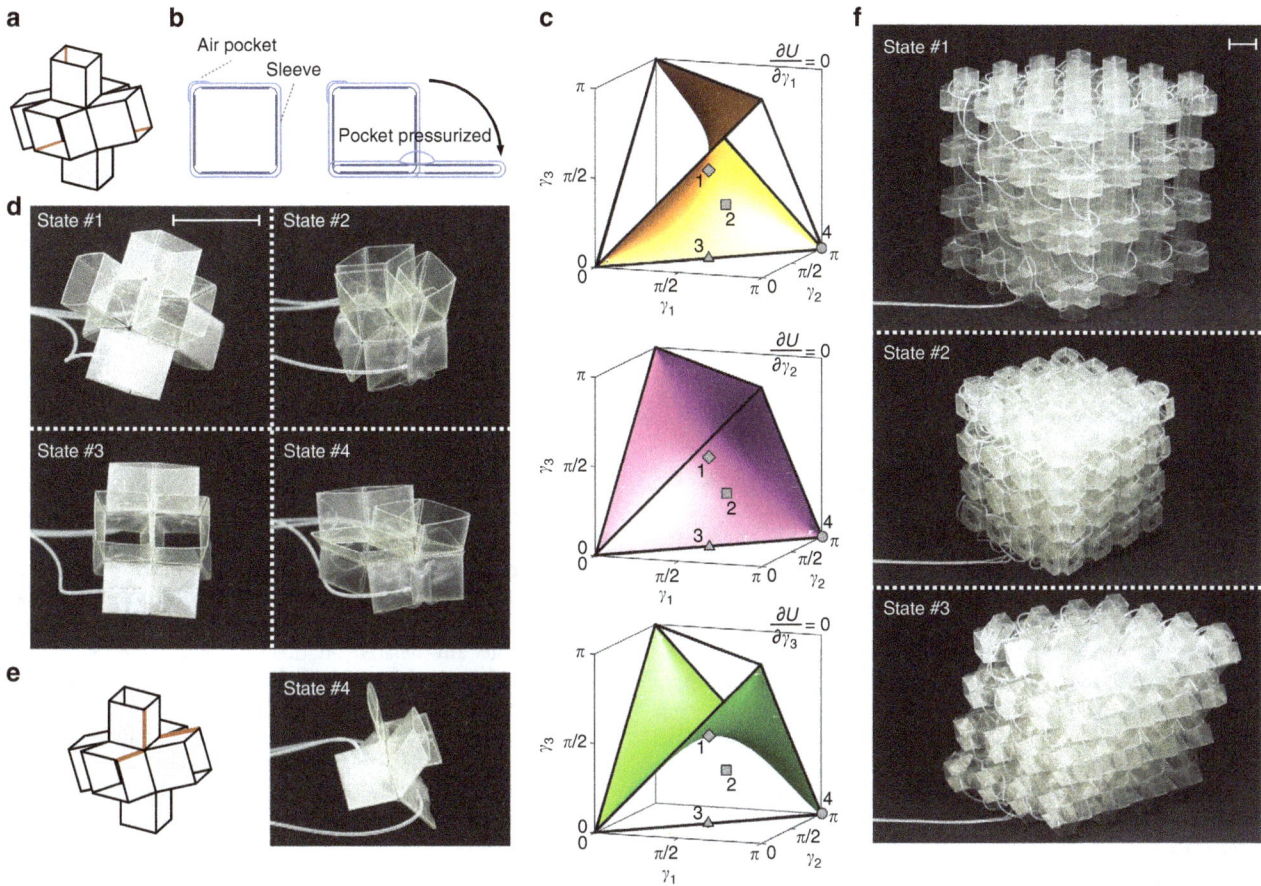

Figure 4 | Actuation of the unit cell and the corresponding mechanical metamaterial. (**a**) To freely transform the entire unit cell, inflatable air pockets are placed on the hinges highlighted in orange (see the 'Methods' section). (**b**) An internal pressure in the air pockets results in a moment in the hinges, causing the extruded rhombus to flatten. (**c**) Surfaces for which $\partial U/\partial \gamma_i = 0$. When moving between two states connected by a path that remains on one of these surfaces, the corresponding γ_i angle does not have to be actuated. (**d**) Configurations obtained by actuating the unit cell (with 3 actuators). (**e**) Improved actuation strategy to reach state #4. As expected state #4 does not fold completely flat, but instead deforms into the state with lowest strain energy for which $\phi_1 = \phi_2 = \phi_3 = 2\pi/3$. (**f**) Actuation of the mechanical metamaterial (with 96 actuators). Note that all structures are actuated by connecting the air pockets to three separate syringes through transparent tubes. Scale bars, 3 cm.

extend this approach to the $4 \times 4 \times 4$ mechanical metamaterial. First, we note that, although the metamaterial still has three degrees of freedom, three air pockets do not provide enough force to simultaneously transform all of its unit cells. Therefore, we distributed 96 air pockets on the outer extruded rhombi of the mechanical metamaterial and actuate each degree of freedom by simultaneously inflating 32 air pockets (Supplementary Movie 5). The snapshots shown in Fig. 4f demonstrate that the shape of our metamaterial can be altered by pressurizing the air pockets. In fact, we could shape all the unit cells of the metamaterial into states #2 and #3. However, the number of actuators was not sufficient to generate high enough forces to achieve state #4.

Finally, we note that since the state domain in Fig. 2b is convex and there is only one energy minima, fully depressurizing the pockets returns the metamaterial to the fully expanded state #1. This means that constant actuation is required to maintain a configuration in a state different from state #1. An interesting approach to mitigate this issue could be to identify different extruded unit cells whose energy is characterized by multiple energy minima, enabling multiple stable configurations. Such multistability, which has already been explored for other origami patterns such as the Miura-ori[6,8], the square twist[9] and even paper folding bags[23], might prove useful to further improve actuation and deployment of the proposed structures.

Load carrying capacity. While so far we have focused on the large geometric changes that can be induced in the mechanical metamaterial, such modifications can also be harnessed to alter its mechanical properties. For example, the stiffness of the metamaterial varies significantly for different configurations, as internal contact arises when the unit cells are deformed to configurations that lie on the faces, edges or vertices of the regular tetrahedron (Fig. 2b,c). Such contact constraints further deformation and therefore effectively increases the material's stiffness in certain directions, which can be harnessed to increase the structure's load carrying capacity. Note that the structure can only carry load when the folding motion is fully constraint, and therefore not during a transformation between states. To demonstrate this effect, we performed uniaxial compression tests on a single unit cell shaped into five different configurations (Fig. 5). Since these tests were performed on a single unit cell, they cannot directly be used to predict the response of the metamaterial, but we expect the metamaterial to show qualitatively a similar increase in load carrying capacity, arising from internal contact.

In these tests, the unit cell is first shaped into a specific configuration and placed between the two plates of a uniaxial compression machine (Fig. 5a,b). Then, it is further compressed in vertical direction while ensuring that it remains in the same

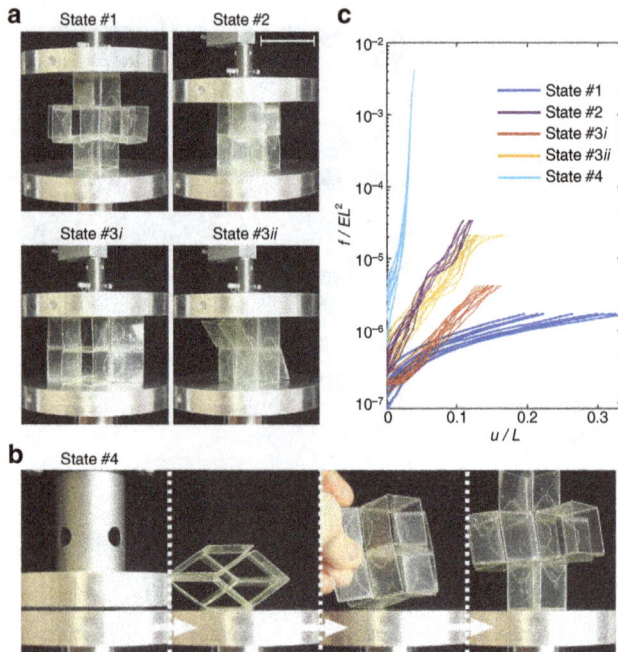

Figure 5 | Uniaxial compression of an extruded cube unit cell pre-folded into four different states. (**a**) Snapshots of the loaded unit cell for states #1, #2 and #3. (**b**) The unit cell is folded into state #4 and then compressed by applying 10,000 N. Remarkably, the fully expanded state can be recovered after removal of the load. (**c**) Force-displacement curves under uniaxial compression of a single unit cell at different configurations. Note that the initial force at zero displacement indicates the force required to maintain the unit cell into its pre-deformed (folded) state. Scale bar, 4 cm.

configuration. In Fig. 5c we report the evolution of the normalized measured force, $f/(EL^2)$, as a function of the normalized applied compressive displacement, u/L, in which $E = 2.6$ GPa is the Young's Modulus of PET[24]. The results indicate that for all configurations except for state #1 a non-zero force f is measured before applying any compressive displacement (that is, at $u/L = 0$). This is the force required to shape and maintain the unit cell into the specific configuration and, as expected, is found to be maximum for state #4. Moreover, the experimental data also shows that the increase in force measured as a function of the applied displacement depends on the shape of the unit cell, indicating that the stiffness of the structure is highly affected by both its configuration and the direction of loading. More specifically, the lowest increase in measured force was observed for state #1, since all the deformation is focused at the hinges and no contact occurs. In contrast, the fully flat state #4 showed the largest increase in force upon compression in the direction perpendicular to the flattened rhombi, since it is fully compacted and therefore behaves similar to the bulk material. Note that, although the observed behaviour was not fully elastic, even after applying 10,000 N to the unit cell in state #4 no permanent damage was observed in the hinges, and with some additional manipulations, the fully expanded configuration was recovered after the removal of the applied load (Fig. 5b and Supplementary Movie 6). This suggests that the strains induced by folding the hinges are not large enough to cause any permanent damage such as fracture or creases (see Supplementary Fig. 3 for details). It is expected that the fully expanded configuration completely recovers without additional manipulations when ideal elastic hinges are used. Finally, comparison between the force-displacement response measured for states #3i and #3ii, indicates that the load carrying capacity of

the unit cell depends greatly on the orientation of the unit cells with respect to the direction of loading.

Discussion

In summary, we have introduced a 3D programmable mechanical metamaterial whose shape, volume and stiffness can be actively controlled, making it ideally suited for the design of deployable and reconfigurable devices and structures. While in this study we focussed on a design based on an extruded cube, many other 3D unit cells with multiple degrees of freedom can be constructed, starting from any convex polyhedra with equal edges and extruding its edges in the direction normal to the faces. Moreover, depending on the characteristic size of the unit cell, different and remote types of actuation can be used to deform the structure, including heat[1,25–28], swelling[29] and magnetic fields[30,31]. In fact, our choice of using tethered pneumatic actuation was motivated mainly by the fact that such actuators are easy and inexpensive to fabricate and can typically generate reasonable forces[32].

By exploiting origami's scale-free geometric character, our approach can be extended to the micro- and nano-scale, as well as to the meter-scale. Since the transformation modes described for this material operate independently of the object's macro-scale external geometry, this transformable metamaterial can be machined into any desired architecture. For example, it can be used to design reconfigurable tubular stents with millimeter-scale features that can easily fit through small openings while in their flat state, as well as centimeter-scale foldable chairs and meter-scale deployable domes (Supplementary Movie 7).

Although in this study we have demonstrated the concept at the centimeter-scale, recent developments in micro-scale fabrication and actuation open exciting opportunities for miniaturization of the proposed metamaterial. In fact, origami-inspired metamaterials at the micro-scale could be manufactured by using self assembly[33] or stress within thin films[34–36], and by taking advantage of recent developments in hinge construction at small scale for laminate-based mechanisms[12,13,25–28,34]. This represents a significant advantage for the proposed structures over structures composed of rods connected by rotational joints, which are challenging to fabricate at a very small scale. Therefore, we believe that our approach can result in simplified routes for the design of transformable structures and devices over a wide range of length scales.

Methods

Fabrication of the unit cell. The unit cells were fabricated from thin polymeric sheets using an efficient stepwise layering and laser-cutting technique, as shown in Fig. 3a. To fabricate each of the six extruded rhombi that together form a unit cell, we started from a nearly inextensible polyethylene terephthalate sheet with thickness of $t = 0.25$ mm, covered with a double-sided tape layer (3 M VHB Adhesive Transfer Tapes F9460PC) with a thickness of $t = 0.05$ mm. Cutting slits were introduced into the bilayer using a CO2 laser system (VLS 3.50, Universal Laser Systems), after which a second, thinner polyethylene terephthalate layer ($t = 0.05$ mm) was bonded to the tape. A second cutting step with low power was then performed to machine only the top layer. Finally, additional slits were introduced through all three layers. As shown in Fig. 3b, the extruded rhombi could then be formed by removing the parts from the layered sheet, and bonding their ends together. Note that no glue was used in this step, but only the double-sided tape already incorporated into the parts. The unit cell was then formed by attaching together six of the extruded rhombi, again using the double-sided tape incorporated into the parts. For this study, we fabricated unit cells with $L = 30$ mm and almost rigid faces with a thickness of $t_f = 0.35$ mm. Moreover, the hinges have a thickness of $t_h = 0.05$ mm and a width of $w_h = 1.5$ mm. Note that we tested multiple values for t_f/t_h, and found that a thickness ratio $t_f/t_h = 7$ provided a good balance between flexibility of the hinges and rigidity of the faces.

Fabrication of the inflatable actuators. To actuate the unit cells and the metamaterial, we fabricated air pockets and embedded them within sleeves, so that they could be easily applied to the hinges. The air pockets were formed by placing two polyvinyl chloride sheets of thickness $t = 0.075$ mm in a hot press and sealing them for 200 s at 175°C. A rectangular piece (14×24 mm) of PTFE-Coated

Fibreglass Fabric was placed between the polyvinyl chloride sheets during the sealing process and then removed through a small opening, creating the internal pocket. To allow inflation, a small Polyethylene (PE) tube was inserted and glued in the same opening using a Cyanoacrylate based glue. Finally, the units cells were actuated by positioning the sleeves around the extruded rhombi with the air pockets aligned to the hinges, and by inflating the air pockets using syringes attached to the PE tubes.

Compression tests. We characterized the response of the unit cells under uniaxial compression using a single-axis Instron (model 5544A, Instron, MA, USA). The response of the unit cells configured in states #1 to #3 were tested using a 100 N load cell at a compression rate of 5 mm min^{-1}, while state #4 was tested using a 50,000 N load cell at a rate of 0.5 mm min. Each test was repeated nine times and three different unit cells were used. We loaded the samples until the point where debonding between the three layers forming each face started to occur. Furthermore, we ensured that there was enough friction between the sample and the testing machine so that no folding occurred.

References

1. Hawkes, E. *et al.* Programmable matter by folding. *Proc. Natl Acad. Sci. USA* **107**, 12441–12445 (2010).
2. Martinez, R. V., Fish, C. R., Chen, X. & Whitesides, G. M. Elastomeric origami: programmable paper-elastomer composites as pneumatic actuators. *Adv. Funct. Mater.* **22**, 1376–1384 (2012).
3. Wei, Z. Y., Guo, Z. V., Dudte, L., Liang, H. Y. & Mahadevan, L. Geometric mechanics of periodic pleated origami. *Physical Review Letters* **110**, 215501 (2013).
4. Schenk, M. & Guest, S. D. Geometry of miura-folded metamaterials. *Proc. Natl Acad. Sci. USA* **110**, 3276–3281 (2013).
5. Felton, S., Tolley, M., Demaine, E., Rus, D. & Wood, R. A method for building self-folding machines. *Science* **345**, 644–646 (2014).
6. Silverberg, J. L. *et al.* Using origami design principles to fold reprogrammable mechanical metamaterials. *Science* **345**, 647–650 (2014).
7. Lv, C., Krishnaraju, D., Konjevod, G., Yu, H. & Jiang, H. Origami based mechanical metamaterials. *Sci. Rep.* **4**, 5979 (2014).
8. Waitukaitis, S., Menaut, R., Chen, B. G.-G. & van Hecke, M. Origami multistability: From single vertices to metasheets. *Phys. Rev. Lett.* **114**, 055503 (2015).
9. Silverberg, J. L. *et al.* Origami structures with a critical transition to bistability arising from hidden degrees of freedom. *Nat. Mater.* **14**, 389–393 (2015).
10. Yasuda, H. & Yang, J. Re-entrant origami-based metamaterials with negative poisson's ratio and bistability. *Phys. Rev. Lett.* **114**, 185502 (2015).
11. Miura, K. Method of packaging and deployment of large membranes in space. *The Institute of Space and Astronautical Science Report* **618**, 1–9 (1985).
12. Kuribayashi, K. *et al.* Self-deployable origami stent grafts as a biomedical application of ni-rich tini shape memory alloy foil. *Mater. Sci. Eng. A* **419**, 131–137 (2006).
13. Kuribayashi-Shigetomi, K., Onoe, H. & Takeuchi, S. Cell origami: self-folding of three-dimensional cell-laden microstructures driven by cell traction force. *PLOS ONE* **7**, e51085 (2012).
14. Song, Z. *et al.* Origami lithium-ion batteries. *Nat. Commun.* **5**, 3140 (2014).
15. Tachi, T. & Miura, K. Rigid-foldable cylinders and cells. *J. Int. Assoc. Shell Spatial Struct.* **53**, 217–226 (2012).
16. Cheung, K. C., Tachi, T., Calisch, S. & Miura, K. Origami interleaved tube cellular materials. *Smart Mater. Struct.* **23** (2014).
17. Strobl, H. Special snapology. Available at: http://www.knotology.eu/PPP-Jena2010e/start.html (2010).
18. Goldman, F. *Using the Snapology Technique to Teach Convex Polyhedra* (CRC Press, 2011).
19. Pardo, J. Flexiball. Available at: http://paper-life.ru/images/origami/modulnoe/flexiball/flexiball.pdf (2005). Accessed on 21 July 2015.
20. Hoberman, C. Deployable structures based on polyhedra having parallelogram faces. *Provisional US Patent Application* **62/023**, 240 (2014).
21. Demaine, E. D. & O'Rourke, J. *Geometric Folding Algorithms* (Cambridge University Press, 2007).
22. Huffman, D. Curvature and creases: a primer on paper. *IEEE Trans. Comput.* **C-25**, 1010–1019 (1976).
23. Balkcom, D. J., Demaine, E. D., Demaine, M. L., Ochsendorf, J. A. & You, Z. *Folding Shopping Bags* (CRC Press, 2009).
24. Bin, Y., Oishi, K., Yoshida, K. & Matsuo, M. Mechanical properties of poly(ethylene terephthalate) estimated in terms of orientation distribution of crystallites and amorphous chain segments under simultaneous biaxial stretching. *Polym. J.* **36**, 888–898 (2004).
25. Leong, T. G. *et al.* Tetherless thermobiochemically actuated microgrippers. *Proc. Natl Acad. Sci. USA* **106**, 703–708 (2009).
26. Liu, Y., Boyles, J. K., Genzer, J. & Dickey, M. D. Self-folding of polymer sheets using local light absorption. *Soft Matter* **8**, 1764–1769 (2012).
27. Laflin, K. E., Morris, C. J., Muqeem, T. & Gracias, D. H. Laser triggered sequential folding of microstructures. *Appl. Phys. Lett.* **101**, 131901 (2012).
28. Kim, J., Hanna, J. A., Hayward, R. C. & Santangelo, C. D. Thermally responsive rolling of thin gel strips with discrete variations in swelling. *Soft Matter* **8**, 2375–2381 (2012).
29. Kim, J., Hanna, J. A., Byun, M., Santangelo, C. D. & Hayward, R. C. Designing responsive buckled surfaces by halftone gel lithography. *Science* **335**, 1201–1205 (2012).
30. Judy, J. & Muller, R. Magnetically actuated, addressable microstructures. *J. Microelectromech. Syst.* **6**, 249–256 (1997).
31. Yi, Y. & Liu, C. Magnetic actuation of hinged microstructures. *J. Microelectromech. Syst.* **8**, 10–17 (1999).
32. Niiyama, R., Rus, D. & Kim, S. Pouch motors: printable/inflatable soft actuators for robotics. *IEEE International Conference on Robotics and Automation (ICRA)* 6332–6337 (2014).
33. Whitesides, G. M. & Grzybowski, B. Self-assembly at all scales. *Science* **295**, 2418–2421 (2002).
34. Shenoy, V. B. & Gracias, D. H. Self-folding thin-film materials: from nanopolyhedra to graphene origami. *MRS Bull.* **37**, 847–854 (2012).
35. Cho, J.-H. *et al.* Nanoscale origami for 3d optics. *Small* **7**, 1943–1948 (2011).
36. Xu, S. *et al.* Assembly of micro/nanomaterials into complex, three-dimensional architectures by compressive buckling. *Science* **347**, 154–159 (2015).

Acknowledgements

This work was supported by the Materials Research Science and Engineering Center under NSF Award No. DMR-1420570. K.B. also acknowledges support from the National Science Foundation (CMMI-1149456-CAREER) and the Wyss institute through the Seed Grant Program.

Author contributions

J.T.B.O., J.C.W., C.H. and K.B. proposed and designed the research; J.T.B.O. and K.B. performed the analytical calculations; J.T.B.O., T.A.J., Y.S. and S.A.B. fabricated the models; J.T.B.O. and T.A.J. performed the experiments; J.T.B.O., J.C.W., C.H. and K.B. wrote the paper with key input from G.W.

Additional information

9

Ultralow-power switching via defect engineering in germanium telluride phase-change memory devices

Pavan Nukala[1], Chia-Chun Lin[1], Russell Composto[1] & Ritesh Agarwal[1]

Crystal-amorphous transformation achieved via the melt-quench pathway in phase-change memory involves fundamentally inefficient energy conversion events; and this translates to large switching current densities, responsible for chemical segregation and device degradation. Alternatively, introducing defects in the crystalline phase can engineer carrier localization effects enhancing carrier–lattice coupling; and this can efficiently extract work required to introduce bond distortions necessary for amorphization from input electrical energy. Here, by pre-inducing extended defects and thus carrier localization effects in crystalline GeTe via high-energy ion irradiation, we show tremendous improvement in amorphization current densities (0.13–$0.6\,\mathrm{MA\,cm^{-2}}$) compared with the melt-quench strategy ($\sim 50\,\mathrm{MA\,cm^{-2}}$). We show scaling behaviour and good reversibility on these devices, and explore several intermediate resistance states that are accessible during both amorphization and recrystallization pathways. Existence of multiple resistance states, along with ultralow-power switching and scaling capabilities, makes this approach promising in context of low-power memory and neuromorphic computation.

[1] Department of Materials Science and Engineering, University of Pennsylvania, Philadelphia, Pennsylvania 19104, USA. Correspondence and requests for materials should be addressed to R.A. (email: riteshag@seas.upenn.edu).

Phase-change materials (PCMs), which rapidly and reversibly transform from crystalline to amorphous phase, are viable alternatives to the relatively slow non-volatile flash memory technology[1] with random access capability. One of the problems with PCMs is the use of high programming currents (and current densities) during the crystal–amorphous transformation (RESET) achieved conventionally via the melt-quench pathway[2]; and reducing the active device volumes has been pursued as a potential solution to mitigate this problem[3–8]. While reports on phase-change line and bridge devices demonstrated lesser RESET currents by shrinking the volume of the PCM directly[4,5,8], works on PCM devices with carbon nanotube electrodes[6,7] showed very low RESET currents ($\sim 5\,\mu A$) by minimizing the contact areas and hence the active device volumes $(35 \times 3 \times 3\,nm)^6$. Alternative approaches to reducing the RESET currents include lowering the melting point of the PCM by doping them with nitrogen or silicon[9,10]. Although these approaches illustrate intelligent device designs based on geometry and chemical doping, none of them have been able to reduce the high amorphization current densities ($\sim 50\,MA\,cm^{-2}$)[7] that are responsible for device degradation issues owing to chemical segregation and heat[11], thus precluding the widespread commercialization of PCM technology.

In the melt-quench pathway, work required to amorphize a crystal is extracted from the heat generated by input electrical energy via inelastic carrier-phonon scattering; and this involves several fundamentally inefficient energy conversion events[12] manifested as large writing current densities. Engineering carrier localization effects by pre-inducing extended defects in crystalline PCM enhances carrier–lattice coupling[13], enabling efficient exchange of energy back and forth between the carriers and the lattice[14,15]. The input electrical energy, hence, can directly perform work on the lattice with minimal loss in the form of heat (wasteful energy) and introduce critical bond distortions required for amorphization[16], potentially lowering the writing current densities.

Here we select crystalline GeTe (*R3m*), a simple binary PCM system, to demonstrate drastic improvement in amorphization current (and power) densities achieved by introducing carrier localization near the Fermi level (E_F). All known phases of crystalline GeTe show *p*-type metallic conduction owing to the presence of large concentration (order of $10^{20}\,cm^{-3}$) of intrinsic Ge vacancies (*p*-type dopants)[17]. The carriers near the Fermi level (E_F), which participate in transport, are delocalized and couple weakly with the lattice. By pre-inducing extended defects using high-energy He$^+$ ion irradiation, we show for GeTe devices in the crystalline phase that the carriers at E_F can be localized, and hence strongly couple with the lattice[14,15]. These devices transformed to an amorphous phase via the defect-based pathway[13,18], at current densities (j_s) of $0.13–0.5\,MA\,cm^{-2}$ significantly lower than $j_s = 50\,MA\,cm^{-2}$ observed in the melt-quench pathway[6,7]. Furthermore, we illustrate scaling of switching currents with device volumes, and reversible and repeatable low-power switching from defect-engineered crystalline states to amorphous phase.

Results

Inducing carrier localization in single-crystalline GeTe nanowires.
Single-crystalline GeTe nanowires were synthesized using the vapour–liquid–solid mechanism[19], and multiple electrode devices were fabricated using electron-beam lithography, and encapsulated with 30 nm of SiO_2 (Fig. 1a, inset; Methods)[13]. All devices were irradiated using 2 MeV He$^+$ ions, in a Rutherford backscattering set-up, at different dosages, and beam currents not exceeding 30 nA (ref. 20; Methods; Supplementary Note 1). Our stopping and range of ions in matter calculations

(Supplementary Note 1) showed that there is a 2% loss in the incident energy upon the penetration of He$^+$ ions through our nanowires, ensuring extended defect formation via knock-on damage and dismissing any role of ion implantation (He bubble or void formation)[21]. To understand the effect of pre-induced defects on the transport characteristics in nanowire devices, we performed temperature-dependent resistivity measurements after ion-beam exposure at various dosages. Resistivity was evaluated as $\rho = R_{NW} A/l_d$, where R_{NW} is the resistance of the nanowire obtained by subtracting the contact resistance measured in a multiple probe configuration (Fig. 1a, inset) from the total device resistance. l_d and A are the length and cross-sectional area of the nanowire device, respectively (see Supplementary Note 2, raw data shown in Supplementary Fig. 1). At dosages up to $700\,\mu C\,cm^{-2}$, the resistivity of all our devices increased linearly with increasing temperature above 30 K and exhibited a saturation value (ρ_0) below 30 K. This behaviour is typical of metals, where dominant contribution to resistivity at low temperatures is from temperature-independent carrier-defect scattering and carrier-phonon scattering at high temperatures[13]. ρ_0 depends on the defect density, and conversely can be used as a measurable metric for the same in the metallic state. As illustrated for representative devices (labelled nanowires 1–4) in Fig. 1a, ρ_0 increased with increasing dosage (from 0 to $700\,\mu C\,cm^{-2}$), consistent with increasing pre-induced defect density in the material.

Another quantity that is sensitive to the defect concentration in metallic state is the slope of temperature–resistivity plots in the linear increase regime (temperature coefficient of resistivity, TCR), which decreases with increasing defect density[22,23], nevertheless remaining positive. Positive TCR is a characteristic of transport via delocalized carriers in a metal, which can be described by Boltzmann transport equation[24]. In all our devices (representative devices nanowires 1–4 shown in Fig. 1b), although TCR showed an initial increase with dosage up to $50\,\mu C\,cm^{-2}$ (for reasons see Supplementary Note 3 and Supplementary Fig. 2), subsequently up to $700\,\mu C\,cm^{-2}$ it showed a decreasing trend to a less positive value, consistent with increasing defect density. More importantly, TCR remained positive, suggesting that at these low dosages ($<700\,\mu C\,cm^{-2}$) carrier localization effects are insignificant.

At higher dosages ($1,800–3,600\,\mu C\,cm^{-2}$), however, the resistivity of all the tested devices showed a non-linear decrease with increasing temperature (TCR is negative but cannot be uniquely defined), which is a signature of localized carriers at E_F participating in transport[25]. This demonstrates an electronic transformation of crystalline GeTe from a metallic state to dirty metallic (ρ_0 is finite) or insulating states ($\rho_0 \to \infty$)[15]. The exact dosage at which this transformation occurred varied from device to device, but all devices showed localization effects in transport above a dosage of $3,600\,\mu C\,cm^{-2}$. For instance, the device nanowire 1 (Fig. 1c) transformed to an insulating state demonstrating variable range hopping (VRH) conduction (Fig. 1c, inset) at a dosage of $1,800\,\mu C\,cm^{-2}$. Devices nanowires 2 and 3 (Fig. 1d) transformed to dirty metal state at $3,600\,\mu C\,cm^{-2}$, demonstrating a power law conduction (σ varying as $T^{0.5}$), a characteristic of disordered metals showing weak localization effects (Fig. 1d, inset)[25]. The stability of these defect-engineered states was tested by heating the devices to 200 °C and monitoring the change in resistance with time; and they showed no change in resistance for 36 h (Supplementary Fig. 3), suggesting that they are thermodynamically stable (Supplementary Note 4).

To understand the structural nature of the radiation induced defects, we performed transmission electron microscopy (TEM) on nanowire devices assembled on TEM compatible platform[18]

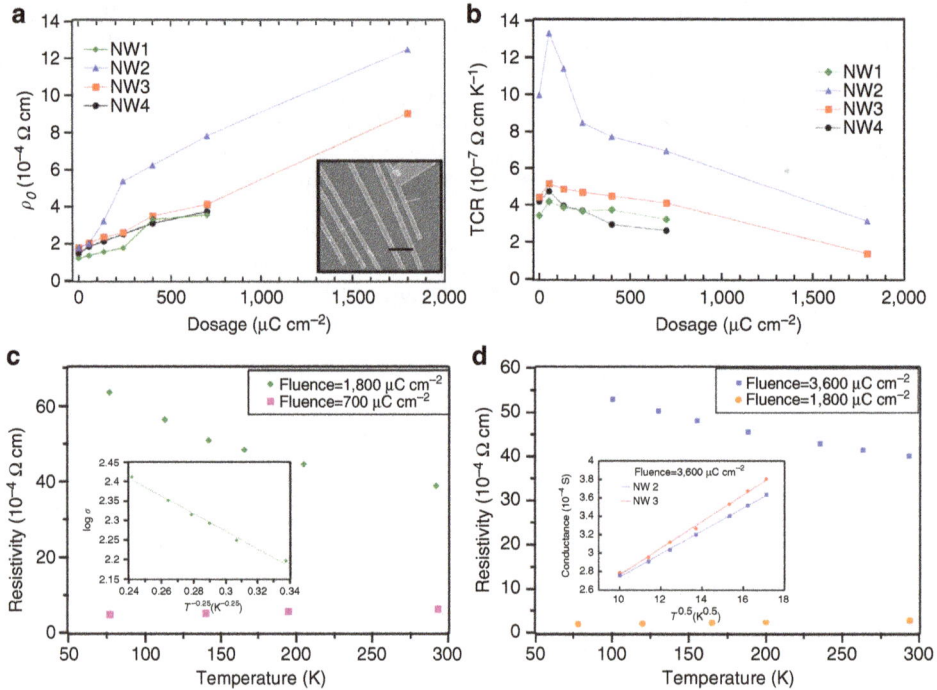

Figure 1 | Transport measurements on crystalline GeTe nanowires irradiated at various dosages. (**a**) Saturation resistivity (ρ_0) plots as a function of dosage on four representative nanowires (NW 1-4), showing an increase in ρ_0 with dosage, in the metallic state. Inset: Scanning electron microscope image of NW 1, a representative multiple probe nanowire devices for transport measurements. Scale bar, 2 μm. (**b**) Temperature coefficient of resistivity (TCR) plots as a function of dosage on four representative nanowires (NW 1, 2, 3 and 4), showing an initial increase in TCR followed by a subsequent decrease with dosage in the metallic state. (**c**) Temperature-resistance plots for NW 1 at dosages 700 μC cm^{-2} (magenta) and 1,800 μC cm^{-2} (green), signifying a metal-insulator transition. Inset: variable range hopping (VRH) conduction behaviour observed at 1,800 μC cm^{-2}, dosage confirming the insulating state. (**d**) Temperature-resistance plots for NW 2 at 1,800 μC cm^{-2} (orange) and 3,600 μC cm^{-2} (blue), signifying a metal-dirty metal transition. Inset: power law conduction behaviour observed for NWs 2 and 3 at 3,600 μC cm^{-2} confirming dirty metallic nature.

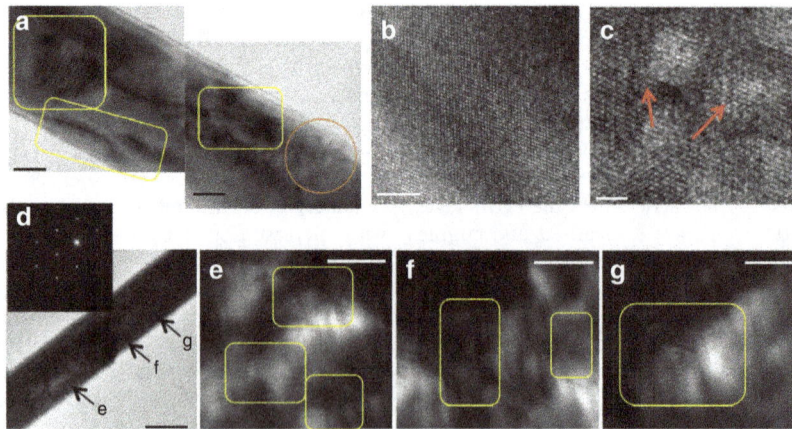

Figure 2 | Structural analysis on crystalline GeTe nanowires irradiated at various dosages. (**a**) Bright-field TEM image showing stacking faults (yellow rectangles) and dislocation loops (orange circles) induced randomly in a nanowire irradiated with a dosage of 45 μC cm^{-2}. Scale bar, 50 nm (**b,c**) High-resolution TEM images of a nanowire before (**b**) and after (**c**) ion irradiation with a dosage of 100 μC cm^{-2} showing defect tetrahedra (red arrows). Scale bar, 5 nm in **b**, and 2 nm in **c**. (**d**) Bright-field TEM image of a nanowire ion irradiated with large fluences (1,800 μC cm^{-2}, inset). Selected area electron diffraction showing that the nanowire is still single crystalline with the satellite spots corresponding to the existence of defects. Scale bar, 200 nm. (**e-g**) Zoomed in, dark-field TEM images of different regions marked in **d**, all showing many intersecting defect templates, a structural feature that corresponds to electron localization. Scale bar, 20 nm (**e,f,g**).

and exposed to different dosages of He$^+$ ion irradiation. At modest dosages (40–100 μC cm^2), where the devices remain metallic, we observed the formation of dislocation loops, two-dimensional defects (stacking faults) and defect tetrahedra—formed due to vacancy/interstitial supersaturation following knockout of atoms (Fig. 2a–c)[26]. The spatial distribution of

these defects along the nanowire is sporadic. For irradiation at higher dosages (1,800 μC cm^{-2}), the entire nanowire became replete with intersecting two-dimensional defects (as illustrated in different regions of a representative nanowire in Fig. 2d–g), still being single crystalline (Fig. 2d, inset), and this stage corresponds to electronic states where carrier localization dominates transport

(Fig. 1c,d). It is important to emphasize that our devices are capped with a conformal coating of SiO_2 (Methods), which prevents these pre-induced defects from annihilating at the surface of the nanowires (the only possible sink) at high temperatures, contributing to their thermal stability (Supplementary Fig. 3). Furthermore, we did not see any evidence of He bubble trapping in GeTe, another common defect observed in materials irradiated with He^+ ions[26] (Supplementary Notes 1 and 5; Supplementary Fig. 4) owing to the high energy of the incident ion beam[21].

Amorphization behaviour of defect-engineered nanowire devices. To verify the idea that pre-inducing defects and engineering carrier localization effects is beneficial for power reduction for amorphization, we studied the switching (crystal–amorphous) and volume scaling properties of the devices as a function of radiation dosage. We amorphized our devices in crystalline phase exposed to different dosages of ion irradiation, by applying a train of voltage pulses (50 ns) of increasing amplitude, separated by 1 s (to allow complete thermalization between pulses), until resistance increased abruptly by at least two orders of magnitude (see Supplementary Note 6 and Supplementary Fig. 5 for pulse shapes, and dynamic currents measured from a pulse). The crystal–amorphous transformation in all these devices occurred via a defect-based mechanism, where heat shock from an electrical pulse first creates extended defects (full and partial dislocations)[13,18]. The pre-induced defects, along with the defects created by electrical pulses, migrate with the hole-wind force and accumulate at a region of local inhomogeneity creating a defect template with intersecting defects, along which amorphization takes place beyond a critical defect density[13,18]. To understand the size scaling of RESET currents (i_s) in this defect-based mechanism, and the influence of pre-induced defects on them, we compared RESET current densities (j_s) as a function of device length (l_d) at various dosages (Fig. 2; Supplementary Fig. 6a). These plots encompass complete information on size (length and cross-sectional area) dependence of i_s. At dosages up to $700\,\mu C\,cm^{-2}$, where no carrier localization effects were present, we observed that j_s increased with increasing dosage for any particular device length (Supplementary Fig. 6a–d). This illustrates that pre-induced defects can be detrimental for i_s (j_s), if they do not induce any significant localization effects in transport (see Supplementary Note 7 for analysis of switching behaviour at low dosages).

However, at higher dosages ($>1,800\,\mu C\,cm^{-2}$), where localization effects dominated transport (Fig. 1a–d), j_s (and i_s) were drastically lowered (Fig. 3). At a dosage of $3,600\,\mu C\,cm^{-2}$, i_s of the device (referred as D1) with active volume as large as $100 \times 100 \times 750\,nm$ (see Supplementary Note 8) was as low as $26\,\mu A$ (j_s: $0.26\,MA\,cm^{-2}$); and for a smaller device ($80 \times 80 \times 320\,nm$, referred as D2), i_s was $8\,\mu A$ (j_s: $0.13\,MA\,cm^{-2}$). The current densities and power densities for amorphizing these defect-engineered devices were ~ 300 and 10^5 times smaller, respectively, than those required by the melt-quench pathway[6] (i_s: $5\,\mu A$, j_s: $50\,MA\,cm^{-2}$; Supplementary Note 8; Supplementary Table 1). With volume scaling also demonstrated on these devices, the absolute power required for switching very small volumes of active PCM[6] is significantly lowered in this approach, and this can potentially mitigate issues such as thermal cross-talk and chemical segregation[11]. Furthermore, the importance of carrier localization and the role of lattice–carrier coupling in efficiently converting input electrical energy into the work required for amorphization resulting in significant lowering of switching powers are highlighted by these results.

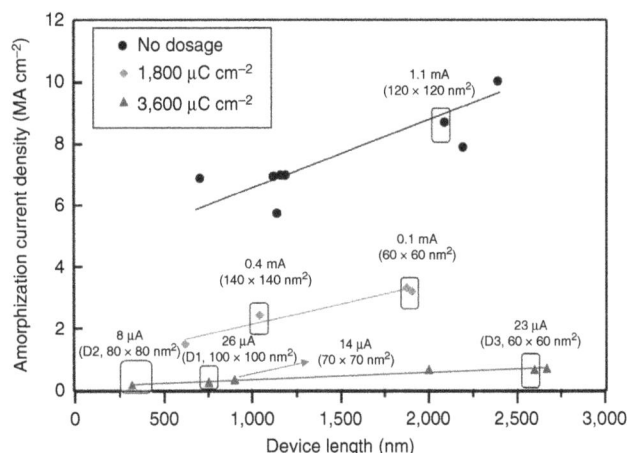

Figure 3 | Scaling behaviour of switching properties of GeTe nanowire devices with pre-induced defects at different dosages. Plot showing amorphization current density (j_s) as a function of device length for devices engineered into states where localized carriers dominate transport when nanowire devices were irradiated at very high dosages (1,800 and $3,600\,\mu C\,cm^{-2}$). Upon comparison with non-irradiated devices (black data points), these devices showed a drastic reduction in switching current densities for amorphization, enabling very low-current switching for large devices. The switching currents (i_s) and device cross-section dimensions are indicated on the plot, and the device volumes and comparison with devices switching via melt quench is shown in Supplementary Table 1. Three particular devices named D1, D2 and D3 defect engineered at a dosage of $3,600\,\mu C\,cm^{-2}$ were selected for further analysis. Data shown in Figs 4–6 are on device D1.

Recrystallization behaviour of defect-engineered nanowire devices. To verify the reversible switching behaviour on the defect-engineered devices that displayed very low j_s (exposed at $3,600\,\mu C\,cm^{-2}$ in Fig. 3), we examined the amorphous–crystal (SET) transformation via threshold switching[27] by applying voltage controlled d.c. I–V sweeps. When the compliance current (I_c) in the circuit was set to $50\,\mu A$, threshold switching followed by recrystallization through Joule heating of the amorphous phase occurred in all the devices at $<1\,V$ (see Fig. 4a for data on D1 and Supplementary Fig. 7 for data on D2). More importantly, as illustrated on D1 (Fig. 4b), the recrystallized phase after a few RESET/SET cycles showed similar resistance and temperature dependence of resistance, as the starting defect-engineered insulating crystalline phase. To demonstrate that the RESET state is stable, and does not fail via defect annihilation upon repeated switching, we cycled D1 between RESET ($26\,\mu A$, 100 ns) and SET (1 V, 250 ns) for 20,000 cycles (Fig. 4c). We observed no appreciable changes in the resistance of the RESET state, validating the robustness of this low-power switching strategy. The band of resistance values in the amorphous phase, however, is a function of pulse width and amplitude. The combination of engineering carrier localization in crystalline phase of PCM, switching to the amorphous phase via the defect-based pathway, and recrystallizing the amorphous phase via Joule heating (pulsed mode or d.c.) is thus encouraging for ultralow-power phase-change memory operation. It must be noted that rigorous endurance tests demonstrating cyclability conforming to commercial standards (10^8–10^9 cycles) is a future endeavour.

Demonstration of metastable intermediate states. The defect-engineered crystalline states where carrier localization dominates transport, structurally corresponds to the entire nanowire device being replete with intersecting extended defects, or defect

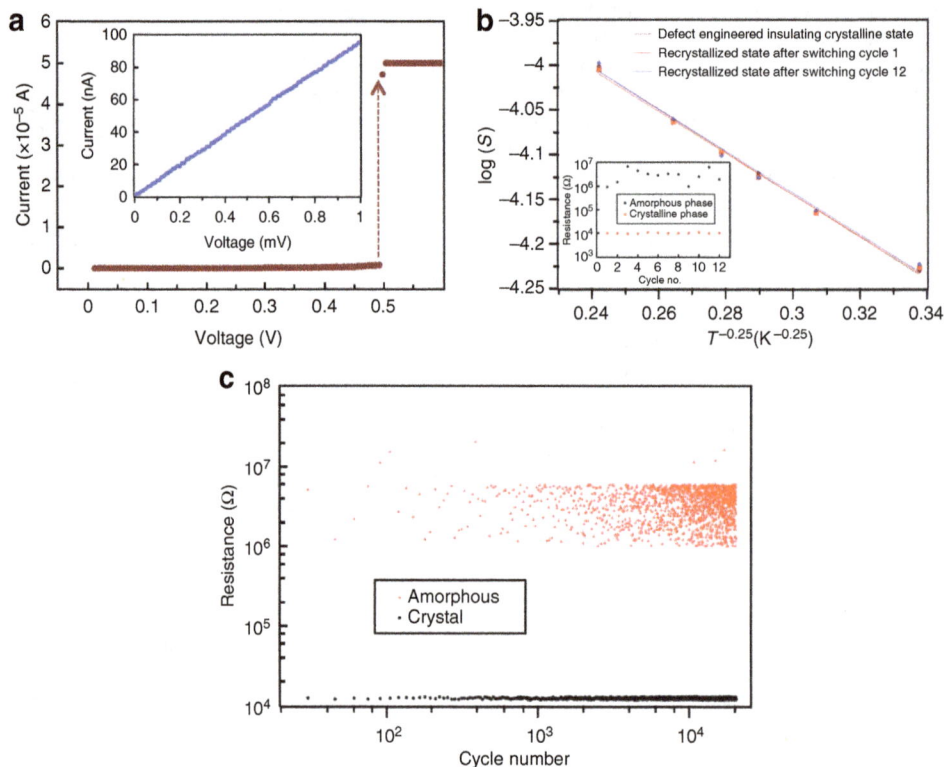

Figure 4 | Recrystallization behaviour of defect-engineered GeTe nanowires switched to amorphous phase. (**a**) I–V sweep from 0 to 0.5 V on D1 (shown in Fig. 3) with compliance current (I_c) set at 50 μA. Recrystallization followed by threshold switching occured at 0.5 V. Inset: low-bias resistance measurement on the recrystallized phase (~10 kΩ). (**b**) Comparisons of the temperature dependence of conductivity measurements between as-engineered insulating crystalline phase, recrystallized phase after one cycle of switching and after 12 cycles of switching. All the recrystallized phases showed similar transport behaviour suggesting reliable and repeatable switching. Inset: data showing cycling between amorphous and crystalline phase. (**c**) Repeated cycling between RESET (26 μA, 100-ns pulse) and SET (1 V, 250-ns pulse) states for 20,000 cycles on D1.

templates (Fig. 2a–d). Upon the application of electrical pulses to this state, more defects will migrate with momentum and energy transferred from the carriers, and accumulate at one (or more) of these templates increasing the local defect concentration[13,18]. A critical defect concentration locally would lead to the collapse of long-range order or 'nucleation' of the amorphous phase. However, the question remains whether some intermediate metastable resistance states in crystalline phase can be accessed as the defect concentration at a template increases continuously towards a critical value at which amorphization occurs. To check whether defects can be controllably accumulated and whether the intermediate resistance states can be accessed, we reduced the pulse width of the voltage pulses used for programming defect-engineered insulating crystalline phase (resistance of 10 kΩ) to an amorphous phase, from 50 to 20 ns (Fig. 5a). Programming device D1 with 50-ns pulses abruptly nucleated the amorphous phase, suggesting that the energy transferred from 50-ns pulses is sufficient to migrate and accumulate defects beyond critical concentration at a region in the defect template. However, with 20-ns pulses, less energy is transferred to the defects, resulting in controlled defect accumulation and access to several intermediate metastable states in the crystalline phase (see Supplementary Fig. 8 for TEM data) whose resistance increases with increasing defect concentration at the template (thermal stability of intermediate and amorphous phases is discussed in Supplementary Note 9 and Supplementary Fig. 9). Similar intermediate states were observed in all the other devices exposed to a dosage of 3,600 μC cm^{-2} upon programming them with 20-ns pulses (Supplementary Fig. 10). For the discussion that follows we will refer to the representative electronic states in

D1 in the crystalline phase with resistances of ~10, 40 and 70 kΩ as states 1, 2 and 3, respectively (Fig. 5a).

To understand whether these intermediate resistance states can be reversibly obtained starting from the amorphous phase, we recrystallized the amorphized device D1, via d.c. I–V sweeps, setting a very low compliance current (I_c) of 5 μA. As shown in Fig. 5b, upon a voltage sweep from 0 to 1 V (green data), the amorphous phase first transformed to an intermediate resistance state (70 kΩ, state 3). Upon a second voltage sweep from 0 to 1 V on state 3, we observed a sudden drop in current at 0.02 V (0.1 μA), followed by a threshold-switching event to another intermediate state (red data in Fig. 5b) with resistance of 40 kΩ, state 2. Another voltage sweep from 0 to 1 V on state 2 showed a similar drop in current at 0.02 V, followed by a switching event to the starting electronic state, state 1 (10 kΩ, blue data in Fig. 5b). To ensure reliability in the formation of all the demonstrated states, we switched these devices for 160 cycles, where every cycle involved the following steps: switching state 1 to a high-resistance amorphous phase by the application of a 100 ns, 26 μA pulse, and switching back to state 1 from the amorphous phase via sweeping d.c. voltage from 0 to 1 V, multiple times, if necessary (depending on the value of I_c). We changed the I_c between cycles to confirm the dependence of formation of intermediate states on I_c (Fig. 5c). When I_c was set to 50 μA, we observed only two states: a high-resistance amorphous state (>1 MΩ), and state 1, a low-resistance crystalline state (~10 kΩ). However, when I_c was 10 μA, we consistently observed amorphous state first transforming into an intermediate resistance state (~40 kΩ) with the first voltage sweep, and then to state 1 with another voltage sweep from 0 to 1 V. Upon further reducing I_c to 5 μA, we observed amorphous phase to state 1 transformation in every cycle requiring three voltage sweeps from 0 to 1 V, with the first two voltage sweeps accessing two intermediate resistance states (between 35 and 80 kΩ with some variability), and the final sweep transforming these intermediate states to state 1. It must be noted that the variability of resistance of the intermediate states (Fig. 5c) upon cycling (with $I_c = 5\,\mu A$), may not currently conform to commercial standards for multistate memory applications, and need to be addressed and improved in further studies. Nevertheless, these findings provide a proof of concept that intermediate states can reliably obtained starting from both crystalline phase and an amorphous phase, with the controlling

parameters being the pulse width and I_c (Joule heating following threshold switching), respectively.

From these results, the mechanism for amorphous–crystal transformation in the defect-based pathway becomes clear. Following amorphization, most of the nanowire has a background density of pre-induced defects, and a local region has a higher defect concentration (defect template). The amorphous region is a part of this template where the defect concentration exceeds a critical value, and it cuts across the cross-section of the nanowire[13,18] (Supplementary Fig. 11a–c for TEM images). Transformation from the amorphous phase to any crystalline state involves threshold switching followed by recrystallization of the amorphous region, and subsequently a reduction of defect concentration in the rest of the template through homogenization of defects (to the background concentration) via Joule heating. Thus following recrystallization, the degree of defect homogenization can be controlled via I_c, and this provides access to several intermediate resistance states in the crystalline phase (Supplementary Note 10).

Amorphization of intermediate resistance states via d.c. current. Another subtle feature in d.c. I–V switching behaviour of intermediate states (states 2 and 3 of D1) is the sudden drop in current (increase in resistance) at very low currents (0.1 μA in Fig. 5b, red and blue curves). This event corresponds to intermediate states first transforming to an amorphous phase, a permanent structural change (Supplementary Fig. 12a,b). It is consistent with the understanding that the momentum (and energy) transfer from carriers to defects at 0.1 μA is sufficient to migrate more defects to the already existing defect-template region in the intermediate states, thus increasing the defect concentration beyond a critical limit to nucleate the amorphous phase. These

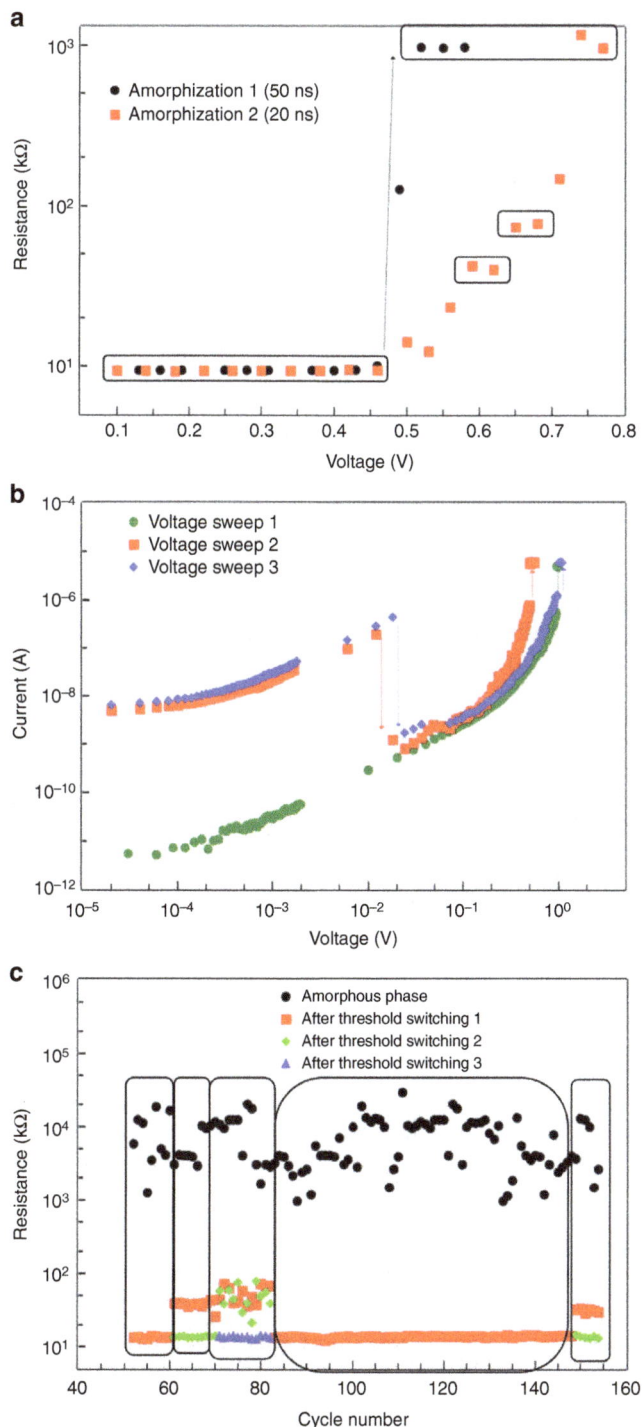

Figure 5 | Accessing intermediate resistance states on GeTe nanowire devices defect engineered into insulating crystalline phase.
(**a**) Programming curve on D1 for the RESET operation. When 50-ns pulses were applied, the transformation to the amorphous phase was sudden. With 20-ns pulses, the transformation happened gradually accessing several intermediate resistance states. Low-voltage, 20-ns pulses accumulates defects at a local region creating an intersecting defect template. Subsequent controlled addition of defects with higher amplitude pulses to this template increases the resistance of the device creating intermediate resistance states. States 1, 2 and 3, and amorphous phase (increasing order of resistance) are boxed. (**b**) Voltage sweep from 0 to 1 V (green) demonstrating a threshold-switching event of the amorphous phase to state 3 at <1 V with compliance current (I_c) set at 5 μA. A second sweep starting from state 3 (red), revealing a drop in current at 0.01 V corresponding to amorphization event, and the amorphous phase subsequently transformed to state 2 after a threshold-switching event to state 2. Another voltage sweep from 0 to 1 V starting with state 2 (blue), again showing a drop in the current at 0.01 V, signifying amorphization (Supplementary Fig. 12a)—and the amorphous phase subsequently threshold switched and transformed to state 1. The arrows in the figure correspondingly indicate carrier-wind force assisted amorphization and threshold-switching events (**c**). Repeatable switching measurements, with every cycle consisting of a 100-ns, 26-μA pulse transforming state 1 to amorphous phase, followed by I–V sweeps until state 1 is eventually retrieved; and between every cycle I_c was randomly set to 50, 10 or 5 μA. For the first 60 cycles, I_c was set to 50 μA, for the next 10 cycles $I_c = 10$ μA. From 70 to 82 cycles, $I_c = 5$ μA, followed by 50 μA from 83 to 150 cycles. Further up to 160 cycles, $I_c = 10$ μA. When $I_c = 50$ μA, amorphous phase always switched to state 1 directly, and when $I_c = 5$ and 10 μA intermediate metastable states became accessible. Here the intermediate resistance states were created by controllably removing defects from the defect-templated region.

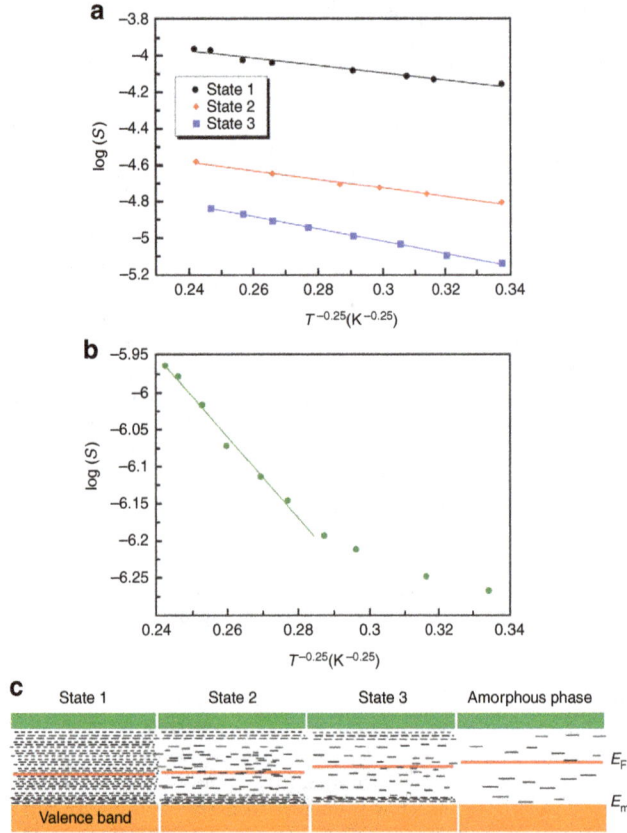

Figure 6 | Transport measurements on various observed resistance states in defect-engineered GeTe nanowires. (**a**) $T^{-0.25}$ versus log(S) plots, where S is the conductance—showing that states 1, 2 and 3 exhibit VRH conduction behaviour with slopes (A) becoming steeper from 1 to 3. A in states 1, 2 and 3 are -2, -2.4 and $-3.4\,K^{0.25}$, respectively. (**b**) Conduction behaviour of the amorphous phase plotted as log S versus $T^{-0.25}$, showing VRH behaviour at high temperatures ($A = -5.1\,K^{0.25}$), and a deviation from VRH at low temperatures most likely due to the formation of paired carrier traps and E_F pinning. (**c**) Schematic band diagrams showing the relative position of Fermi level (E_F) and density of single-electron trap states at E_F ($N(E_F)$) in all the observed states. From state 1 to the amorphous phase, Fermi level progressively moves into deeper traps until E_F gets pinned to the mid-gap in the amorphous phase, and the single-electron trap density ($N(E_F)$) decreases at the cost of increasing charged (paired) traps.

results demonstrate that crystal–amorphous transformation in the defect-templated pathway is a solid-state transformation, assisted by carrier-wind force and strong carrier–lattice coupling (Supplementary Note 11).

Transport properties of various observed resistance states. To understand the differences in various observed resistance states (states 1–3 and amorphous phase) from an electronic viewpoint, we performed temperature-dependent resistance measurements on D1 prepared in these states (Fig. 6a,b). States 1, 2 and 3 (~ 10, 40 and $70\,k\Omega$, respectively) showed VRH conduction, where conductance S depends on temperature as $S = S_0 \exp(-AT^{-0.25})$, with A, the temperature-independent prefactor, increasing from 2 to 2.4 to $3.4\,K^{0.25}$ from states 1 to 3 (Fig. 6a). $A = [\alpha^3/kN(E_F)]^{0.25}$, where α is the inverse of carrier localization length, k is the Boltzmann constant and $N(E_F)$ is the density of single-electron trap states at the E_F (ref. 28). The increase in value of A from states 1 to 3 could be a result of decrease in localization

length (or increase in α) or a decrease in $N(E_F)$ or both. On the other hand, the conduction characteristics of the amorphous phase (Fig. 6b) showed VRH behaviour at high temperatures ($> 150\,K$), with $A = 5.1\,K^{0.25}$; and deviated from VRH behaviour (log(S) proportional to $T^{-0.25}$) at low temperatures (Supplementary Fig. 13). This behaviour is different from that of typical melt-quench amorphous phase, which shows activated conduction via emission of carriers into delocalized states at high temperatures and VRH at low temperatures[29] (Supplementary Note 12). It will be an interesting future direction to rigorously study the nature of the observed amorphous phase and understand device-related issues such as resistance drift in comparison with the conventional melt-quench amorphous phase.

The schematics of band structure shown in Fig. 6c explain the observed conduction characteristics (Fig. 6a,b) of various representative states. In state 1, E_F is above the mobility edge (E_m) surrounded by a large density of single-electron traps $N(E_F)$. Progression from state 1 to state 3 by adding more defects either shifts the E_F towards the mid-gap, which reduces $N(E_F)$ and localization length, or creates more paired trap centres, which reduces $N(E_F)$. Finally, in the amorphous phase, E_F is pinned to the mid-gap[30,31], with $N(E_F)$ at its minimum value. In the reverse process (SET), by controlled homogenization (removal) of defects from the defect-template region via Joule heating, E_F moves back towards the mobility edge, and $N(E_F)$ starts to increase, accessing all the crystalline intermediate resistance states. Pulse width (Fig. 5a) and compliance current (Fig. 5c) can be used to control the addition and homogenization of defects respectively. It is important to note here that multiple resistance states in PCM reported in earlier works[30–34] were created by controlling the relative volumes of the amorphous and crystalline states, the only two physically different states, and are fundamentally different from the multiple resistance states in this work.

Discussion

We demonstrated that defect-engineered crystalline GeTe with dominant carrier localization effects transforms to an amorphous phase at current and power densities significantly lower than some of the best reported low-power devices operated through melt-quench strategy[6]. These results emphasize the importance of carrier–lattice coupling in the insulating state in carrying out energy efficient amorphization. Our devices displayed good reversibility and endurance, suggesting new strategies for potentially mitigating the issue of chemical segregation and heat-based device degradation problems, common in melt-quench approach[11]. The recent discoveries of localization effects[35] and defect-templated amorphization[18] in Ge–Sb–Te alloys suggests that defect-engineering approach could be applied to other well-known PCMs too. In addition, with our demonstration of scaling of switching properties in the defect-based approach and multistate switching, we believe that nanoscale PCM structures (mushroom, sidewall, pillar or confined architectures)[5,36] engineered into electronic states in crystalline phase that show localization behaviour in transport will be promising for ultralow-power memories and novel computation strategies[29,30].

Methods

Synthesis of GeTe nanowires. GeTe nanowires were synthesized using metal catalyst-mediated vapour–liquid–solid process, where bulk GeTe powder (99.9%, Alfa Aesar, melting temperature, 724 °C) was placed at the centre of a tube furnace. Silicon oxide substrate evaporated with Au film (8 nm) and subsequently annealed at 720 °C for 10 min was placed on the downstream side of the furnace (~ 15 cm away from the middle). The furnace was heated to 400 °C at a carrier gas (Ar) flow rate of 100 s.c.c.m. and a pressure of 10 torr, and maintained so for 10 h before the furnace was slowly cooled to room temperature.

Device fabrication. Multiple electrode devices were fabricated using electron-beam lithography. Nanowires were dry transferred onto an insulating substrate with pre-patterned markers. Three layers of PMMA 495 A4 and three layers of PMMA 950 A2 were spin coated and baked (180 °C) following which electron-beam lithography, metallization (Ti/Au: 50/100 nm) and lift-off procedure were performed. The devices were subsequently annealed at 350 °C, and a 30-nm SiOx film was conformally deposited by plasma-enhanced chemical vapour deposition.

Ion irradiation. The devices thus fabricated were irradiated with 2 MeV He$^+$ ions, (beam area: 5×5 mm) in a tandem accelerator (NEC minitandem ion accelerator). Substrates were aligned perpendicular to the beam, and ion bombardment was performed until a cumulative prescribed dosage was reached, with ion current maintained below 30 nA (as measured from picoammeter). Dosage was calculated as It/A, where I is the instantaneous ion current, t is the time of exposure and A is the area of the beam area (5×5 mm). We verified using SRIM (Stopping and Range of Ions in Matter) software that He$^+$ ion exposure and knock-on damage is uniform throughout the thickness of the nanowire devices (Supplementary Note 1).

Electrical testing. Temperature–resistance measurements were performed in the Lakeshore TTPX cryogenic probe station, and resistance measurements were carried out via I–V sweeps at very low bias (-2 to 2 mV) using Keithley 2,602 (I–V analyser/source meter). Switching and endurance tests were performed using Keitheley 3,401 for pulse generation, 2,602 (I–V analyser) for resistance measurement after the application of the pulse and Keithley 2,700 as the data acquisition system. The shapes of the voltage pulses generated and dynamic current response produced were verified using a 500 MHz Tektronix DPO3052 digital oscilloscope. Applied voltage pulse was measured by connecting the device in parallel to the 50 Ω input channel 1 of the oscilloscope. The current response was measured by measuring the voltage drop across a 50-Ω resistor connected in series with the device, and in parallel with a second 50 Ω input channel of the oscilloscope.

References

1. Wong, H.-S.P. *et al.* Phase change memory. *Proc. IEEE* **98**, 2201–2227 (2010).
2. Bez, R. & Pirovano, A. Non-volatile memory technologies: emerging concepts and new materials. *Mater. Sc. Semicond. Process.* **7**, 349–355 (2004).
3. Pirovano, A. *et al.* in *IEDM '03 Technical Digest IEEE International Electron Devices Meeting*, 699–702 (Washington, DC, USA, 2003).
4. Chen, Y. C. *et al.* in *IEDM '06 International Electron Devices Meeting*, 1–4 (San Francisco, CA, USA, 2006).
5. Lankhorst, M. H., Ketelaars, B. W. & Wolters, R. A. Low-cost and nanoscale non-volatile memory concept for future silicon chips. *Nat. Mater.* **4**, 347–352 (2005).
6. Xiong, F., Liao, A. D., Estrada, D. & Pop, E. Low-power switching of phase-change materials with carbon nanotube electrodes. *Science* **332**, 568–570 (2011).
7. Xiong, F. *et al.* Self-aligned nanotube–nanowire phase change memory. *Nano Lett.* **13**, 464–46902 (2012).
8. Lee, S.-H., Jung, Y. & Agarwal, R. Highly scalable non-volatile and ultra-low-power phase-change nanowire memory. *Nat. Nanotechnol.* **2**, 626–630 (2007).
9. Hwang, Y. N. *et al.* in *IEDM '03 Technical Digest. IEEE International Electron Devices Meeting*, 893–896 (Washington, DC, USA, 2003).
10. Qiao, B. *et al.* Effects of Si doping on the structural and electrical properties of Ge$_2$Sb$_2$Te$_5$ films for phase change random access memory. *Appl. Surf. Sci.* **252**, 8404–8409 (2006).
11. Kim, C. *et al.* Direct evidence of phase separation in Ge$_2$Sb$_2$Te$_5$ in phase change memory devices. *Appl. Phys. Lett.* **94**, 193504 (2009).
12. Moran, M. J., Shapiro., H. N., Boettner, D. D. & Bailey., M. B. *Fundamentals of Engineering Thermodynamics* 8th edn (John Wiley & Sons, 2014).
13. Nukala, P. *et al.* Direct observation of metal-insulator transition in single-crystalline germanium telluride nanowire memory devices prior to amorphization. *Nano Lett.* **14**, 2201–2209 (2014).
14. Mott., N. F. *Metal-Insulator Transitions* 2nd edn (Taylor & Francis, 1990).
15. Mott., N. F. & Davis, E. A. *Electronic Processes in Non-Crystalline Materials* Vol. 2 (Oxford Univ. Press Inc., 1979).
16. Kolobov, A. V., Krbal, M., Fons, P., Tominaga, J. & Uruga, T. Distortion-triggered loss of long-range order in solids with bonding energy hierarchy. *Nat. Chem.* **3**, 311–316 (2011).
17. Edwards, A. H. *et al.* Electronic structure of intrinsic defects in crystalline germanium telluride. *Phys. Rev. B* **73**, 045210 (2006).
18. Nam, S. W. *et al.* Electrical wind force-driven and dislocation-templated amorphization in phase-change nanowires. *Science* **336**, 1561–1566 (2012).
19. Lee, S.-H., Ko, D.-K., Jung, Y. & Agarwal, R. Size-dependent phase transition memory switching behaviour and low writing currents in GeTe nanowires. *Appl. Phys. Lett.* **89**, 223116 (2006).
20. Composto, R. J., Walters, R. M. & Genzer, J. Application of ion scattering techniques to characterize polymer surfaces and interfaces. *Mater. Sci. Eng. Rep.* **38**, 107–180 (2002).
21. Iwakiri, H., Yasunaga, K., Morishita, K. & Yoshida, N. Microstructure evolution in tungsten during low-energy helium ion irradiation. *J. Nucl. Mater.* **283**, 1134 (2000).
22. Mooij, J. H. Electrical conduction in concentrated disordered transition metal alloys. *Phys. Stat. Sol.* **17**, 521–530 (1973).
23. Park, M.-A., Savran, K. & Kim, Y.-J. Weak localization and the Mooij rule in disordered materials. *Phys. Stat. Sol.* **237**, 500–506 (2003).
24. Harris, S. *An Introduction to the Theory of Boltzmann Equation* (Dover Publications, 1999).
25. Lee, P. A. & Ramakrishnan, T. V. *et al.* Disordered electronic systems. *Rev. Mod. Phys.* **57**, 287–337 (1985).
26. Yu, K. Y. *et al.* Radiation damage in helium ion irradiated nanocrystalline Fe. *J. Nucl. Mater.* **2012**, 140–146 (2012).
27. Pirovano, A., Lacaita, A. L., Benvenuti, A., Pellizzer, F. & Bez, R. Electronic switching in phase-change memories. *IEEE Trans. Electron Devices* **51**, 452–459 (2004).
28. Zallen, R. *The Physics of Amorphous Solids* (John Wiley & Sons, 1983).
29. Longeaud, C., Luckas, J. & Wuttig, M. Some results on the germanium telluride density of states. *J. Phys. Conf. Ser.* **398**, 012007 (2012).
30. He, Q. *et al.* Continuous controllable amorphization ratio of nanoscale phase change memory cells. *Appl. Phys. Lett.* **104**, 223502 (2014).
31. Kuzum, D., Jeyasingh, R. G., Lee, B. & Wong, H. S. Nanoelectronic programmable synapses based on phase change materials for brain-inspired computing. *Nano Lett.* **12**, 2179–2186 (2012).
32. Wright, C. D., Hosseini, P. & Diosdado, J. A. V. Beyond von-Neumann computing with nanoscale phase-change memory devices. *Adv. Funct. Mater.* **23**, 2248–2254 (2013).
33. Jie, F. *et al.* Design of multi-states storage medium for phase change memory. *Jpn J. Appl. Phys.* **46**, 5724–5727 (2007).
34. Skelton, J. M., Loke, D., Lee, T. H. & Elliott, S. R. Understanding the multistate SET process in Ge-Sb-Te-based phase-change memory. *J. Appl. Phys.* **112**, 064901 (2012).
35. Siegrist, T. *et al.* Disorder-induced localization in crystalline phase-change materials. *Nat. Mater.* **10**, 202–208 (2011).
36. Hudgens, S. & Johnson, B. Overview of phase-change chalcogenide nonvolatile memory technology. *MRS Bull.* **29**, 829–832 (2004).

Acknowledgements

This work was supported by NSF (DMR-1002164 and 1210503), Penn-MRSEC (DMR05-20020), Materials Structures and Devices Center at MIT, NSF/MRSEC-DMR 11-20901, and partially from the NSF Polymer Program DMR09-07493 and NSF Materials World Network DMR-1210379. Ion-irradiation and electron microscopy were performed at the Nanoscale Characterization Facility at the University of Pennsylvania.

Author contributions

P.N. and R.A. conceived the concepts, designed experiments and co-wrote the manuscript. P.N. carried out synthesis, device fabrication, TEM characterization, transport measurements and data analysis. C.C.L. carried out ion-irradiation experiments. C.C.L. along with R.C. carried out SRIM calculations, and subsequent analysis on ion penetration in GeTe nanowire devices. P.N., C.C.L., R.A. and R.C. discussed all the data and their interpretations.

Additional information

Competing financial interests: The authors declare no competing financial interests.

10

Bio-recognitive photonics of a DNA-guided organic semiconductor

Seung Hyuk Back[1,*], Jin Hyuk Park[2,*], Chunzhi Cui[2,*] & Dong June Ahn[1,2,3]

Incorporation of duplex DNA with higher molecular weights has attracted attention for a new opportunity towards a better organic light-emitting diode (OLED) capability. However, biological recognition by OLED materials is yet to be addressed. In this study, specific oligomeric DNA–DNA recognition is successfully achieved by tri (8-hydroxyquinoline) aluminium (Alq_3), an organic semiconductor. Alq_3 rods crystallized with guidance from single-strand DNA molecules show, strikingly, a unique distribution of the DNA molecules with a shape of an 'inverted' hourglass. The crystal's luminescent intensity is enhanced by 1.6-fold upon recognition of the perfect-matched target DNA sequence, but not in the case of a single-base mismatched one. The DNA–DNA recognition forming double-helix structure is identified to occur only in the rod's outer periphery. This study opens up new opportunities of Alq_3, one of the most widely used OLED materials, enabling biological recognition.

[1]KU-KIST Graduate School of Converging Science and Technology, Korea University, Seoul 02841, Korea. [2]Department of Chemical and Biological Engineering, Korea University, Seoul 02841, Korea. [3]Center for Theragnosis, Biomedical Research Institute, Korea Institute of Science and Technology, Seoul 02792, Korea. * These authors contributed equally to this work. Correspondence and requests for materials should be addressed to D.J.A. (email: ahn@korea.ac.kr).

Novel display materials have gained keen attraction recently in the fields of electronics and photonics research especially owing to the rapid evolution of smart communication devices[1–3]. Among the various display materials available, organic semiconductors or metal-organic compounds are considered to be very promising, and they have therefore been intensely investigated[4–6]. An alumina quinoline, tri (8-hydroxyquinoline) aluminium (Alq_3), first reported approximately three decades ago, which emits in the green and blue spectra, is a material of central interest[7–9]. Alq_3 is currently used in a multitude of organic light-emitting diodes (OLEDs)[10–13] that are used in various displays. Since it was first reported, enormous improvements have been made in its light emission efficiency, to provide higher display quality[14–18]. One peculiar approach incorporates a biological material into the light-emitting device, often called a BioLED[19]. An example is DNA in the form of a thin film introduced within a conventional electroluminescent cell incorporating an Alq_3 layer[20]. Utilized in the device were double-strand DNAs (dsDNAs) extracted from natural organisms and complexed with cationic surfactants; the device provides ∼30-fold increase in luminescence intensity[21]. This phenomenon was attributed to the contribution of the DNA layer to the electron blocking effect, thus reducing significant loss of electrons and enhancing electron–hole recombination in the cell[20]. Luminescent dyes entrapped within dsDNA thin films reported[22] also exhibited higher intensity owing to less non-radiative relaxation. This novel capability of DNA is noteworthy as a gadget in light-emitting devices. The value of the devices can be recognized even higher as they incorporate the 'bio-recognition function'. Current BioLEDs now face a new journey to the realm of biological recognition.

To this end, this study presents a critical step endowing an OLED material with a biological recognition function. We demonstrate for the first time that only specific DNA–DNA recognition triggers photoluminescent enhancement reflected by Alq_3, the most widely used OLED material.

Results

Optical properties analyses of DNA-guided Alq_3 rods. We first observed the characteristic alteration when Alq_3 particles incorporating single-strand DNA (ssDNA) moieties interacted with specific target DNA (tDNA) molecules. Crystallization of Alq_3

has been conventionally executed with the aid of surfactants and recently become successful using ssDNA molecules only[23]. With guidance from ssDNA, we fabricated prismatic hexagonal rod crystals composed of Alq_3. In this study, the oligomeric ssDNA used for crystal guidance was a 27-mer sequence of anthrax lethal factor. Figure 1a shows a schematic illustration of the recognition of specific tDNA by the light-emitting Alq_3 rod crystallized by ssDNA. Figure 1b,c provide colour charge-coupled device (CCD) images of the ssDNA-guided Alq_3 (ssDNA-Alq_3) rods before and after treatment with tDNA molecules, respectively. We observed the ssDNA-Alq_3 rods emitting green luminescence. Interestingly, the intensity of the green luminescence of the ssDNA-Alq_3 rods was markedly enhanced after interaction with specific tDNA molecules. For quantitative analysis of the intensity enhancement, we measured the photoluminescence (PL) spectra of the Alq_3 rods. As shown in Fig. 1d, a broad PL peak was observed at ∼512 nm when samples were excited with a laser at 365 nm, which corresponds to the main absorption band of Alq_3. The PL spectra were yellowish-green, composed of both α and δ phases[8,9,24]. After interaction with specific tDNA molecules, the PL peak intensity increased ∼1.6-fold, which is concordant with the results of the CCD analysis. Interestingly, when treated with single-base (1-mer) mismatched tDNA molecules that are less specific, the Alq_3 rods showed little enhancement of PL intensity. In addition, PL excitation (PLE) spectrum analysis confirmed the enhancement of PL intensity, as shown in Fig. 1e. The intensity with excitation at 365 nm and emission at 512 nm was clearly higher following treatment with specific target molecules.

Crystal structure analyses upon interaction with DNA. To further explore the PL enhancement of the Alq_3 rods after interaction with specific tDNA molecules, we selected four crystal samples of ssDNA-Alq_3, ssDNA-Alq_3 treated with specific tDNA, ssDNA-Alq_3 treated with 1-mer mismatched tDNA and dsDNA-Alq_3 (dsDNA-guided Alq_3 rods crystallized by the use of dsDNA molecules from the start). X-ray diffraction (XRD) patterns were observed, as shown in Fig. 2a, to examine structural changes in the Alq_3 crystals. The XRD pattern of the ssDNA-Alq_3 rod showed typical α-phase peaks for Alq_3 at 11.40° and 12.81°, along with a δ-phase peak at 11.79°. Hence, the ssDNA-Alq_3 rods fabricated in this study contain both α- and δ-phases[8,9,25–27], which is consistent with the yellowish-green luminescence

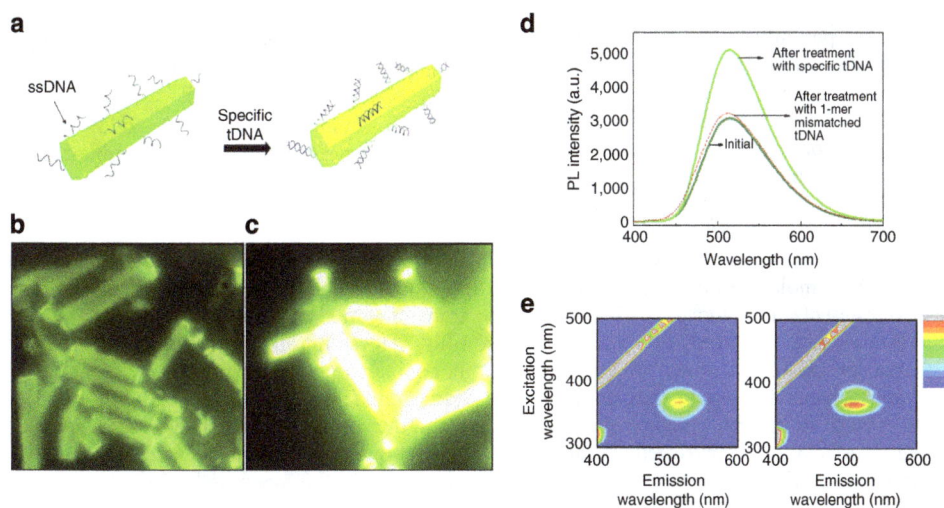

Figure 1 | Optical effects of DNA-guided Alq_3 rods. (a) Schematic illustration of the recognition of specific tDNA by Alq_3 rod crystals. **(b,c)** Colour CCD images of the samples before and after interaction with tDNA, respectively. **(d)** PL spectra of the initial Alq_3 crystals (indicated by dark green line), and after interaction with specific tDNA (green line) and 1-mer mismatched DNA (red dotted line), respectively, with excitation at 365 nm. **(e)** PLE spectra of the samples before and after interaction with tDNA, respectively.

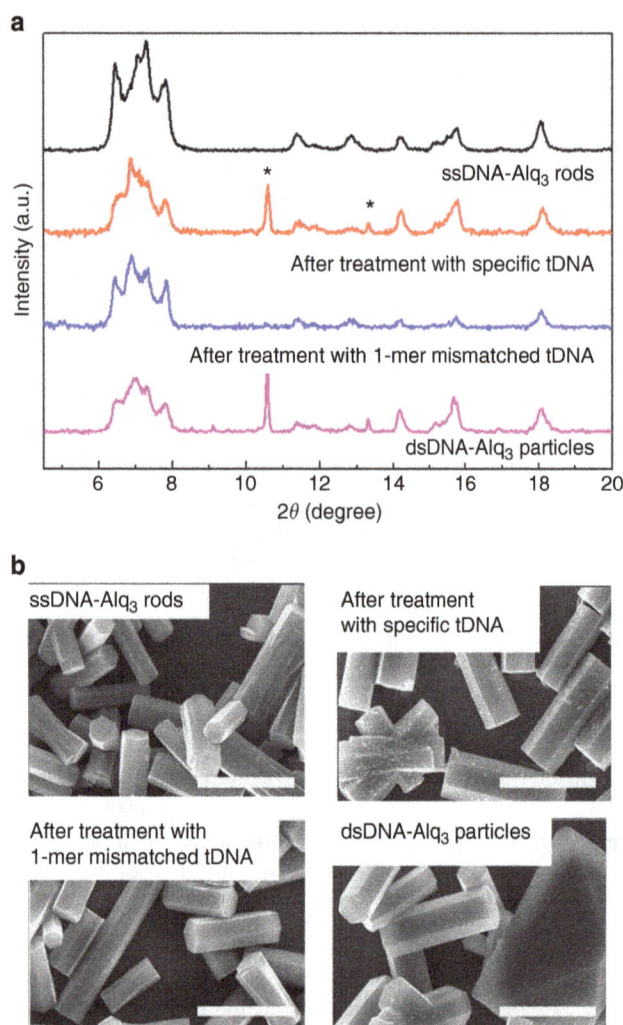

Figure 2 | Crystal structure and morphology of Alq$_3$ rods upon interaction with DNA. (**a**) XRD patterns and (**b**) SEM images of ssDNA-Alq$_3$ rods, ssDNA-Alq$_3$ rods treated with specific tDNA, ssDNA-Alq$_3$ rods treated with 1-mer mismatched tDNA and dsDNA-Alq$_3$ particles in which dsDNA formed *a priori* (scale bars, 10 μm).

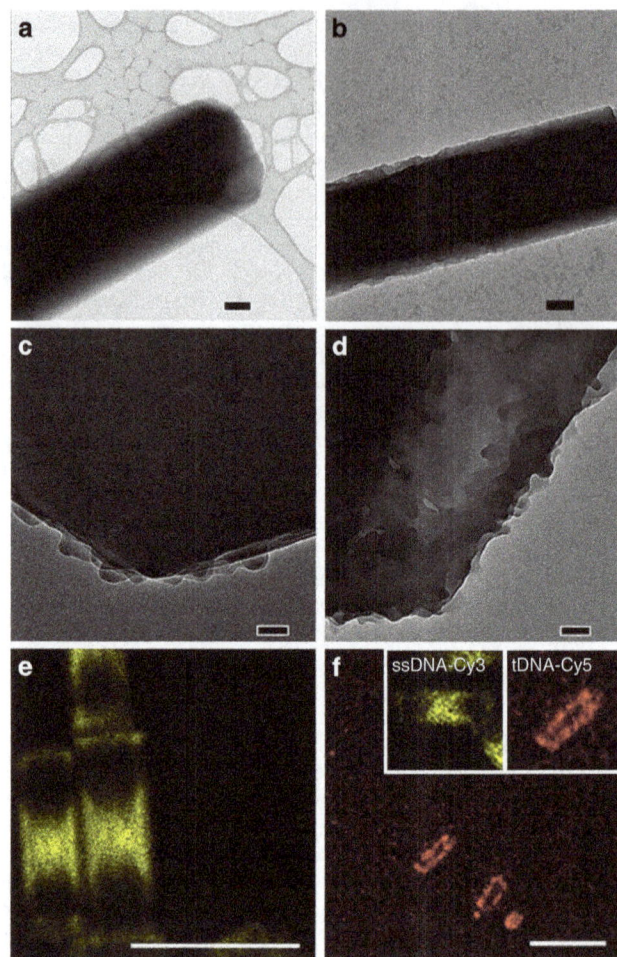

Figure 3 | Observation of inner core and crust layers in Alq3 rod with distributions of DNA molecules. HR-TEM images of a ssDNA-Alq$_3$ rod (**a,c**) before and (**b,d**) after being treated with the specific tDNA: (**a,b**) top view (scale bar, 200 nm) and (**c,d**) near edge (scale bar, 50 nm). CLSM images captured at the centre plane with excitation by a 555-nm laser of (**e**) the ssDNA-Cy3 molecules used for crystal guidance (scale bar, 20 μm) and (**f**) the perfect-match tDNA-Cy5 molecules (scale bar, 20 μm) recognized by the surrounding peripheral ssDNA. (Insets represent respectively the ssDNA-Cy3 and tDNA-Cy5 distributions after hybridization.) The filters used for observing the Cy3 and Cy5 dyes ranged from 300 to 630 nm and 630 to 800 nm, respectively.

observed in the PL analyses. Upon interaction with specific tDNA molecules, the rods showed an XRD pattern almost identical to that of the initial Alq$_3$ rods, except that two additional peaks appeared at 10.59° and 13.31°. The source of these two peaks was identified as the dsDNA forming the helical structure[28,29]. These two peaks were also evident in the dsDNA-Alq$_3$ sample, in which the double-helix DNA structure was formed before the crystallization. In a very clear comparison, these two peaks were completely absent in the case of ssDNA-Alq$_3$ rods treated with 1-mer mismatched tDNA molecules, indicating that nearly no helical dsDNA was present. The morphological features of the four Alq$_3$ crystals were observed by field-emission scanning electron microscopy (SEM), as shown in Fig. 2b. The ssDNA-Alq$_3$ rods showed a regular prismatic hexagonal shape with a smooth surface, similar to the morphology of Alq$_3$ rods fabricated using the surfactant, cetyltrimethylammonium bromide (Supplementary Fig. 1)[30]. Upon interaction with specific tDNA, the initially smooth surface of the Alq$_3$ rods became rough. A rough surface was not observed either following treatment with 1-mer mismatched tDNA or in the sample of dsDNA-Alq$_3$ rods. Therefore, the Alq$_3$ rods only showed a significant enhancement in PL following interaction with specific

tDNA, accompanied by the following interesting features: diffraction peaks indicative of double-helix DNA and surface roughening.

Layers in a single Alq$_3$ rod and DNA distribution. To observe the roughened surface more closely, we employed high-resolution transmission electron microscopy (HR-TEM) of the ssDNA-Alq$_3$ rods, before and after interaction with specific tDNA molecules. In comparison with the prismatic hexagonal shape of the ssDNA-Alq$_3$ rods (Fig. 3a), interaction with specific tDNA resulted in formation of an 120-nm thick crust layer surrounding the inner core of the rod of which the whole thickness is 800 nm, as shown in Fig. 3b. In the magnified HR-TEM images of regions near the rod edges as shown in Fig. 3c,d, the surface crust layer was observable to a small degree in ssDNA-Alq$_3$ rod, but increased markedly upon specific tDNA interaction. We can estimate, from the HR-TEM image (Fig. 3b), the volume ratio of the crust layer in the Alq$_3$ rod to be 50 vol% (Supplementary Fig. 2). It is noted

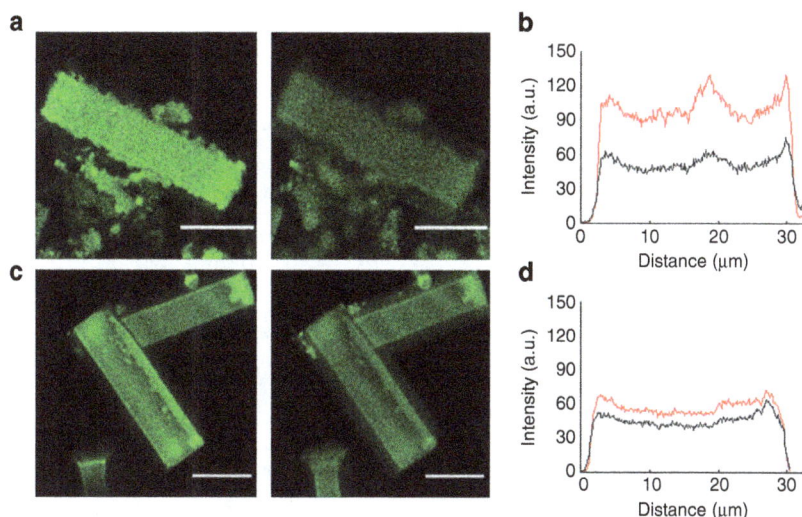

Figure 4 | Depth-wise identification of localized PL of Alq₃ molecules. (**a**) CLSM images of the upper plane (left) and the 2.5-µm inner plane (right) of a tDNA-recognized ssDNA-Alq₃ rod, (**b**) longitudinal PL profiles in which the intensity of the upper plane (indicated by red line) is ∼2-fold higher than that of the inner plane (black line). (**c**) Images and (**d**) PL profiles of ssDNA-Alq₃ rods are also shown. The images and profiles were obtained with excitation by a 405-nm laser; the filter used for observation of Alq₃ molecules ranged from 300 to 550 nm (scale bars, 10 µm).

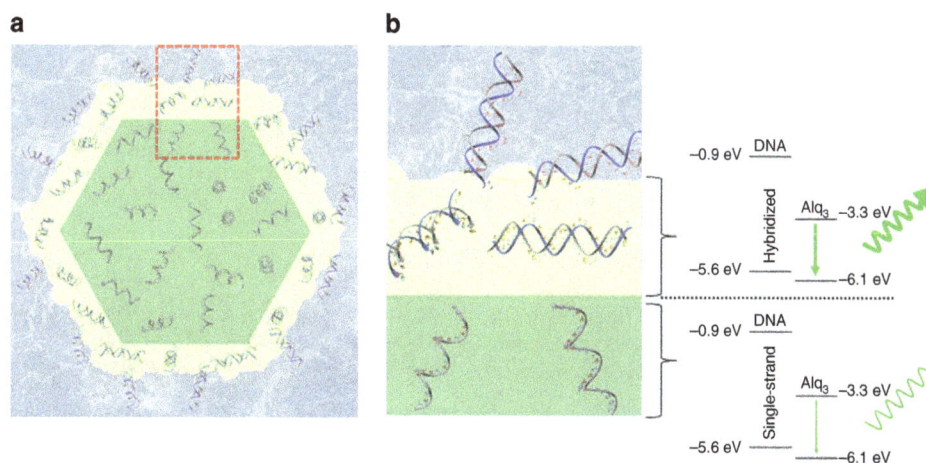

Figure 5 | Cross-sectional schematic illustration and energy-level diagram. (**a**) Cross-section of a ssDNA-guided Alq₃ rod recognizing a specific tDNA at the crust layer and (**b**) enlarged picture indicating double-helix DNA that exerts less non-radiative dissipation.

that such a crust layer was completely absent in the reference case, the cetyltrimethylammonium bromide-guided Alq₃ rod (Supplementary Fig. 1). Hence, it is rational that the increased crust layer was induced by the recognition of specific tDNA by the ssDNA present in the Alq₃ rod. To confirm the positioning of the DNA molecules, we used ssDNA and specific tDNA molecules labelled with Cy3 and Cy5 fluorescent dyes, respectively. The distribution of the DNA molecules was visualized by tracing the corresponding dye moieties using a confocal laser scanning microscope (CLSM) capable of excitation with a 555-nm laser with a variable-wavelength filter. Strikingly to observe in Fig. 3e, the ssDNA-Cy3 molecules showed a very unique distribution in the shape of an 'inverted' hourglass over the Alq₃ rod. DNA molecules have recently been shown to play a role in crystallization[23] and to act as an alternative to surfactants. A conventional concept of crystallization regarding the role of surfactant molecules (that is, wrapping around particulate seeds and resulting in subsequent crystallization) was found to be invalid for the DNA molecules used for the crystal guidance, at least in the present case. A detailed understanding of why the ssDNA molecules are distributed in this unique manner is

presently lacking. How the ssDNA molecules having the inverted hourglass distribution interact with the tDNA molecules is investigated next. Tracing of the dye moieties showed that the tDNA-Cy5 molecules were only present on the rod's outer periphery, which emitted red fluorescence as shown in Fig. 3f. Interestingly, the unique distribution of ssDNA molecules is nearly maintained. Thus, we can infer that recognition of the tDNA molecules by the ssDNA molecules occurred only in the region limited to the roughened surface but not in the inner core. Hence, the abovementioned 120-nm thick surface crust, shown in Fig. 3b, was induced by specific DNA–DNA recognition.

Depth-wise identification of localized photoluminescence. In the next stage, we investigated precisely where in the Alq₃ rods the enhanced PL originates, that is, the inner core, the surface crust, or the entire rod. To compare the PL intensity of the crust layer and inner core of a single Alq₃ rod, we again employed CLSM, but with excitation at 405 nm, to excite the Alq₃ molecules. From the rod having thickness of 11 µm, we acquired two-dimensional cutting-plane images, one from the uppermost surface layer and the other at the depth of the 2.5-μm below the surface, and thus below the crust. Specific DNA–DNA recognition occurred and Fig. 4a shows a comparison of the uppermost surface layer (left) and the 2.5-μm inner planes (right), corresponding to the crust layer and the inner core, respectively. We observed that the upper plane was markedly brighter than the inner plane. We quantified this phenomenon using a profiling analysis of the rod crystals in a longitudinal direction, as shown in Fig. 4b; the PL intensity of the upper plane (indicated by red line) was ∼2-fold higher than that of the inner plane (black line). In contrast, the upper plane was a little brighter than the lower in the ssDNA-Alq$_3$ rod, as shown in Fig. 4c,d. Therefore, careful comparison of the PL intensity of the crust layer and the inner core, in addition to analyses of structural variation and molecular profiling, enables us to conclude that specific DNA–DNA recognition caused the increase in the surface crust layer, the very region responsible for the PL enhancement. Thus, Alq$_3$, one of the most widely used OLED materials, becomes capable of DNA–DNA recognition.

This novel phenomenon can be explained by the following mechanism: Fig. 5 suggests a cross-sectional schematic illustration of the Alq$_3$ rod recognizing the specific tDNA and the energy-level diagrams for both crust layer and inner core. No shift at the main absorption peak of the Alq$_3$ rods before and after the recognition was found, which was also true for ssDNA and hybridized rods (Supplementary Fig. 3). This indicates that specific DNA–DNA recognition did not induce any changes in the bandgap of the Alq$_3$ rods. As illustrated in the energy diagram, DNA has a wide bandgap such that the lowest unoccupied molecular orbital energy level was −0.9 eV and the highest occupied molecular orbital level was −5.6 eV, whereas the lowest unoccupied molecular orbital and highest occupied molecular orbital levels of the yellowish-green Alq$_3$ were −3.3 and −6.1 eV, respectively[20,21]. We analysed fluorescence lifetimes to evaluate the effect of the DNA–DNA interaction on luminescence enhancement. The average lifetime was measured to be 1.28 ns for the ssDNA-Alq$_3$ rods and evidently increased to 1.63 ns after specific DNA–DNA interaction, indicating that DNA recognition evokes enhanced prevention of non-radiative relaxation of Alq$_3$ molecules (Supplementary Fig. 4 and Supplementary Table 1). Hence, the PL intensity is enhanced only in the limited region of the surface crust layer, where specific recognition event occurs.

Discussion

In summary, the Alq$_3$, a molecular organic semiconductor, crystallized and functionalized with ssDNA molecules is found capable of recognizing biological interaction, for the first time. The crystal's luminescent intensity is enhanced by 1.6-fold upon label-free recognition of perfect-matched tDNA sequence, but not in the case of single-base mismatched one. Specific DNA–DNA interaction induces double-helix DNA structure on the crystal's surface crust layer that is analysed to be responsible for longer fluorescence lifetime and the increase in the luminescent response. This study signifies a new direction of OLED materials toward unprecedented bio-recognitive photonic functions and applications.

Methods

Fabrication of DNA-guided Alq$_3$ rods. Commercial Alq$_3$ powder was dissolved in tetrahydrofuran at a concentration of 1 mg ml^{-1}, to form a stock solution. The stock solution (2 ml) was injected into 20 ml of various aqueous DNA solutions at a concentration of 0.5 μM, with vigorous stirring (∼800 r.p.m.) for 10 min. The mixture was stored at room temperature (RT) overnight to allow formation of visible precipitate. The ssDNA used in this fabrication had a sequence of

NH$_2$-5′-ATC CTT ATC AAT ATT TAA CAA TAA TCC-3′; this ssDNA, hybridized with its complementary sequence, was used as the dsDNA.

Hybridization of DNA molecules. The ssDNA used in this experiment was the amine-terminated anthrax lethal factor probe DNA sequence (NH$_2$-5′-ATC CTT ATC AAT ATT TAA CAA TAA TCC-3′). The fabricated ssDNA-Alq$_3$ rods were reacted with complementary tDNA (3′-TAG GAA TAG TTA TAA ATT GTT ATT AGG-5′) at a concentration of 0.5 μM at 52 °C for 30 min and then returned to RT. The 1-mer mismatched tDNA sequence used in this study was 3′-TAG GAA TAG TTA CAA ATT GTT ATT AGG-5′.

Characterization of the Alq$_3$ crystal samples. The surface morphology of the Alq$_3$ rods was analysed using a field-emission SEM (Hitachi, S-4300) using an acceleration voltage of 15 kV. Powdered samples of Alq$_3$ rods were cast on an ultrathin carbon-coated Cu grid or a holey carbon-coated Cu grid and images were captured using an HR-TEM (Tecnai G2, Fei) with an acceleration voltage of 200 kV. The powder XRD (Bruker, D8 Advance with DaVinci) patterns were captured at a voltage of 40 kV, a current of 40 mA and Cu-Kα radiation (λ = 1.540 Å). The scanning rate was 0.02° s^{-1}, and the 2θ range was captured from 2° to 20°. A fluorescence spectrophotometer (Hitachi, F-7000) was used for measuring PL and PLE spectra excited by a Xe lamp. A CLSM (Carl Zeiss, LSM700) was used for measuring z-sectioning fluorescence images of the isolated single Alq$_3$ rod, ssDNA molecules, and tDNA molecules. For this analysis, Alq$_3$ rods were fabricated with ssDNA-Cy3 (NH$_2$-5′-ATC CTT ATC AAT ATT TAA CAA TAA TCC-3′-Cy3) and tDNA-Cy5 (3′-TAG GAA TAG TTA TAA ATT GTT ATT AGG-5′-Cy5) to visualize the interactions with ssDNA. The Alq$_3$ molecules were guided by ssDNA and were excited at 405 nm and detected with a 300–550 nm filter. The ssDNA-Cy3 molecules present in the Alq$_3$ rod were excited at 555 nm and detected with a 300–630 nm filter. In addition, the tDNA-Cy5 molecules recognized by the Alq$_3$ rods were excited at 555 nm and detected with a 630–800 nm filter. The Alq$_3$ rods were analysed using a z-stack of images collected at 100 nm intervals through the ×20, ×40 and ×100 objective lenses. Fluorescence lifetimes of the solution-phase samples were obtained with an Edinburgh Instruments FL920 Fluorescence Lifetime spectrometer equipped with 376.6-nm pulsed-diode laser at RT. Quantum yields were measured using IESP-150B (Sumitomo Heavy Industries Advanced Machinery Co. Ltd.) equipped with a Xe lamp (CW500W).

References

1. Rogers, J. A., Someya, T. & Huang, Y. Materials and mechanics for stretchable electronics. *Science* **327**, 1603–1607 (2010).
2. Sekitani, T. *et al.* Stretchable active-matrix organic light-emitting diode display using printable elastic conductors. *Nat. Mater.* **8**, 494–499 (2009).
3. Tsukazaki, A. *et al.* Repeated temperature modulation epitaxy for p-type doping and light-emitting diode based on ZnO. *Nat. Mater.* **4**, 42–46 (2005).
4. Li, H., Eddaoudi, M., O'Keeffe, M. & Yaghi, O. M. Design and synthesis of an exceptionally stable and highly porous metal-organic framework. *Nature* **402**, 276–279 (1999).
5. Coropceanu, V. *et al.* Charge transport in organic semiconductors. *Chem. Rev.* **107**, 926–952 (2007).
6. Allard, S., Forster, M., Souharce, B., Thiem, H. & Scherf, U. Organic semiconductors for solution-processable field-effect transistors (OFETs). *Angew. Chem. Int. Ed. Engl.* **47**, 4070–4098 (2008).
7. Tang, C. W. & VanSlyke, S. A. Organic electroluminescent diodes. *Appl. Phys. Lett.* **51**, 913–915 (1987).
8. Cölle, M., Gmeiner, J., Milius, W., Hillebrecht, H. & Brütting, W. Preparation and characterization of blue-luminescent tris (8-hydroxyquinoline)-aluminum (Alq$_3$). *Adv. Funct. Mater.* **13**, 108–112 (2003).
9. Cölle, M. & Brütting, W. Thermal, structural and photophysical properties of the organic semiconductor Alq$_3$. *Phys. Status. Solidi A* **201**, 1095–1115 (2004).
10. Chen, C. & Shi, J. Metal chelates as emitting materials for organic electroluminescence. *Coord. Chem. Rev.* **171**, 161–174 (1998).
11. Kido, J. & Okamoto, Y. Organo lanthanide metal complexes for electroluminescent materials. *Chem. Rev.* **102**, 2357–2368 (2002).
12. Liao, S.-H. *et al.* Hydroxynaphthyridine-derived group III metal chelates: wide band gap and deep blue analogues of green Alq$_3$ (tris (8-hydroxyquinolate) aluminum) and their versatile applications for organic light-emitting diodes. *J. Am. Chem. Soc.* **131**, 763–777 (2008).
13. Chi, Y. & Chou, P.-T. Transition-metal phosphors with cyclometalating ligands: fundamentals and applications. *Chem. Soc. Rev.* **39**, 638–655 (2010).
14. Knupfer, M., Peisert, H. & Schwieger, T. Band-gap and correlation effects in the organic semiconductor Alq$_3$. *Phys. Rev. B* **65**, 033204 (2001).
15. Pohl, R. & Anzenbacher, P. Emission color tuning in Alq$_3$ complexes with extended conjugated chromophores. *Org. Lett.* **5**, 2769–2772 (2003).
16. Wang, X.-Y. & Weck, M. Poly (styrene)-supported Alq$_3$ and BPh$_2$q. *Macromolecules* **38**, 7219–7224 (2005).

17. Ravi Kishore, V. V. N., Narasimhan, K. L. & Periasamy, N. On the radiative lifetime, quantum yield and fluorescence decay of Alq in thin films. *Phys. Chem. Chem. Phys.* **5**, 1386–1391 (2003).

18. Lunt, R. R., Benziger, J. B. & Forrest, S. R. Relationship between crystalline order and exciton diffusion length in molecular organic semiconductors. *Adv. Mater.* **22**, 1233–1236 (2010).

19. Hagen, J. A., Li, W., Steckl, A. J. & Grote, J. G. Enhanced emission efficiency in organic light-emitting diodes using deoxyribonucleic acid complex as an electron blocking layer. *Appl. Phys. Lett.* **88**, 171109 (2006).

20. Steckl, A. J. DNA-a new material for photonics? *Nat. Photon.* **1**, 3–5 (2007).

21. Singh, T. B., Sariciftci, N. S. & Grote, J. G. Bio-organic optoelectronic devices using DNA. *Org. Electron.* **223**, 73–112 (2010).

22. Kawabe, Y., Wang, L., Horinouchi, S. & Ogata, N. Amplified spontaneous emission from fluorescent-dye-doped DNA-surfactant complex films. *Adv. Mater.* **12**, 1281–1283 (2000).

23. Cui, C., Park, D. H., Kim, J., Joo, J. & Ahn, D. J. Oligonucleotide assisted light-emitting Alq3 microrods: energy transfer effect with fluorescent dyes. *Chem. Commun.* **49**, 5360–5362 (2013).

24. Fukushima, T. & Kaji, H. Green-and blue-emitting tris (8-hydroxyquinoline) aluminum (III)(Alq3) crystalline polymorphs: preparation and application to organic light-emitting diodes. *Org. Electron.* **13**, 2985–2990 (2012).

25. Suzuki, F., Fukushima, T., Fukuchi, M. & Kaji, H. Refined structure determination of blue-emitting tris(8-hydroxyquinoline) aluminum(III) (Alq3) by the combined use of cross-polarization/magic-angle spinning 13C solid-state NMR and first-principles calculation. *J. Phys. Chem.* **117**, 18809–18817 (2013).

26. Muccini, M. *et al.* Blue luminescence of facial tris(quinolin-8-olato) aluminum(III) in solution, crystals, and thin films. *Adv. Mater.* **16**, 861–864 (2004).

27. Brinkmann, M. *et al.* Correlation between molecular packing and optical properties in different crystalline polymorphs and amorphous thin films of mer-tris (8-hydroxyquinoline) aluminum (III). *J. Am. Chem. Soc.* **122**, 5147–5157 (2000).

28. Drew, H. R. & Dickerson, R. E. Structure of a B-DNA dodecamer: III. geometry of hydration. *J. Mol. Biol.* **151**, 535–556 (1981).

29. Lu, X. J. & Olson, W. K. 3DNA: a software package for the analysis, rebuilding and visualization of three-dimensional nucleic acid structures. *Nucleic Acids Res.* **31**, 5108–5121 (2003).

30. Chen, W., Peng, Q. & Li, Y. Alq3 nanorods: promising building blocks for optical devices. *Adv. Mater.* **20**, 2747–2750 (2008).

Acknowledgements

This work was supported by the National Research Foundation (MSIP 2014023305 and 2015M3C1A3002152), KU-KIST Graduate School of Converging Science and Technology (R1309521), LOTTE CHEMICAL CORPORATION and a Korea University Grant. Prof S.W. Han at KAIST and Dr S. Kim at KIST are acknowledged for their assistance in the measurements of fluorescence lifetime and quantum yield.

Author contributions

D.J.A. designed and supervised the project. S.H.B. contributed to diffraction pattern and microscopy analyses. J.H.P. contributed to confocal molecular profiling analyses. C.C. contributed to DNA-guided crystallization, recognition and optical analyses. S.H.B., J.H.P., C.C. and D.J.A. suggested the concept and wrote the manuscript.

Additional information

11

Effective energy storage from a triboelectric nanogenerator

Yunlong Zi[1,*], Jie Wang[1,*], Sihong Wang[1,*], Shengming Li[1], Zhen Wen[1], Hengyu Guo[1] & Zhong Lin Wang[1,2]

To sustainably power electronics by harvesting mechanical energy using nanogenerators, energy storage is essential to supply a regulated and stable electric output, which is traditionally realized by a direct connection between the two components through a rectifier. However, this may lead to low energy-storage efficiency. Here, we rationally design a charging cycle to maximize energy-storage efficiency by modulating the charge flow in the system, which is demonstrated on a triboelectric nanogenerator by adding a motion-triggered switch. Both theoretical and experimental comparisons show that the designed charging cycle can enhance the charging rate, improve the maximum energy-storage efficiency by up to 50% and promote the saturation voltage by at least a factor of two. This represents a progress to effectively store the energy harvested by nanogenerators with the aim to utilize ambient mechanical energy to drive portable/wearable/implantable electronics.

[1] School of Materials Science and Engineering, Georgia Institute of Technology, Atlanta, Georgia 30332, USA. [2] Beijing Institute of Nanoenergy and Nanosystems, Chinese Academy of Sciences, Beijing 100083, China. * These authors contributed equally to this work. Correspondence and requests for materials should be addressed to Z.L.W. (email: zhong.wang@mse.gatech.edu).

With the rapid development of mobile and portable electronics, more and more efforts have been dedicated to looking for sustainable mobile energy source at the power levels of micro-to-milli watts for these electronics. Currently, the major approach for powering these electronics is by using energy storage units such as batteries[1-3] and capacitors[4,5]. However, the main drawback of these energy storage units is the limited lifetime, so that they cannot drive the electronics sustainably. Alternatively, to drive small electronics sustainably through harvesting ubiquitous ambient small-scale energy, nanogenerators[6] based on piezoelectric, pyroelectric and triboelectric effects have been developed, which are dramatically different from traditional generators. Among nanogenerators, triboelectric nanogenerators (TENG)[7-10] have attracted attention due to their high output and high energy conversion efficiency. Hence, our study here mainly focuses on TENGs. The fundamental working mechanism of a TENG is based on coupling of triboelectrification and electrostatic induction[10-12]. First, at least one pair of triboelectric layers made from materials with distinct electron affinities get into physical contact to create triboelectric charges. Second, as triggered by external mechanical force, the relative motion between the triboelectric layers breaks the balanced electrostatic charge distribution on electrodes. As a result, the potential difference between electrodes is built and free electrons flow through external circuits to achieve new equilibrium. When the triboelectric layers move back, the free electrons flow back to return to the original electrostatic equilibrium. Under periodical mechanical motions such as vibrations[13-15], human walking[16,17] and ocean waves[18,19], pulsed alternating current (AC) output is delivered via the TENG to external circuit. Due to the nature of variable frequency and irregular amplitude of the pulsed AC output, TENGs cannot be directly used to drive most electronic devices. An energy storage unit is required to store the energy harvested by nanogenerators and to provide a regulated and manageable output. Consequently, self-charging power systems[20-22] have been developed by hybridizing a nanogenerator and an energy storage unit, and the latter is charged by the former through a full-wave bridge rectifier.

However, most of the previous research on TENGs mainly focused on output performance under external load resistances[12,17,23-26]; while only a few papers[20,22,27,28] explored the process of using a TENG to charge an energy storage unit, from which we obtained initial understandings on the charging characteristics. From these studies, it has been demonstrated that the charging rate decays quickly after several charging cycles. The saturation voltage, which is the highest achievable voltage of the energy storage unit, is much smaller than the open-circuit voltage of the TENG, resulting in a low energy-storage efficiency regardless of the energy conversion efficiency of the TENG. Therefore, studies are required to further understand the charging process and to improve the energy storage performances for TENGs. According to a recent study on the standards of TENGs, the average output power is mainly determined by the encircled area of its operation cycle in the built-up voltage V – transferred charge Q plot[12]. Therefore, the corresponding operation cycle for charging the energy storage units should be studied by considering the charging process of the TENG to achieve the maximized energy-storage efficiency.

In this work, we first analysed the operation cycle of using a TENG to directly charge a battery/capacitor through a bridge rectifier by our recently proposed V–Q plot[12]. A sliding freestanding-triboelectric-layer (SFT) mode TENG was fabricated to experimentally measure the V–Q plots of the direct charging cycle. Then a rationally designed cycle was proposed to improve the charging rate and the saturation voltage. The voltage of the energy storage unit was demonstrated to be a key parameter for optimizing the stored energy per cycle. As a proof of concept, a motion-triggered switch was added to the fabricated TENG to achieve this designed charging cycle. Furthermore, the enhanced performances including the enhanced charging rate, the improved energy-storage efficiency and the promoted saturation voltage were achieved experimentally. Finally, the self-charging power system operated under this designed charging cycle was used to drive a commercial calculator in the sustainable mode. Our work represents a paradigm shift in strategies and experimental methods to achieve effective energy harvesting and storage by TENGs as well as other nanogenerators.

Results

Theoretical analysis of the direct charging cycle. Conventional integration of a TENG and an energy storage device was achieved through a full-wave bridge rectifier, as shown in the inset of Fig. 1a. We utilize the V–Q plot as the analytical tool for TENG energy storage, as shown in Fig. 1, in which V is the potential difference between the two electrodes, and Q is the amount of charge transferred between the electrodes. Several important parameters of the TENG are defined in Table 1. The parameters for TENG energy storage are given in Supplementary Table 1. We use the most commonly utilized minimum achievable charge reference state[29], so both Q and V at $x=0$ position are set to be 0 initially, where x is the relative displacement between triboelectric layers, which is used to specify the operation of the TENG. The cycle for maximized energy output with infinite load resistance, which has the largest possible energy output per cycle E_m, is also plotted in Fig. 1a,c using the dashed lines (with equations given in Supplementary Note 1)[12]. Here we assume that the charging voltage V_C, which is the voltage of the energy storage unit, does not change significantly during one cycle of the TENG operation (as stated in Supplementary Note 2).

Therefore, starting from $(Q, V) = (0, 0)$, the direct charging cycles of the TENG can be described with the following steps (Fig. 1 a,b). For the first cycle, the first step is to change from status I to II. The initial status I of the system is at $(Q, V) = (0, 0)$ and $x=0$. Because the initial voltage V supplied by the TENG is lower than the charging voltage V_C, all of the diodes in the rectifier are at 'off' state, which makes the external circuit condition close to the open-circuit condition. When the operation of the TENG starts, V begins to increase with x. No charge transfer occurs during this step, until V reaches V_C as shown in status II.

The second step is to change from status II to III. Two of the four diodes in the rectifier are turned on to enable the charging process. Then x continues to increase until $x=x_{max}$, during which the TENG charges the battery/capacitor at $V=V_C$ (See Supplementary Note 3 and Supplementary Fig. 1 for the detailed process) until status III is achieved. Please note that in status III the charges cannot be fully transferred to the other electrode ($Q<Q_{SC,max}$), since $V=V_C$ should be maintained during the charging process in this step.

The third step is to change from status III to IV: x starts to decrease. Since the reversed-direction current is forbidden, V decreases without charge transfers, until $V=-V_C$ as shown in status IV. During this step all of the diodes in the rectifier are off because the net voltage $|V|$ is lower than the threshold voltage at V_C.

The fourth step is to change from status IV to V. The other two diodes in the rectifier are turned on to enable the charging process again. x continues to decrease until $x=0$, as shown in status V, during which the TENG charges the battery/capacitor with $V=-V_C$. Similar to status III, in status V the charge cannot be fully transferred back to the original electrode ($Q>0$), since $V=-V_C$ should be maintained during the charging process in this step. Here the first charging cycle is completed.

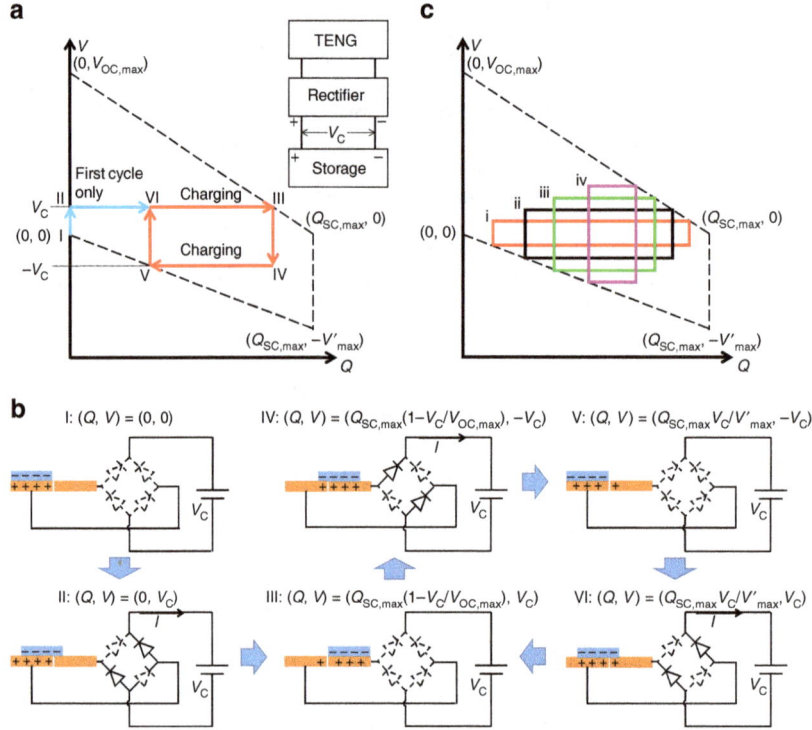

Figure 1 | The V–Q plot and the physical process of the direct charging cycle. (a) The V–Q plot. The inset of **a** shows the circuit diagram of the system. The dashed lines in **a** represent the cycle with the maximized energy output with infinite resistance. **(b)** The schematic diagrams of the physical process. In **b**, the diodes drawn by solid and dashed lines are for the on and off states, respectively. **(c)** The direct charging cycles (the second cycle and thereafter) for different charging voltages (voltage iv > iii > ii > i).

Table 1 | The definitions of important parameters of the TENG.

Symbol	Definitions
$Q_{SC,max}$	The maximum short-circuit transferred charge
$V_{OC,max}$	The maximum open-circuit voltage at $Q = 0$
V'_{max}	The maximum achievable absolute voltage at $Q = Q_{SC,max}$
E_m	The largest possible energy output per cycle

For the second cycle and thereafter (starting from status V), x starts to increase, all of the diodes in the rectifier are off (since $|V| < V_C$) and V increases without charge transfers, until $V = V_C$ as shown in status VI. x continues to increase and the corresponding two diodes are turned on to allow the charging process continue until $x = x_{max}$, the same as status III. The process to complete one cycle is the same as steps 3 and 4 of the first cycle, which comes back to status V. Therefore, the second cycle and thereafter follow the sequence of $V \rightarrow VI \rightarrow III \rightarrow IV \rightarrow V$, which is the steady-state charging cycle.

The corresponding V–Q cycle is plotted in Fig. 1a, with the (Q, V) coordinates corresponding to statuses calculated in Supplementary Note 4. The stored energy per cycle $E_{C,direct}$, which is proportional to the average output power, can be calculated as the encircled area of each cycle[12]:

$$E_{C,direct} = \begin{cases} V_C Q_{SC,max}(2 - 2V_C/V_{OC,max} - V_C/V'_{max}), & \text{for the first cycle} \\ 2V_C Q_{SC,max}(1 - V_C/V_{OC,max} - V_C/V'_{max}), \\ \qquad \text{for the second cycle and thereafter} \end{cases} \quad (1)$$

Since the charging of an energy storage device requires many cycles of operation of the TENG, we will only consider the equation for the steady-state cycles. The amount of charge ($Q_{C,direct}$) flowing into the energy storage unit per cycle can be

calculated as $E_{C,direct}/V_C$, which equals to the total length of the sides that are parallel to the Q axis in the V–Q plot. The direct charging cycles corresponding to different charging voltages V_C are plotted in Fig. 1c, which shows the decrease of $Q_{C,direct}$ due to the increase of V_C (which is further discussed in Supplementary Note 5).

From equation (1), with varied V_C, we can derive that the maximum value of $E_{C,direct}$ is achieved when $V_C = \frac{1}{2}\frac{V_{OC,max}V'_{max}}{V_{OC,max} + V'_{max}}$, which is:

$$E_{C,direct,max} = \frac{Q_{SC,max}}{2}\frac{V_{OC,max}V'_{max}}{V_{OC,max} + V'_{max}} \quad (2)$$

We define the energy storage efficiency η as the percentage of the stored energy per cycle in the largest possible energy output per cycle E_m. Therefore, for the direct charging cycle, the maximum energy-storage efficiency is:

$$\eta_{direct} = \frac{E_{C,direct,max}}{E_m} = \frac{V_{OC,max}V'_{max}}{(V_{OC,max} + V'_{max})^2} \leq 25\% \quad (3)$$

And the largest possible V_C to operate the direct charging cycle is at the saturation voltage:

$$V_{Sat,direct} = \frac{V_{OC,max}V'_{max}}{V_{OC,max} + V'_{max}} \quad (4)$$

Experimental demonstration of the direct charging cycle. A SFT mode TENG is fabricated to measure the direct charging cycle experimentally. The TENG is schematically illustrated in the inset of Fig. 2a. The top static part of the TENG is made by attaching two parallel aluminium (Al) electrodes on the bottom surface of an acrylic board. The moving part of the TENG is made by attaching a piece of fluorinated ethylene propylene (FEP) film on the top surface of another acrylic board. The FEP

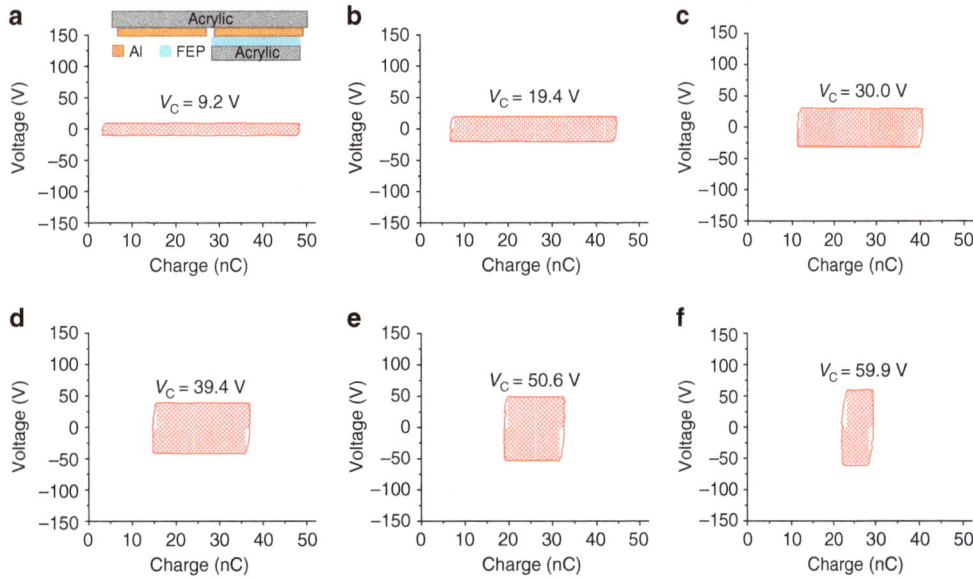

Figure 2 | The V–Q plots of the direct charging cycle for a commercial capacitor with charging voltage V_C increased from 9.2 V to 59.9 V. For these plots, V_C are (**a**) 9.2 V, (**b**) 19.4 V, (**c**) 30.0 V, (**d**) 39.4 V, (**e**) 50.6 V and (**f**) 59.9 V. The inset of **a** shows the TENG structure. The shaded areas are the stored energy per cycle $E_{C,direct}$.

film and Al electrodes are the triboelectric layers. The fabrication process is described in detail in the Methods section. In this structure, $V_{OC,max} = V'_{max}$ since the capacitance between the two electrodes is kept identical[29,30]. The measured cycle for maximized energy output with infinite load resistance is shown in Supplementary Fig. 2b, where $V_{OC,max} = V'_{max} \approx 140$ V, and $Q_{SC,max} \approx 50$ nC. The circuit diagram for this measurement is shown in Supplementary Fig. 2a. A commercial capacitor (0.73 µF) and fabricated lithium-ion batteries connected in series are used as the energy storage devices, respectively. For the capacitor, the charging voltage V_C increases gradually during the charging process, and the V–Q curves of direct charging cycle with different V_C are plotted in Fig. 2. For the batteries, each battery can supply a voltage of about 3–4 V, and different number of batteries are connected to achieve different V_C. The fabrication method of the batteries and the measurement set-up are described in the Methods section and Supplementary Fig. 3, respectively. The V–Q curves of the direct charging cycle for batteries are plotted in Supplementary Fig. 4. From these V–Q curves, we notice that all of the curves in the direct charging cycle for both batteries and the capacitor have a rectangular shape. The total length of the two sides parallel to the Q axis can be directly used as $Q_{C,direct}$, as demonstrated in Supplementary Note 6 and Supplementary Fig. 5. $Q_{C,direct}$ decreases with increasing V_C, and the saturation voltage $V_{Sat,direct}$ is measured at almost 70 V, which are consistent with the calculated results from equations (1) and (4).

Theoretical analysis of the designed charging cycle. From the equations and experimental results presented above, we can make three observations on the direct charging cycle. First, the charge flowing into the energy storage unit per cycle $Q_{C,direct}$, which is proportional to the charging rate, quickly decreases with the increased V_C. Second, the maximum value of the energy storage efficiency η is only 25%. Finally, the saturation voltage $V_{Sat,direct}$ is much smaller than $V_{OC,max}$ and V'_{max}. To improve these charging performances, we rationally design a charging cycle. A switch is added in parallel to the TENG, which is controlled to instantaneously turn on only at $x=0$ and $x=x_{max}$, and turn off at all the other x. Starting from $(Q, V) = (0, 0)$, the process of the designed charging cycle can be described as follows (Fig. 3a,b):

Steps 1 and 2 (status I–III) are the same as those in the first cycle of the direct charging cycle. Similarly, in status III the charge cannot be fully transferred to the other electrode $(Q < Q_{SC,max})$.

Step 3 (status III–IV): while in status III, x reaches x_{max}, and then the switch is turned on. The charge transfers in short-circuit condition to achieve $(Q, V) = (Q_{SC,max}, 0)$, which is status IV.

Step 4 (status IV–V): the switch is turned off as x begins to decrease; at the same time all of the diodes in the rectifier are off since $|V| < V_C$. Therefore, V decreases without any charge transfers until it reaches $-V_C$, which is status V.

Step 5 (status V–VI): the second half-cycle of the charging starts as x continues decreasing, until $x=0$ at status VI. Also, in status VI the charge cannot be fully transferred back to the original electrode $(Q > 0)$.

Step 6 (status VI to I): the switch is turned on again to make (Q, V) return to $(0, 0)$, that is, status I.

Therefore, the repeated process for the designed charging cycle is through I→II→III→IV→V→VI→I. As concluded from steps 3 and 6, the function of the switch is to enable the complete charge transfer by creating an instantaneous short-circuit condition. Consequently, in the next half-cycle, there are more charges available to charge the energy storage unit, which is the key to enhance the charging performances by using the designed charging cycle.

By marking the coordinates of corresponding statuses (as calculated in Supplementary Note 4) in the V–Q plot (Fig. 3a), we observe that there are three regions of energy in the encircled area in the designed charging cycle. Area 1 represents the energy that can be stored in both the direct and the designed charging cycles; area 3 represents the energy released through the switch; and the energy of area 2 is the part that can only be stored in the designed charging cycle. Thus, the stored energy per cycle in the designed charging cycle should only include energies in areas 1 and 2. During steps 1–3 and steps 4–6, the energies stored are derived as $V_C Q_{SC,max}(1 - V_C/V_{OC,max})$ and $V_C Q_{SC,max}(1 - V_C/V'_{max})$, respectively. Therefore, the total stored energy per cycle is:

$$E_{C,designed} = V_C Q_{SC,max}(2 - V_C/V_{OC,max} - V_C/V'_{max})$$

Note that this equation is only valid for $V_C \leq \min \{V_{OC,max},$

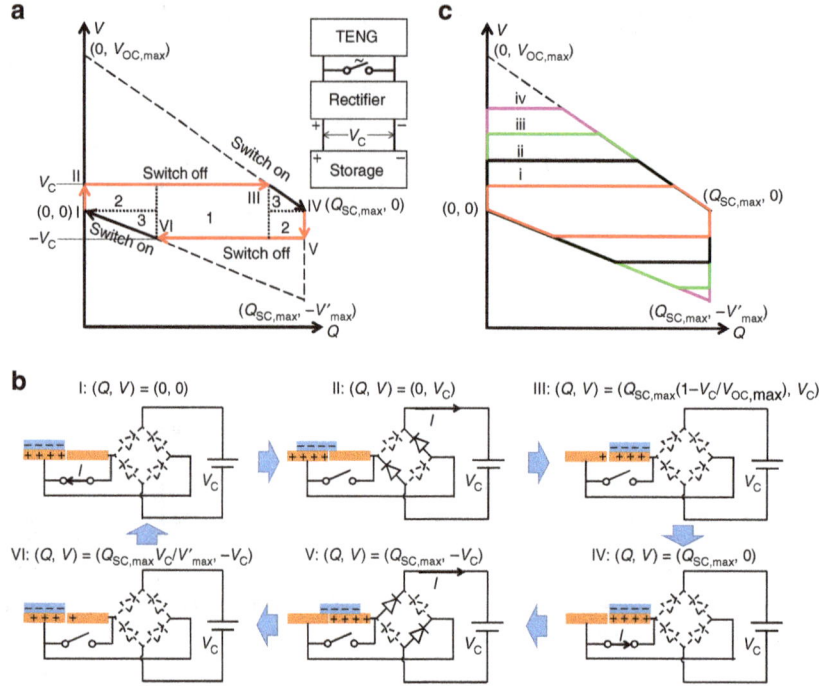

Figure 3 | The V–Q plot and the physical process of the rationally designed charging cycle. (a) The V–Q plot. The inset of **a** shows the circuit for the designed charging cycle. The dashed lines in **a** represent the cycle of maximized energy output with infinite resistance. (**b**) The schematic diagrams of the physical process. In **b**, the diodes drawn by solid and dashed lines are for the on and off states, respectively. (**c**) The designed charging cycle for different charging voltages, with charging voltage intensity iv > iii > ii > i.

V'_{max}. Considering the symmetric roles of $V_{OC,max}$ and V'_{max} (as explained in Supplementary Note 7), we can simply assume $V_{OC,max} \geq V'_{max}$. If $V'_{max} < V_C \leq V_{OC,max}$, since during step 4 the voltage cannot achieve $-V_C$, there is no energy stored during steps 4–6. Then the equation becomes:

$$E_{C,designed} = V_C Q_{SC,max}\left(1 - V_C/V_{OC,max}\right)$$

Therefore:

$$E_{C,designed} = \begin{cases} V_C Q_{SC,max}\left(2 - V_C/V_{OC,max} - V_C/V'_{max}\right), V_C \leq V'_{max} \leq V_{OC,max} \\ V_C Q_{SC,max}\left(1 - V_C/V_{OC,max}\right), V'_{max} < V_C \leq V_{OC,max} \end{cases}$$

$$(5)$$

Similarly, the other characteristic parameters of the designed charging cycle can be calculated as follows:

The maximum stored energy per cycle (achieved at $V_C = \frac{V_{OC,max}V'_{max}}{V_{OC,max}+V'_{max}}$ or $V_C = \frac{V_{OC,max}}{2}$):

$$E_{C,designed,max} = \max\left\{Q_{SC,max}\frac{V_{OC,max}V'_{max}}{V_{OC,max}+V'_{max}}, \frac{Q_{SC,max}V_{OC,max}}{4}\right\}$$

$$(6)$$

The maximum energy storage efficiency:

$$\eta_{designed} = \frac{E_{C,designed,max}}{E_m} = \max\left\{\frac{2V_{OC,max}V'_{max}}{\left(V_{OC,max}+V'_{max}\right)^2}, \frac{V_{OC,max}}{2\left(V_{OC,max}+V'_{max}\right)}\right\} \leq 50\%$$

$$(7)$$

And the saturation voltage is:

$$V_{Sat,designed} = \max\left\{V_{OC,max}, V'_{max}\right\}$$

$$(8)$$

As derived by these equations of the designed charging cycle: the amount of charge flowing to the battery/capacitor per cycle of the designed charging cycle ($Q_{C,designed}$) is larger than that of the direct charging cycle ($Q_{C,direct}$), corresponding to the same V_C (Supplementary Note 5); the maximum energy storage efficiency

has been increased up to 50%; and the saturation voltage $V_{Sat,designed}$ is the same as the larger one in $V_{OC,max}$ and V'_{max}, which is at least twice the value of $V_{Sat,direct}$. Therefore, this designed cycle represents an effective strategy to greatly enhance the performances of using TENG to charge an energy storage unit.

Experimental demonstration of the designed charging cycle. We demonstrate this designed charging cycle experimentally by adding a motion-triggered switch on the fabricated SFT mode TENG, which is similar to the design of the pulsed nanogenerator as reported previously[31], as shown in Fig. 4. The top static part is kept the same (Fig. 4a). Two parallel titanium (Ti) ribbons are attached on the acrylic board of the moving part (Fig. 4b). These two Ti ribbons are connected to the two Al electrodes through copper wires, respectively. Two Ti bars (Fig. 4c) are fixed in the positions of $x=0$ and $x=x_{max}$. The configuration of the integrated structure is shown in Fig. 4d and the detailed fabrication process is described in the Methods section. When $x=0$ or $x=x_{max}$, both of the two Ti ribbons contact with one of the Ti bars, and then the two electrodes are connected, which represents the switch on. Otherwise, the electrodes are not connected, which represents the switch off.

The V–Q curves of these TENGs operated in the designed charging cycle are measured. Figure 5 presents the V–Q plots of the designed charging cycle for the same capacitor used in the experiments for the direct charging cycle, and the plots for the batteries in series are shown in Supplementary Fig. 6. Similarly, the total length of the two sides that are parallel to Q axis in V–Q plot can be used directly as $Q_{C,designed}$. The stored energy per cycle $E_{C,designed}$ is marked as the shaded area in all the plots in Fig. 5, in which the red and grey shaded areas represent areas 1 and 2 in Fig. 3a, respectively. Compared with the direct charging cycle, the designed charging cycle can store more energy per cycle for the

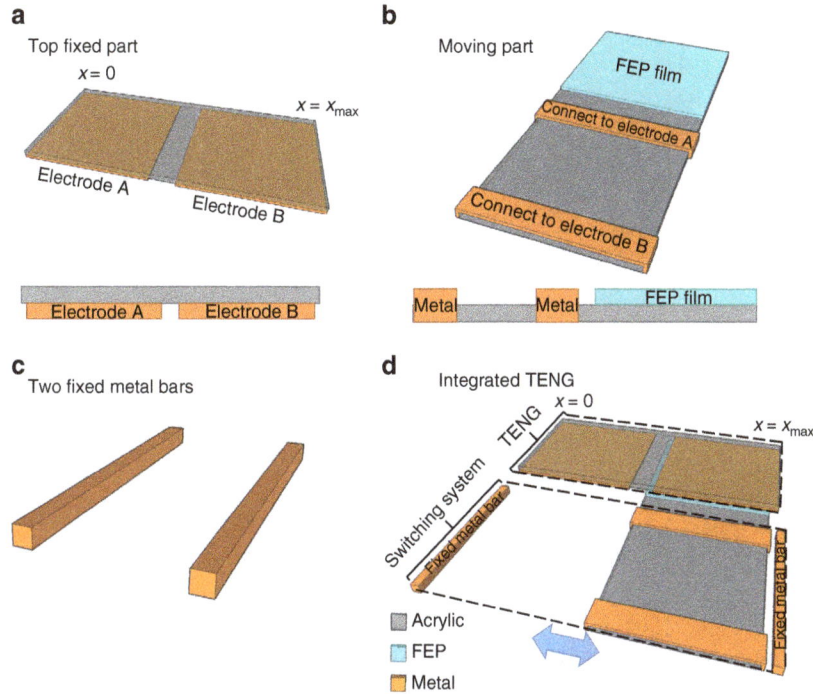

Figure 4 | The TENG with the motion-triggered switch fabricated for the designed charging cycle. This TENG is composed of **a** the static part, (**b**) the moving part and (**c**) the two fixed metal bars. (**d**) The TENG as integrated by the three parts in **a-c**. The insets in **a,b** show the front view of the static part and the right-side view of the moving part.

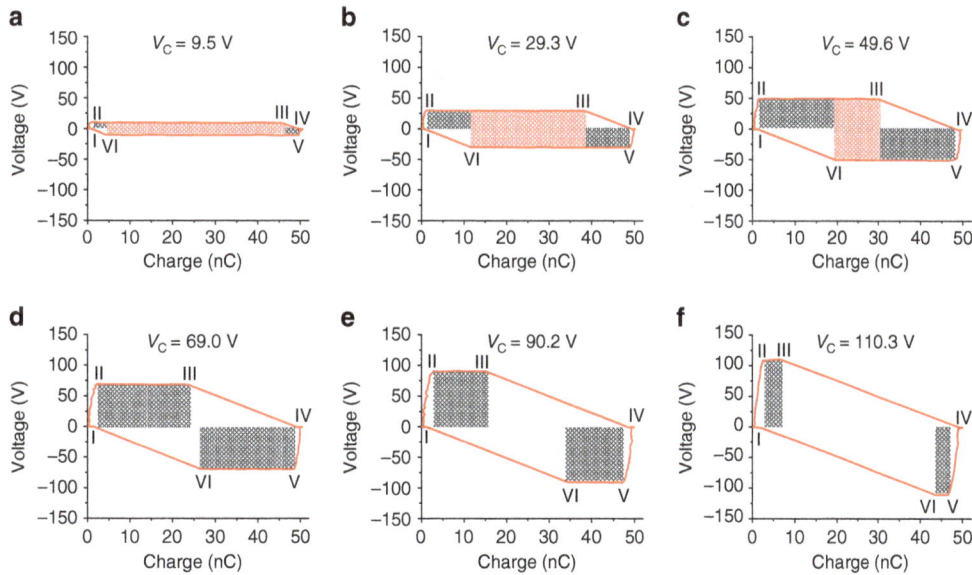

Figure 5 | The V–Q plots of the designed charging cycle for a commercial capacitor with charging voltage increased from 9.5 V to 110.3 V. For these plots, V_C are (**a**) 9.5 V, (**b**) 29.3 V, (**c**) 49.6 V, (**d**) 69.0 V, (**e**) 90.2 V and (**f**) 110.3 V. The six statuses in the designed charging cycle are marked in the plots. The shaded areas are the stored energy per cycle by the designed charging cycle $E_{C,designed}$, in which the red shaded areas are the part of the energy that can also be stored in the direct charging cycle with the same V_C, while the grey shaded areas are the 'extra' energy that can only be stored in the designed charging cycle.

same V_C, and V_{Sat} can be enhanced to be close to 140 V. To further understand the V–Q plots in Fig. 5d,f when V_C is larger than $V_{Sat,direct}$ but smaller than $V_{Sat,designed}$, the corresponding charging cycles are discussed in Supplementary Note 8, and the theoretically derived V–Q plots are displayed in Supplementary Fig. 7. All of these experimental results are consistent with our theoretical derivations.

Quantitative comparisons between the charging cycles. To further reveal the advantages of the designed charging cycle in energy storage and verify our theoretical derivations above, quantitative comparisons between experimental results of the direct and the designed charging cycles are performed, as shown in Fig. 6. Here the plots for the capacitor are used since it can easily achieve different V_C during the charging process.

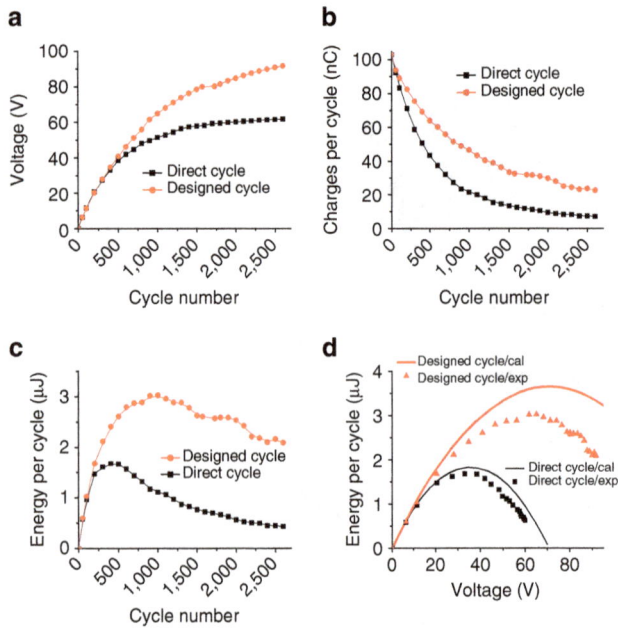

Figure 6 | Quantitative comparisons between the charging cycles.
(a) The charging voltage V_C, (b) the amount of charge Q_C flowing to the capacitor per cycle and (c) the stored energy per cycle E_C changes versus the number of the charging cycles for both the direct and the designed charging cycles. (d) The experimental (dots) and calculated (lines) results (as calculated from equations (1) and (5)) of the stored energy per cycle E_C versus the charging voltage V_C for both the direct and the designed charging cycles.

Figure 6a–c show the changes of the charging voltage V_C, the charge Q_C flowing to the capacitor per cycle and the stored energy per cycle E_C versus the number of the charging cycles, respectively. For both types of cycles, with the increase of the number of the charging cycles, V_C increases since more and more charges flow into the capacitor; Q_C decreases due to the increases of V_C, as discussed in Supplementary Note 5. Compared with the direct charging cycle, V_C increases faster, Q_C decreases slower and hence E_C is significantly promoted in the designed charging cycle. The reason of slower decay of Q_C in the designed charging cycle is during the switch-on operations (statuses III to IV and VI to I), the charges are fully transferred to $Q_{SC,max}$ or 0, so that in the next half-cycle there are more charges available to flow into the capacitor (as further discussed in Supplementary Note 5). We also notice that the higher saturation voltage V_{Sat} can be achieved in the designed charging cycle, since for the direct charging cycle there is no energy stored once V_C approaches $V_{Sat,direct}$, as discussed in Supplementary Note 8. The curves of E_C versus V_C for both the direct and the designed charging cycles are also plotted in Fig. 6d for both experimental and calculated results (as calculated from equations (1) and (5)). From this we can conclude that, the designed charging cycle can supply a larger E_C under the same V_C, and a higher V_{Sat} can be achieved.

Application of the designed charging cycle. To further demonstrate the promoted energy storage in the designed charging cycle, we use a TENG working in the designed charging cycle to power a commercial calculator (working voltage 1.5 V) in sustainable mode using the circuit in Fig. 7a. To provide a constant voltage to the calculator, the average input power from the TENG needs to equal to the power consumption in the output circuit[20]. The capacitor is charged to let $V_C = 19.2$ V initially. As indicated by the dashed frame in Fig. 7a, the output circuit, which is connected to the capacitor in parallel, includes a power

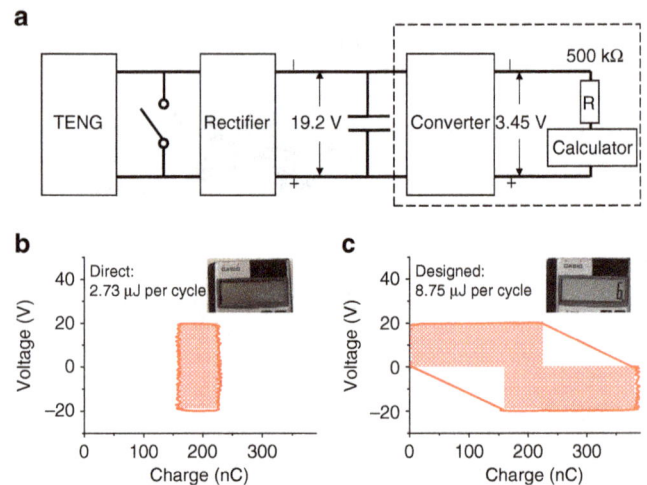

Figure 7 | Demonstration of a calculator sustainably powered by the self-charging power system through the designed charging cycle.
(a) Schematic diagram of the operation circuit; (b,c) the V–Q plots of the direct and the designed charging cycle for the TENG. The shaded areas represent the energy stored per cycle for both cycles. The power required by units in dashed frame in a is satisfied by that supplied by TENG through the designed charging cycle. The inset pictures show the calculator's display screen as charged by the direct and designed charging cycles, respectively.

converter (from Linear Technology) to regulate the voltage to 3.45 V, the calculator and a divider resistor of 500 kΩ to provide the calculator a voltage of ~1.5 V. The measured current required to drive the output circuits is ~1.5 μA (Supplementary Fig. 8), which indicates that an average output power of ~28.8 μW needs to be provided by the TENG. If we set the working period as 0.3 s for the TENG, the required stored energy per cycle is ~8.65 μJ. To achieve that, we fabricate a TENG that can generate ~780 nC per cycle in maximum ($Q_{SC,max} \approx 390$ nC). The measured V–Q plots of both the direct and the designed charging cycles at $V_C = 19.2$ V are shown in Fig. 7b,c. The measured E_C for the direct charging cycle is 2.73 μJ, which is much smaller than the required energy per cycle of 8.65 μJ. Consequently, the capacitor's energy operated in the direct charging cycle is consumed quickly, and the calculator turned off quickly, as shown in the inset of Fig. 7b. As a comparison, the measured E_C for the designed charging cycle is 8.75 μJ, which is larger than the required energy per cycle of 8.65 μJ. Therefore, as shown in the insets of Fig. 7c, with the same $V_C = 19.2$ V, the TENG operating in the designed charging cycle can drive the calculator sustainably. The calculator working in sustainable mode powered by the self-charging power system through the designed charging cycle is also shown in Supplementary Movie 1.

Discussion
In summary, a rationally designed route for more effective energy storage of the random-pulsed output power generated by TENGs was developed theoretically and experimentally. The step-by-step charging processes of both the direct and the designed charging cycles were illustrated by V–Q plots, which revealed the advantages of the designed charging cycle in enhanced charging rate, improved energy-storage efficiency (up to 50%), and promoted saturation voltage (by at least a factor of two). As a proof of concept, we designed a TENG with motion-triggered switch to achieve the designed charging cycle. The results from both approaches were compared quantitatively, and the results confirmed the advantages of the designed charging cycle, as

predicted by theoretical derivations. A commercial calculator was demonstrated to be sustainably powered by the TENG-based self-charging power system through the designed charging cycle. This designed charging cycle provides a route to enhance the energy harvesting and storage for TENGs as well as other nanogenerators, which represents a solid progress towards effectively utilizing ambient mechanical energy as a sustainable power source for electronics.

Methods

Fabrication and operation of the TENG. The static part of the TENG was fabricated by attaching Al foils on a 10.1×7 cm acrylic board to form two rectangular shape electrodes each with a size of 5×7 cm and the spacing of 1 mm. A 5×7-cm 50-μm FEP film attached along one side of a 5×15-cm acrylic board was used as the moving part. For the designed charging cycle, two Ti metal ribbons (1 cm in width for each, 5 cm in spacing) were attached on the other side of the acrylic board in parallel in the moving part, which were connected to two electrodes, respectively. Two Ti bars (7 cm in length) were also prepared. To operate the TENG in the designed charging cycle, the moving part was mounted on a linear motor and the static part and the two Ti bars were mounted on 3 three-dimensional stages, respectively, and the FEP surface and one Al electrode were placed face to face, as shown in Fig. 4d. The linear motor was controlled to move periodically between two Al electrodes with a displacement of 5.1 cm. For the direct charging cycle, the two Ti bars were removed. The TENG used to demonstrate the sustainable mode was fabricated in the same procedure with twice of all the sizes. The voltage and transferred charge were measured by two Keithley 6,514 system electrometers simultaneously. The measurement circuits for V–Q plots are illustrated in Supplementary Fig. 3.

Fabrication of the lithium-ion battery. Lithium-ion rechargeable coin cell batteries were fabricated by using $LiCoO_2$/carbon black/binder mixture on Al foil (1 cm in diameter) as the anode, polyethylene (PE) as separator (2 cm in diameter), the graphite/carbon black/binder mixture on Al foil (1.5 cm in diameter) as the cathode. The electrolyte (1 M $LiPF_6$ in 1:1:1 ethylene carbonate/dimethyl carbonate/diethyl carbonate) was injected inside between the anode and cathode before the coin cell was pressed firmly. The charging-discharging curves of these batteries were tested as shown in Supplementary Fig. 9, which exhibits a plateau voltage of ~ 3.8 V. In our experiments, each battery can supply voltage of ~ 3–4 V.

References

1. Dunn, B., Kamath, H. & Tarascon, J. M. Electrical energy storage for the grid: a battery of choices. *Science* **334**, 928–935 (2011).
2. Huskinson, B. *et al.* A metal-free organic-inorganic aqueous flow battery. *Nature* **505**, 195–198 (2014).
3. Peng, Z. Q., Freunberger, S. A., Chen, Y. H. & Bruce, P. G. A reversible and higher-rate Li-O$_2$ battery. *Science* **337**, 563–566 (2012).
4. El-Kady, M. F., Strong, V., Dubin, S. & Kaner, R. B. Laser scribing of high-performance and flexible graphene-based electrochemical capacitors. *Science* **335**, 1326–1330 (2012).
5. Zhu, Y. W. *et al.* Carbon-based supercapacitors produced by activation of graphene. *Science* **332**, 1537–1541 (2011).
6. Wang, Z. L. & Song, J. Piezoelectric nanogenerators based on zinc oxide nanowire arrays. *Science* **312**, 242–246 (2006).
7. Fan, F.-R., Tian, Z.-Q. & Lin Wang, Z. Flexible triboelectric generator. *Nano Energy* **1**, 328–334 (2012).
8. Zhu, G. *et al.* Linear-grating triboelectric generator based on sliding electrification. *Nano. Lett.* **13**, 2282–2289 (2013).
9. Wang, Z. L., Chen, J. & Lin, L. Progress in triboelectric nanogenerators as a new energy technology and self-powered sensors. *Energy Environ. Sci.* **8**, 2250–2282 (2015).
10. Wang, Z. L. Triboelectric nanogenerators as new energy technology and self-powered sensors – Principles, problems and perspectives. *Faraday Discuss.* **176**, 447–458 (2014).
11. Wang, Z. L. Triboelectric nanogenerators as new energy technology for self-powered systems and as active mechanical and chemical sensors. *ACS Nano* **7**, 9533–9557 (2013).
12. Zi, Y. *et al.* Standards and figure-of-merits for quantifying the performance of triboelectric nanogenerators. *Nat. Commun.* **6**, 8376 (2015).
13. Chen, J. *et al.* Harmonic-resonator-based triboelectric nanogenerator as a sustainable power source and a self-powered active vibration sensor. *Adv. Mater.* **25**, 6094–6099 (2013).
14. Yang, J. *et al.* Triboelectrification-based organic film nanogenerator for acoustic energy harvesting and self-powered active acoustic sensing. *ACS Nano* **8**, 2649–2657 (2014).
15. Fan, X. *et al.* Ultrathin, rollable, paper-based triboelectric nanogenerator for acoustic energy harvesting and self-powered sound recording. *ACS Nano* **9**, 4236–4243 (2015).
16. Yang, W. *et al.* Harvesting energy from the natural vibration of human walking. *ACS Nano* **7**, 11317–11324 (2013).
17. Xie, Y. *et al.* Grating-structured freestanding triboelectric-layer nanogenerator for harvesting mechanical energy at 85% total conversion efficiency. *Adv. Mater.* **26**, 6599–6607 (2014).
18. Chen, J. *et al.* Networks of triboelectric nanogenerators for harvesting water wave energy: a potential approach toward blue energy. *ACS Nano* **9**, 3324–3331 (2015).
19. Wang, X. *et al.* Triboelectric Nanogenerator based on fully enclosed rolling spherical structure for harvesting low-frequency water wave energy. *Adv. Energy Mater.* **5**, 15011467 (2015).
20. Wang, S. *et al.* Motion charged battery as sustainable flexible-power-unit. *ACS Nano* **7**, 11263–11271 (2013).
21. Wang, J. *et al.* A flexible fiber-based supercapacitor–triboelectric-nanogenerator power system for wearable electronics. *Adv. Mater.* **27**, 4830–4836 (2015).
22. Luo, J. *et al.* Integration of micro-supercapacitors with triboelectric nanogenerators for a flexible self-charging power unit. *Nano Res.* **8**, 3934–3943 (2015).
23. Zhou, Y. S. *et al.* Manipulating nanoscale contact electrification by an applied electric field. *Nano. Lett.* **14**, 1567–1572 (2014).
24. Zi, Y. *et al.* Triboelectric–pyroelectric–piezoelectric hybrid cell for high-efficiency energy-harvesting and self-powered sensing. *Adv. Mater.* **27**, 2340–2347 (2015).
25. Han, M. *et al.* Magnetic-assisted triboelectric nanogenerators as self-powered visualized omnidirectional tilt sensing system. *Sci. Rep.* **4**, 4811 (2014).
26. Seung, W. *et al.* Nanopatterned textile-based wearable triboelectric nanogenerator. *ACS Nano* **9**, 3501–3509 (2015).
27. Niu, S. *et al.* Optimization of triboelectric nanogenerator charging systems for efficient energy harvesting and storage. *IEEE Trans. Electron Devices* **62**, 641–647 (2015).
28. Pu, X. *et al.* Efficient charging of Li-ion batteries with pulsed output current of triboelectric nanogenerators. *Adv. Sci.* **3**, 1500255 (2015).
29. Niu, S. *et al.* Theory of freestanding triboelectric-layer-based nanogenerators. *Nano Energy* **12**, 760–774 (2015).
30. Wang, S., Xie, Y., Niu, S., Lin, L. & Wang, Z. L. Freestanding triboelectric-layer-based nanogenerators for harvesting energy from a moving object or human motion in contact and non-contact modes. *Adv. Mater.* **26**, 2818–2824 (2014).
31. Cheng, G., Lin, Z.-H., Lin, L., Du, Z.-l. & Wang, Z. L. Pulsed nanogenerator with huge instantaneous output power density. *ACS Nano* **7**, 7383–7391 (2013).

Acknowledgements

Y.Z., J.W. and S.W. contributed equally to this work. This research was supported by the National Science Foundation (DMR-1505319), the Hightower Chair foundation and the 'thousands talents' programme for pioneer researcher and his innovation team, China, National Natural Science Foundation of China (Grant No. 51432005). We would like to thank Dr Ken C. Pradel for his help in revising the manuscript.

Author contributions

Y.Z., J.W., S.W. and Z.L.W. conceived the idea, discussed the data and prepared the manuscript. Y.Z. and S.W. did the theoretical derivations and calculations on the direct charging cycle and the designed charging cycle. Y.Z. and J.W. did the experiments and analysed the data. S.L., Z.W. and H.G. helped the experiments. All the authors helped revise the manuscript.

Additional information

12

In-line three-dimensional holography of nanocrystalline objects at atomic resolution

F.-R. Chen[1], D. Van Dyck[2] & C. Kisielowski[3]

Resolution and sensitivity of the latest generation aberration-corrected transmission electron microscopes allow the vast majority of single atoms to be imaged with sub-Ångstrom resolution and their locations determined in an image plane with a precision that exceeds the 1.9-pm wavelength of 300 kV electrons. Such unprecedented performance allows expansion of electron microscopic investigations with atomic resolution into the third dimension. Here we report a general tomographic method to recover the three-dimensional shape of a crystalline particle from high-resolution images of a single projection without the need for sample rotation. The method is compatible with low dose rate electron microscopy, which improves on signal quality, while minimizing electron beam-induced structure modifications even for small particles or surfaces. We apply it to germanium, gold and magnesium oxide particles, and achieve a depth resolution of 1-2 Å, which is smaller than inter-atomic distances.

[1] Department of Engineering and System Science, National Tsing-Hua University, 101 Kuang-Fu Road, Hsin Chu 300, Taiwan. [2] EMAT, Department of Physics, University of Antwerp, 2020 Antwerpen, Belgium. [3] Lawrence Berkeley National Laboratory, The Molecular Foundry and Joint Center for Artificial Photosynthesis, One Cyclotron Road, Berkeley California 94720 USA. Correspondence and requests for materials should be addressed to F.R.C. (email: fchen1@me.com)

n the late 1950s Richard Feynman pointed out[1] that 'It would be very easy to make an analysis of any complicated chemical substance; all one would have to do would be to look at it and see where the atoms are.' In principle, the latest generation aberration-corrected transmission electron mocroscopes can achieve this goal[2–4] but for a variety of reasons one is still far away from a reliable method that would meet Feynman's challenge of extracting the three-dimensional (3D) position of all the atoms in an object, to understand its physical and chemical properties[5–13]. A most noticeable bottleneck is the large accumulated electron dose required to produce tilt series of atomic resolution images, because electron dose rates are commonly chosen large (10^4–$10^5\,\text{e\AA}^{-2}\,\text{s}^{-1}$) to achieve a needed resolution around 1 Å and single atom sensitivity. Any such single image can exhibit uncontrolled electron beam-induced surfaces alterations or even bulk modifications, in particular if particles are small[14–17]. Therefore, only a few favourable cases allowed for an extraction of atom positions in the beam (z) direction with high precision. They included the study of a graphene double layer[18] that can tolerate extraordinary large electron dose rates, tomography of embedded nanocrystals where a sacrificial matrix protects the nanoparticles[19] or a 3D structure determination by comparing experimental images with theoretical expectations that typically include assumptions of debatable validity[20,21]. On the other hand, small crystalline particles are known to exhibit drastically altered bulk or surface properties such as their catalytic activity. Consequently, there is a strong need for a tomographic technique with atomic resolution that can maintain the pristine structure of small objects, which requires imaging with small electron dose rates.

In this study we present a self-consistent approach to recover the 3D atomic structure of nanocrystalline particles from single projections by exploiting the dynamic nature of electron scattering and pursuing a quantitative interpretation of the electron exit wave reconstructed from focal series of high-resolution images. The exit wave contains not just the usual intensity, but the entire field information, amplitude and phase, which is the same as ' holography'. In particular, this reconstruction method allows capturing images with choosable dose rates that can be adjusted to maintain structural integrity during the imaging process without compromising spatial atomic resolution and single atom sensitivity[22–26]. Currently, exit waves can be reconstructed from images captured with dose rates reduced to a level of roughly 1 atto-Ampere per Å ($6\,\text{e\AA}^{-2}\,\text{s}^{-1}$). Moreover, it is pointed out that the reconstructed electron wave in an image plane is not identical to the wave at the exit surface of the crystal, because crystals exhibit a shape and surfaces are often not flat at the atomic scale but exhibit roughness. Therefore, focus values with respect to a common image plane change locally.

If crystalline objects are imaged along a zone axis orientation in an electron microscope, the incident electrons are trapped in the strong electrostatic potential of the atomic columns in beam direction. This trapping of electrons is commonly described by electron channelling[27,28], which has all ingredients for a full 3D quantification of image contrast, as the scattering process critically depends on column length and its chemical composition but only weakly on electron channelling in neighbouring columns as long as samples are thin and the column spacing exceeds $\sim 1\,\text{Å}$ in the image plane. Therefore, the exit wave of a crystalline object in zone orientation can be analysed column by column. Within each column, successive atoms are aligned with an equidistant spacing set by the crystal lattice and they act as lenses that focus and defocus the propagating electron wave periodically with increasing column length. Thus, the electron wave inside the column oscillates periodically with sample thickness and contains depth information. Element-specific contrast changes can be observed in high-performance microscopes, because atoms are discrete objects with a characteristic scattering power yielding element-specific phase changes. Our procedure to extract quantitative information from exit waves is described in the Methods section and partly in previous publications[18,23]. It addresses the challenge how to extract for every column its mass and its distance to a common reference plane with single-atom sensitivity and a precision in beam direction that exceeds interatomic distances. Once focus and mass of all atomic columns are deduced from a single-projected exit wave, the 3D structure of the crystalline particle can be reconstructed. Conveniently, atomicity provides an internal and self-consistent calibration standard that can be used to recover any sample shape from only one projection at truly atomic resolution. In addition, it is shown for exit waves recorded with the TEAM 0.5 microscope[24] how a chosen electron dose rate affects the absolute phase values for the detection single atoms. In summary, we show analysing electron channelling along atom columns allows for a full 3D quantification of the atomic structure as long as the atom columns are homogeneously occupied. Concerning defects, our method is readily available to investigate edge-on dislocations or planar defects such as twin boundaries in 3D, which will be demonstrated later in an experimental exit wave of a gold particle. A point defect such as a vacancy can be detected as an atomic step in the surface with respect to the neighbouring columns if single projections are used.

Results

Mass and focus circles for 3D reconstruction. The exit wave $\Psi_e(\mathbf{r},\mathbf{z})$, at a particular image plane, can be expressed analytically as[23]:

$$\Psi_e(\mathbf{r},t)=\Psi(\mathbf{r},0)+\Phi_{1s}(\mathbf{r})\left(e^{-iEt}-1\right)\left(1e^{-i\alpha\Delta f}\right) \qquad (1)$$

where t is the mass thickness of the sample, $\Psi(\mathbf{r},0)$ is the incident wave, $\Phi_{1s}(\mathbf{r})$ is the 1s eigenstate of the projected electrostatic potential of the atom column with eigenenergy E. α is a constant and Δf is the focus difference given by the distance of the image plane to the exact exit surface defined by the last atom in a column. The exit wave is complex and can be represented in an Argand plot. Figure 1a shows the Argand plot of pixels from the centre of atom columns. It is clear that the two factors ($e^{-iEt}-1$) and ($1-e^{-i\alpha\Delta f}$) from equation describe two circles called 'mass (Et) circle' (black circle) and 'defocus ($\alpha\Delta f$) circle' (red circles), respectively. It is seen that information about the column mass and its local focus are orthogonal. If the sensitivity of a microscope suffices to isolate the contrast contribution from scattering at single atoms, the 'mass' values will be discrete and their regular spacing will give an ultimate mass calibration. Similarly, discrete focus values must be detected if the spatial resolution in beam direction is smaller than the periodic distance between successive atoms in a column.

The Argand plot (Fig. 1a) explains in a natural manner how shape changes can be separated from column thicknesses (masses) by applying local wave propagation. The blue points on the black mass circle represent columns of different thickness characterized by the angle θ' between any blue point and the incident (or vacuum) wave at (1,0). If the column mass increases, θ' increases proportional to the projected thickness (mass) of the column. To convert the column mass θ' into the number of atoms in a column, the phase θ' must be divided by the phase change per single atom θ, which roughly increases with the atomic number $Z^{2/3}$ (ref. 26) as used in the Methods and Table 1. Experimentally, however, the scattering power of atoms measured by θ is not a constant, because it is modulated by thermal vibrations and by electron beam-induced sample excitations[26]. Nevertheless, the complex wave values located directly on the mass circle represent columns where the exit surface of the sample coincides with an image plane so that there is no focus difference between them. However, shape or surface roughness at the bottom of a sample must locally create finite focus deviations, because the altered geometry moves away the exit wave from the image plane along the red defocus circle. The arc length between the red and blue dots in Fig. 1a is a measurement of this local defocus and wave propagation must be used to refocus all atomic columns into the same image plane for a quantitative analysis. It is worth noting that the intensity of the exit wave always decreases as the electron wave is propagated away from the common image plane (Fig. 1b). We use the criterion of maximum propagation intensity (MPI) to determine local shape changes at the bottom of the sample by measuring the local defocus values and further refine them with the Big-Bang scheme[18] to a nominal precision of better than 1 Å. Once column thickness and surface shape at the bottom of the sample are known, their linear combination creates a tomogram. The detailed procedures and analyses are given in Methods.

Analysis of experimental exit wave functions. Intentionally, we prepared a semiconductor, a metal and a ceramic sample by different techniques, namely by ion milling, thin film deposition and by electron beam-assisted processing, to obtain differently

shaped objects. Amplitude and phase of their exit wave are shown in Fig. 2. Figure 2a displays a Ge [110] sample prepared from a bulk Ge single crystal by mechanical polishing and successive ion milling[29], and one expects the creation of surface steps forming a wedge-shaped sample with shallow angles. The gold [110] sample was grown by physical vapour deposition on germanium[30]. After growth, the Ge substrate was etched away creating a free-standing metal sample with flat bottom and rounded top (Fig. 2b). It is emphasized that in this case twin boundaries and a dislocation core are included in the analysis to make the point that extended defects can be analysed by our procedure. Finally, the MgO sample originates from a polished MgO [001] single crystal, which was prepared in [100] cross-section[31] and exposed to the high brightness electron beam at 300 kV for several minutes. *In-situ* observations revealed that the high-energy electron beam removes all sample preparation induced surface roughness and forces the formation of the stacked cube structure with one global [100] zone axis orientation (Fig. 2c). All images of Fig. 2 show crystals are suspended in the high vacuum of the electron microscope and the support films are not visible in the field of view. Moreover, we did not find any evidence for an attachment of residual gas molecules from the high column vacuum to the surfaces of the samples, as a cold N_2 trap was used. The expected

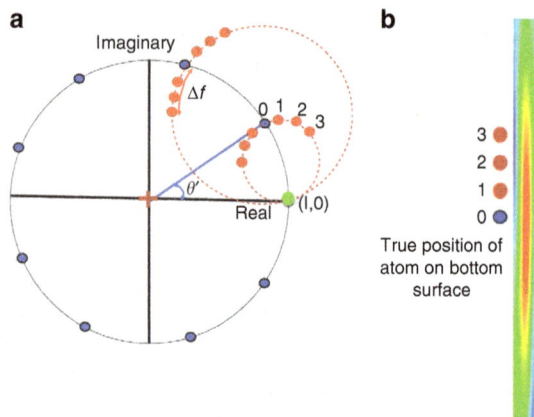

Figure 1 | Schematic representation of the exit wave function in Argand space and propagation intensity. (a) The complex values of pixels in the centre of each atomic column are represented as blue dots. The red dots correspond to pixels of the wave function if propagated across a distance Δf away from the exit surface of the sample. Black circle is called 'mass circle' and the red dashed circle is called 'defocus circle'. (b) The propagation intensity of one atomic column. The true position of atom at the bottom exit surface is at the position of maximum intensity (blue point).

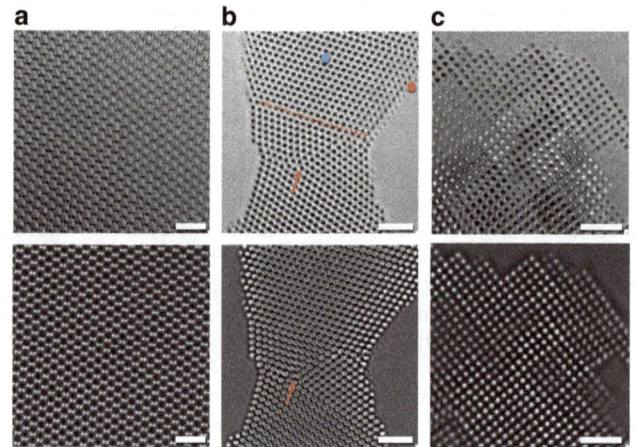

Figure 2 | In-line holograms acceleration voltage = 300 kV. Dose rates are given in Table 1. (a) Ge[110] and (b) Au[110] were reconstructed from focal series of 35 images with a focus step of 2 nm and Cs = 0 mm. (c) MgO[100] was reconstructed from focal series of 80 images with a focus step of 1 nm and Cs = − 0.015 mm. Top and bottom rows show the amplitude and the phase of the reconstructed electron exit wave functions, respectively. Scale bar, 1 nm in all cases. It is noteworthy that there are edge-on twin boundaries present in the imaged gold sample (b, one is highlighted by a dash line). At the intersection of three twin boundaries, a present end-on dislocation marked by an arrow. In addition, in phase image of b, many atom columns assume 'donut' shapes because mass and focus information are convoluted.

Table 1 | Phase change/atom, accuracy in the mass phase and dose rate for Ge, Au and MgO.

Material/Z (a.m.u.)	Peak to valley phase change/atom (rad)(experimental measurement)	Phase change per atom (rad) (calculated)	Thickness accuracy (nm), 2σ	Phase accuracy (rad), 2σ	Dose rate * 10³ (e/A⁻²s⁻¹)
Germanium /32 (72.63)	0.11 Extrapolated by $Z^{2/3}$	0.16	0.12	0.07	15
gold/79 (179)	0.21	0.40	0.24	0.13	46
Magnesium oxide/20 (40)	0.08	0.09	0.22	0.04	1.3

geometrical features including the presence of surface steps that, however, are not obvious in these images, except for contrast fluctuations in the amplitude image of the MgO [100] sample (Fig. 2c), which suggests the presence of MgO cubes. This information is simply masked by mixing sample shape with column length in the experimental images as described earlier. For their deconvolution we apply MPI and column mass measurements (Methods) to each atom column of the electron exit waves (Fig. 2), which provides the histograms of column mass and defocus values (Fig. 3). It is seen that all histograms reveal discrete sets of peaks, which are periodically spaced. Image simulations confirm that the incremental steps between adjacent peaks correspond to the addition of single atoms or molecules to atom columns with a periodic spacing in the beam direction set by the crystal structures of the materials. In a second process, we determine the confidence levels of these measurements by fitting Gaussians to the accumulation points (Fig. 3). This allows for an extraction of error bars that are given as 64.2% probability values (2σ values) of the measurements and are listed in Table 1.

To convert column mass values into radians, we determined experimentally the phase changes of the electron wave caused by scattering at one gold atom and one MgO molecule (Methods). Table 1 lists these phase values. It is seen that the phase of the exit wave is changed by 0.21 ± 0.07 rad by passing through a single gold atom in an atomic column or by 0.08 ± 0.02 rad if it is passing through a single MgO molecule. A value of 0.11 rad for scattering at one Ge atom is estimated using a $Z^{2/3}$ dependence. Remarkably, it is also seen that error bars increase with increasing dose rates, suggesting that measurements with a best element differentiation can only be performed if dose rates are kept low. In addition, column locations in the beam direction can be determined to a precision 1.2–2.4 Å on an absolute scale. Consequently, depth resolution has reached interatomic distances in thickness reconstructions from single projections. It is now straightforward to create 3D tomograms from these measurements, as the focus values describe the exit surface profile of the sample and the column length is given by the number of atoms of known spacing along the column length. In this manner, we have created the tomograms (Fig. 4) that show all geometrical properties that one expects to be imprinted by the chosen sample preparation procedure. The number of atoms in each tomogram is a time average that is dose rate dependent and equals 35,389 for Ge, 4,883 for Au and 10,750 (Mg, O) for MgO.

Specifically, the tomogram of Ge[110] in Fig. 4a shows a wedge-shaped sample prepared by ion milling with a low incidence angle ($\sim 6°$). Usually such wedges are formed by irregularly spaced terraces on both sides of the sample[31], which is confirmed by this tomogram. The Au [110] sample is dome shaped with a reasonably flat bottom at the side of the crystal initially attached to the germanium substrate. In addition,

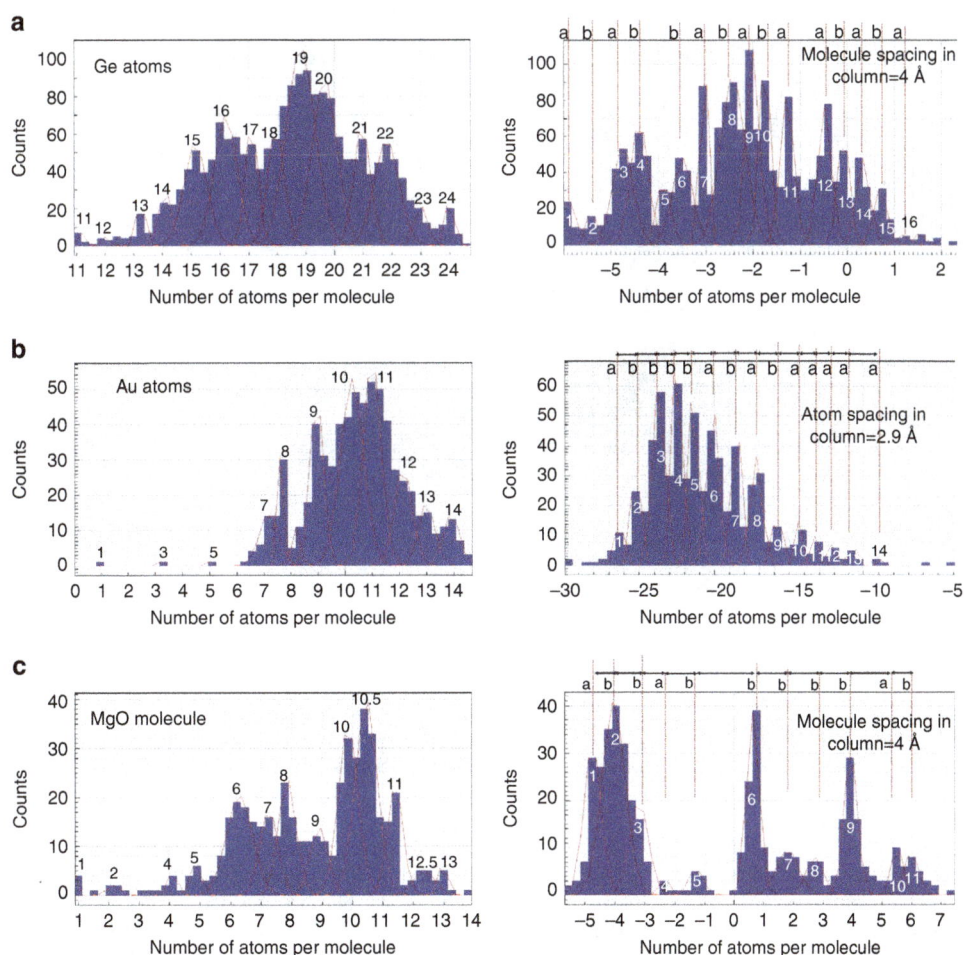

Figure 3 | Column mass and focus histograms. (**a**) Germanium and (**b**) gold values are given in terms of single atoms. (**c**) MgO graphs refer to single molecules. Gaussian functions (red lines) are fitted to their width, which is given in form of an averaged 2*σ error bar in Table 1. In focus graphs, the number of atoms/molecules are converted into focus values by multiplication with their listed spacing in beam direction. For germanium and gold, **a** and **b** sites refer to the existing (110) surface corrugation. In case of MgO [100], a surface corrugation is absent but surfaces are either terminated by Mg or by O atoms.

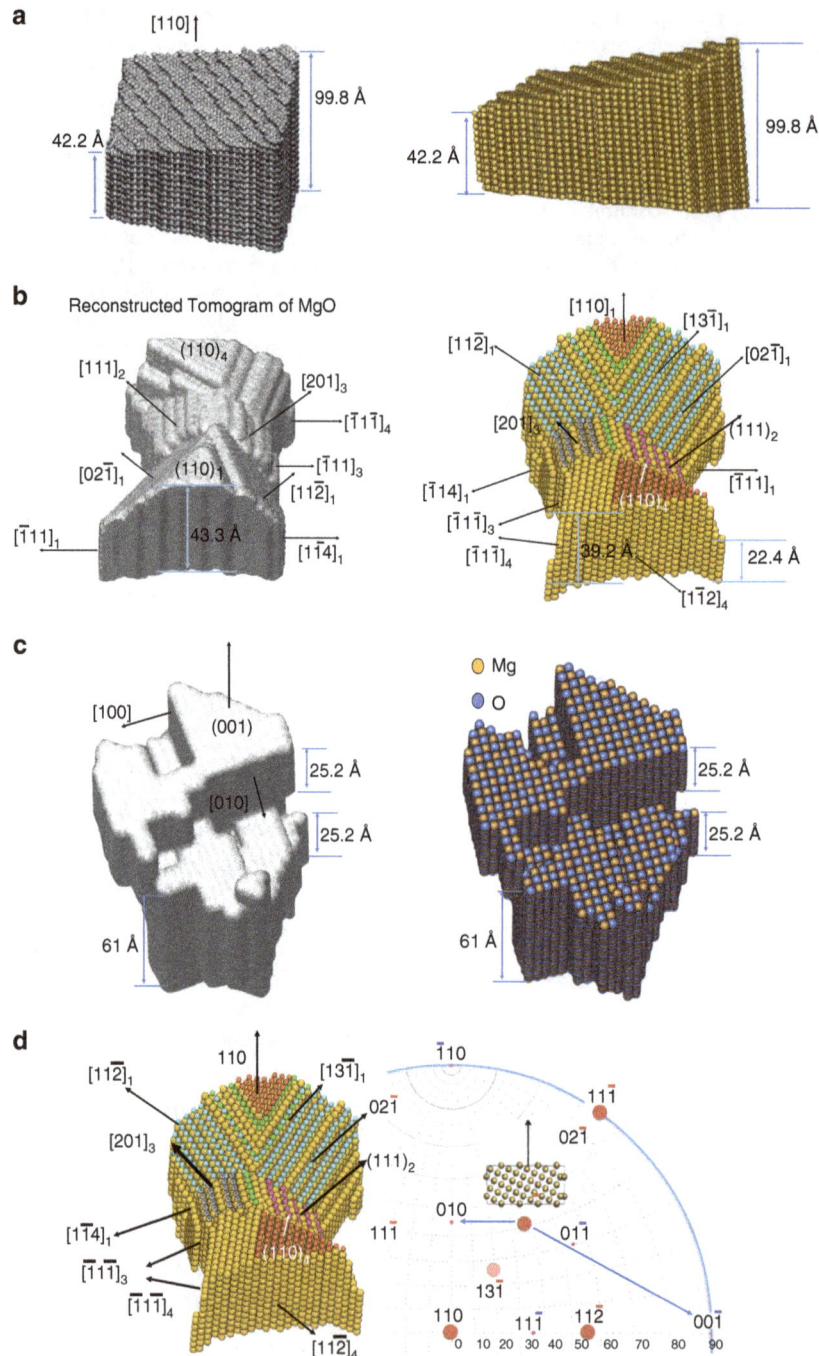

Figure 4 | Atomic resolution tomograms. (**a**) Surface shape and atomic structure views of the Ge [110] sample. (**b**) Surface shape and atomic structure views of the Au[110] sample. The facets are highlighted with different colours. (**c**) Surface shape and atomic structure views of the MgO [100] sample. Orange atoms: Mg, blue atoms: O (the size of the atoms is intentionally enlarged to render the shape of the particle). (**d**) The Wulf net shows the relationship of the high-energy facets (red dots) with low indexed facets for four grains indexed with $i = 1\ldots 4$. In grain 1, a [02$\bar{1}$] surface facet can be formed by low energetic [010] and [00$\bar{1}$] (red symbols) surfaces as shown by the insets and observed in the tomogram.

pronounced facets and surface reconstructions are seen. It was established[32] that high beam currents rapidly alter all surfaces of gold crystals during the acquisition of high-resolution images. The process transforms the material into a thermodynamically more stable form that can be recognized by the exposure of (111) surfaces. Such atom rearrangements are driven by the low surface energy $\gamma_{Au}(111)$[33] and are visible in Fig. 4b. Consequently, the tomogram depicts a crystal structure averaged over the 60 seconds acquisition of the focus series, while the surfaces of the crystal were altered from image frame to image frame. The misleading impression that a static situation would be considered only exists, because surface diffusion is fast compared with the image acquisition time[34] and the loss of single atoms from atom columns is hard to detect. The MgO [100] particle does not exhibit the typical wedge shape that is characteristic for sample initially prepared by ion milling. Instead, a shape transformation took place into the stack of cubes shown in the tomogram of Fig. 4c during the prolonged exposure of the sample to a high electron beam current of $\sim 50,000\,\mathrm{e\AA}^{-2}\mathrm{s}^{-1}$ before the experiment. The focus series itself was recorded with a reduced beam current of $1,300\,\mathrm{e\AA}^{-2}\mathrm{s}^{-1}$. Similar to the Au sample, an exposure of the material to high beam currents triggers the formation of a thermodynamically more stable shape that, however, creates cubes with exposed (100) surfaces in MgO, because the surface energy $\gamma_{MgO}(100)$ is the lowest[35]. From the tomogram, it is seen that the edge length of such cubes varies between 2.5 and 6 nm. The sample thickness at the highlighted vacuum/MgO interface exceeds 6 nm and any {100}/vacuum interface is atomically flat.

Discussion

Successful reconstructions of sample shapes from single projections date back to the early 1990s (ref. 11) when the shape of a silicon crystal was reconstructed from a single high-resolution image. However, this approach did not allow to identify single atoms or surface steps, because lens aberrations were ignored and a procedure was lacking to separate column length from surface profile. Ever since, progress was steady[12,18,19,21,23] and has now reached the point that atom locations can indeed be determined in 3D so that Feynman's challenge can be met. Instead, debates evolve around the implementation of most suitable methods, the validity of recovered values, the necessity of recording many projections and the need to control of beam–sample interactions.

The development of exit wave reconstruction methods is a key element for the achieved progress, as it allows to describe the dynamic electron channelling in an Argand plot, which is transparent and anchored in physics[36–39]. The Argand plot (Fig. 1) explains in a natural manner how the vertical position of a column can be separated from the column mass by applying local wave propagation and how the column mass can be quantified. It also allows to understand how point defects affect the mass of a column. A vacancy, for example, will reduce the column mass by a single atomic step that can be predicted and measured if the related phase change/atom exceeds the noise level. Certainly, grain boundaries and dislocation cores can be included in the analysis (Fig. 2b). Moreover, our tomograms (Fig. 4) show that the thickness of the analysed crystalline objects exceeds now 10 nm, which makes the tool generally applicable for investigations of nanocrystals and catalysts. In general, the sample thickness is limited to a full oscillation period (or extinction distance) of the channelling electron wave in the order of tens of

nanometres. It is noted that beside electron microscopy, 3D atom probe (3D-AP) is also available for atom counting[40]. Two complementary differences between 3D-AP technique are that our technique accounts for every atom, while $\sim 50\%$ of all atoms can escape an AP observation and AP can detect much smaller impurity levels.

The measured phase change per Au atom and per MgO molecule (Table 1) of 0.21 ± 0.07 and 0.08 ± 0.02 rad, respectively, can directly be compared with multislice calculations using the electron scattering factors by Doyle and Turner[41], and a reasonable Debye–Waller factor of $0.5\,\mathrm{\AA}^2$, which accounts for a Gaussian distribution of averaged atom displacement by ~ 8 pm (ref. 41). We calculate an expected phase change of 0.40 rad for a gold atom and 0.09 rad for a MgO molecule (Table 1). By comparison, these theoretical expectations exceed the measurements by factors of 2.5 and 1.1, and the larger discrepancy occurs for the Au atoms where the images were recorded with the largest electron dose rate of $45,000\,\mathrm{e\AA}^{-2}\mathrm{s}^{-1}$ (Table 1).

It is instructive to investigate the impact of different damping functions on the signal strength using an Argand plot. Contribution from damping functions such as a poor modulation transfer function of a camera or mechanical vibrations, for example, simply reduce the diameter of the Argand circle but do not affect the phase change per atom as long as phase changes are measured from the centre of an Argand circle (Fig. 5a). The exit wave for the red Argand plot (Fig. 5a) is reconstructed from simulated images of a Au [001] crystal with a Debye–Waller factor of $0.5\,\mathrm{\AA}^2$, whereas the green Argand plot is obtained with the same parameter set, except for an additional mechanical damping of the contrast transfer function by 50 pm, which coincides with the information limit resolution of TEAM 0.5. It is seen that the phase change per atom is maintained if measured from the origin of the Argand circle even though its diameter is largely reduced. Consequently, we do not correct for a poor camera performance or mechanical vibrations, because such corrections only boost high-frequency noise but leave phases unaffected if described in an Argand plot. Instead, we fit Argand circles to the data points and translate the origin of the circle to (0,0). On the other hand, damping processes such as electron beam-induced atom vibrations can soften the scattering potential and reduce the phase values for scattering at single atoms. If we model electron beam-induced object excitations of 45 pm by using a larger Debye–Waller factor of $16\,\mathrm{\AA}^2 = 8\pi^2(45)\ \mathrm{pm}^2$

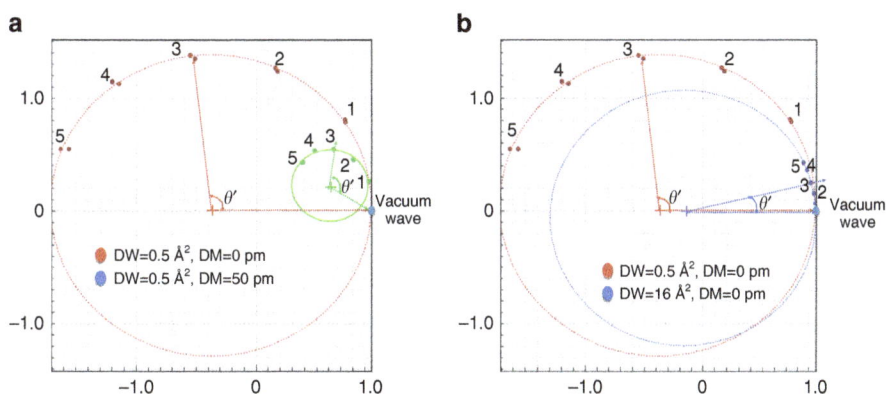

Figure 5 | Effects of mechanical damping and potential softening to Argand plots. (**a**) Red Argand plot shows exit waves reconstructed from simulated images of a Au [001] crystal with Debye–Waller (DW) factor of $0.5\,\mathrm{\AA}^2$ with no mechanical damping (DM), whereas the green Argand plot is obtained from the same crystal but with mechanical damping (DM = 50 pm) of the contrast transfer function. The numbers in the plot correspond to number of atom. (**b**) In our case, phases are reduced due to reversible electron beam-induced object excitations in the image formation process. This electron beam stimulation effect is modelled as a higher Debye–Waller factor of $16\,\mathrm{\AA}^2$ with DM = 0 pm.

(ref. 41), the phases are greatly reduced, which leads to the blue circle in Fig. 5b. This description is consistent with the view that reversible electron beam-induced object excitations contribute to the image formation process. As such excitations can cause large displacements and decrease logarithmically with decreasing dose rates[26], low dose rate electron microscope becomes advantageous or even mandatory if it is needed to maintain the pristine structure of small particles[42], surfaces or even molecules[43].

To reduce electron beam-induced sample alterations, electron dose rates were dropped to 1,300 eÅ^{-2}s^{-1} for the acquisition of the exit wave of MgO. It is advantageous that focal series of images were recorded, because they can be used to study and track electron beam-induced object changes by splitting the data set into different subsets that can be reconstructed separately. For example, Fig. 6a is a reconstruction from the first 35 images of the MgO data set, Fig. 6b is reconstructed from images 36 through 70 and Fig. 6c makes use of the entire focus series. Arrows in Fig. 6a,b mark locations where the electron beam visible changed the sample/vacuum interface. An estimated 5–10% of the all surface sites are affected by the chosen dose rate and one has to assume that such alterations also occur on the top and bottom of the sample where they escape a direct observation. In fact, it was recently reported that structural integrity of oxide catalysts smaller than 5 nm can only be maintained if dose rates are kept well below 1,000 eÅ^{-2}s^{-1} (ref. 44). It is also seen in Fig. 6 that the displaced entities on the MgO [100] surfaces occupy regular lattice sites and do not distort the MgO lattice if attached somewhere else. Therefore, it is compelling that they are elements of the crystal structure. At the vacuum/crystal interface, we commonly measure a contrast change of 0.07–0.09 rad (Fig. 6c). Smaller phase changes around 0.04 rad could be detected but occur rarely. If the contrast change of 0.04 rad is assigned to single Mg or O atoms, the most commonly occurring phase change of 0.08 rad represents the presence of MgO molecules aligned in beam direction. In fact, a phase value of 0.08 rad is very close to the calculated phase change of 0.09 rad for a molecule (Table 1). The common occurrence of the molecular unit also explains the discrete nature of the histogram in Fig. 2c. Therefore, unlike the case of Ge[110] and Au[110], MgO molecules are the basic unit that differentiates column masses and entire MgO molecules are commonly displaced by the electron beam or added to different sites of the crystal structure. In addition, the calibration of the mass phase is consistent with a contrast dependence proportional to $Z^{2/3}$ (refs 25,26). In addition, it seems energetically unfavourable to attach single Mg or O atoms to the [100] surfaces of the MgO compound, because the bonding is ionic and electric charge would occur locally on a crystal surface that is of minimal surface energy[5] if neutral. If carbon contaminants were present, they would also modulate the column masses with values around 0.04 rad. The rare occurrence of such column mass modulations proves that the crystal surfaces remain free of carbon contamination. Therefore, we find that electron irradiation can be used to form perfect cubes from MgO samples initially prepared by ion milling and consequently exhibited a wedge shape similar to that of the Ge[110] crystal.

Jia et al.[21] reports on the 3D reconstruction of an almost identical MgO [100] cube from a single image. In this work, a dose rate is estimated from the reported counts 3,500 (ref. 21), exposure time 0.5 s and the photons per electron conversion rate 2 photons per electron at 300 keV to be approximately between 50,000 and 100,000 eÅ^{-2}s^{-1}, which roughly equals the total dose of our experiment. By comparison, our experiment spreads this dose over a recording time of 120 s so that the dose rate is two orders of magnitude lower. Otherwise, their experimental conditions are very similar but the conclusions differ. It is

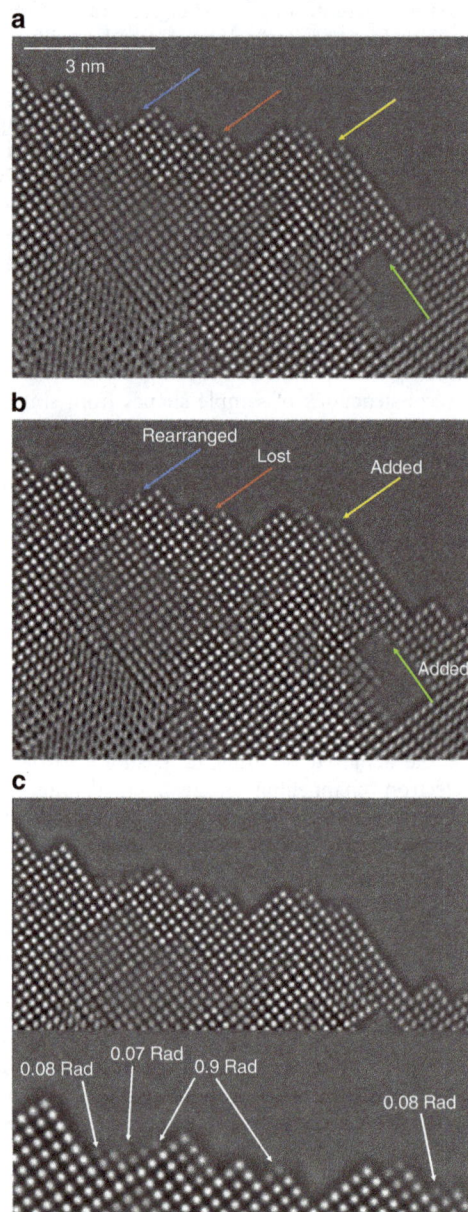

Figure 6 | Reconstructed exit waves from different time series.
(**a**) Reconstructed exit wave from the first 35 time series images. (**b**) Reconstructed exit waves from the 35th to the 70th images. The edge atoms are obviously altered via electron beam–sample interaction. (**c**) Reconstructed exit waves using the images of whole time series. The phase value from the edge atoms can be read to be ∼0.08 rad. Scale bar, 3 nm in all cases.

remarkable that the authors[21] determine a sample thickness of only 1.5–2.0 nm from an area close to the vacuum/MgO interface that is similar to ours and postulate the presence of residual gas contaminants on the MgO[100] surfaces to match the image contrast quantitatively with multislice calculations. As our self-consistent approach neither recovers sample areas of such small thickness nor provides evidence for the presence of surface contaminations, it is likely to be that the existing differences relate to electron beam-induced sample excitations. They greatly affect absolute values in such a manner if experiments are directly compared with theory as described in ref. 21.

In our experiments, structural rearrangements occur on the gold sample surfaces from image to image[32,45] due to the choice of a high dose rate for the recording. One outstanding drawback of using high beam currents is a greatly increased error bar on the determination of the phase changes (Table 1), which prohibits a reliable element differentiation. Nevertheless, meaningful information can be extracted for the mono-elemental gold crystal. As the contrast is high and single gold atoms can be distinguished, structural feature showing equilibrium configurations at atomic resolution become visible in the tomogram. The formation of equilibrium structures is driven by surface energies γ that increase in the order $\gamma_{Au}(111) < \gamma_{Au}(100) < \gamma_{Au}(110)$ for face-centred gold crystals[33]. In the tomogram of Fig. 4b, the existence of large (111) facets confirms this view. This line of argumentation can be extended by considering that any other surface that can be reconstructed by exposing combinations of low energetic surfaces. For example, a missing row reconstruction occurs, because it increases low energy contributions from (111) surfaces instead of simply exposing a high energetic (110) surface and combinations of (111) with (100) surfaces yield similar effects. In Fig. 4d, the pronounced exposure of a (02−1) surface is highlighted that is formed by a combination of two {100} surfaces. However, the existing equilibrium configurations are distorted by the dynamic surface alterations induced by the large beam current. Nonetheless, the result strengthens our claim that structure recovery from single projection has emerged as a robust tool to

determine atom positions in 3D at atomic resolution, which addresses Feynman's challenge.

The possibility to perform inline 3D holography at very low dose rates provides a powerful new tool for *in-situ* observation of structural transformation dynamics in small particles. This is particularly clear from the observation of the MgO particle, as both the Ge and the MgO samples were prepared by ion milling to form a wedge-shaped sample with random terraces. From the reconstructed tomogram, however, one can see that the Ge sample has kept this form but the MgO sample was transformed into a cubic shape with cubic protrusions terminated by (100) terraces.

Our working hypothesis is that the illuminating electrons transfer kinetic energy to the atoms in the sample, which cause them to vibrate. Certainly, near head-on collisions displace the atoms; however, as argued in refs 24,25, most of them are displaced in a metastable position from which they can return to their original site. When the dose rate is sufficiently small, the transferred energy can be dissipated before being accumulated and the average kinetic energy of the atoms stays below threshold limits. In that case, all atoms remain close to their original sites but the apparent Debye–Waller factor can be larger than expected from the thermal motion alone[24-26]. Thus, low dose rates still transform samples locally but in a stationary manner with an average structure that can still be observed in high resolution electron microscopy. In contrast, dose rates that exceed the

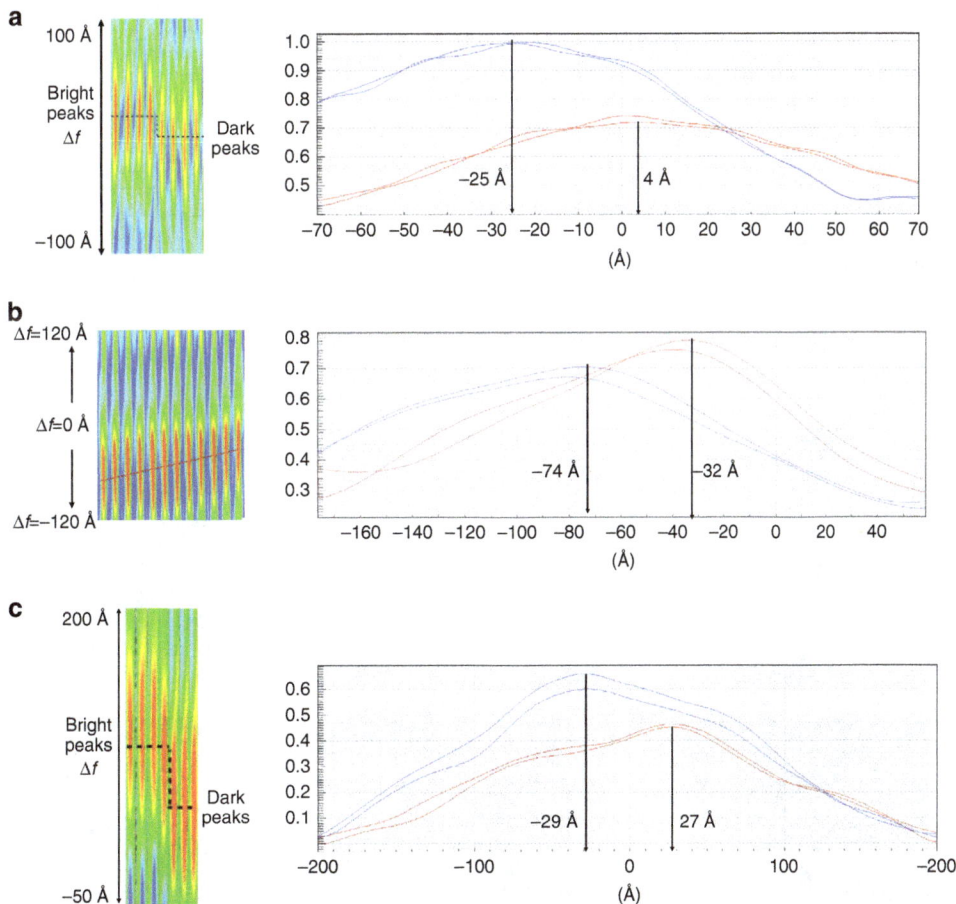

Figure 7 | Intensity of the column wave as function of the defocussing distance. (**a**) Cross-section of propagated intensity (ellipsoid shape, left of panel) and intensity profiles (right of panel) from dark (blue) and bright (red) columns of modulus of Ge exit wave (Fig. 2a). (**b**) Cross-section of propagated intensity (ellipsoid shape) and intensity profiles from the centre (blue dot) to the edge (red dot) columns of the modulus of Au exit wave (Fig. 2b). (**c**) Cross-section of propagated intensity (ellipsoid shape) and intensity profiles from dark (blue) and bright (red) columns of modulus of MgO exit wave (Fig. 2c). In three cases, two columns at two different positions are analysed. The position of the atom locates at the maximum (indicated), which can be determined with a precision of the order of 0.1 nm.

threshold cause additional collisions before the atoms return to their equilibrium sites and cause permanent damage and eventually destroys the material.

Conversely, permanent radiation damage can be retarded by reducing the dose rate below a certain threshold, which keeps the sample in a stationary regime between creation and annihilation of atom displacements that can still be observed by high resolution electron microscopy. In the case of the MgO sample, the atoms are constantly moving under the irradiation so that the object goes through many different unstable states. However, at the end all the atoms move to stable (100) planes where they are bound in a much more stable position around which they then constantly vibrate.

We argue it is a general phenomenon that most of the surface atoms are more easily displaced than bulk atoms. Under the electron beam, nano-objects are constantly transformed going through several intermediate structures until the atoms are finally grouped in more stable planes where they can then stay in a kind of steady state. Thus, irrespective of the original shape of the sample, it will always evolve into a stable structure that is stationary in the electron beam. Only when the dose rate exceeds a certain threshold, the sample will be damaged irreversibly. This hypothesis holds a very promising method to create nanocrystalline objects with a well-controlled shape by choosing appropriate dose rates.

Our results demonstrate that arbitrary 3D structures of nanomaterials can be recovered up to the level of single atoms from only one projection if the crystal structure is known and the material is homogeneous. However, it is noted that the electron dose rate used for imaging is important in two aspects. First, modification of the sample take place that can be controlled and, second, it induces a softening of the scattering potential that leads to an underestimation of the mass in atomic column. The method allows investigating defects such as edge-on dislocation or planar defects such as stacking fault and grain boundaries. Certainly, the addition of two additional projections of the same sample make any assumptions obsolete and fully resolves the internal structure of any crystalline object[12].

Methods

Procedures for 3D reconstruction. As a result of electron channelling in the atomic columns, the exit wave function consists of sharp peaks superimposed on a constant background. For crystalline materials, basically, the exit wave function can be analysed column-by-column. The procedure of 3D reconstruction can be subdivided into several steps as follows: (1) determination of the true 'z' height (focus) of the exit surface of a column from an image plane with the maximum propagation intensity (MPI) criterion by wave propagation along 'defocus circles' as shown in Fig. 1; (2) refining the 'z' height using the Big-Bang scheme[18]; (3) correcting the focus of each column wave by back propagation, that is, propagating the red dot to the blue dot (focus-corrected wave, FCW), which is located on the 'mass circle' as shown in Fig. 1; and (4) extracting the wave values in the 'valleys' between the atomic columns (background or valley wave).

(5) The phase of the exit wave of a column can suffer from a phase shift caused by the mean inner potential of the crystal and by atom vibrations, which results in a phase offset that causes an error in the determination of the column mass. This can be corrected using peak to valley ratios by dividing its complex peak value with the complex value of its neighbouring valley wave for every column. The phase of this normalized wave θ' can then be used as input to determine the column mass.

(6) Fitting a mass circle to the normalized FCW (blue dots). The centre of the mass circle is displayed as a red cross in Fig. 1.

(7) Measuring the phase of the vacuum wave from the centre of the mass circle. Usually, the vacuum wave is very close to (1,0) as indicated as a green dot. The phase of the vacuum wave is used as a reference for zero mass.

(8) Measuring the phase of the normalized FCW (blue dot) from the centre of mass circle. The column mass is given by the phase angle θ' between the normalized FCW and the vacuum wave. See Fig. 1.

(9) The column mass in units of the number of atoms can be deduced by dividing the θ' by a standard phase change θ per one atom (column 2 and 3 in Table 1). The value of the phase change θ per atom (molecule) is sensitive to the electron dose rate and will be described in Methods.

Determination of the defocus of a column. We back propagate the exit wave function numerically by convolution with an inverse defocus operator. In this process, every spherical wave is refocused backwards into its point of origin where the intensity is maximal. This position can be obtained by monitoring the intensity of the back-propagated wave as a function of the propagated distance. The exit wave function was propagated (defocused) in both positive and negative directions, and for each column the intensity at every focus positions was recorded and forms a 3D $(x, y, \Delta f)$ intensity stack along the focus axis. We applied this to the cases of Ge [110], MgO [100] and Au [110]. Some cross-sections of defocused intensity stacks are displayed in Fig. 7a–c, respectively. All local column intensities of the waves show an elongated ellipsoidal shape (left in Fig. 7). The intensity profile (right in Fig. 7) along the centre of the ellipsoid are also given to show the difference in 'z' height of the exit surface. However, the ellipsoid always exhibits a 'flat' intensity maximum, which limits the depth precision.

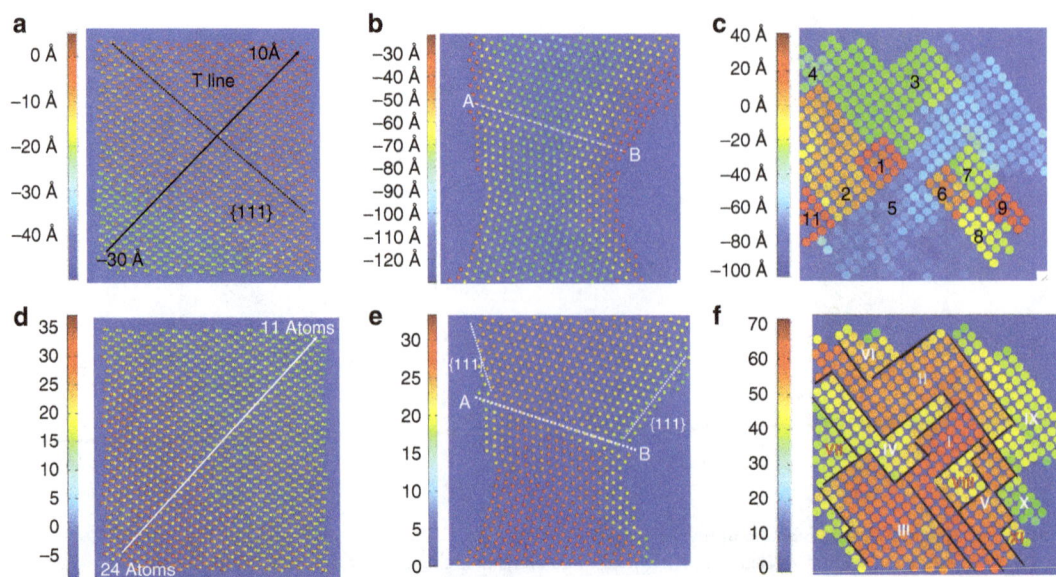

Figure 8 | Focus maps and mass maps. (**a**) Focus map for Ge. (**b**) Focus map for Au bridge. (**c**) Focus map for MgO, which shows focus patches that correspond to the peaks indicated in the histogram (fig. 3c). (**d**) Mass map for Ge. (**e**) Mass map for Au bridge. (**f**) Mass map for MgO. The humps indicated in the mass histogram of MgO (fig. 3c) corresponds to the mass patches.

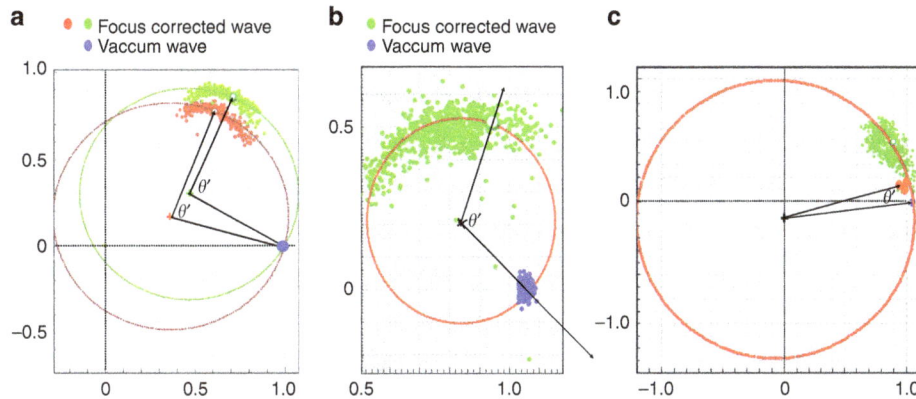

Figure 9 | Plot of the mass circle and the phase of θ'. (a) Ge. **(b)** Au bridge. **(c)** MgO. The green dots and red dots are FCWs , and the blue dots are the vacuum wave. The green and red dots in Ge case are the FCW from right and left columns of the dumbbell pairs. The red dots in the MgO case are the FCW from the edge.

In a second step, we improved this precision by the 'Big-Bang' procedure[18], which basically acts as follows. Because of the electron channelling, the exit wave function is sharply peaked at the centre of every column. When this wave propagates in free space towards the plane of observation, these peaks acts as point sources of spherical waves. It was recently outlined[23] theoretically how the position of the exit plane for each atomic column in 3D can be determined very accurately from fitting of the intersection of this spherical wave in the plane of observation. Here, the three sets of experimental data are refined in this manner, yielding a precision of about 0.1 nm.

Figure 8 shows the focus maps that are obtained by correcting the focus in every atomic column and the corresponding focus histograms are shown in Fig. 3a–c. For Ge [110], we observe a focus gradient across the image from bottom left (~ -30 Å) to the top right (~ 10 Å). In the focus map of Au, the focus value along the line A–B line shows a difference of ~ 26–30 Å between the edge and centre parts. The focus map of the MgO exhibits a number of flat focus patches numbered from 1 to 11. This histogram is discrete at the Angstrom level, which suggests that the addition of single atoms to a column can be measured by a related change of focus.

Thereafter, the experimental exit wave function of every column is then back propagated column so as to be used as the wave ψ for a determination of the mass of the corresponding columns. Next, another wave φ is determined in the 'valley' between the atomic columns. Using the valley wave φ, the FCW ψ can then be 'normalized' as follows.

$$\Psi(\text{norm}) = \psi/\varphi \qquad (2)$$

where Ψ (norm) is the normalized FCW so as to corrected by the mean inner potential of the crystal. Figure 9 shows the Argand plots for the cases of Ge [110], MgO [100] and Au [110] after focus correction, which reduces the data scatter significantly so that the expected circular arc can be recognized. The position of the point along this circular curve is now only a function of the column mass.

Determination of the column mass. The last step then consists in determining the column mass for every individual column with respect to the vacuum wave. The green dots and the blue dots in the Fig. 9 are the FCWs and the vacuum waves, respectively, which are fitted with a 'mass circle'. The red and green dots in the Fig. 9a are the FCWs of Ge. The two branches correspond with the left and right columns of the dumbbell and this difference is caused by sample tilt. The red dots in the Fig. 9c are the FCW functions of single MgO molecules from the edge of the crystal.

From the channelling theory, the projected mass of a column is proportional to the angle θ' measured from the centre of a mass circle. The radius of the mass circle depends on the electron dose rate and other damping functions as described in the main text. As the exit wave functions can only be determined apart from a constant phase offset, we use the vacuum wave to define (1,0). The column mass is then given by the difference between the phase of Ψ (norm) and the vacuum wave. This procedure applies to MgO and Au where vacuum values are visible in the images but not to the case of the Ge where the material fills the entire field of view (Fig. 2a). In that case, we have set the vacuum wave to the theoretical position (1,0). The green and red circles in Fig. 9a–c are the fitted mass circles for Ge, MgO and Au, respectively. The mass associated with a column can be calibrated in terms of number of atoms by

$$n = \theta'/\theta \qquad (3)$$

where θ is the phase change/atom given in Table 1 and Methods.

Concerning an absolute calibration of the phase change per atom in a column, which is commonly influenced by electron beam-induced and thermal vibrations,

nature offers its own proper yardstick (M-II). Figure 8d–f show the mass maps of the investigated samples (mass histograms are depicted in Fig. 3d–f). Discrete values are observed, because the contrast contribution from scattering at single atoms can be isolated in high-performance microscopes. The number of atoms in each detected column is used to form the histograms and the maps. The Ge mass map shows a mass gradient of ~ 24 atoms in the bottom left corner towards the top right corner ~ 11 atoms along the diagonal direction. A mass profile along line A–B of the Au[110] sample reveals a difference of seven to eight atoms between the edge and centre of the sample. The mass map of the MgO shows the number of MgO molecules in each patch # I to # X. Along [001] projection, the Mg and O atoms alternate in each column. In our analysis, the Mg and O atoms are treated as a pair of atoms and the number of MgO molecules are deduced from the mass circle in Fig. 9c.

Based on the focus and mass maps shown in the Fig. 8, we built the 3D tomography of Ge, MgO and Au samples that are shown in Fig. 4. It is worth mentioning that with our present method we cannot unambiguously determine the termination of the {100} surfaces with either Mg or O atoms from only one projection even though patches of relevant checkerboard patterns exist in the images and a projection along another crystallographic orientation that shows the Mg and O alternating layers would be helpful.

The phase change per atom θ (rad per atom). For an absolute calibration of the phase change per atom in a column, which is commonly influenced by electron beam-induced and thermal vibrations, nature offers its own proper yardstick.

The phase change per atom θ for Ge, Au and MgO is calculated along [001] crystallographic orientation with multislice simulation using MacTempas (totalresolution.com) for different thicknesses. The imaging parameters are those for the 300-keV TEAM0.5 microscope with g_{max} set at $2\,\text{Å}^{-1}$. The Debye–Waller factor is $0.5\,\text{Å}^2$. The simulation were carried out for thickness up to eight atoms with increments of one atom (or molecule for MgO). The peak-to-valley phase values were measured from electron exit wave functions that were reconstructed from focus series of simulated images for the different thicknesses. The peak-to-valley phase values θ (rad per atom) are as follows: 0.16 for Ge (a.m.u. $= 72.6$), 0.09 for MgO (a.m.u. $= 40.3$) and 0.4 for Au (a.m.u. $= 197$). It was reported[26] that the phase/atom scales with the atomic number $Z^{2/3}$. A comparison with experimental values is given in Table 1, where it is seen that deviations from the expectations increase with increasing dose rate. Part of the deviation comes from the potential softening by e-beam excitations and part from an imperfect recovery of the background phase.

References

1. Feynman, R. P. There is plenty of room at the bottom. *Eng. Sci.* **23**, 22–36 (1960).
2. Girit, Ç. Ö. *et al.* Graphene at the edge: stability and dynamics. *Science* **323**, 1705–1708 (2009).
3. Krivanek, O. L. *et al.* Atom-by-atom structural and chemical analysis by annular dark-field electron microscopy. *Nature* **464**, 571–574 (2010).
4. Suenaga, K. *et al.* Visualizing and identifying single atoms using electron energy-loss spectroscopy with low accelerating voltage. *Nat. Chem.* **1**, 415–418 (2009).
5. Chen, D. *et al.* The properties of SIRT, TVM, and DART for 3D imaging of tubular domains in nanocomposite thin-films and sections. *Ultramicroscopy* **147**, 137–148 (2014).

6. Hovden, R. *et al.* Breaking the Crowther limit: combining depth-sectioning and tilt tomography for high-resolution, wide-field 3D reconstructions. *Ultramicroscopy* **140**, 26–31 (2014).

7. Scott, M. C. *et al.* Electron tomography at 2.4-ångström resolution. *Nature* **483**, 444–447 (2012).

8. Midgley, P. A., Weyland, M., Thomas, J. M. & Johnson, B. F. G. Z-contrast tomography: a technique in three-dimensional nanostructural analysis based on Rutherford scattering. *Chem. Commun.* **10**, 907–908 (2001).

9. Borisevich, A. Y., Lupini, A. R. & Pennycook, S. J. Depth sectioning with the aberration-corrected scanning transmission electron microscope. *Proc. Natl Acad. Sci. USA* **103**, 3044–3048 (2006).

10. Behan, G., Cosgriff, E. C., Kirkland, A. I. & Nellist, P. D. Three-dimensional imaging by optical sectioning in the aberration-corrected scanning transmission electron microscope. *Phil. Trans. R. Soc. A Math. Phys. Eng. Sci.* **367**, 3825–3844 (2009).

11. Kisielowski, C. *et al.* An approach to quantitative high resolution electron microscopy of crystalline materials. *Ultramicroscopy* **58**, 131–155 (1995).

12. Jinschek, J. R. *et al.* 3-D reconstruction of the atomic positions in a simulated gold nanocrystal based on discrete tomography: prospects of atomic resolution electron tomography. *Ultramicroscopy* **108**, 589–604 (2008).

13. Goris, B. *et al.* Three-dimensional elemental mapping at the atomic scale in bimetallic nanocrystals. *Nano Lett.* **13**, 4236–4241 (2013).

14. Smith, D. J., Petfordlong, A. K., Wallenberg, L. R. & Bovin, J.-O. Dynamic atomic-level rearrangements in small gold particles. *Science* **233**, 872–875 (1986).

15. McBride, J. R., Pennycook, T. J., Pennycook, S. J. & Rosenthal, S. J. The possibility and implications of dynamic nanoparticle surfaces. *ACS Nano* **10**, 8358–8365 (2013).

16. Egerton, R. F., Li, P. & Malac, M. Rediation damage in TEM and SEM. *Micron* **35**, 399–409 (2004).

17. Specht, P. *et al.* Quantitative contrast evaluation of an industry-style rhodium nanocatalyst with single atom sensitivity. *ChemCatChem* **3**, 1034–1037 (2011).

18. Van Dyck, D., Jinschek, J. R. & Chen, F. R. Big-Bang tomography as a new route to atomic resolution electron tomography. *Nature* **486**, 243–246 (2012).

19. Van Aert, S., Batenburg, K. J., Rossell, M. D., Erni, R. & Van Tendeloo, G. Three-dimensional atomic imaging of crystalline nanoparticles. *Nature* **470**, 374–377 (2011).

20. Gontard, L. C. *et al.* Aberration-corrected imaging of active sites on industrial catalyst nanoparticles. *Angew. Chem.* **46**, 3683–3685 (2007).

21. Jia, C. L. *et al.* Determination of the 3D shape of a nanoscale crystal with atomic resolution from a single image. *Nat. Mater.* **13**, 1044–1049 (2014).

22. Botton, G. A., Calderon, H. A. & Kisielowski, C. Preface. Electron beam irradiation effects, modifications and control. *Micron* **68**, 140 (2015).

23. Chen, F.-R., Kisielowski, C. & Van Dyck, D. 3D reconstruction of nanocrystalline particles from a single projection. *Micron* **68**, 59–65 (2015).

24. Kisielowski, C. *et al.* Instrumental requirements for the detection of electron beam-induced object excitations at the single atom level in high-resolution transmission electron microscopy. *Micron* **68**, 186–193 (2015).

25. Van Dyck, D., Lobato, I., Chen, F. -R. & Kisielowski, C. Do you believe that atoms stay in place when you observe them in HREM? *Micron* **68**, 158–163 (2015).

26. Kisielowski, C. *et al.* Real-time, sub-ångstrom imaging of reversible and irreversible conformations in rhodium catalysts and graphene. *Phys. Rev.* **B88**, 024305 (2013).

27. Sinkler, W. & Marks, L. D. A simple channelling model for HREM contrast transfer under dynamical conditions. *J. Microsc.* **194**, 112–123 (1999).

28. Van Aert, S., Geuens, P., Van Dyck, D., Kisielowski, C. & Jinschek, J. R. Electron channelling based crystallography. *Ultramicroscopy* **107**, 551–558 (2007).

29. Alloyeau, D., Freitag, B., Dag, S., Wang, L.-W. & Kisielowski, C. Atomic-resolution three-dimensional imaging of germanium self-interstitials near a surface: aberration-corrected transmission electron microscopy. *Phys. Rev. B* **80**, 014114 (2009).

30. Westmacott, K. H., Hinderberger, S., Radetic, S. T. & Dahmen, U. PVD growth of fcc metal films on single crystal Si and Ge substrates. *Mat. Res. Soc. Symp. Proc.* **562**, 93–102 (1999).

31. Barna, A., Pecz, B. & Menyhard, M. Amorphisation and surface morphology development at low-energy ion milling. *Ultramicroscopy* **70**, 161–171 (1998).

32. Martin, A. V., Ishizuka, K., Kisielowski, C. & Allen, L. J. Phase imaging and the evolution of a gold-vacuum interface at atomic resolution. *Phys. Rev. B* **74**, 172102 (2006).

33. Shi, H. & Stampfl, H. Shape and surface structure of gold nanoparticles under oxidizing conditions. *Phys. Rev. B* **77**, 094127 (2008).

34. Schneider, S., Surrey, A., Poh, D., Schultz, L. & Rellinghaus, B. Atomic surface diffusion on Pt nanoparticles quantified by high-resolution transmission electron microscopy. *Micron* **63**, 52–56 (2014).

35. Saylor, D. M. & Rohrer, G. S. Measuring the influence of grain-boundary misorientation on thermal groove geometry in ceramic polycrystals. *J. Am. Cer. Soc.* **82**, 1529–1536 (1999).

36. Coene, W., Thust, A., Van Dyck, D. & Op de Beeck, M. Maximum-likelihood method for focus-variation image reconstruction in high resolution transmission electron microscopy. *Ultramicroscopy* **64**, 109–135 (1996).

37. Sinkler, W. & Marks, L. D. A simple channelling model for HREM contrast transfer under dynamical conditions. *J. Microsc.* **194**, 112–123 (1999).

38. Hsieh, W. K., Chen, F. R., Kai, J. J. & Kirkland, A. I. Resolution extension and exit wave reconstruction in complex HREM. *Ultramicroscopy* **98**, 99 (2004).

39. Wang, A., Chen, F.-R., Van Aert, S. & Van Dyck, D. Direct structure inversion from exit waves part I: theory and simulations. *Ultramicroscopy* **110**, 527–534 (2010).

40. Perea, D. E. *et al.* Determining the location and nearest neighbours of aluminium in zeolites with atom probe tomography. *Nat. Commun.* **6**, 7589 (2015).

41. Doyle, P. A. & Turner, P. S. Relativistic Hartree–Fock X-ray and electron scattering factors. *Acta Crystallogr.* **A24**, 390–397 (1968).

42. Kisielowski, C. *et al.* Real-time, sub-ångstrom imaging of reversible and irreversible conformations in rhodium catalysts and graphene. *Phys. Rev. B* **88**, 024305 (2013).

43. Trueblood, K. N. *et al.* Atomic displacement parameter nomenclature. *Acta Crystallogr.* **A52**, 770–781 (1996).

44. Haber, J. A., Anzenburg, E., Yano, J., Kisielowski, C. & Gregoire, J. M. Multiphase nanostructure of a quinary metal oxide electrocatalyst reveals a new direction for OER electrocatalyst design. *Adv. Energy Mater.* **5**, 1043207 (2015).

45. Kisielowski, C. *et al.* Detection of single atoms and buried defects in three dimensions by aberration-corrected electron microscope with 0.5-angstrom information limit. *Microsc. Microanal.* **14**, 469–477 (2008).

Acknowledgements

Electron microscopy was performed with the TEAM 0.5 microscope at the Molecular Foundry, NCEM, which is supported by the Office of Science, Office of Basic Energy Sciences, of the U.S. Department of Energy under contract number DE-AC02-05CH11231. The MgO substrate was kindly provided by Wangfeng Li from the University of Delaware. D.V.D. acknowledges the financial support from the Fund for Scientific Research - Flanders (FWO) under Project Numbers VF04812N and G.0188.08. F.-R.C. thanks the support from NSC 96-2628-E-007-017-MY3 and NSC 101-2120-M-007-012-CC1.

Author contributions

C.K. provided the high-resolution transmission electron microscopy image series recorded with the TEAM0.5. He restored the electron exit wave functions and developed procedures for contrast quantification. D.V.D. developed the theory and F.-R.C. developed the programme codes to test theory and analysis of experimental data. All authors drafted the paper.

Additional information

Competing financial interests: The authors declare no competing financial interests.

13

Tellurium as a high-performance elemental thermoelectric

Siqi Lin[1], Wen Li[1], Zhiwei Chen[1], Jiawen Shen[1], Binghui Ge[2] & Yanzhong Pei[1]

High-efficiency thermoelectric materials require a high conductivity. It is known that a large number of degenerate band valleys offers many conducting channels for improving the conductivity without detrimental effects on the other properties explicitly, and therefore, increases thermoelectric performance. In addition to the strategy of converging different bands, many semiconductors provide an inherent band nestification, equally enabling a large number of effective band valley degeneracy. Here we show as an example that a simple elemental semiconductor, tellurium, exhibits a high thermoelectric figure of merit of unity, not only demonstrating the concept but also filling up the high performance gap from 300 to 700 K for elemental thermoelectrics. The concept used here should be applicable in general for thermoelectrics with similar band features.

[1] Key Laboratory of Advanced Civil Engineering Materials of Ministry of Education, School of Materials Science and Engineering, Tongji University, 4800 Caoan Road, Shanghai 201804, China. [2] Beijing national laboratory for condensed matter physics, Institute of physics, Chinese academy of science, Beijing 100190, China. Correspondence and requests for materials should be addressed to W.L. (email: liwen@tongji.edu.cn) or to Y.P. (email: yanzhong@tongji.edu.cn).

Thermoelectric devices, which enable a direct conversion between heat and electricity based on either Seebeck or Peltier effects, have attracted increasing interest as a sustainable and emission free solution to the imminent global energy crisis and environment pollution for a few decades[1]. The performance of a thermoelectric material is determined by the dimensionless figure of merit, $zT = S^2\sigma T/(\kappa_E + \kappa_L)$, where S, σ, κ_E, κ_L, and T are the Seebeck coefficient, electrical conductivity, electronic thermal conductivity, lattice thermal conductivity and absolute temperature, respectively.

Because the electrical properties including S, σ and κ_E couple with each other strongly, a simple improvement in one of these three parameters usually leads to a compensation in the other two, resulting in the difficulty for enhancing zT. Minimizing the lattice thermal conductivity (κ_L), the only one independent material property, has been proven to be effective through nanostructuring[2–6], liquid phonons[7,8] and lattice unharmonicity[9,10] in the recent 15 years.

Alternatively, recent band engineering efforts aiming to obtain a high number of degenerated valleys (N_v) (refs 11–18), a low carrier inertial mass[19] and a weak scattering[20,21] has also led to great success in increasing the figure of merit zT (ref. 22). Taking the strategy of increasing N_v by converging two different valence (or conduction) bands in the k-space as an example, which has been well-demonstrated in p-type PbTe (ref. 15) and other IV–VI (ref. 23) semiconductors, zT has found to be increased significantly.

A straightforward understanding on how band convergence leading to high thermoelectric performance, is the increased conducting channels for high electrical conductivity, without affecting the Seebeck coefficient that is determined by the position of Fermi level and scattering mechanism[11]. Being slightly different from the band convergence where two or more band branches having similar energy but unnecessarily the same k-space location, nested bands have not only similar energy but also the same k-space location. In spite of the difference in k-space location between band convergence and nestification, the aligned bands in both cases should equally contribute to the transport of charge carriers, leading to a superior electrical performance for high thermoelectric efficiency.

Band nestification often occurs in well-known simple semiconductors such as group IV elements and III–V compounds, particularly in p-type conduction, this is mainly due to the splitting of degenerate bands by spin-orbit interaction[24]. This interesting band feature has led these materials to be playing important roles in the electronic industry for many decades, and many of these semiconductors have actually been considered as thermoelectrics since they are known[25], although the relative low atomic mass for the constitute elements and the simple crystal structure may lead to a high lattice thermal conductivity[25–27].

As an important member among the elemental semiconductors, trigonal Te with the P3$_1$21 space group undergoes a transition to a topological insulator phase[28]. However, it has been much less considered as a thermoelectric material.[29–31] Available experimental results are limited to the electrical properties[29–32] and low temperature thermal conductivity only[33,34]. Providing its intrinsically nested valence bands[30,32,35,36], which are very similar to those of group IV and III–V semiconductors, as superior electronic performance can then be reasonably expected in tellurium. In addition, the relatively heavy atomic mass and the complexity in crystal structure in tellurium, as compared with the well-studied group IV and III–V semiconductors, should lead to a much lower lattice thermal conductivity[27].

Guided by the concept of nested bands for high thermoelectric performance, this work focuses on the thermoelectric performance

of polycrystalline tellurium, a constitute element commonly used for producing conventional thermoelectrics including PbTe and Bi$_2$Te$_3$. The thermoelectric figure of merit, zT, as high as 1.0, achieved in a material as simple as elemental tellurium, demonstrating the validity of the concept. The achieved zT of ~1.0 here is actually the highest among the reported element-based thermoelectrics[37] including those are heavily alloyed, such as SiGe (refs 38,39) and BiSb (refs 40–42). Furthermore, the obtained high zT fills the gap of elemental thermoelectrics showing high zT in the temperature range of 300–700 K as shown in Fig. 1, revealing the importance of tellurium as a thermoelectric material when the application circumstance strictly disallows precipitation, segregation or volatilization.

Results

Band structure of tellurium. It was reported[35] as early as in 1950s, followed by a few other researchers in 1970s (refs 30,32,36,43–45). Very recently, the detailed band structure has been given by theoretical calculations[28,46]. Regardless the different sources, the most important similarity, among the majority of the literatures, is that the nested valence bands at the H point in the Brillouin zone[30,32,36,43–46] due to the spin-orbit coupling in tellurium. The four-orbital degenerated valence band at point H was split into two upper valence bands H4 and H5 and the lower doubly degenerate band H6 (refs 28,46). Because the energy separation, either calculated[36,43,46] (https://www.materialsproject.org/materials/mp-19/) or measured[47], between the two upper valence bands (H4 and H5) is as small as 0.1 eV or less, they both contribute to the hole transport concurrently. On the other hand, the third valence band (H6) has a much lower energy[36,43,46] and therefore is not influential to the electrical properties in the temperature and doping ranges studied here. Further due to the spin-orbital coupling, the first valence band, H4, may exhibit a weak camel's back in shape along the H → K direction[36,46]. However, this work focuses on the transport properties at temperatures > 300 K, the resulting Fermi distribution broadening is significantly larger in energy than the difference between the extremums of the camel's back H4 band,

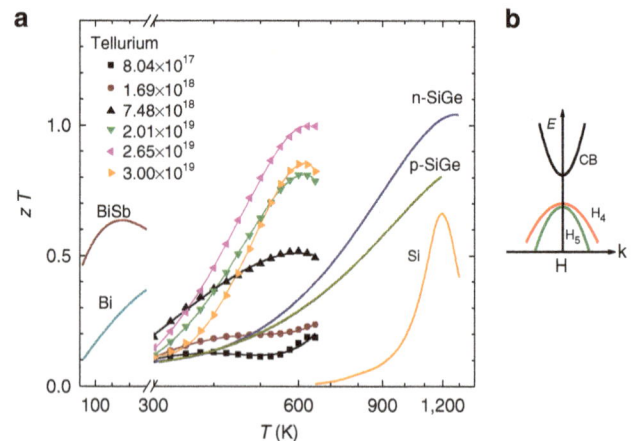

Figure 1 | Survey of zT for elemental thermoelectrics. Temperature-dependent figure of merit (zT) for p-type ploycrystalline tellurium with different carrier concentrations shown in a unit of cm^{-3} (**a**). Both low temperature Bi/Bi-Sb alloys[40] and high temperature Si (ref. 37)/Si-Ge alloys[39] are included for comparison. p-type tellurium studied here shows a highest zT n the temperature range from 300 to 700 K, largely relies on its inherently nested valence bands (H4 and H5) as shown in **b**. The overlying lines in **a** are included to guide the eye.

leading to an unobservable effect on the electrical properties. Therefore, the band structure of Te can be approximated as the inset of Fig. 1 schematically, without including either the low energy H6 band or the camel's back feature for band H4.

It is then clear that the bands H4 and H5 are effectively nested in tellurium, giving a rise to the conducting channels for holes. This inherent band feature, and its effect on the thermoelectric performance is essentially similar with those caused by band convergence[11,22,48], in which the converged bands unnecessarily have band extremum at the same k-space location. In this way, the two upper valence bands, with a valley degeneracy of 2 each, accumulate the hole pockets to a total number of 4 approximately, being comparable with $4\sim6$ that obtained in n-SiGe, Bi_2Te_3 and n-PbTe thermoelectrics.

Carrier concentration-dependent transport properties. The transport properties were measured on single phased polycrystalline tellurium samples, where the trigonal structure and an average grain size of $\sim100\,\mu m$ are determined by X-ray diffraction and transmission electron microscope analyses (Supplementary Fig. 1). The transport properties in the directions along and perpendicular to the applied pressure of the hot press, are found to be nearly isotropic (Supplementary Fig. 2). Indeed, the experimental thermoelectric figure of merit, zT, of elemental tellurium is found to be as high as unity, being the highest among the reported element-based thermoelectrics including those are heavily alloyed such as SiGe (refs 38,39) and BiSb (refs 40–42). This further leads to a fill-up to the gap of elemental thermoelectrics showing high zT in the temperature range from 300 to 700 K as shown in Fig. 1. The measured zT shows

a good reproducibility (Supplementary Fig. 3) and comparability (Supplementary Fig. 4) to that measured by a different technique. The observed discrepancy on zT between the experimental results and the *ab initio* calculations[46] is largely due to the difference on estimating the electronic thermal conductivity.

In nested bands on the transport properties, the Hall carrier concentration dependence is given in Fig. 2. According to the above discussion on the band structure, a two-band (H4 and H5) model is used to understand the transport properties. Furthermore, due to the small band gap in tellurium, the band may be slightly nonparabolic and therefore needs to take into the first order of nonparabolicity (Kane band) into account[49]. The two Kane band model enables a reasonable prediction (dashed curves) on the hall carrier concentration-dependent Seebeck coefficient (Fig. 2a), Hall mobility (Fig. 2b) and power factor (Fig. 2c). The *ab initio* calculated[46] Seebeck coefficient is slighter higher, which is presumably due to the fact that this method does not take into account the reduction of the band gap with increasing temperature[30]. As a result, the *ab initio* calculated[46] power factor is also higher than the measurement (Supplementary Fig. 5). Similarly, this discrepancy can also be seen from the temperature-dependent transport properties as discussed below.

The Kane band model has shown similar success on understanding the transport properties of *n*-type PbTe (refs 19,50,51). It should be noted that the two-band model takes the band nonparabolicity, temperature-dependent effective mass and band gap[30] into account. This model assumes a dominant charge carrier scattering by acoustic phonons, as evidenced in Fig. 2d by the observed temperature dependence of $T^{-1.5}$ on the Hall mobility (μ_H), because any

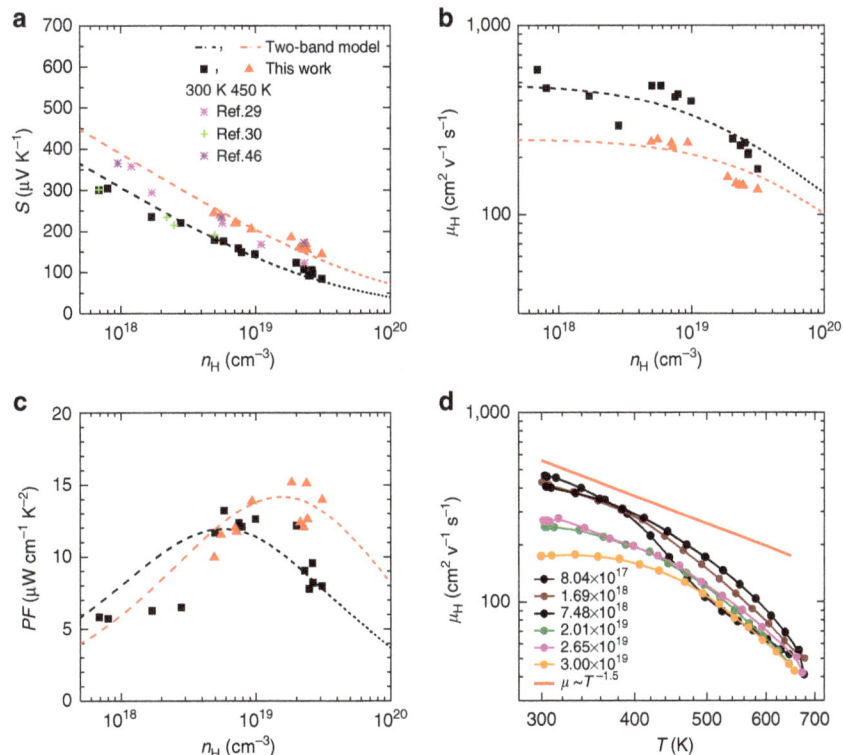

Figure 2 | Transport properties for tellurium. Hall carrier concentration-dependent Seebeck coefficient (**a**) Hall mobility (**b**) and power factor (**c**) at 300 K (blue, green, purple and pink) and 450 K (red), and the temperature-dependent Hall mobility (**d**) with a comparison to literature results[29,30,46]. The dashed curves in (**a-c**) show the prediction based on a two-band Kane model with a scattering mechanism by acoustic phonons as evidenced from the temperature-dependent mobility (red line in **d**). Overlying lines in **d** are plotted to guide the eye and the carrier concentrations for the samples are shown in a unit of cm^{-3} (**d**).

other scattering mechanisms such as by grain boundaries, polar-optical phonons, and ionized impurities predict a dependence of $\mu_H \sim T^p$ with $p \geq -0.5$. The even faster decrease in the Hall mobility than the $\mu_H \sim T^{-1.5}$ relationship at temperatures higher than 500 K, can be understood by the increased effective mass according to the two-band model.

This model further tells the optimal carrier concentration (n_{opt}) that allows a maximum power factor to be achieved (Fig. 3c), and the resulting n_{opt} is found to be in the range of $1-3 \times 10^{19}$ cm^{-3}, depending the temperature and density of state effective mass[51]. The obtained n_{opt} is consistent with the *ab initio* calculation[46], and is comparable with that of narrow band gap ($E_g < 0.5$ eV) thermoelectric semiconductors such as n-PbTe (ref. 51) and Bi_2Te_3 (ref. 52).

Temperature-dependent transport properties. The temperature-dependent Seebeck coefficient (S) and resistivity (ρ) are shown in Fig. 3a and Fig. 3b, respectively. The doping effectiveness of arsenic can be seen from the significant decrease in both resistivity and Seebeck coefficient when the doping concentration increases. The positive sign of the Seebeck coefficient indicates the p-type conduction, which is consistent with our Hall coefficient measurements. Majority of the samples studied here show degenerated semiconducting behaviour, meaning a continuous increase in both resistivity and Seebeck coefficient with increasing temperature. The decrease in ρ and

S at high temperatures can be ascribed to the existence of minority carriers, which is normally seen in narrow band gap semiconductors. The existence of minority carriers also lead to an increase in the high-temperature thermal conductivity, particularly in lightly doped samples.

The temperature-dependent total thermal conductivity and (κ) its lattice contribution (κ_L) are shown in Fig. 3c and Fig. 3d, respectively. It can be seen from Fig. 3c that due to the effective doping by arsenic, the resulting reduced resistivity (Fig. 3b) leads to an increased electronic thermal conductivity (κ_E) and therefore the total thermal conductivity. The electronic thermal conductivity can be determined by the Wiedeman–Franz law ($\kappa_E = LT/\rho$), where the temperature-dependent Lorenz factor (L) is estimated by the two-Kane-band model. The lattice thermal conductivity is determined by subtracting the electronic contribution from the total thermal conductivity via ($\kappa_L = \kappa - \kappa_E$). The room temperature lattice thermal conductivity for unintentionally doped tellurium is ~ 1.6 W m^{-1} K^{-1}, which is comparable with conventional thermoelectric lead chalcogenides[21,50,53,54] and bismuth/antimony tellurides[52,55], but significantly lower than those of group IV or group III–V semiconudctors[38,56]. The low lattice thermal conductivity is presumably due to the heavy atomic mass and relatively complex crystal structure[27]. Importantly, all the heavily doped samples show a nearly identical temperature dependence, and follow a nice T^{-1} decrease with increasing temperature, indicating a dominant phonon scattering by Umklapp process. According to

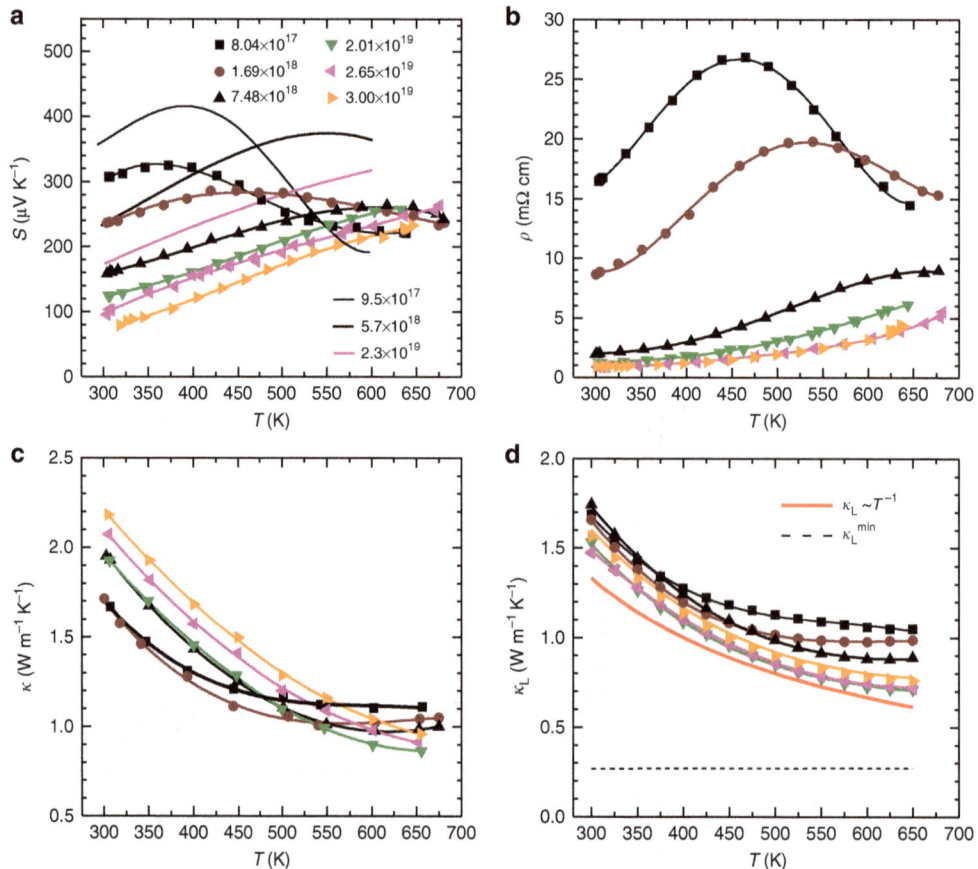

Figure 3 | Transport properties as a function of temperature. Seebeck coefficient (**a**) resistivity (**b**) total thermal conductivity (**c**) and lattice thermal conductivity (**d**) for p-tellurium. The *ab initio* calculated Seebeck coefficient[46] is included as solid curves for comparison in **a**. Majority of the samples studied here show a degenerate p-type semiconducting behaviour and a dominant phonon scattering by Umklapp process (red curve in **d**). The black dashed line in **d** shows the estimated minimal lattice thermal conductivity (κ_L^{min}) according to the Cahill model. Overlying lines are used to guide the eye and the carrier concentrations for the samples are shown in a unit of cm^{-3} in **a**.

the Cahill model[57], the measured sound velocities (2,287 ms^{-1} for longitudinal and 1,410 ms^{-1} for transverse ones, respectively), enable a determination of minimal lattice thermal conductivity (κ_L^{min}), which is also shown in Fig. 3d. It can be seen that there should be a reasonably big room for a reduction on κ_L for an even higher zT through well-demonstrated strategies in thermoelectrics such as solid solution[38,58] and nanostructuring[2–5].

Discussion
In summary, the inherently nested valence bands in tellurium enable an approximate hole pockets of 4, leading to a reasonably high power factor. In combination with its acceptably low thermal conductivity, elemental semiconducting tellurium surprisingly shows a high figure of merit, $zT = 1.0$, and therefore nicely fills up the high performance gap from 300 to 700 K for elemental thermoelectrics. The guiding principle here introduces a pure electronic effect for discovering high thermoelectric performance materials, application of other well-demonstrated independent strategies such as alloying or nanostructuring for decreasing the lattice thermal conductivity, is expected to lead to an even higher zT.

Methods
Synthesis. Polycrystalline Te samples were prepared by melting high purity element (>99.99%) at 823 K for 8 h, followed by quenching in cold water and annealing at 673 K for 3 days. Dopants including phosphorus (P), arsenic (As), antimony (Sb) and bismuth (Bi) were used to tune the carrier concentration. It is found that As-doping is the most effective to achieve a high enough carrier concentration that is needed for realizing the high zT at high temperatures, and is therefore focused on in this study. The ingot materials were ground into fine powders and hot pressing[59] at 673 K for 20 min under a uniaxial pressure of 90 MPa. The obtained dense pellet samples were a ∼ 12 mm in diameter and 1.5 mm in thickness.

Structural characterization. The phase impurity was characterized by X-ray diffraction (Dandong Haoyuan Instrument Co. LTD). The samples for the transmission electron microscope observation were prepared by mechanical polishing, dimpling and ion milling with liquid nitrogen. STEM images were taken with a JEOL ARM 200 equipped with a probe corrector. The obtained pellet samples showed a density higher than 98% of the theoretical one, where microvoids with a size of microns can be occasionally observed.

Transport property measurements. To be less involved in measurement uncertainties due to the possible hysteresis and the sample dimension determinations, the electrical transport properties including resistivity, Seebeck coefficient and Hall coefficient were simultaneously measured on the same pellet sample during both heating and cooling. The Seebeck coefficient was obtained from the slope of the thermopower versus temperature gradients of 0–5 K (ref. 60). The resistivity and Hall coefficient (R_H) were measured using the van der Pauw technique under a reversible magnetic field of 1.5 T. For a comparison, the Seebeck coefficient and resistivity for two high performance samples were also measured using a ULVAC ZEM-3 system. The thermal diffusivity (D) was measured through laser flash method with the Netzsch LFA457 system. The heat capacity (C_p) was assumed to be the Dulong–Petit limit and be temperature independent, which is consistent with the literature result at room temperature[61]. The thermal conductivity was calculated via $\kappa = dC_pD$, where d is the density measured using the mass and geometric volume of the pellet. All the transport property measurements were performed under vacuum in the temperature range of 300–650 K. The sound velocity was measured using an ultrasonic pulse-receiver (Olympus-NDT) equipped with an oscilloscope (Keysight). The uncertainty for each measurement of transport property (including S, σ and κ) is ∼ 5%.

References
1. Bell, L. E. Cooling heating, generating power, and recovering waste heat with thermoelectric systems. *Science* **321**, 1457–1461 (2008).
2. Hsu, K. F. et al. Cubic AgPb$_m$SbTe$_{2+m}$: Bulk thermoelectric materials with high figure of merit. *Science* **303**, 818–821 (2004).
3. Biswas, K. et al. High-performance bulk thermoelectrics with all-scale hierarchical architectures. *Nature* **489**, 414–418 (2012).
4. Poudel, B. et al. High-thermoelectric performance of nanostructured bismuth antimony telluride bulk alloys. *Science* **320**, 634–638 (2008).
5. Pei, Y., Lensch-Falk, J., Toberer, E. S., Medlin, D. L. & Snyder, G. J. High thermoelectric performance in PbTe Due to large nanoscale Ag$_2$Te precipitates and La doping. *Adv. Funct. Mater.* **21**, 241–249 (2011).
6. Kim, S. I. et al. Dense dislocation arrays embedded in grain boundaries for high-performance bulk thermoelectrics. *Science* **348**, 109–114 (2015).
7. Liu, H., Shi, X., Xu, F., Zhang, L. & Zhang, W. Copper ion liquid-like thermoelectrics. *Nat. Mater.* **11**, 422–425 (2012).
8. Qiu, W. et al. Part-crystalline part-liquid state and rattling-like thermal damping in materials with chemical-bond hierarchy. *Proc. Natl Acad. Sci. USA* **111**, 15031–15035 (2014).
9. Zhao, L. D. et al. Ultralow thermal conductivity and high thermoelectric figure of merit in SnSe crystals. *Nature* **508**, 373–377 (2014).
10. Morelli, D. T., Jovovic, V. & Heremans, J. P. Intrinsically minimal thermal conductivity in Cubic I-V-VI$_2$ semiconductors. *Phys. Rev. Lett.* **101**, 035901 (2008).
11. Pei, Y. et al. Convergence of electronic bands for high performance bulk thermoelectrics. *Nature* **473**, 66–69 (2011).
12. Liu, W. et al. Convergence of conduction bands as a means of enhancing thermoelectric performance of n-Type Mg$_2$Si$_{1-x}$Sn$_x$ solid solutions. *Phys. Rev. Lett.* **108**, 166601 (2012).
13. Pei, Y. et al. Stabilizing the optimal carrier concentration for high thermoelectric efficiency. *Adv. Mater.* **23**, 5674–5678 (2011).
14. Xie, H. et al. Beneficial contribution of alloy disorder to electron and phonon transport in half-heusler thermoelectric materials. *Adv. Funct. Mater.* **23**, 5123–5130 (2013).
15. Pei, Y., LaLonde, A. D., Heinz, N. A. & Snyder, G. J. High thermoelectric figure of merit in PbTe alloys demonstrated in PbTe–CdTe. *Adv. Energy Mater.* **2**, 670–675 (2012).
16. Zhang, Q. et al. Heavy doping and band engineering by potassium to improve the thermoelectric figure of merit in p-Type PbTe, PbSe, and PbTe$_{1-y}$Se$_y$. *J. Am. Chem. Soc.* **134**, 10031–10038 (2012).
17. Zhao, L. D. et al. All-scale hierarchical thermoelectrics: MgTe in PbTe facilitates valence band convergence and suppresses bipolar thermal transport for high performance. *Energy Environ. Sci.* **6**, 3346–3355 (2013).
18. Pei, Y., Wang, H., Gibbs, Z. M., LaLonde, A. D. & Snyder, G. J. Thermopower enhancement in Pb$_{1-x}$Mn$_x$Te alloys and its effect on thermoelectric efficiency. *NPG Asia Mater.* **4**, e28 (2012).
19. Pei, Y., LaLonde, A. D., Wang, H. & Snyder, G. J. Low Effective, Mass Leading to high thermoelectric performance. *Energy Environ. Sci.* **5**, 7963–7969 (2012).
20. Liu, X. et al. Low electron scattering potentials in high performance Mg$_2$Si$_{0.45}$Sn$_{0.55}$ Based thermoelectric solid solutions with band convergence. *Adv. Energy Mater.* **3**, 1238–1244 (2013).
21. Wang, H., Pei, Y., LaLonde, A. D. & Snyder, G. J. Weak electron-phonon coupling contributing to high thermoelectric performance in n-type PbSe. *Proc. Natl Acad. Sci. USA* **109**, 9705–9709 (2012).
22. Pei, Y., Wang, H. & Snyder, G. J. Band engineering of thermoelectric materials. *Adv. Mater.* **24**, 6125–6135 (2012).
23. Hoang, K., Mahanti, S. D. & Kanatzidis, M. G. Impurity clustering and impurity-induced bands in PbTe-, SnTe-, and GeTe-based bulk thermoelectrics. *Phys. Rev. B* **81**, 115106 (2010).
24. Vurgaftman, I., Meyer, J. R. & Ram-Mohan, L. R. Band parameters for III-V compound semiconductors and their alloys. *J. Appl. Phys.* **89**, 5815–5875 (2001).
25. Wood, C. Materials for thermoelectric energy conversion. *Rep. Prog. Phys.* **51**, 459–539 (1988).
26. Rosi, F. D., Hockings, E. F. & Lindenblad, N. E. Semiconducting materials for thermoelectric power generation. *R.C.A. Rev.* **22**, 82–121 (1961).
27. Goldsmid, H. J. *Thermoelectric refrigeration* (Plenum Press, 1964).
28. Agapito, L. A., Kioussis, N., Goddard, W. A. & Ong, N. P. Novel family of chiral-based topological insulators: elemental tellurium under strain. *Phys. Rev. Lett.* **110**, 176401 (2013).
29. Fukuroi, T., Tanuma, S. & Tobisawa, S. Electrical properties of antimony-doped tellurium crystals. *Sci. Rep.Res. Inst.Ser. A* **4**, 283–297 (1952).
30. Grosse, P. *Die Festkörpereigenschaften von Tellur* (Springer, 1969).
31. Fukuroi, T., Tanuma, S. & Tobisawa, S. On the electro-magnetic properties of single crystals of Tellurium. I: electric resistance, Hall effect Magneto-resistance, and Thermo-electric Power. *Sci. Rep.Res. Inst.* **Ser. A 1**, 373–386 (1949).
32. Gerlach, E. & Grosse, P. (eds) *The Physics of Selenium and Tellurium* (Springer, 1979).
33. Adams, A., Baumann, F. & Stuke, J. Thermal conductivity of selenium and tellurium single crystals and phonon drag of tellurium. *Phys. Status Solidi (b)* **23**, K99–K104 (1967).
34. Devyatkova, E. D., Moizhes, B. Y. & Thermal, A. S. I. Conductivity of Telltuium containing different concentration of additional element in the temperature range 80-480K. *Sov. Phys. Solid State* **1**, 555–569 (1959).
35. Reitz, J. R. Electronic Band Structure of Selenium and Tellurium. *Phys. Rev.* **105**, 1233 (1957).

36. Doi, T., Nakao, K. & Kamimura, H. The valence band structure of Tellurium. I. The k · p perturbation method. *J. Phys. Soc. Jpn* **28**, 36–43 (1970).

37. Fulkerson, W., Moore, J. P., Williams, R. K., Graves, R. S. & McElroy, D. L. Thermal Conductivity, electrical resistivity, and seebeck coefficient of silicon from 100 to 1300°K. *Phys. Rev.* **167**, 765–782 (1968).

38. Vining, C. B. in *CRC Handbook of Thermoelectrics*. (ed. Rowe, D. M.) Ch. 28 (CRC Press, 1995).

39. Dismukes, J. P., Ekstrom, L., Steigmeier, E. F., Kudman, I. & Beers, D. S. Thermal and Electrical properties of heavily doped Ge-Si alloys up to 1300°K. *J. Appl. Phys.* **35**, 2899–2907 (1964).

40. Yim, W. M. & Amith, A. Bi-Sb Alloys for magneto-thermoelectric and thermomagnetic cooling. *Solid State Electron.* **15**, 1141–1165 (1972).

41. Horst, R. & Williams, L. Potential figure of merit of the BiSb alloys. in *Proceedings of the Third International Conference on Thermoelectrics* **183**, 139–173 (1980).

42. Mahan, G. D. in *Solid State Physics*. Vol. 51 (eds Ehrenreich, H. & Spaepen, F.) 81–157 (Academic Press, 1998).

43. Doi, T., Nakao, K. & Kamimura, H. The valence band structure of Tellurium. II. The infrared absorption. *J. Phys. Soc. Jpn* **28**, 822–826 (1970).

44. Nakao, K., Doi, T. & Kamimura, H. The valence band structure of Tellurium. III. The landau levels. *J. Phys. Soc. Jpn* **30**, 1400–1413 (1971).

45. Joannopoulos, J. D., Schlüter, M. & Cohen, M. L. Electronic structure of trigonal and amorphous Se and Te. *Phys. Rev. B* **11**, 2186–2199 (1975).

46. Peng, H., Kioussis, N. & Snyder, G. J. Elemental tellurium as a chiralp-type thermoelectric material. *Phys. Rev. B* **89**, 195206 (2014).

47. Caldwell, R. S. & Fan, H. Optical properties of tellurium and selenium. *Phys. Rev.* **114**, 664–675 (1959).

48. Li, W. *et al.* Band and scattering tuning for high performance thermoelectric $Sn_{1-x}Mn_xTe$ alloys. *Journal of Materiomics*. doi:10.1016/j.jmat.2015.09.001 (2015).

49. Kane, E. Band structure of indium antimonide. *J. Phys. Chem. Solids* **1**, 249–261 (1957).

50. LaLonde, A. D., Pei, Y. & Snyder, G. J. Reevaluation of $PbTe_{1-x}I_x$ as high performance n-type thermoelectric material. *Energy Environ. Sci.* **4**, 2090–2096 (2011).

51. Pei, Y., Gibbs, Z. M., Balke, B., Zeier, W. G. & Snyder, G. J. Optimum carrier concentration in n-type PbTe thermoelectrics. *Adv. Energy Mater.* **4**, 1400486 (2014).

52. Schemer, H. & Scherrer, S. *Bismuth Telluride, Antimony Telluride and their Solid*. (eds Rowe, D. M.) Ch. 19 (CRC Press, 1995).

53. Pei, Y., LaLonde, A., Iwanaga, S. & Snyder, G. J. High thermoelectric figure of merit in heavy-hole dominated PbTe. *Energ. Environ. Sci.* **4**, 2085–2089 (2011).

54. Jian, Z. *et al.* Significant band engineering effect of YbTe for high performance thermoelectric PbTe. *J. Mater. Chem. C* **3**, 12410–12417 (2015).

55. Champness, C., Chiang, P. & Parekh, P. Thermoelectric properties of Bi_2Te_3-Sb_2Te_3 alloys. *Can. J. Phys.* **43**, 653–669 (1965).

56. Pei, Y. & Morelli, D. Vacancy phonon scattering in thermoelectric In_2Te_3-InSb solid solutions. *Appl. Phys. Lett.* **94**, 122112 (2009).

57. Cahill, D. G., Watson, S. K. & Pohl, R. O. Lower limit to the thermal conductivity of disordered crystals. *Phys. Rev. B* **46**, 6131 (1992).

58. Steele, M. & Rosi, F. Thermal conductivity and thermoelectric power of Germanium-Silicon alloys. *J. Appl. Phys.* **29**, 1517–1520 (1958).

59. LaLonde, A. D., Ikeda, T. & Snyder, G. J. Rapid consolidation of powdered materials by induction hot pressing. *Rev. Sci. Instrum.* **82**, 025104 (2011).

60. Zhou, Z. H. & Uher, C. Apparatus for Seebeck coefficient and electrical resistivity measurements of bulk thermoelectric materials at high temperature. *Rev. Sci. Instrum.* **76**, 023901 (2005).

61. DeSorbo, W. Concerning the low temperature specific heat of Tellurium. *J. Chem. Phys,* **21**, 764 (1953).

Acknowledgements

This work is supported by the National Natural Science Foundation of China (Grant No. 51422208, 11474219 and 51401147), the national Recruitment Program of Global Youth Experts (1000 Plan), the programme for professor of special appointment (Eastern Scholar) at Shanghai Institutions of Higher Learning, the Pujiang Project of Shanghai Science and Technology Commission (13PJ1408400), the fundamental research funds for the central universities and the Bayer-Tongji Eco-Construction and Material Academy (TB20140001). The authors thank Professors Xun Shi and Lidong Chen from the Shanghai Institute of Ceramics, CAS for their support on Seebeck coefficient and resistivity measurements for comparison.

Author contributions

S.L., W.L. and Y.P. designed the work. S.L. and J.C. prepared the samples and carried out the thermoelectric property measurements. S.L. and Z.C. Analysed the experimental data and established the model for the transport properties. B.G. performed the transmission electron microscope observation. All the authors wrote and edited the manuscript.

Additional information

14

Screen-printed flexible MRI receive coils

Joseph R. Corea[1], Anita M. Flynn[1], Balthazar Lechêne[1], Greig Scott[2], Galen D. Reed[1,3], Peter J. Shin[1,3], Michael Lustig[1] & Ana C. Arias[1]

Magnetic resonance imaging is an inherently signal-to-noise-starved technique that limits the spatial resolution, diagnostic image quality and results in typically long acquisition times that are prone to motion artefacts. This limitation is exacerbated when receive coils have poor fit due to lack of flexibility or need for padding for patient comfort. Here, we report a new approach that uses printing for fabricating receive coils. Our approach enables highly flexible, extremely lightweight conforming devices. We show that these devices exhibit similar to higher signal-to-noise ratio than conventional ones, in clinical scenarios when coils could be displaced more than 18 mm away from the body. In addition, we provide detailed material properties and components performance analysis. Prototype arrays are incorporated within infant blankets for *in vivo* studies. This work presents the first fully functional, printed coils for 1.5- and 3-T clinical scanners.

[1] Department of Electrical Engineering and Computer Sciences, University of California, Berkeley, California 94720, USA. [2] Department of Electrical Engineering, Stanford University, Stanford, California 94305, USA. [3] Department of Bioengineering, University of California, San Francisco, California 94722, USA. Correspondence and requests for materials should be addressed to M.L. (email: mlustig@eecs.berkeley.edu) or to A.C.A. (email: acarias@eecs.berkeley.edu).

Magnetic resonance imaging (MRI) is a widely used non-invasive imaging modality that provides an unsurpassed variety of high-resolution soft-tissue contrast and functional information[1,2]. Unlike computed tomography, MRI scans do not expose patients to harmful ionizing radiation, and are considered safe[3,4]. Unfortunately MRI data acquisition is inherently slow and signal-to-noise ratio (SNR) starved. This limits the spatial resolution, diagnostic image quality and typically results in long acquisition times that are prone to motion artefacts. Recent advances in MRI such as parallel imaging[5,6] and compressed sensing[7] enable reduction in scan time by collecting less data and using advanced reconstruction techniques. However, these reductions are ultimately limited by the SNR obtained during the shorter scan. SNR can be increased by the use of contrast agents[8,9] and higher-field scanners[10], but better receive coils[11,12] often provide more significant gains. A typical MRI paradigm consists of placing the patient in the large static field, giving rise to a net magnetic moment, using radio frequency (RF) pulses to excite the magnetization, which emits tiny amounts of RF energy at a characteristic resonant frequency proportional to the magnetic field, switching gradient magnetic fields to encode spatial information by manipulating the resonant frequency in three-dimensional space, and setting the spatial resolution and receiving a current signal from receive coils placed in close proximity to the body via Faraday induction throughout the duration of the encoding time. These received RF signals, which represent partial coded information, are amplified, digitized and stored. This procedure is repeated with different gradient waveforms until enough information is collected to form an image. It is the rate in which gradients switch that sets the amount of data collected each time. This rate is fundamentally limited by physiological constraints leading to relatively long exams that are uncomfortable, limit patient access, increase cost and more importantly, make the acquisitions susceptible to motion artefacts. The use of coil arrays provides additional SNR and scan acceleration by parallel imaging[5,6], mitigating the above shortcomings.

Surface receive coil and arrays are typically built to acquire images with the highest possible SNR for a specific area of the body[13]. Currently, the manufacturing process for commercial coils relies on the use of high-quality electronic components such as porcelain capacitors, thick copper traces and low-loss substrates. The electrical elements of each coil are packaged with medical grade, fire resistant materials that contribute to the size and weight of a given array. Figure 1a shows typical coil arrays used for head and chest imaging on an adult. If the same chest array were used on a smaller person or a child, there would be large gaps between the coil elements and the body, squandering much of the SNR gained from high-quality components. This problem is aggravated in small children, often requiring general anaesthesia to restrict motion during the exam. Therefore, good coil fit to obtain high SNR is often critical in shortening scan time and reducing complications[14,15]. For example, Fig. 1b illustrates the importance of coil placement on SNR. It compares a cervical spine image obtained by a printed flexible coil (right) that fits perfectly against the neck to one obtained using a conventional surface coil (left) mimicking a worse case scanning condition, placing it on the patient table 8 cm away from the base of the neck. This imaging case, while extreme, clearly highlights the importance of coil placement during a scan, showing a large loss in SNR from poor coil placement. When a receive coil is placed close to the body, its sensitivity to tissue signal is markedly increased. At the same time, the coil is strongly affected by the conductivity of human tissue, which can be modelled as additional resistive losses. The latter effect increases with field strength (frequency) and coil size to the point that in

most typical clinical imaging scenarios, the losses due to the sample dominate the intrinsic losses in the system (shown in Supplementary Fig. 1 using Supplementary Equations 1 and 2)[16,17]. This provides an opportunity in which novel solution processed electronic materials, which previously have been dismissed due to higher loss, can still perform adequately for receive coils without compromising image SNR. At the same time, these materials could provide significant added value of flexibility, lightness and mass manufacturing ability. In the past, several works have focused on adding flexibility and conformity to MRI receive arrays using a conductor sewn into fabric[18], a mercury-based conductor[19,20] and semi-flexible copper tape[21]. The advantage that printing has over previous techniques is the scalability and adaptability it possesses, qualities necessary to become a commonly used technology. Here, we report a powerful new approach that uses printing for the design and fabrication of MRI receive coils. Advances in electronic materials processed from solution have resulted in the demonstration of flexible electronic devices such as light emitting diodes, thin film transistors and photovoltaic devices[22-26]. Flexible electronics applications targeting the consumer electronics market are very exciting, but when devices are in contact with the human body advantages given by flexibility add considerable functionality[27-29]. Our method addresses the imaging limitations by enabling highly flexible, lightweight devices that conform to the human body, much like bespoke garments.

Results

Fabricating MRI coils using screen-printing on flexible substrates. Printing can be tailored by using different inks, substrates and techniques enabling custom pattern design[22]. Inkjet printing has previously been used to deposit metal layers for a single-element receive coil designed for a high-frequency small animal system; however, inkjet printing does not scale well[30]. We use screen-printing because coils require thick, low resistance conductive traces over a large area (that is, body size) at a high throughput, something not easily achieved with inkjet printing. The coils demonstrated here are screen-printed onto lightweight flexible substrates as illustrated in Fig. 1c. In screen-printing, ink is forced through a pre-patterned mesh onto a substrate[23] (Fig. 1c). To take advantage of flexibility and demonstrate feasibility, receive arrays were integrated with a baby blanket, as shown in Fig. 1d and used to scan volunteers. We envision that this technology could enable tightly fitting customized garments with integrated MRI coils, as illustrated in Fig. 1e. These could enable paediatric patients to receive shorter MRI exams with increased comfort and image quality.

Printed MRI receive coils designed for 1.5 and 3.0 T scanners. In the simplest sense, receive coils are formed by loops of wire integrated with capacitors. The resonant frequency of a conductive loop is determined by its inductance and capacitance, which both depend on the geometry and materials used. The size of the loop is typically predetermined, fixing the inductance. Therefore, tuning capacitors, C_t, are added to the loop to tune the desired resonant frequency. To minimize cable losses a matching capacitor, C_m, is added to match the input impedance to $50\,\Omega$ (refs 13,31). The schematic representation of a typical MRI coil is shown in Fig. 2a. We use octagonal coils with a diameter of 8.7 cm (ref. 32), with a conductor width of 0.5 cm and four capacitors evenly spaced throughout the loop as shown in Fig. 2b. To fabricate the coils, we print the coils layer by layer from solution, illustrated in Fig. 2c. The first layer of conductive ink is printed onto a thin flexible substrate, typically polyethylene terephthalate, forming the metal loop of the coil. The coil is

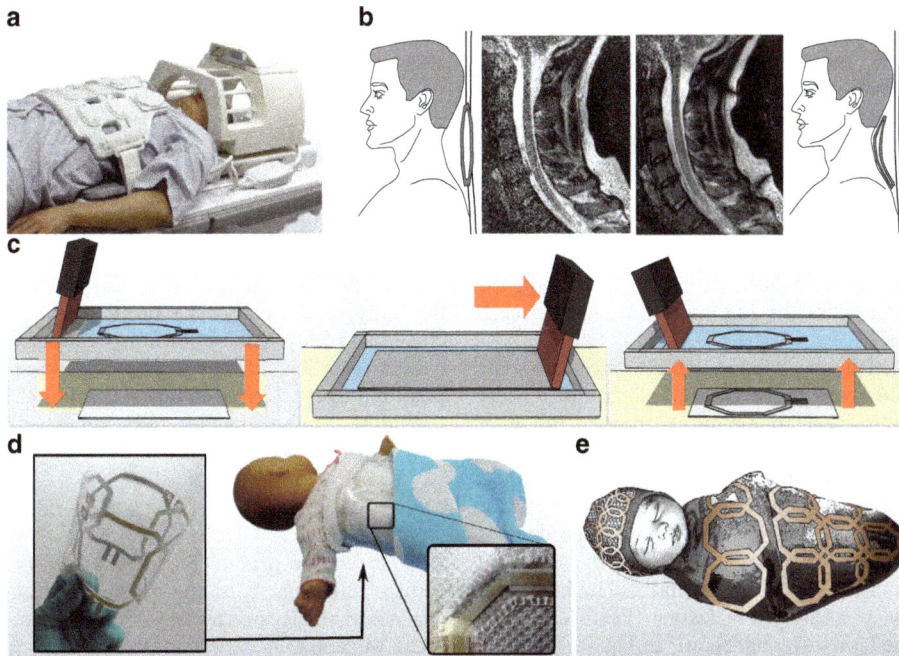

Figure 1 | RF receive coil arrays proximity to body results in better image SNR. (**a**) Conventional MRI receive arrays on the chest and head of a patient. (**b**) Cervical spine images of volunteer showing low-SNR when using a coil placed 8 cm away from the spine (left) and high SNR when placed against the skin (right). (**c**) Schematic representation of fabrication process of flexible printed coils. The screen is patterned with emulsion (blue) and shows the coil design. Ink (grey) is transferred to the substrate (white) during the screen-printing process. (**d**) Photograph of a printed flexible four-channel coil array fabricated on plastic film and integrated into an infant blanket. The inset shows how a printed coil is stitched into the fabric. (**e**) Concept drawing of an infant swaddle and hat with an integrated printed receive coil array.

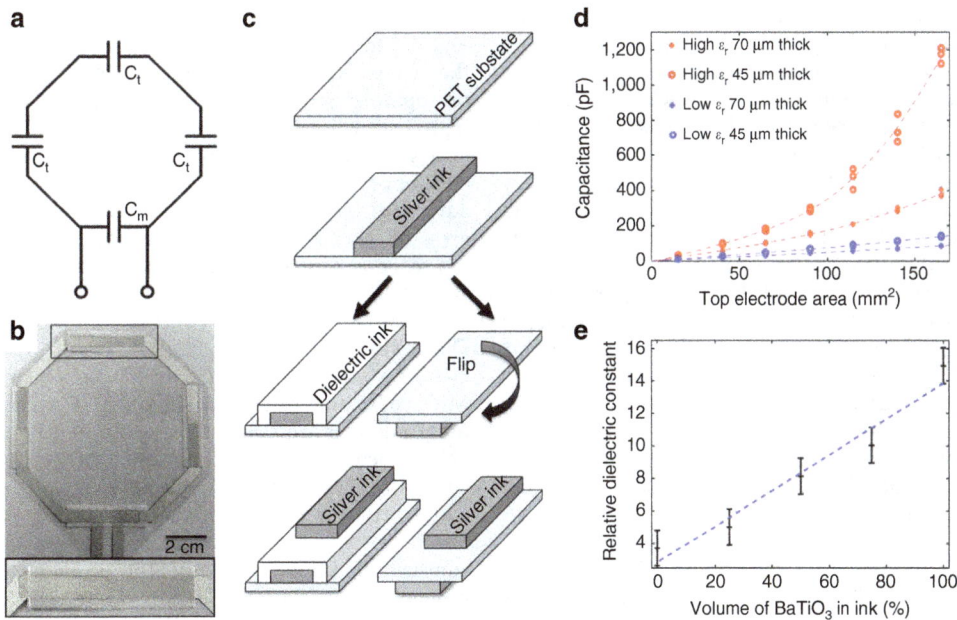

Figure 2 | Fabrication method and characterization of printed receive coils. (**a**) Schematic of a printed coil showing tuning, C_t, and matching, C_m, capacitors. (**b**) Photograph of a printed coil. Inset highlights top-down view of printed capacitor. (**c**) Coil printing process flow showing two optional possible processes: printed dielectric or using the substrate as a dielectric. (**d**) Dependence of capacitance with top electrode area, dielectric thickness and ink composition. (**e**) Relative dielectric constant, measured at 127 MHz, as the volume of barium titanate in the ink is increased. High dielectric constant is achieved with barium titanate ink, while low dielectric constant is achieved with ultraviolet-curable ink. Error bars show standard deviation.

completed with matching and tuning capacitors by printing a dielectric layer and the top electrode metal layer. The metal ink is a conductive silver micro-flake solution (Creative Materials 118-09 A/B). The insulating dielectric ink is a mixture of barium titanate (Conductive Compounds BT-101) and a ultraviolet-

curable resin-based ink (Creative Materials 116-20). When tuning a coil, it is desirable to control the capacitance to reach the Larmor frequencies used in MRI systems, ~ 64 MHz (1.5 T) and 127 MHz (3.0 T). In our printed process, the area of the top electrode along with the thickness and composition of the dielectric layer can modify the capacitance. The dependence of capacitance with printed dielectric thickness (40 and 75 µm), composition of dielectric ink ($\varepsilon_r = 15$ and $\varepsilon_r = 4$) and top electrode area is summarized in Fig. 2d. We have found that relative dielectric constant increases linearly with the concentration of barium titanate in the ink, as shown in Fig. 2e. This experimental window allowed us to achieve capacitances ranging from 2 to 1,200 pF, matching the ranges needed for tuning coils to different frequencies with similar coil geometry. We concluded that the most effective strategy for coarse tuning the capacitance is the control of the composition of the dielectric ink while changing the area of top electrode provides the fine-tuning needed to reach the specific frequencies.

Currently, the materials for printed devices are constantly evolving and better-suited inks are becoming available for a wider variety of substrates. Fortunately, our printing process lends itself very well to rapid prototyping with these new materials without the need for completely redesigning the printing process. For example, it is possible to use the substrate as the dielectric for the printed capacitors without a significant change in coil design. To demonstrate this case, we created a coil with an improved conductive silver ink (Silver micro-flake, Dupont 5064H) printed on both sides of a low-loss substrate, polyether ether ketone (PEEK), forming capacitors wherever the layers overlap, shown in Fig. 2c.

Signal-to-noise ratio of printed flexible coils. The viability of using our printed flexible coils in a clinical setting was first evaluated by characterizing image SNR, using a $NiCl_2$-doped saltwater (0.68 S m^{-1} at 3T) phantom as a model for human tissue. In our study, we fabricated five different types of coils as follows: an all-printed flexible coil; a copper coil in which printed capacitors replaced the conventional capacitors; a coil with printed silver conductors integrated with low-loss porcelain capacitors; a coil with improved printed silver conductors utilizing the low-loss PEEK substrate as the dielectric for capacitors; and a semi-flexible control coil composed of copper conductors and low-loss porcelain capacitors. All coils had the same geometry, and the control coil was not placed in any mechanical enclosure, allowing us to flex and measure SNR with all five types of coils. The SNR for each type of coil was calculated based on measured quality factor (Q) and by the method described in Hayes et al.[33] The loaded Q values (measured in close proximity to the phantom) were 6.7 (fully printed), 7.6 (printed capacitors), 9.7 (printed conductors, discrete capacitors), 9.5 (PEEK dielectric, printed conductors), and 11.4 (non-printed). Unloaded Q values are shown in Supplementary Tables 1 and 2. A diagram illustrating the experimental imaging set up and relative SNR in cross-section through the phantom are shown in Fig. 3a. We found that at 3 T and close to the phantom, surface coils present $79 \pm 3\%$, $86 \pm 3\%$, $93 \pm 3\%$ and $96 \pm 3\%$ relative SNR corresponding to fully printed, printed capacitors, PEEK capacitors and printed conductor, respectively. SNR values are normalized with respect to the control coil, and the SNR predicted from bench tests is shown as a dot in the bar graph of Fig. 3a. The fully printed coils show slightly higher relative SNR when used at 3-T compared with the 1.5-T system, as shown in Fig. 3b. This difference is attributed to the larger role that coil loss plays at lower frequencies as described in Darrasse et al.[16]. The fully printed and control coils were placed at increasing offsets from the phantom and the SNR was measured in actual scanned images. As expected and shown in Fig. 3c (tabulated in Supplementary Tables 3–6), SNR decreases as the distance from the phantom increases for both types of coils. We have shown that the SNR of control coil is surpassed by the printed coils at phantom offsets of 18 mm using PEEK as dielectric and 28 mm

using printed dielectric, when the printed coils are kept in close contact with the phantom. In our experiments, all coils could show additional losses from the cables used to connect to them when there is impedance mismatch, since we do not place preamplifiers directly on the coils. The calculated SNR considering the use of preamplifiers is shown as thin dotted lines in Fig. 3c (calculated with Supplementary Equations 3 and 4)[34]. When taking these losses into account, we find that printed coils based on PEEK show higher SNR when compared with control coils placed at 21 mm from the phantom. In addition, we show that SNR performance is minimally changed by the flexibility of the coils. The normalized SNR profiles around a curved and flat phantom of the same volume and composition are compared in Fig. 3d. The printed coils were wrapped around the phantom with a 22-mm radius of curvature. When placed in close proximity to the sample, coils are heavily loaded; therefore, changes in tuning due to the different geometry have a negligible effect on image SNR.

The fabrication process is scalable to larger area coverage. The focus of this work is on the design, fabrication and performance analysis of single surface coils. However, most clinical coils today come in arrays. We therefore believe it is important to demonstrate that our printing process is scalable to printing coil arrays. We therefore developed a proof of concept; a simple four-channel receive array to demonstrate array capacity of the process. The array was designed for 3 T and composed of four overlapped coil elements shown in Fig. 4a. This array was used as a proof of concept to image the cervical spine and knee, areas of the body where curvature can be a limiting factor. The cervical spine image is shown in Fig. 4b. The improved coverage and sensitivity of flexible coil arrays are illustrated by comparing knee images in Fig. 4c, taken with a single element, with Fig. 4d, taken with the array. Even though our prototype array did not focus on optimizing array geometry, coupling or fit, it produced high-quality images. Utilizing printing as a technique, other arrays with more elements can be built that include strain relief cuts, pre-curved substrates, more conformable materials or more advanced topography to better address even the most complex areas of the body—however, this advanced approach is beyond the scope of this preliminary work.

Discussion
Here, we present the first fully functional, printed and flexible MRI coils, and array for 1.5-and 3-T clinical scanners. Our unique designs achieved a remarkable 80–93% of the control coils SNR depending on materials and construction. While current custom-built conventional coils have less intrinsic loss compared with current printed materials, it is impractical for custom traditional arrays to be built for each patient. The ease of adjustability in the printing process lends itself well to new geometries and materials, as was shown with our coil using the substrate as the dielectric. While printed coils with printed dielectric capacitors exhibit lower SNR in a one-to-one comparison with control coils, printed coils which use the PEEK substrate as the dielectric are less lossy. In fact, they near the performance of the control coils, in a typical sample-loaded, sample-noise dominated regime. At the same time, when the printed coils conform to patients and can be placed in close proximity to the body, they provide similar or better image quality than conventional ones that do not necessarily fit as well. The printed array integrated into a baby blanket is extremely lightweight and provides new opportunities for conformity and comfort with a mass manufacturing technique.

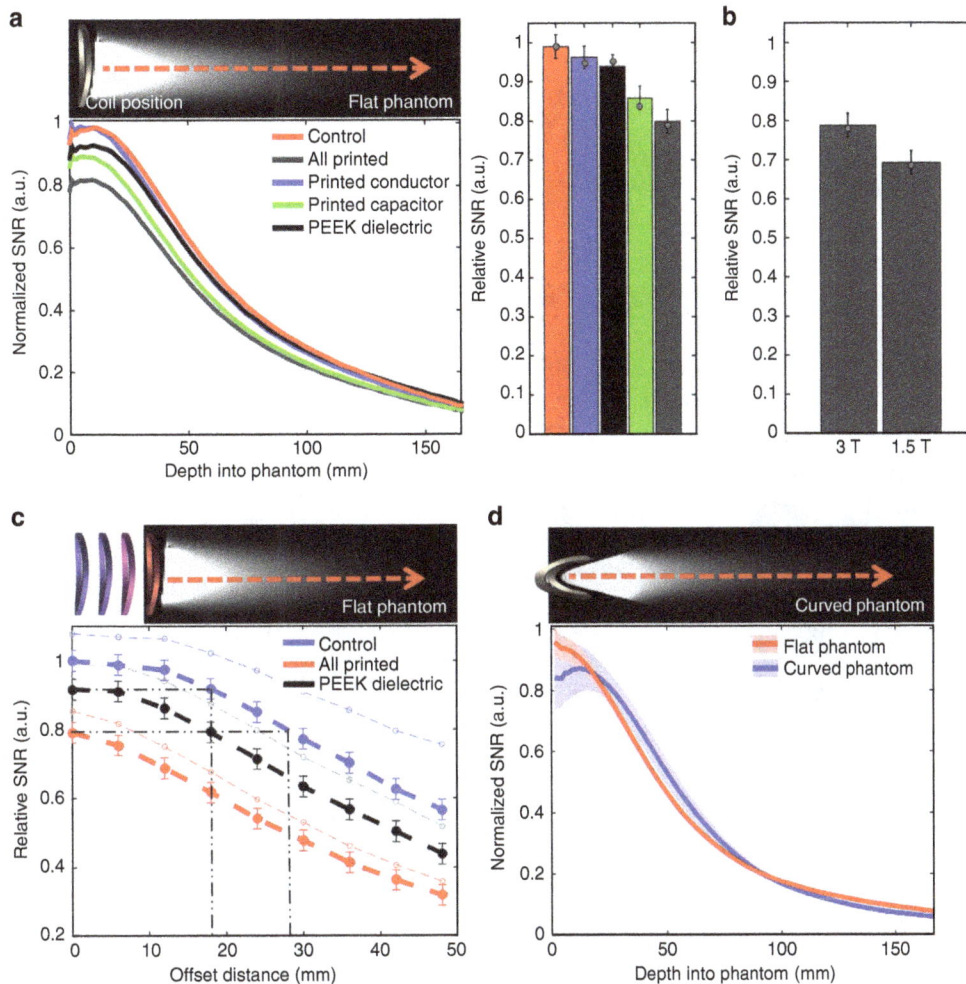

Figure 3 | 1.5- and 3-T scanner receive coil SNR characterization. (a) Normalized SNR versus depth into the phantom for coils fabricated with different permutations of printed components at 3 T, with schematic showing coil position 3 mm away from conductive fluid. Bar graph summarizes trends shown in relative SNR for each coil type. Dot on bar graph shows predicted SNR extracted from bench top quality factor measurements. **(b)** Relative measured (bars) and bench top-predicted (dots) image SNR of printed coils at 1.5 and 3 T. **(c)** Relative SNR for control and printed coils versus increasing coil offsets from the top surface of the phantom. Black dashed-dotted line highlights the position offset where the control coil shows equal SNR to the printed coils when the printed one has no offset from the top surface of the phantom. Light thin lines represent calculated best case performance when preamplifiers are added to the coil. Error bars show standard deviation. **(d)** Average normalized SNR profile for printed coils flexed around the surface of a curved saltwater phantom (blue) and placed on a flat phantom (red) at 3 T. Wide coloured bands indicate the s.d. across several coils.

Methods

Coil fabrication. For coils with a printed dielectric, the first metal layer of the conductive coil was screen-printed, using an ASYS APM101 screen printer, onto a 75-μm-thick polyethylene terephthalate film using a silver micro-flake ink, with flake size of 7 μm, purchased from Creative Materials (118-19A/B). The metal layer was annealed at 125 °C for 15 min before the deposition of the dielectric material. Two types of dielectric inks were used when printing the tuning and matching capacitors, a ultraviolet-curable resin (Creative Materials 116-20) and a BaTiO₃ ink (Conductive Compounds BT-101). A 60-μm-thick layer of ultraviolet-curable resin was used for 3-T coils and cured with a mercury arc lamp, with 24 W cm⁻² power flux for 3 s. Coils designed for 1.5-T scanners required higher dielectric constant. For these, 30-μm-thick layer of BaTiO₃ ink was used as the dielectric layer of the capacitors. After deposition, the BaTiO₃ ink was annealed on a hot plate at 125 °C for 15 min. The top electrode of the capacitors was formed with a 30-μm-thick layer of silver micro-flake ink. The finished coil was further annealed at 125 °C for 15 min on a hot plate.

For coils with the substrate as the dielectric, the two metal layers on opposite sides of a 75-μm-thick PEEK film were printed using a silver micro-flake ink, purchased from Dupont (5064H). These layers were annealed at 140 °C for 15 min.

The control coils were fabricated using a 70-μm-thick etched copper on 75-μm-thick Pyralux AP low-loss substrate. Advanced Technical Ceramics 100B low-loss porcelain-based capacitors were soldered onto the copper traces to form tuned coils.

Coil and component electrical characterization. Printed test capacitors were 5 mm wide and had an overlap ranging from 1 to 30 mm in length. Capacitors were mounted with plastic clamps on a copper PCB (printed circuit board) test fixture over a 30 × 30 mm² opening in the board. For each measurement, the experimental fixture with capacitors was tested and calibrated on an Agilent E5061B ENA network analyzer with open, short and 50 Ω loads on an identically shaped calibration board. All coils were tuned with the phantom used to image via an S11 measurement on an Agilent E5061B ENA network analyzer. The printed tuning capacitors set the correct resonant frequency while the matching capacitors set the impedance. The capacitance values were varied by using different dielectric inks or changing the size of the metal electrodes. The optimization process was repeated until the coils resonated at the Larmor frequency and displayed 50 Ω impedance [31]. Coil Q, was measured using an Agilent E5061B ENA network analyzer with two broadband magnetic field probes separated by 30 cm and facing each other to minimize the |S21| noise floor to approximately − 90 dB (ref. 33). During all measurements, care was taken to ensure coils and test apparatus were at least 50 cm away from conductive material to prevent artificial loading of the coil. To measure loaded Q, coils were taped to a cubic phantom 7 litre in volume, filled with solution of 3.356 g l⁻¹ NiCl₂*6H₂O and 2.4 g l⁻¹ NaCl for conductivity of 0.68 Sm⁻¹ at 123–127 MHz. Unloaded Q was measured using the magnitude of the S21 response with 1601 points averaged 16 times, centred at the Larmor frequency with the network analyzer set to a frequency span of 25 MHz, while loaded Q was measured with a span of 100 MHz.

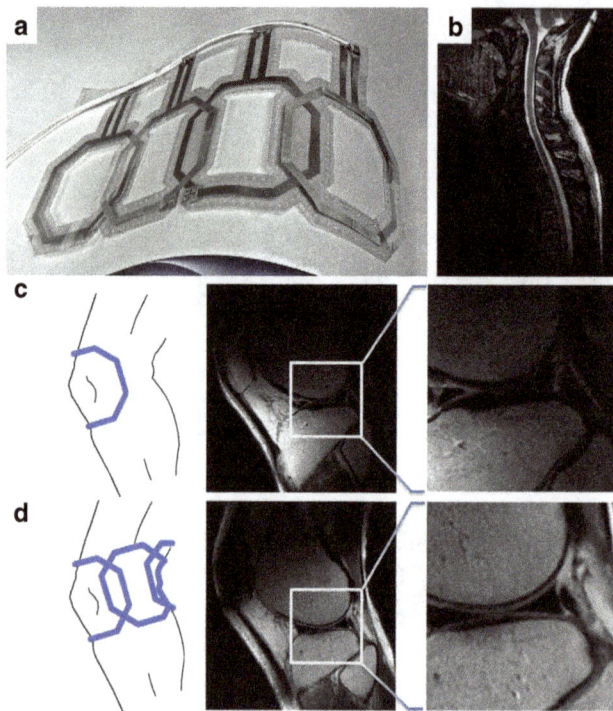

Figure 4 | *In vivo* imaging with flexible coil array at 3 T. (a) Proof of concept, prototype of printed flexible four-channel receive array.
(b) Sagittal cervical spine MRI image showing excellent penetration due to the conformity of the array. **(c)** Single-element MRI image of a knee.
(d) Scan showing the expected improved penetration using a four-channel array wrapped around the leg of a volunteer. Highlighted areas show region of interest with higher SNR from increased field of view from array.

MRI Imaging. Single-channel cervical spine images were acquired with a turbo spin echo sequence on a 3-T Siemens scanner with an echo time (TE) of 112 ms, repetition time (TR) of 3500 ms and flip angle (FA) 90°. The field of view was 200×200 mm^2 with resolution of 436 lines in phase encodes and readout directions with slice thickness of 4 mm. To compensate for imaging intensity variation due to the coil sensitivity, the image was normalized with respect to a uniform body coil image. SNR measurements were performed on the scanner by placing the coils on the same 7 litre phantom used to measure loaded Q. Coil bending tests were performed on different 7 litre phantoms containing a solution of $3\,\mathrm{g\,l}^{-1}$ of CuSO$_4$ and $3\,\mathrm{g\,l}^{-1}$ NaCl. To interface with the scanner, all coils were clamped into a test fixture that had a PIN (p-type/intrinsic/n-type) diode to deactivate the coil during the transmit phase of each scan. This fixture was connected via half-wavelength long coaxial cables to an interface box made by Stark Contrast (Erlangen Germany), which housed preamplifiers connected to the scanner. The half-wavelength coaxial cable contained a cable-trap circuit tuned to the Larmor frequency. Image reconstruction was unchanged from that used in conventional coils.

The scans used to measure SNR were two-dimensional gradient echo sequences on 3-T Siemens and 1.5-T General Electric scanners with a TE of 10 ms, TR of 438 ms and FA of 25°. Field of view was 200×200 mm^2 with resolution of 256 phase encodes and readouts. Slice thickness was 5 mm. The same prescan settings were used for all experiments, reducing variations due to differing magnet shimming, analogue gain and digital gain. Images of phantoms had SNR measured by dividing signal (that is, pixel in phantom), by an estimate of the noise sd. The noise was estimated from an image area with no signal containing at least 2,800 points at least 5 pixels from the edge of the image to avoid effects from the scanner's low-pass filter. The noise only area did not contain any ringing or streaking artefacts in the phase encode direction. To maintain a uniform offset for the experiments, different thicknesses of polycarbonate sheet were inserted between the phantom and each coil.

Fabrication and characterization. The four-channel array was fabricated by printing neighbouring coils on alternating sides of the substrate. The leads of each coil, along with a PIN diode, formed the dynamic disable circuit which detunes the coils during transmit. The PIN diode was attached using copper rivets pressed to the silver ink traces to form the electrical contact. The amount of coil overlap in the array was determined using two single coils and connecting each to one port of a network analyzer. Coils were overlapped until |S21| between the coils was

minimized. The array was connected to low input impedance preamplifiers to take advantage of preamplifier decoupling, to reduce the amount of cross-channel coupling. Anatomy images of the spine taken with the printed array used a T2-weighted spin echo sequence: TE 114.8 ms, TR 3,500 ms, FA 90° and two averages on a 3-T General Electric scanner. The scan sequence for imaging a volunteer's knee with a printed single-channel coil and with the four-channel array was a T2-weighted turbo spin echo with sequence parameters of TE of 39 ms, TR 3,000 ms, FA 150° and 1 average.

All experimental procedures were approved by the local ethical committee, Committee for protection of human subjects, University of California at Berkeley.

References

1. Lauterbur, P. C. Image formation by induced local interactions - examples employing nuclear magnetic-resonance. *Nature* **242**, 190–191 (1973).
2. Nishimura, D. *Principles of Magnetic Resonance Imaging* (Stanford Univ., 2010).
3. Wright, G. A. Magnetic resonance imaging. *IEEE Sig. Proc. Mag.* **14**, 56–66 (1997).
4. Hoult, D. I. & Lauterbur, P. C. Sensitivity of the zeugmatographic experiment involving human samples. *J. Magn. Reson.* **34**, 425–433 (1979).
5. Pruessmann, K. P., Weiger, M., Scheidegger, M. B. & Boesiger, P. SENSE: sensitivity encoding for fast MRI. *Magn. Reson. Med.* **42**, 952–962 (1999).
6. Griswold, M. A. *et al.* Generalized autocalibrating partially parallel acquisitions (GRAPPA). *Magn. Reson. Med.* **47**, 1202–1210 (2002).
7. Lustig, M., Donoho, D. & Pauly, J. M. Sparse MRI: the application of compressed sensing for rapid MR imaging. *Magn. Reson. Med.* **58**, 1182–1195 (2007).
8. Caravan, P., Ellison, J. J., McMurry, T. J. & Lauffer, R. B. Gadolinium(III) chelates as MRI contrast agents: structure, dynamics, and applications. *Chem. Rev.* **99**, 2293–2352 (1999).
9. Na, H. B., Song, I. C. & Hyeon, T. Inorganic nanoparticles for MRI contrast agents. *Adv. Mater.* **21**, 2133–2148 (2009).
10. Ocali, O. & Atalar, E. Ultimate intrinsic signal-to-noise ratio in MRI. *Magn. Reson. Med.* **39**, 462–473 (1998).
11. Roemer, P. B., Edelstein, W. A., Hayes, C. E., Souza, S. P. & Mueller, O. M. The NMR phased array. *Magn. Reson. Med.* **16**, 192–225 (1990).
12. Wiesinger, F., Boesiger, P. & Pruessmann, K. P. Electrodynamics and ultimate SNR in parallel MR imaging. *Magn. Reson. Med.* **52**, 376–390 (2004).
13. Vaughan, J. T. & Griffiths, J. R. *RF Coils for MRI* (John Wiley and Sons Ltd., 2012).
14. Frush, D. P., Bisset, G. S. & Hall, S. C. Pediatric sedation in radiology: the practice of safe sleep. *Am. J. Roentgenol.* **167**, 1381–1387 (1996).
15. Mason, K. P. *et al.* High dose dexmedetomidine as the sole sedative for pediatric MRI. *Pediatr. Anesth.* **18**, 403–411 (2008).
16. Darrasse, L. & Ginefri, J. C. Perspectives with cryogenic RF probes in biomedical MRI. *Biochimie* **85**, 915–937 (2003).
17. Suits, B. H., Garroway, A. N. & Miller, J. B. Surface and gradiometer coils near a conducting body: the lift-off effect. *J. Magn. Reson.* **135**, 373–379 (1998).
18. Nordmeyer-Massner, J. A., De Zanche, N. & Pruessmann, K. P. Stretchable coil arrays: application to knee imaging under varying flexion angles. *Magn. Reson. Med.* **67**, 872–879 (2012).
19. Malko, J. A., McClees, E. C., Braun, I. F., Davis, P. C. & Hoffman, Jr. J. C. A flexible mercury-filled surface coil for MR imaging. *AJNR Am. J. Neuroradiol.* **7**, 246–247 (1986).
20. Rousseau, J., Lecouffe, P. & Marchandise, X. A new, fully versatile surface coil for MRI. *Magn. Reson. Imaging* **8**, 517–523 (1990).
21. Adriany, G. *et al.* A geometrically adjustable 16-channel transmit/receive transmission line array for improved RF efficiency and parallel imaging performance at 7 Tesla. *Magn. Reson. Med.* **59**, 590–597 (2008).
22. Arias, A. C., MacKenzie, J. D., McCulloch, I., Rivnay, J. & Salleo, A. Materials and applications for large area electronics: solution-based approaches. *Chem. Rev.* **110**, 3–24 (2010).
23. Krebs, F. C. Fabrication and processing of polymer solar cells: A review of printing and coating techniques. *Sol. Energy Mater. Sol. Cells* **93**, 394–412 (2009).
24. Muller, C. D. *et al.* Multi-colour organic light-emitting displays by solution processing. *Nature* **421**, 829–833 (2003).
25. Yan, H. *et al.* A high-mobility electron-transporting polymer for printed transistors. *Nature* **457**, 679–U671 (2009).
26. Pierre, A. *et al.* All-printed flexible organic transistors enabled by surface tension-guided blade coating. *Adv. Mater.* **26**, 5722–5727 (2014).
27. Schwartz, G. *et al.* Flexible polymer transistors with high pressure sensitivity for application in electronic skin and health monitoring. *Nat. Commun.* **4**, 1859 (2013).
28. Lochner, C. M., Khan, Y., Pierre, A. & Arias, A. C. All-organic optoelectronic sensor for pulse oximetry. *Nat. Commun.* **5**, 5745 (2014).
29. Kim, D. H. *et al.* Stretchable and foldable silicon integrated circuits. *Science* **320**, 507–511 (2008).
30. Mager, D. *et al.* An MRI receiver coil produced by inkjet printing directly on to a flexible substrate. *IEEE Trans. Med. Imaging* **29**, 482–487 (2010).

31. Mispelter, J. l., Lupu, M. & Briguet, A. *NMR Probeheads for Biophysical and Biomedical Experiments: Theoretical Principles & Practical Guidelines* (Imperial College Press, Distributed by World Scientific, 2006).
32. Vasanawala, S. *et al.* in *International Socieity of Magentic Resosnance and Medicine 19* (Montreal, 2011).
33. Hayes, C. E. & Axel, L. Noise performance of surface coils for magnetic-resonance imaging at 1.5-T. *Med. Phys.* **12,** 604–607 (1985).
34. American Radio Relay League. in *Radio Amateur's Library Publication No 6 v* (ARRL, 2002).

Acknowledgements

The work was partially funded by the National Institutes of Health under R21 EB015628 and R01 EB019241 grants, the Hellman Family Fund, Sloan Research Fellowship, Okawa Research Grant, the Bakar Fellowship and GE Healthcare. In addition, we would like to thank Professor Dan Vigneron at the University of California San Francisco,- Professor John Pauly and Shreyas Vasawanawala at Stanford University for the time allowed on their scanners. We would also like to thank Thomas Grafendorfer, Dr James Tropp, Dr Fraser Robb and Dr Victor Taracila along with GE Healthcare, for guidance with electrical testing of coils. We like to thank Mark Davis for his help with illustrations, and Professor Ali Niknejad, Dr Martin Uecker, Frank Ong, John Tamir, Donjin Seo, Dr Joseph Cheng, Filip Maksimovic and Yasser Khan for helpful discussions.

Author contributions

A.C.A., J.R.C., A.M.F., M.L., and G.S. conceptualized the work. J.R.C., A.M.F., and B.L. carried out device fabrication and characterization. Scanning experiments were designed and performed by J.R.C., M.L., G.D.R., G.S., and P.J.S. J.R.C., B.L., and M.L. volunteered as test subjects for human scanning experiments. All authors discussed the results and commented on the manuscript.

Additional information

15

Widely tunable two-colour seeded free-electron laser source for resonant-pump resonant-probe magnetic scattering

Eugenio Ferrari[1,2,*], Carlo Spezzani[1,3,*], Franck Fortuna[4], Renaud Delaunay[5], Franck Vidal[6], Ivaylo Nikolov[1], Paolo Cinquegrana[1], Bruno Diviacco[1], David Gauthier[1], Giuseppe Penco[1], Primož Rebernik Ribič[1], Eleonore Roussel[1], Marco Trovò[1], Jean-Baptiste Moussy[7], Tommaso Pincelli[8], Lounès Lounis[6,9], Michele Manfredda[1], Emanuele Pedersoli[1], Flavio Capotondi[1], Cristian Svetina[1,10], Nicola Mahne[1], Marco Zangrando[1,11], Lorenzo Raimondi[1], Alexander Demidovich[1], Luca Giannessi[1,12], Giovanni De Ninno[1,13], Miltcho Boyanov Danailov[1], Enrico Allaria[1] & Maurizio Sacchi[6,14]

The advent of free-electron laser (FEL) sources delivering two synchronized pulses of different wavelengths (or colours) has made available a whole range of novel pump–probe experiments. This communication describes a major step forward using a new configuration of the FERMI FEL-seeded source to deliver two pulses with different wavelengths, each tunable independently over a broad spectral range with adjustable time delay. The FEL scheme makes use of two seed laser beams of different wavelengths and of a split radiator section to generate two extreme ultraviolet pulses from distinct portions of the same electron bunch. The tunability range of this new two-colour source meets the requirements of double-resonant FEL pump/FEL probe time-resolved studies. We demonstrate its performance in a proof-of-principle magnetic scattering experiment in Fe-Ni compounds, by tuning the FEL wavelengths to the Fe and Ni 3*p* resonances.

[1] ELETTRA—Sincrotrone Trieste, Area Science Park, 34149 Trieste, Italy. [2] Dipartimento di Fisica, Università degli Studi di Trieste, 34127 Trieste, Italy. [3] Laboratoire de Physique des Solides, Université Paris-Sud, CNRS-UMR 8502, Bât. 510, 91405 Orsay, France. [4] Centre de Sciences Nucléaires et de Sciences de la Matière, Université Paris-Sud, CNRS UMR 8609, Bât. 104-108, 91405 Orsay, France. [5] Laboratoire de Chimie Physique Matière et Rayonnement, Sorbonne Universités, UPMC Univ Paris 06, CNRS UMR 7614, 75005 Paris, France. [6] Institut des NanoSciences de Paris, Sorbonne Universités, UPMC Univ Paris 06, CNRS UMR 7588, 75005 Paris, France. [7] Service de Physique de l'Etat Condensé, DSM/IRAMIS/SPEC, CNRS UMR 3680, CEA Saclay, 91191 Gif-sur-Yvette, France. [8] Dipartimento di Fisica, Università degli Studi di Milano, 20133 Milano, Italy. [9] Ecole Normale Supérieure, PSL Research University, 75231 Paris, France. [10] Graduate School of Nanotechnology, Università degli Studi di Trieste, 34127 Trieste, Italy. [11] Istituto Officina dei Materiali, Consiglio Nazionale delle Ricerche, 34149 Trieste, Italy. [12] ENEA, Centro Ricerche Frascati, Via E. Fermi 45, 00044 Frascati, Italy. [13] Laboratory of Quantum Optics, University of Nova Gorica, 5001 Nova Gorica, Slovenia. [14] Synchrotron SOLEIL, L'Orme des Merisiers, Saint-Aubin, B.P. 48, 91192 Gif-sur-Yvette, France. * These authors contributed equally to this work. Correspondence and requests for materials should be addressed to E.A. (email: enrico.allaria@elettra.eu) or to M.S. (email: maurizio.sacchi@insp.jussieu.fr).

Free-electron laser (FEL) sources covering the wide spectral range from extreme ultraviolet to hard X-rays represent a breakthrough in photon science, with applications in physics, chemistry and biology. Many aspects of the spectral and temporal characteristics of the FEL pulses can be tailored to specific experimental needs by an accurate control of the lasing process, in the so-called beam by design approach[1]. The ability to run the FEL source in two-colour configuration, that is, to create two synchronized FEL pulses of differing wavelengths, has enormous potential for femtosecond time-resolved studies[2,3] as it opens up unique opportunities for studying the dynamic response in atomic, molecular and solid state systems by selectively tuning electron resonances in atoms. As a consequence it has engendered major research[4–8] and development[9–15] efforts at all FEL facilities worldwide, with the ambition of attaining wide-ranging colour tunability and timing control.

Various two-colour schemes have been proposed, both for seeded[2,9,10,14] and for self-amplified spontaneous emission (SASE)[11–13] FEL sources. Initial configurations delivered two short FEL pulses with a controlled temporal separation in the range of a few hundred femtoseconds and a small photon wavelength separation ($\sim 1\%$). Such configurations, where a single electron bunch generates the two FEL pulses, have served users for experiments both at seeded[2] and at SASE[11–13] facilities. In the case of SASE, differing photon wavelengths are obtained by dividing the radiator in two slightly detuned sections[11]. In the case of external seeding, the FEL wavelength separation is controlled by acting on the seed laser wavelength and by taking advantage of a residual controllable energy chirp on the electron beam[2,10]. For self-seeding schemes, it has been demonstrated[14] that two seeded FEL pulses can be generated using two distinct Bragg diffraction lines in the self-seeding crystal recombined within the taper-tuned undulators. The possibility of producing two colours with a wider spectral separation (up to 30%) has been demonstrated recently at the SACLA hard X-ray SASE source by using the capabilities of a variable gap undulator[13].

Until now, no configuration that generates two pulses with independently tunable wavelengths over a wide spectral range had been designed for externally seeded FELs. A whole new class of pump–probe experiments that require both pump and probe to be element selective is created by combining the full coherence of seeded FELs with a broad and independent tunability of the two colours.

Over the last decade, time-resolved studies made frequent use of short X-ray pulses as a probe that is coupled to an optical laser pump. Femto-slicing at synchrotrons[16–22], high harmonic generation in gases[23–26] and FEL sources[27–32] deliver extreme ultraviolet and X-ray pulses with sub-100-femtoseconds duration that have been used for studying the ultrafast dynamics of magnetic[16–27,29,31] and structural[20,28–30] order in optical-laser pump/X-ray probe experiments. Tuning the wavelength to an atomic resonance provides the probe with element selectivity, which is of considerable interest especially for magnetic studies.

Developing FEL sources that can produce two pulses with independently selectable wavelengths for the pump and the probe and with a well-defined time separation obviously widens the potential of FEL radiation for studying the dynamics of a process and makes it possible to associate the pump energy to a specific electronic excitation of a given element. One field that will surely profit from this new tool is magnetization dynamics in $3d$-transition-metal and rare-earth based oxides and compounds[18,19,22,33–37]: the presence of highly localized $3d$ and $4f$ orbitals and of mediated exchange interactions suggests that associating the pump energy to a specific electronic excitation will

influence the magnetization dynamics profoundly, compared with using a non-resonant pump.

In the proof-of-principle time-resolved scattering experiment on Fe–Ni compounds described here, we use the new two-colour configuration of the externally seeded FERMI FEL source to generate, from the same electron bunch, two synchronized pulses with up to 30% spectral separation. The pump FEL pulse excites the Fe $3p \rightarrow 3d$ transition resonantly, while the second FEL pulse, tuned to the Ni $3p \rightarrow 3d$ resonance, probes the ultrafast Ni magnetization dynamics. The experiment successfully reveals the potential of this new source for investigating structural, electronic and magnetization dynamics in the fields of condensed matter as well as atomic and molecular physics.

Results

Two-colour seeded FEL with wide wavelength tunability. The experiment was performed at the FERMI facility[38,39], which is a seeded FEL operated in the high-gain harmonic generation (HGHG) mode[40,41]. The chosen configuration (see Methods) provided a relatively long (~ 1 ps) electron bunch interacting with a short (~ 100 fs) ultraviolet laser pulse (seed laser) in the first undulator section called the modulator (Mod in Fig. 1a). As a consequence of this interaction, the electron beam energy is modulated with a periodicity imposed by the seed laser wavelength λ_{seed}. Following a magnetic chicane that works as a dispersive section (DS in Fig. 1a), the energy modulation is converted into a density modulation (bunching), which has strong harmonic components. Finally, in a second long undulator section called the radiator (Rad in Fig. 1a), the bunched electrons generate coherent FEL emission at one of the harmonics of the seed laser which is selected by setting the undulator gap. The advantages of HGHG with respect to SASE FEL stem from the fine control of the initial bunching, making it possible to generate FEL pulses with a high degree of longitudinal coherence[42]. Moreover, since only electrons interacting with the seed laser are bunched, this scheme provides a good control of the FEL temporal properties[43].

Two FEL pulses with a controlled delay can be produced by seeding the same electron bunch with two seed pulses[2]. Since in the HGHG seeding process the final FEL wavelength is determined mainly by λ_{seed} and it must be close to one of its harmonics, a way for delivering two-colour FEL pulses with very different wavelengths ($>10\%$ separation) relies on seeding the electron beam with two laser pulses and on sustaining the amplification process at both wavelengths independently (Fig. 1a).

To achieve this, some constraints have to be dealt with. Both seed wavelengths λ_{seed_1} (for the probe) and λ_{seed_2} (for the pump) have to modulate the electron energy in the interaction region efficiently so their separation must be within the modulator working bandwidth. The two seed pulses modulate the energy in distinct regions of the electron beam. For each region, the dispersive section converts the energy modulation into an electron density modulation that carries all the harmonic components of the corresponding seed wavelength, either λ_{seed_1} or λ_{seed_2}. The electron beam is now ready for the amplification of one of these harmonics, selected by the resonance condition of the radiator (undulator gap). A large separation between the two colours can be obtained by dividing the radiator into two subsections (Rad_1 and Rad_2 in Fig. 1a), one resonant at $\lambda_{FEL\_1} = \lambda_{seed\_1}/m$ and the other at $\lambda_{FEL\_2} = \lambda_{seed\_2}/n$, with m and n integers. Since the radiator bandwidths are markedly narrower than the modulator one, we can emit efficiently the pump (or the probe) beam from one radiator subsection only, while suppressing its amplification in the other, selectively

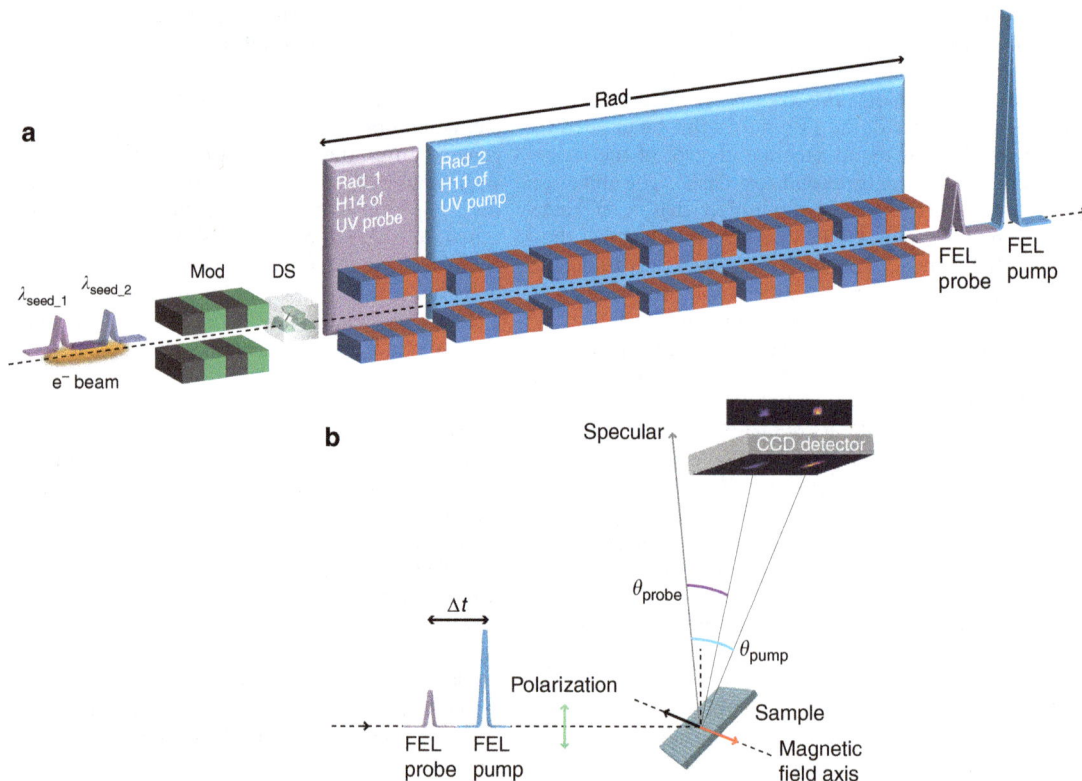

Figure 1 | Schematic setup for a two-colour double resonance FEL experiment. (**a**) Two-colour seeded FEL source configuration: the modulator (Mod), dispersive (DS) and radiator (Rad) sections of the FEL source are outlined. In the modulator section, two ultraviolet (UV) laser pulses of wavelength λ_{seed_1} and λ_{seed_2} delayed by Δt interact with the same electron bunch, imposing an energy modulation that is converted into density modulation in DS. The first radiator subsection Rad_1 is tuned to the 14th harmonic of λ_{seed_1} and the second subsection Rad_2 is tuned to the 11th harmonic of λ_{seed_2}, generating the FEL probe and pump pulses, respectively (see Methods). (**b**) Magnetic scattering experiment: the two linear p-polarized FEL pulses reach the magnetic grating sample and diffract at different angles according to their wavelengths. The diffracted intensities are recorded by a two-dimensional detector (CCD camera). The wavelength separation between pump and probe is detected as a spatial separation at the CCD, while their time separation Δt is defined by the delay between the two seed pulses.

(see Methods). Finally, constraints on the temporal separation Δt between the two FEL pulses are set by the need to avoid interference between the laser seeds (lower limit) and by the electron bunch duration (upper limit). In the example reported below, we spanned delays ranging from 300 to 800 fs.

The Fe-$3p$ resonant-pump and Ni-$3p$ resonant-probe test experiment (Fig. 1b) used two FEL pulses tuned to $\lambda_{FEL\_2} = 23.2$ nm and $\lambda_{FEL\_1} = 18.7$ nm, corresponding to the 11th harmonic of $\lambda_{seed\_2} = 255$ nm and to the 14th harmonic of $\lambda_{seed\_1} = 261.5$ nm, respectively. To this purpose, a special configuration of the FERMI seed laser was implemented based on the combined use of two ultraviolet pulses originating from a common infra-red source through two separated generation channels. One made use of an optical parametric amplifier (OPA) for producing the 255 nm seed, the other of a third harmonic generation (THG) setup for the 261.5 nm seed. This approach made the twin seeding possible at two different ultraviolet wavelengths, one of them tunable via the OPA (see Methods). We tested different distributions of the six undulator modules over the Rad_1 and Rad_2 radiator subsections (see Fig. 1a). We obtained different power distributions between pump and probe by going from one module in Rad_1 and five in Rad_2 to three modules in each subsection. All the configurations provided satisfactory stable conditions for producing two-colour FEL pulses. Since in our test experiment the pump is required to be more energetic than the probe, five of the six available radiator modules were tuned to produce 23.2 nm pulses (Rad_2 in Fig. 1a), while the remaining module (Rad_1) was tuned to the probe wavelength. It was

important, in this configuration, that Rad_1 was the first of the undulator modules, to prevent the smearing of the electron density modulation along the radiator section to degrade its performance. We verified also that one can switch readily the FEL pump and probe wavelengths, by reversing the time delay between OPA and THG generated seed pulses and inverting the gap settings of the Rad_1 and Rad_2 radiator subsections.

Figure 2a shows the spectral distribution of the two ultraviolet seed laser pulses and Fig. 2b shows the FEL pulse energy as a function of the modulator gap, when using only the ultraviolet -probe or only the ultraviolet -pump seeds. The two curves of Fig. 2b, which are normalized to the same amplitude, illustrate at each wavelength the extreme sensitivity of the FEL intensity to the modulator setting. A modulator gap of 19.94 mm optimizes the FEL pump emission when seeding at λ_{seed_2}, while a gap of 19.60 mm is best when seeding at λ_{seed_1} to produce the FEL probe pulse. The gap can be used as an adjustable parameter for the fine control of the relative efficiency in the generation of the pump and probe FEL pulses, thanks to the $\sim 3\%$ resonance bandwidth of the modulator (Supplementary Fig. 1). In our case, a good compromise was found at a gap of 19.75 mm, which made it possible to generate both $\lambda_{FEL\_1} = 18.7$ nm and $\lambda_{FEL\_2} = 23.2$ nm pulses, albeit with a reduced intensity. For the Fe–Ni experiment, the FERMI FEL source was characterized by pulse energies of up to $\sim 10\,\mu J$ at the pump wavelength and $\sim 1\,\mu J$ at the probe wavelength using these parameters. Once converted into a fluence F at the sample surface (see Methods), these values were sufficient to reach, in our experiment, the damage threshold

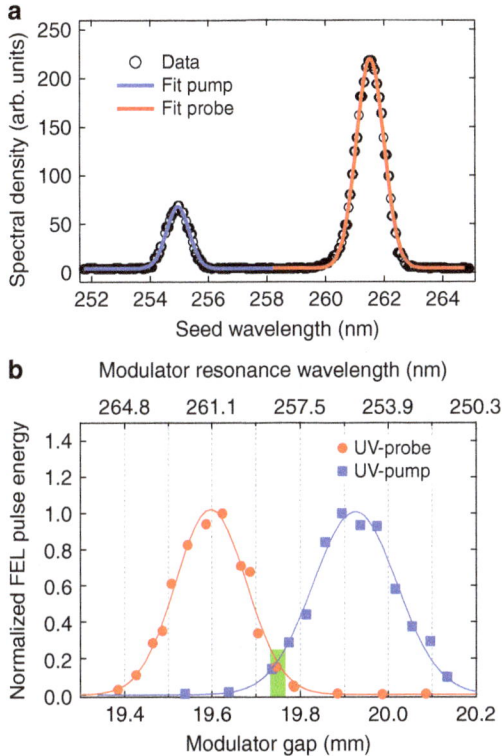

Figure 2 | Seed pulses and modulator setting. (**a**) Spectral properties of the ultraviolet (UV) laser twin-seed source. Lines are Gaussian fits to the $\lambda_{seed\_1} = 261.5$ nm (red line) and $\lambda_{seed\_2} = 255$ nm (blue line) probe and pump contributions, respectively. (**b**) Modulator gap dependence of the FEL output for the two seed wavelengths. Circles and squares refer to seeding at 261.5 and 255 nm, respectively. Each point is the average of 100 consecutive FEL shots. Lines represent Gaussian fits to the intensity distributions. The curves are normalized to the same average maximum, showing that tuning the modulator gap to 19.75 mm (vertical green bar) makes it possible to seed with two colours simultaneously, preserving a fraction of the maximum pulse energy.

and single shot detection conditions for the pump and the probe pulses, respectively.

Resonant-pump/resonant-probe magnetic scattering experiment. We tested the two-colour twin-seeded FEL source by studying the resonant-pump/resonant-probe magnetization dynamics in Fe–Ni samples, using the IRMA reflectometer[44] installed at the DiProI beamline[45,46]. The samples were a 20-nm-thick permalloy ($Ni_{0.81}Fe_{0.19}$ alloy) film deposited on a Si grating and a 12.5-nm-thick $NiFe_2O_4$ layer epitaxially grown on $MgAl_2O_4(001)$. Both samples were structured as line gratings with a period of ~ 600 nm (see Methods). They worked as dispersive elements, separating different wavelengths at the level of the two-dimensional in-vacuum charge-coupled device (CCD) detector[23,24]. All Bragg peaks generated by the grating samples at different wavelengths fell within the angular acceptance of the detector and could be collected simultaneously (see Fig. 3).

The FEL polarization was set to linear vertical to optimize the sensitivity to the sample magnetization in transverse geometry[23,24,47–49], that is, with the external magnetic field applied normal to the scattering plane and parallel to the lines of the grating sample (see Fig. 1b). After an initial 80 mT magnetic pulse, the scattered intensity was collected in an applied field of 20 mT, guaranteeing the sample magnetic

saturation (see Methods). In the following, the magnetic signal is defined as an asymmetry ratio, that is, as the difference between scattered intensities measured for opposite signs of the applied field divided by their sum, as shown in Fig. 4. At each given delay Δt, the Ni magnetic signal was measured as a function of the pump fluence F (see Methods for the relationship between FEL pulse energy and fluence at the sample). The pump wavelength was tuned either to the Fe-3p resonance ($\lambda_{FEL\_2} = 23.2$ nm) or off-resonance ($\lambda_{FEL\_2} = 25.5$ nm), the latter being obtained simply by tuning the radiator subsection Rad_2 to the 10th harmonic of the λ_{seed_2} seed laser wavelength, instead of the 11th. It is worth underlining that, according to calculations based on tabulated optical constants[50] (see also http://henke.lbl.gov/optical_constants/), the fraction of pump energy absorbed by the sample at 23.2 nm and at 25.5 nm differs by less than 2% for both permalloy and ferrite films.

First, we explored the ultrafast Ni demagnetization while varying the delay Δt between the FEL probe and pump by adjusting the delay between the corresponding seed laser pulses. An example of delay dependence spanning the 300–800 fs range is shown in Fig. 5 where the Ni magnetic signal is reported after a Fe-3p resonant pump pulse with fluence $F = 10$ mJ cm^{-2} (dots and squares refer to Ni-ferrite and permalloy samples, respectively). The asymmetry ratio in the Bragg peak intensity is calculated over a limited detector area of $\sim 100 \times 100$ μm^2 to ensure homogeneous pump fluence and the Ni magnetic signal is normalized to its static value measured with no pump.

The main advantage of this novel two-colour scheme over those developed previously at the FERMI seeded source[2] is its ability to tune both λ_{FEL_1} and λ_{FEL_2} to selected values over a broad range. It is also important to stress that this scheme makes the switching between on- and off-resonance pumping fast and easy. As mentioned before, this can be achieved simply by changing the gap of the Rad_2 radiator subsection for selecting a different harmonic of the λ_{seed_2} wavelength. An example of on/off-resonance pumping is given in Fig. 6. It shows the Ni magnetic signal (normalized to its static value) measured at a fixed time delay of ~ 400 fs for a FEL pump wavelength tuned to the Fe-3p resonance ($\lambda_{FEL\_2} = 23.2$ nm, red circles) or off-resonance ($\lambda_{FEL\_2} = 25.5$ nm, blue squares) as a function of the pump fluence F. The permalloy results (Fig. 6a) do not reveal a measurable effect of the pump wavelength: both curves show the same F-dependence of the Ni magnetic signal, which attains a $\sim 50\%$ reduction at $F \sim 10$ mJ cm^{-2}. On the contrary, pumping at the two on/off-resonance wavelengths results in an apparent difference in Ni demagnetization behaviour when F exceeds ~ 5 mJ cm^{-2} in the case of Ni-ferrite (Fig. 6b).

Although a detailed discussion of the results reported in Figs 5 and 6 is not within the scope of this communication, the observed differences between ferrite and permalloy behaviour can be ascribed to the direct hybridization of delocalized Fe and Ni 3d orbitals in ferromagnetic permalloy versus indirect exchange (via oxygen) of more localized 3d orbitals in ferrimagnetic $NiFe_2O_4$. These early results are intriguing and more studies are under consideration to shed light on the observed pump wavelength dependence.

Discussion

We have developed and tested a new FEL setup capable of delivering two-colour time-delayed pulses with independent wavelength tunability over a wide spectral range (18.7–25.5 nm). Combined with the seeded nature of the FERMI source[39], this provides improved conditions for two-colour FEL experiments that require tuning both the pump and the probe to selected atomic resonances. The potential of this two-colour

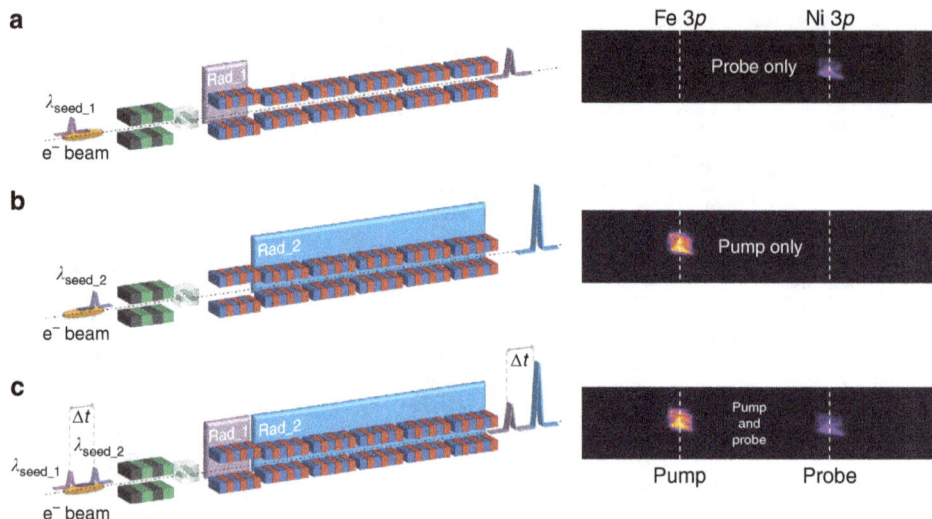

Figure 3 | FEL source configuration and scattering data recording. Diffracted intensity from the 20-nm-thick permalloy grating sample at 46.3° incidence. Data are collected under different seeding conditions (schematics on the left) using a position-sensitive CCD detector (images on the right); the 1,025 × 202 pixel images correspond to 13.84 × 2.73 mm^2 and cover ~1.48° in scattering angle. (**a**) The $\lambda_{seed\_1} = 261.5$ nm laser pulse is sent through the modulator, turning on the Ni-3p resonant FEL emission at $\lambda_{FEL\_1} = 18.7$ nm in Rad_1 (14th harmonic) and no emission from Rad_2. (**b**) The $\lambda_{seed\_2} = 255$ nm laser pulse generates the Fe-3p resonant FEL emission at $\lambda_{FEL\_2} = 23.2$ nm in the radiator section Rad_2 (11th harmonic) and no emission from Rad_1. (**c**) Both seed laser pulses, delayed by Δt, interact with the electron bunch, generating Fe-3p resonant pump and Ni-3p resonant probe FEL pulses, also delayed by Δt.

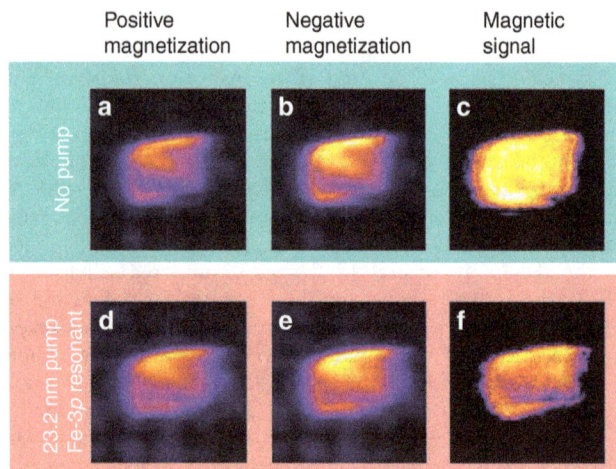

Figure 4 | Magnetic signal in the scattering data. Diffracted intensity at the Ni-3p resonant probe wavelength with no pump (**a**-**c**) and following a Fe-3p resonant pump pulse (**d**-**f**). The pump fluence is $F = 8$ mJ cm^{-2}, the delay Δt is 450 fs. (**a,d**) and (**b,e**) Diagrams refer to a positive and negative saturating magnetic field, respectively. The magnetic signal, expressed as the asymmetry ratio, is shown in **c** and **f**. Each picture is 128 × 128 pixels, corresponding to 1.73 × 1.73 mm^2.

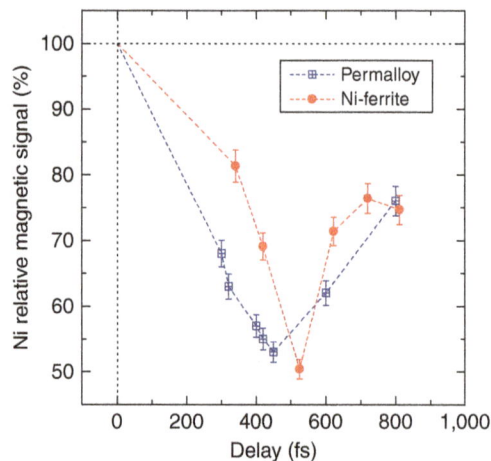

Figure 5 | Time-dependent magnetic signal. Ni demagnetization in the permalloy (blue squares) and the Ni-ferrite (red circles) samples at several delays Δt between probe and pump pulses ($F = 10$ mJ cm^{-2}). Vertical error bars represent s.d. (see Methods). The maximum fluctuation in the pump-probe delay over the measurement duration (± 5 fs, see Methods) is smaller than the point width. Lines are a guide to the eye.

scheme has been demonstrated by a scattering experiment that probes the magnetization dynamics in systems containing two magnetic elements, Fe and Ni. Undoubtedly, it can find original applications in many other fields of condensed matter, atomic and molecular physics.

From a technical point of view, the solution that we propose is based on seeding the same electron bunch with two independent laser pulses and on splitting the FEL radiator into two subsections. On one hand, this solution offers the possibility of selectively tuning the two FEL colours over a very wide range. It may go well beyond the 30% bandwidth demonstrated here, by amplifying different harmonics of the seed wavelengths in each radiator subsection. On the other hand, using two laser seeds that modulate the same electron bunch, and two radiators impose some constraints on the relationship between the λ_{FEL_1} and λ_{FEL_2} wavelengths, both in terms of FEL intensity and of possible gaps in the range of wavelengths that can be spanned.

Figure 7 summarizes the calculated source performance when λ_{FEL_1} and λ_{FEL_2} span the 16–28 nm range, showing that marked intensity variations are present. The colour code represents the relative modulator efficiency for each couple of wavelengths, calculated assuming that the modulator resonance is set to the average value of λ_{FEL_1} and λ_{FEL_2}. Both seed wavelengths are

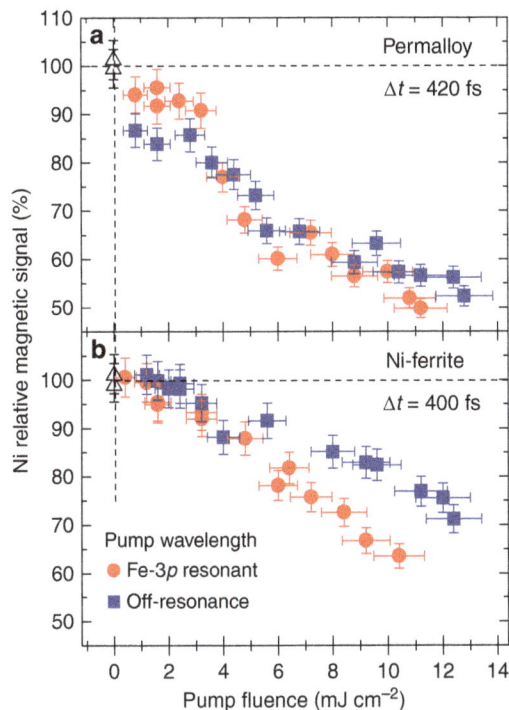

Figure 6 | Resonant versus non-resonant pumping. Pump fluence dependence of the Ni demagnetization in permalloy (**a**) and Ni-ferrite (**b**) at ~400 fs delay, comparing the results for Fe-3p resonant ($\lambda_{FEL\_2} = 23.2$ nm, red circles) and non-resonant ($\lambda_{FEL\_2} = 25.5$ nm, blue squares) FEL pump pulses. The Ni magnetic signal is reported as the asymmetry ratio in the Bragg peak intensity, normalized to the value measured with no pump (open triangles). Error bars represent s.d. (see Methods).

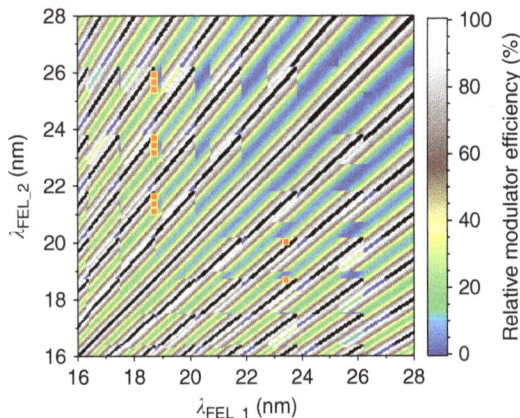

Figure 7 | Calculated seeding efficiency over the 16–28 nm range. The colour code represents the relative modulator efficiency at λ_{FEL_1} and λ_{FEL_2} when the modulator gap is set to resonate with their average value. The calculation uses radiator harmonics from 9 to 16 and λ_{seed} values between 228 and 262 nm. Black dots correspond to (λ_{FEL_1}, λ_{FEL_2}) couples whose λ_{seed} values are within the radiator bandwidth and cannot be produced using the proposed source scheme. Red squares identify the couples of wavelengths explored during the test experiment (Supplementary Figs 2 and 3).

allowed to span the 228–262 nm range covered by the OPA, and radiator harmonics from 9 to 16 are considered. The finite modulator bandwidth defines the maximum intensity that can be obtained for each (λ_{FEL_1}, λ_{FEL_2}) combination, hence the efficiency of the two-colour process. The radiator bandwidth imposes limitations on the independent tunability of λ_{FEL_1} and λ_{FEL_2}: black dots forming diagonal lines in Fig. 7 mark couples of wavelengths whose corresponding λ_{seed} values are close enough to be amplified in both radiator subsections. In this case, four FEL pulses, and not two, would be generated and the proposed two-colour scheme does not work properly.

Figure 7 shows the tuning capabilities and limitations of the adopted two-colour FEL scheme over the 16–28 nm range that broadly covers the wavelengths used in our test experiment. The red squares in Fig. 7 indicate the pairs of FEL wavelengths that were actually explored for the Fe–Ni double-resonant pump–probe measurements (Supplementary Fig. 2 and Supplementary Fig. 3). In principle, a much wider range of λ_{FEL} values extending up to 90 nm can be covered by using the full set of harmonics available at FERMI[39,51]. At wavelengths longer than ~45 nm, though, the limited range of the OPA and the low radiator harmonic numbers introduce gaps in the (λ_{FEL_1}, λ_{FEL_2}) values that can be covered by this two-colour FEL scheme (Supplementary Fig. 4).

The accessible delay range between the pump and the probe is limited by the generation of the two FEL pulses from the same electron bunch of finite temporal length. In our experiment, we spanned the 300–800 fs range and an extension to 200–1,000 fs can be envisaged. This remains a strong constraint on the class of dynamic phenomena that can be addressed. Concerning ultra-fast demagnetization, in particular, many systems of interest feature

response times of the order of 200 fs (refs 21,26,52), at the limit of the accessible range.

Further developments can be envisaged for improving the source characteristics, such as the twin-bunch mode recently demonstrated in SASE configuration[15]. The implementation of a similar scheme at FERMI would provide a more efficient bunching at the two wavelengths, a more efficient coupling in the radiator sections and, in fine, a significant increase in the energy per pulse, which could attain tens of microjoules for both the pump and the probe. Moreover, using two independent bunches would provide additional flexibility for tuning the two λ_{seed} wavelengths and would soften the constraints on the temporal separation between pump and probe pulses. Another significant improvement, already planned at FERMI, implies a second OPA for tuning both λ_{seed} wavelengths independently, as used for computing the tuning range reported in Fig. 7 and Supplementary Fig. 4. The desired resonant condition for both the pump and the probe FEL pulses could be finely matched.

Finally, it is worth remembering that the FERMI radiator section is composed of Apple-II type undulators[53] delivering radiation of selectable polarization, either circular (right/left) or linear (vertical/horizontal). Therefore our two-colour source offers the possibility of choosing the polarization state of each pulse independently, which may be especially important in atomic and molecular physics studies.

The two-colour extreme ultraviolet source that we have developed at FERMI already has potential for many interesting and original studies in magnetization dynamics and beyond. For instance, it can cover the 3p resonances of any couple of elements among Mn, Fe, Co and Ni, making a wide class of relevant magnetic materials accessible to resonant FEL pump/resonant FEL probe experiments. More generally, it enables the excitation of a particular energy and polarization-selected resonance on a well-defined atomic site in a complex system and makes it possible to study its dynamics with the second FEL pulse, by choosing for the probe another electronic subshell or another atomic site. This new source will provide unprecedented opportunities for probing in a highly selective way the dynamics of complex relaxation processes, such as Auger cascades or sequential multiple ionization, and of charge transfer processes in large molecules and clusters.

Methods

Accelerator. The FERMI linac[54] was operated at 1.3 GeV electron beam energy and 700 pC nominal charge. A moderate compression produced almost flat 500 A current electron bunches. The bunch length provided the conditions for an effective twin seeding with temporal separation of up to ~ 900 fs. The longitudinal phase space of the electron beam (energy versus time) was characterized by a chirp with both linear and quadratic components[55] that can be exploited to further enhance the difference between the resonant wavelengths in the two parts of the beam.

Seed lasers. The special twin-seed laser configuration for wide tunability two-colour FEL was based on the standard FERMI seed laser system described earlier[56–58]. The output of a Ti:Sapphire amplifier (5–7 mJ per pulse, 100 fs pulse duration, 784 nm central wavelength) was shared between an infrared OPA and a THG setup. Inside the OPA box, the signal pulses delivered by a two-stage white-light-seeded OPA process were frequency mixed with a residual pump pulse and the generated visible light was further up-converted by second harmonic generation to obtain ultraviolet pulses in the 228–262 nm range. The second ultraviolet pulse, generated in a time-plate-type BBO crystal-based THG setup, was adjusted to generate pulses with a central wavelength of 261.5 nm. The intensities of the two pulses could be varied independently through remotely controlled waveplates. The time delay between the two seed pulses was measured using an optical cross-correlator, where each ultraviolet pulse was cross-correlated with an IR pulse derived from the ultrafast oscillator that seeds the Ti:Sapphire amplifier. A remotely controlled delay stage on the THG path was used to set the time delay between the two seed pulses, before recombining them through a 50% beam splitter. Both seed pulses originate from the same source (laser oscillator and regenerative amplifier) and their relative time delay is very stable[56,57]. It has been verified that once set, the relative time delay between the two ultraviolet pulses (hence between the two FEL pulses) remains stable within less than ± 5 fs over a time span of 2 h. This includes both short-term timing jitter and slow timing drifts. The adjustment and long-term stabilization of the spatial coincidence and collinearity of the two seed beams inside the FEL undulator, which are essential for obtaining the coincidence of the two FEL pulses on the sample, were obtained by using a dedicated feedback loop based on independent steering optics for each beam.

Undulators. The modulator is a 100 mm period 3 m long planar undulator with $\sim 3\%$ nominal resonance bandwidth. The radiator comprises six independent 55 mm period 2.42 m long undulators based on the APPLE–II design[53] that provide adjustable polarization[59,60]. The radiator was divided into two subsections, Rad_1 and Rad_2, set to resonate with harmonic 14 of λ_{seed_1} and harmonic 11 of λ_{seed_2}, respectively (Fig. 1a). The $\lambda_{seed\_2} = 255$ nm OPA seed produces a localized bunching at all the harmonics including the 11th that matches the resonance in Rad_2, generating 23.2 nm coherent emission which is amplified along the radiator. However, the beam has also bunching at the 18.2 nm 14th harmonic close to the resonant wavelength of Rad_1 (18.7 nm), which may produce unwanted emission. Similarly, the $\lambda_{seed\_1} = 261.5$ nm THG seed induces a bunching at 18.7 nm (14th harmonic), which generates the FEL probe pulse in Rad_1, but also at 23.8 nm (11th harmonic) which may excite emission from Rad_2 tuned at 23.2 nm. In both cases, though, the separation between the undesired bunching wavelength and the radiator resonant wavelength is $> 2\%$, that is, larger than the $\sim 0.7\%$ gain bandwidth measured for the radiators (Allaria et al. FEL-1 current status and recent achievements, FERMI Machine Advisory Committee, Sincrotrone Trieste, April 2014, unpublished). It is the narrow bandwidth of the radiators compared with the modulator that makes it possible to produce time-delayed single-frequency pump and probe FEL pulses from the same electron bunch.

Samples. The 20 nm permalloy film was sputter-deposited from a $Fe_{19}Ni_{81}$ target onto a commercial Si grating (605 nm period, 190 nm groove depth), with 3 nm Al buffer and capping layers. Room temperature magneto-optical Kerr effect measurements showed 100% remanence and ~ 8 mT coercive field along the grating lines. The 12.5-nm-thick $NiFe_2O_4$ layer was grown on $MgAl_2O_4(001)$ by molecular beam epitaxy in atomic oxygen plasma. A $100 \times 400 \, \mu m^2$ area of the Ni-ferrite layer was ruled by focused ion beam etching with a set of ~ 350 nm wide stripes with a ~ 600 nm period. The magnetic signal at the Fe-3p resonance measured on the patterned area at the FEL source showed an ~ 50 mT coercive field with 100% remanence along the stripes.

Scattering setup. The experiment was performed using the IRMA vertical-scattering-plane reflectometer[44]. A horseshoe electromagnet applied variable fields (± 150 mT) parallel to the sample surface and normal to the scattering plane (Fig. 1b). The FEL beam was refocused at the sample position by two bendable mirrors in Kirkpatrick–Baez configuration, using an extreme ultraviolet imager at the sample position. The final spot size ($\sim 80 \, \mu m$) was estimated by scanning a movable pin-hole while measuring the transmitted intensity. The reflectometer

allowed for a precise alignment of the sample with respect to the FEL beam using a slitted photodiode mounted on the detector arm. The vertically scattered intensity was detected by an in-vacuum CCD camera (2,048 × 2,048 pixels, pixel size $13.5 \times 13.5 \, \mu m^2$) shielded from visible light by a 100-nm-thick Al filter. The CCD was mounted at 90° from the incoming FEL beam and at 535 mm from the sample. The pump fluence F at the sample was evaluated by correcting the pump energy measured at the source for the transport-line transmission (six reflections and a 200-nm-thick Al filter), focal spot size ($\sim 80 \times 80 \, \mu m^2$) and angle of incidence (46.5°). Error bars on fluence (Fig. 6) account for both the pump energy measurement accuracy and for the source intensity fluctuations. The maximum fluence at the sample was ~ 40 and ~ 3.5 mJ cm^{-2} for the pump and the probe, respectively. F values could be adjusted rapidly and continuously by attenuating the pump seed laser. The scattering of the p-polarized FEL radiation was measured near the Brewster extinction condition, reducing non-magnetic contributions and maximizing the magnetic contrast[23,24,47–49]. All the data reported here were collected at 46.5° incidence of the FEL radiation. Magnetization-dependent data were collected following the same protocol for both samples and for all the measurements: (a) application of $+80$ mT pulse of ~ 10 ms duration, exceeding the saturation field; (b) $+20$ mT applied while collecting the scattered intensity at the CCD detector during a given acquisition time (1–10 s per frame); (c) repeat (a,b) for negative field values; (d) repeat the whole (a–c) sequence 50 times. The magnetic signal was then defined as an asymmetry ratio, that is, as the difference divided by the sum of two images collected for opposite signs of the applied field. Data reported in Figs 5 and 6 represent average values taken over a 7×7 pixels area.

References

1. Hemsing, E., Stupakov, G. & Xiang, D. Beam by design: laser manipulation of electrons in modern accelerators. *Rev. Mod. Phys.* **86**, 897–941 (2014).
2. Allaria, E. *et al.* Two-colour pump–probe experiments with a twin-pulse-seed extreme ultraviolet free-electron laser. *Nat. Commun.* **4**, 2476 (2013).
3. Bencivenga, F. *et al.* Multi-colour pulses from seeded free-electron-lasers: towards the development of non-linear core-level coherent spectroscopies. *Faraday Discuss.* **171**, 487–503 (2014).
4. Ciocci, F. *et al.* Two color free-electron laser and frequency beating. *Phys. Rev. Lett.* **111**, 264801 (2013).
5. Marcus, G., Penn, G. & Zholents, A. A. Free electron laser design for four-wave mixing experiments with soft X-ray pulses. *Phys. Rev. Lett.* **113**, 024801 (2014).
6. Campbell, L. T., McNeil, B. W. J. & Reiche, S. Two-colour free electron laser with wide frequency separation using a single monoenergetic electron beam. *New J. Phys.* **16**, 103019 (2014).
7. Chiadroni, E. *et al.* Two color FEL driven by a comb-like electron beam distribution. *Phys. Procedia* **52**, 27–35 (2014).
8. Dattoli, G., Mirian, N. S., DiPalma, E. & Petrillo, V. Two-color free-electron laser with two orthogonal undulators. *Phys. Rev. Spec. Top. Accel. Beams* **17**, 050702 (2014).
9. De Ninno, G. *et al.* Chirped seeded free-electron lasers: self-standing light sources for two-color pump-probe experiments. *Phys. Rev. Lett.* **110**, 064801 (2013).
10. Mahieu, B. *et al.* Two-colour generation in a chirped seeded free-electron laser: a close look. *Opt. Express* **21**, 22728–22741 (2013).
11. Lutman, A. A. *et al.* Experimental demonstration of femtosecond two-color X-ray free-electron lasers. *Phys. Rev. Lett.* **110**, 134801 (2013).
12. Marinelli, A. *et al.* Multicolor operation and spectral control in a gain-modulated X-ray free-electron laser. *Phys. Rev. Lett.* **111**, 134801 (2013).
13. Hara, T. *et al.* Two-colour hard X-ray free-electron laser with wide tunability. *Nat. Commun.* **4**, 2919 (2013).
14. Lutman, A. A. *et al.* Demonstration of single-crystal self-seeded two-color X-ray free-electron lasers. *Phys. Rev. Lett.* **113**, 254801 (2014).
15. Marinelli, A. *et al.* High-intensity double-pulse X-ray free-electron laser. *Nat. Commun.* **6**, 6369 (2015).
16. Stamm, C. *et al.* Femtosecond modification of electron localization and transfer of angular momentum in nickel. *Nat. Mater.* **6**, 740–743 (2007).
17. Boeglin, C. *et al.* Distinguishing the ultrafast dynamics of spin and orbital moments in solids. *Nature* **465**, 458–461 (2010).
18. Radu, I. *et al.* Transient ferromagnetic-like state mediating ultrafast reversal of antiferromagnetically coupled spins. *Nature* **472**, 205–208 (2011).
19. Wietstruk, M. *et al.* Hot-electron-driven enhancement of spin-lattice coupling in Gd and Tb 4f ferromagnets observed by femtosecond X-ray magnetic circular dichroism. *Phys. Rev. Lett.* **106**, 127401 (2011).
20. Mariager, S. O. *et al.* Structural and magnetic dynamics of a laser induced phase transition in FeRh. *Phys. Rev. Lett.* **108**, 087201 (2012).
21. Eschenlohr, A. *et al.* Ultrafast spin transport as key to femtosecond demagnetization. *Nat. Mater.* **12**, 332–336 (2013).
22. Bergeard, N. *et al.* Ultrafast angular momentum transfer in multisublattice ferrimagnets. *Nat. Commun.* **5**, 3466 (2014).
23. La-O-Vorakiat, C. *et al.* Ultrafast demagnetization dynamics at the M edges of magnetic elements observed using a tabletop high-harmonic soft X-ray source. *Phys. Rev. Lett.* **103**, 257402 (2009).

24. La-O-Vorakiat, C. *et al.* Ultrafast demagnetization measurements using extreme ultraviolet light: comparison of electronic and magnetic contributions. *Phys. Rev. X* **2**, 011005 (2012).

25. Mathias, S. *et al.* Probing the timescale of the exchange interaction in a ferromagnetic alloy. *Proc. Natl Acad. Sci. USA* **109**, 4792–4797 (2012).

26. Günther, S. *et al.* Testing spin-flip scattering as a possible mechanism of ultrafast demagnetization in ordered magnetic alloys. *Phys. Rev. B* **90**, 180407 R (2014).

27. Pfau, B. *et al.* Ultrafast optical demagnetization manipulates nanoscale spin structure in domain walls. *Nat. Commun.* **3**, 1100 (2012).

28. Zhang, W. *et al.* Tracking excited-state charge and spin dynamics in iron coordination complexes. *Nature* **509**, 345–348 (2014).

29. Beaud, P. *et al.* A time-dependent order parameter for ultrafast photoinduced phase transitions. *Nat. Mater.* **13**, 923–927 (2014).

30. Clark, J. N. *et al.* Imaging transient melting of a nanocrystal using an X-ray laser. *Proc. Natl Acad. Sci. USA* **112**, 7444–7448 (2015).

31. Först, M. *et al.* Spatially resolved ultrafast magnetic dynamics initiated at a complex oxide heterointerface. *Nat. Mater.* **14**, 883–888 (2015).

32. Wernet, P. *et al.* Orbital-specific mapping of the ligand exchange dynamics of $Fe(CO)_5$ in solution. *Nature* **520**, 78–81 (2015).

33. Ostler, T. A. *et al.* Ultrafast heating as a sufficient stimulus for magnetization reversal in a ferrimagnet. *Nat. Commun.* **3**, 666 (2012).

34. Graves, C. E. *et al.* Nanoscale spin reversal by non-local angular momentum transfer following ultrafast laser excitation in ferrimagnetic GdFeCo. *Nat. Mater.* **12**, 293–298 (2013).

35. Finazzi, M. *et al.* Laser-induced magnetic nanostructures with tunable topological properties. *Phys. Rev. Lett.* **110**, 177205 (2013).

36. Mangin, S. *et al.* Engineered materials for all-optical helicity-dependent magnetic switching. *Nat. Mater.* **13**, 286–292 (2014).

37. Le Guyader, L. *et al.* Nanoscale sub-100 picosecond all-optical magnetization switching in GdFeCo microstructures. *Nat. Commun.* **6**, 5839 (2015).

38. Bocchetta, C. J. *et al.* FERMI@Elettra FEL conceptual design report (Sincrotrone Trieste, 2007).

39. Allaria, E. *et al.* Highly coherent and stable pulses from the FERMI seeded free-electron laser in the extreme ultraviolet. *Nat. Photon.* **6**, 699–704 (2012).

40. Yu, L. H. Generation of intense UV radiation by sub-harmonically seeded single-pass free-electron lasers. *Phys. Rev. A* **8**, 5178–5193 (1991).

41. Yu, L. H. *et al.* High-gain harmonic-generation free-electron laser. *Science* **289**, 932–934 (2000).

42. De Ninno, G. *et al.* Single-shot spectro-temporal characterization of XUV pulses from a seeded free-electron laser. *Nat. Commun.* **6**, 8075 (2015).

43. Gauthier, D. *et al.* Spectrotemporal shaping of seeded free-electron laser pulses. *Phys. Rev. Lett.* **115**, 114801 (2015).

44. Sacchi, M. *et al.* Ultra-high vacuum soft X-ray reflectometer. *Rev. Sci. Instrum.* **74**, 2791–2795 (2003).

45. Pedersoli, E. *et al.* Multipurpose modular experimental station for the DiProI beamline of Fermi@Elettra free electron laser. *Rev. Sci. Instrum.* **82**, 043711 (2011).

46. Capotondi, F. *et al.* Coherent imaging using seeded free-electron laser pulses with variable polarization: first results and research opportunities. *Rev. Sci. Instrum.* **84**, 051301 (2013).

47. Spezzani, C. *et al.* Magnetization and microstructure dynamics in Fe/MnAs/GaAs(001): Fe magnetization reversal by a femtosecond laser pulse. *Phys. Rev. Lett.* **113**, 247202 (2014).

48. Sacchi, M., Panaccione, G., Vogel, J., Mirone, A. & van der Laan, G. Magnetic dichroism in reflectivity and photoemission using linearly polarized light: 3*p* core level of Ni(110). *Phys. Rev. B* **58**, 3750–3754 (1998).

49. Hecker, M., Oppeneer, P. M., Valencia, S., Mertins, H. C. & Schneider, C. M. Soft X-ray magnetic reflection spectroscopy at the 3*p* absorption edges of thin Fe films. *J. Electron Spectros. Relat. Phenomena* **144-147**, 881–884 (2005).

50. Henke, B. L., Gullikson, E. M. & Davis, J. C. X-ray interactions: photoabsorption, scattering, transmission, and reflection at $E = 50$-30,000 eV, $Z = 1$-92. *Atom. Data Nucl. Data Tables* **54**, 181–342 (1993).

51. Allaria, E. *et al.* The FERMI free-electron lasers. *J. Synchrotron Radiat.* **22**, 485–491 (2015).

52. Vodungbo, B. *et al.* Laser-induced ultrafast demagnetization in the presence of a nanoscale magnetic domain network. *Nat. Commun.* **3**, 999 (2012).

53. Sasaki, S. Analyses for a planar variably-polarizing undulator. *Nucl. Instrum. Methods A* **347**, 83–86 (1994).

54. Di Mitri, S. *et al.* Design and simulation challenges for FERMI@ELETTRA. *Nucl. Instrum. Methods A* **608**, 19–27 (2009).

55. Penco, G. *et al.* Experimental demonstration of electron longitudinal-phase-space linearization by shaping the photoinjector laser pulse. *Phys. Rev. Lett.* **112**, 044801 (2014).

56. Danailov, M. B. *et al.* Design and first experience with the FERMI seed laser. *Proceedings of the 33rd International Free Electron Laser Conference (FEL 2011)*, (eds Zhao, Z. & Wang, D.) 183–186 (SINAP, Shanghai, China TUOC4, 2012).

57. Danailov, M. B. *et al.* Towards jitter-free pump-probe measurements at seeded free electron laser facilities. *Opt. Express* **22**, 12869–12879 (2014).

58. Cinquegrana, P. *et al.* Optical beam transport to a remote location for low jitter pump-probe experiments with a free electron laser. *Phys. Rev. Spec. Top. Accel. Beams* **17**, 040702 (2014).

59. Schmidt, T. & Zimoch, D. About APPLE II Operation. *AIP Conf. Proc.* **879**, 404–407 (2007).

60. Allaria, E. *et al.* Control of the polarization of a vacuum-ultraviolet, high-gain, free-electron laser. *Phys. Rev. X* **4**, 041040 (2014).

Acknowledgements

We are grateful to Maya Kiskinova (Sincrotrone Trieste), Jan Vogel (Institut Néel, Grenoble), Giancarlo Panaccione (CNR-IOM, Trieste), Fausto Sirotti (Synchrotron SOLEIL), Nicolas Moisan (LPS, Orsay), Michael Meyer (European XFEL, Hamburg) and Coryn F. Hague (LCPMR, Paris) for useful discussions and suggestions. This research received financial support from the European Community 7th Framework Programme under grant agreement n° 312284, and from CNRS (France) via the PEPS_SASLELX program. The FERMI project at Elettra—Sincrotrone Trieste is supported by MIUR under grants FIRB-RBAP045JF2 and FIRB-RBAP06AWK3.

Author contributions

E.F., C.Sp., G.D.N., M.B.D., E.A. and M.S. devised and coordinated the experiment. E.F., C.Sp., L.G., G.D.N., M.B.D. and E.A. designed and optimized the two-colour FEL source. I.N., P.C., A.D. and M.B.D. reconfigured and operated the twin-seed laser source. E.F., B.D., D.G., G.P., P.R.R., E.R., M.T., L.G., G.D.N. and E.A. operated the FEL source during the experiment. F.F., R.D., F.V., J.B.M. and L.L. fabricated and characterized the samples. M.M., E.P., F.C., C.Sv., N.M., M.Z. and L.R. contributed to the integration of the IRMA experimental chamber at the DIPROI beamline. C.Sp., F.F., R.D., T.P. and M.S. performed the scattering experiment. E.F., C.Sp., F.V., M.B.D., G.D.N., E.A. and M.S. analysed the data and wrote the manuscript, with contributions from all the authors.

Additional information

# 16

Controllable positive exchange bias via redox-driven oxygen migration

Dustin A. Gilbert[1,2], Justin Olamit[1], Randy K. Dumas[1,3], B.J. Kirby[2], Alexander J. Grutter[2], Brian B. Maranville[2], Elke Arenholz[4], Julie A. Borchers[2] & Kai Liu[1]

Ionic transport in metal/oxide heterostructures offers a highly effective means to tailor material properties via modification of the interfacial characteristics. However, direct observation of ionic motion under buried interfaces and demonstration of its correlation with physical properties has been challenging. Using the strong oxygen affinity of gadolinium, we design a model system of Gd_xFe_{1-x}/NiCoO bilayer films, where the oxygen migration is observed and manifested in a controlled positive exchange bias over a relatively small cooling field range. The exchange bias characteristics are shown to be the result of an interfacial layer of elemental nickel and cobalt, a few nanometres in thickness, whose moments are larger than expected from uncompensated NiCoO moments. This interface layer is attributed to a redox-driven oxygen migration from NiCoO to the gadolinium, during growth or soon after. These results demonstrate an effective path to tailoring the interfacial characteristics and interlayer exchange coupling in metal/oxide heterostructures.

[1] Physics Department, University of California, Davis, One Shields Avenue, Davis, California 95616, USA. [2] NIST Center for Neutron Research, Gaithersburg, Maryland 20899, USA. [3] Department of Physics, University of Gothenburg, Gothenburg 412 96, Sweden. [4] Advanced Light Source, Lawrence Berkeley National Laboratory, Berkeley, California 94720, USA. Correspondence and requests for materials should be addressed to K.L. (email: kailiu@ucdavis.edu).

Modification of metal/oxide heterostructures through ionic motion is highly effective in tailoring the interfacial characteristics and consequently their physical and chemical properties[1]. For example, forced oxygen migration has been explored in resistive switching of memristors[2,3] and control over metal–insulator transitions in electrolyte-gated materials[4]; charge-trapping has been demonstrated to substantially enhance the efficiency of magnetoelectric effects[5,6]. Most recently, the role of oxygen migration-induced surface chemistry modification has been highlighted in electrical tuning of interfacial magnetic anisotropy[7-11], which are highly relevant to energy-efficient magnetization switching in magnetic tunnel junctions and other spintronic devices[12,13]. In particular, gadolinium oxide films have been used as a source of ionic oxygen, which can then be driven into a neighbouring ferromagnet (FM) with an electric field[6,10,11]. The strong oxygen affinity and preferred oxidation state of gadolinium (almost exclusively $+3$) plays a key role in mobilizing any off-stoichiometry oxygen. This offers an opportunity to design a system with an inverted construction, whereby, relying on the gadolinium oxygen affinity, a neighbouring oxide film is reduced. To date, however, direct observations of the oxygen migration under buried interfaces and its correlation with the physical properties in metal/oxide heterostructures have been few and far between[10,11,14].

In this work, we demonstrate effective magneto-ionic manipulation of metal/oxide interfaces using an inverted gadolinium-based heterostructure design, manifested through the interface-sensitive exchange bias effect. We report direct evidence of controllable positive exchange bias in bilayer films of GdFe/NiCoO (ferrimagnet/antiferromagnet (AF)) enabled by the redox-driven oxygen migration. The exchange bias phenomenon is central to spin-valve type of spintronic devices[15-17] and to several emerging frontiers such as multiferroics[18,19], chiral ordering and spin texture[20,21], control of quantum magnets[22] and AF spintronics[23,24]. Conventionally, positive exchange bias, most notably in TM/TMF_2 ($TM(F) = $ FM transition metal (fluoride)), is due to the competition between the Zeeman energy and an AF interfacial exchange coupling, and often requires a large cooling field (on the order of a few Tesla) to realize[25]. The present GdFe/NiCoO films exhibit a full range of controllability, both the sign and magnitude of the exchange bias, using a much smaller cooling field range that is an order of magnitude smaller than previous observations[25], along with a variable GdFe composition that affects its Curie temperature. Depth profiling with polarized neutron reflectometry (PNR) directly identifies an FM interfacial layer, a few nanometres in thickness, whose moments are substantially larger than typically expected from uncompensated AF interfacial moment alone. Using element-specific X-ray magnetic circular dichroism (XMCD) spectroscopy, the interfacial layer is shown to result from rotatable NiCo and have exchange coupling both to the adjacent GdFe and NiCoO. Thermodynamic considerations suggest that a Gd-NiCoO redox reaction causes the formation of the interfacial NiCo, which is manifested in the controllable positive exchange bias. These results provide important insights into the emerging field of magneto-ionics.

Results

Magnetometry. Magnetometry measurements were performed on thin films of $Ni_{0.47}Co_{0.53}O$ (20 nm)/Gd_xFe_{1-x} (30 nm), where $x = 0.42$, 0.48, 0.53 and 0.57, identified as samples A–D, respectively (see Methods). Henceforth, for simplicity, $Ni_{0.47}Co_{0.53}O$ and Gd_xFe_{1-x} will be identified as NiCoO and GdFe, respectively. The GdFe Curie temperature T_C was measured to be

around 480 K for samples A and B, and around 350 K for samples C and D. To establish exchange bias, the samples were heated to 420 K (above the AF NiCoO Néel temperature of $T_N = 401$ K)[26,27] in a helium flow furnace and then cooled to room temperature in the presence of an in-plane cooling field $\mu_0 H_{FC}$. Room-temperature hysteresis loops for sample A ($x = 0.42$) are shown in Fig. 1a, after field cooling in 15, 100 and 300 mT, and 1.5 T. A small vertical shift of 1–2% of the saturation magnetization M_S is observed in the loops, resulting from uncompensated pinned moments at the AF interface[28,29]. Two magnetic phases are evident: a single loop at low fields with a small coercivity and a pair of asymmetrically biased subloops at higher fields (zoomed-in views shown in Fig. 1b,c, referred to as phase 1 and 2, respectively). When the sample is field cooled in small $\mu_0 H_{FC}$ (< 100 mT), the phase 1 subloop is biased in the $-H$ direction and the phase 2 subloops are asymmetrically biased to the $+H$ direction. As H_{FC} is increased, the phase 1 subloop gradually shifts from negative to positive bias, while the two subloops of phase 2 collectively shift to the $-H$ direction. These trends illustrate controllable, yet opposite, exchange biases experienced by the two phases under increasing cooling fields. The phase 1 behaviour is similar to the positive exchange bias reported earlier by Yang et al.[27], whereas that of phase 2 is unexpected and different from bifurcated loops reported earlier in systems with macroscopic domains[30,31].

Temperature-dependent hysteresis loops of sample A after field cooling in 15 mT are shown in Fig. 1d (1 μemu = 1 nA m²). With increasing temperature, the phase 1 exchange bias decreases and vanishes just below the NiCoO Néel temperature of 401 K (Fig. 1e)[26,27], whereas the coercivity remains largely unchanged, as shown in Fig. 1f; the phase 2 exchange bias also decreases, vanishing around 420 K (Fig. 1d). The loop squareness, defined as the ratio of the remanent magnetization M_R and M_S, measured up to 2 T, exhibits a non-monotonic temperature dependence, as shown in Fig. 1g: first decreasing to 0 at 420 K, coincident with the disappearance of phase 2, then increasing to unity over 450–520 K. This behaviour is consistent with the magnetization of the NiCo being balanced against the GdFe. The M_S decreases with increasing temperature until 450 K and then increases. For $Gd_{0.42}Fe_{0.58}$ we do not expect a compensation point where the Gd and Fe moments cancel out each other[32]. Similar trends are also seen for sample B ($x = 0.48$).

The results for samples C ($x = 0.53$) and D ($x = 0.57$) are quite different. In these samples, by adjusting the Gd content, the GdFe T_C is tuned below the NiCoO Néel temperature. Room-temperature hysteresis loops for sample C under different H_{FC} are shown in Fig. 2a, which also exhibit two magnetic phases. However, the low anisotropy phase with small coercivity (phase 1, Fig. 2b) is always positively biased, whereas the high anisotropy phase (phase 2, Fig. 2c) exhibits only a single open loop that is always negatively biased. Unlike samples A and B, there is little difference between $\mu_0 H_{FC} = 15$ mT and 1.5 T. Temperature-dependent magnetic characteristics for sample C after field cooling in 15 mT are shown in Fig. 2d,e. Phase 1 disappears above 350 K (Fig. 2e), in agreement with T_C for $Gd_{0.53}Fe_{0.47}$ (ref. 32), suggesting that phase 1 can be attributed to the GdFe; in contrast, the exchange bias in phase 2 exhibits a two-step temperature dependence (Fig. 2f): although substantially suppressed beyond the GdFe T_C of 350 K, it persists until ≈ 400 K, the NiCoO Néel temperature. Thus, phase 2 is not related to the GdFe but rather to a higher T_C phase; the two-step dependence suggests that this phase is in contact with both the NiCoO and GdFe, probably at the interface of the two layers. Furthermore, with increasing temperatures the saturation magnetization decreases, while the loop squareness increases (Fig. 2g). Sample D ($x = 0.57$) behaves similar to sample C, with the only difference being that the room-temperature coercivity of phase 1 was further reduced. The Gd concentration and cooling field dependence of the magnetic properties are summarized in Supplementary Fig. 1.

X-ray magnetic circular dichroism. To further investigate the origin of phase 2 that gives rise to the subloops at high fields, XMCD was used to extract element-specific hysteresis loops on sample A at room temperature after field cooling in 15 mT, as shown in Fig. 3b–e, along with the vibrating sample magnetometer (VSM) loop in Fig. 3a. For each of the elements probed, the XMCD asymmetry (see Methods) is non-zero, indicating the presence of rotatable ferromagnetic moments (Supplementary Fig. 2). The Gd and Fe loops shown respectively in Fig. 3b,c

illustrate the expected ferrimagnet ordering with the Gd moments dominating the Fe ones at room temperature; the correlation with the VSM loop confirms that phase 1 is due to the GdFe. In contrast, the presence of significant uncompensated rotatable Co and Ni moments is rather unexpected. Magnetometry measurements of as-grown NiCoO films (Supplementary Fig. 3) and X-ray photoemission spectroscopy studies of the copper capping layer revealed no appreciable magnetic moments. The XMCD loops for Co and Ni, shown in Fig. 3d,e, respectively, exhibit an interesting dip/bump/dip structure, indicating a complex reversal behaviour where the switching of Co and Ni are at times against the external magnetic field.

From the element-specific hysteresis loops, we can determine the sequence of magnetization reversal from positive saturation,

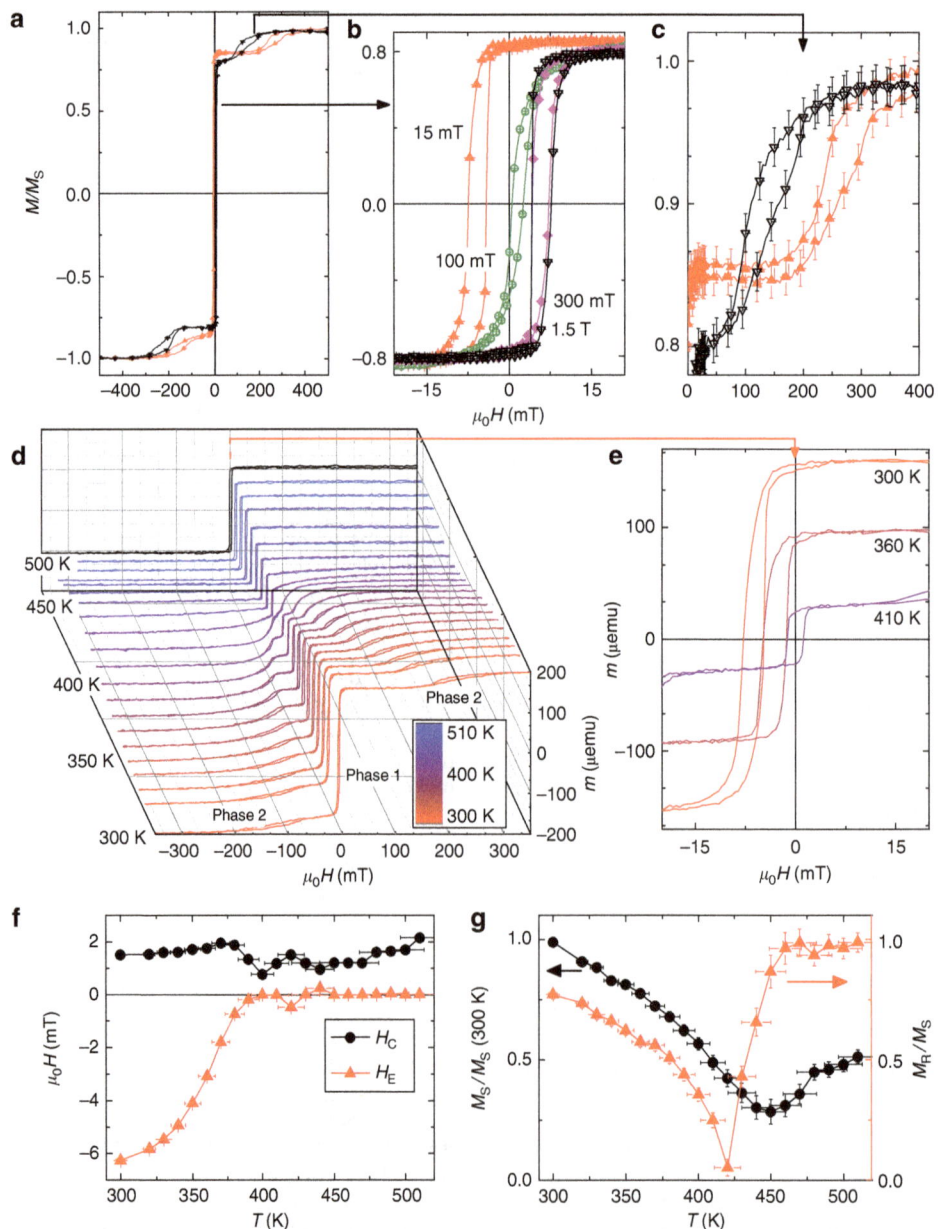

Figure 1 | Magnetometry results for sample A (Gd$_{0.42}$Fe$_{0.58}$/NiCoO). (**a**) Room-temperature hysteresis loops under different cooling fields. (**b,c**) Zoomed-in views of phase 1 and phase 2, respectively, as indicated by arrows. The colored loops in **a**–**c** correspond to cooling fields of 15 mT (red), 100 mT (green), 300 mT (pink), and 1.5 T (blue). Temperature-dependent (**d**) hysteresis loop (300–510 K, marked by the scale bar, with a zoomed-in view of the phase 1 loops shown in **e**), (**f**) coercivity and bias, and (**g**) saturation magnetization and squareness are shown after field cooling in 15 mT. Error bars are determined by the machine sensitivity limits. For clarity, not all data points are shown.

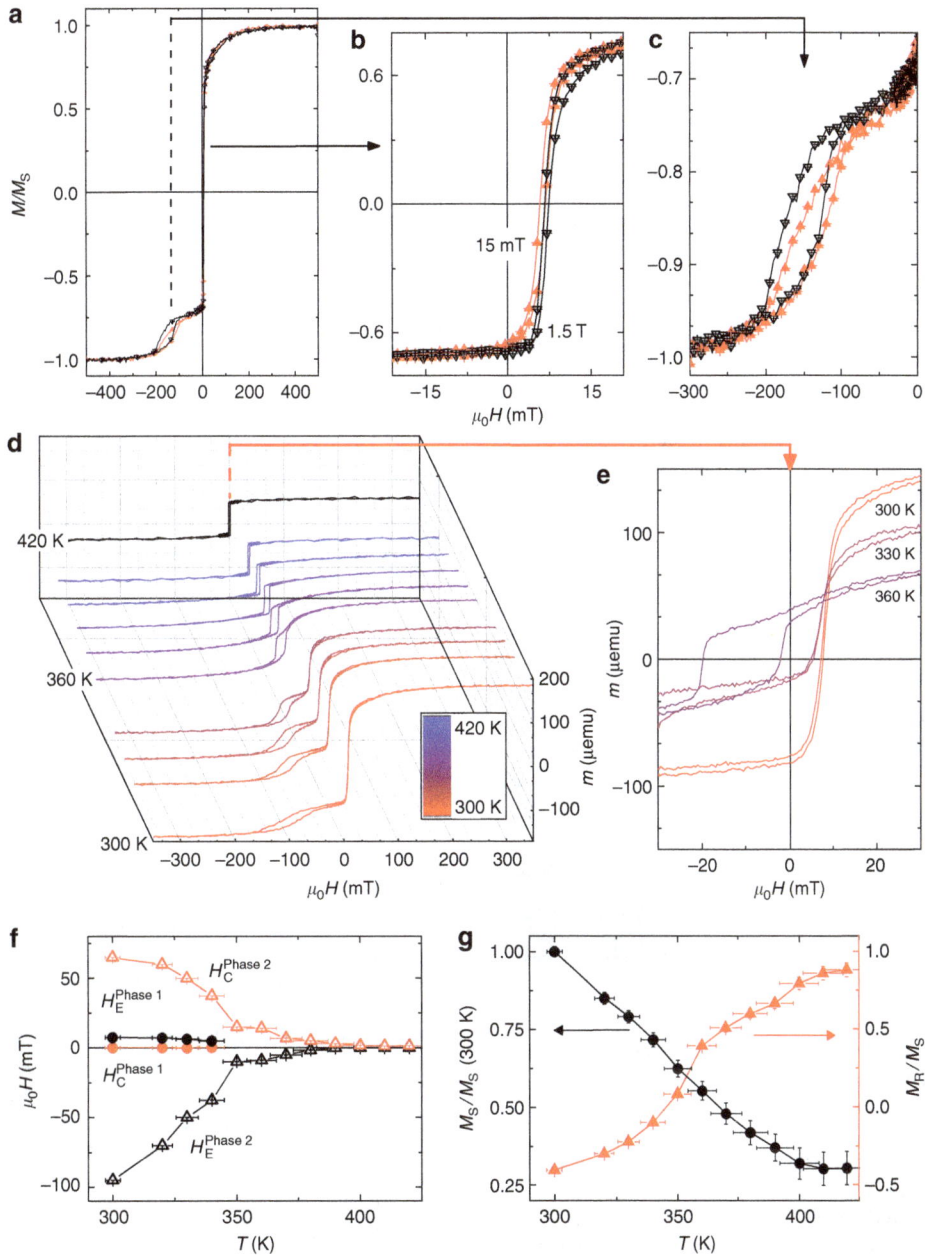

Figure 2 | Magnetometry results for sample C (Gd$_{0.53}$Fe$_{0.47}$/NiCoO). (**a**) Room-temperature hysteresis loops under different cooling fields of 15 mT (red) and 1.5 T (blue). (**b,c**) Zoomed-in views of phase 1 and phase 2, respectively, as indicated by arrows. Temperature-dependent (**d**) hysteresis loop (300–420 K, marked by the scale bar, with a zoomed-in view shown in **e**), (**f**) coercivity and exchange bias, and (**g**) saturation magnetization and squareness are shown after field cooling in 15 mT. Error bars are determined by the machine sensitivity limits. For clarity, not all data points are shown.

as illustrated in Fig. 3f–i for four representative stages. At high positive fields, the Gd, Ni and Co are all parallel to the applied field, whereas the Fe is opposite, AF coupled to Gd (Fig. 3f). As the field is reduced but remains positive (Fig. 3g), the NiCo reverses, presumably to satisfy the AF exchange coupling to the GdFe, giving rise to the apparent phase 2 subloop. As the applied field becomes negative (Fig. 3h), the Gd reverses and so do the Fe, Ni and Co, such that the Gd remains AF coupled to Fe and NiCo. The reversal of the NiCo against the applied field shows that the NiCo is quite strongly exchange coupled to the GdFe. Finally, at large negative fields (Fig. 3i) the NiCo again becomes parallel to the GdFe, as the field breaks their exchange coupling, leading to the second subloop for phase 2.

These magnetization characteristics are also consistent with the temperature-dependent hysteresis loops shown in Fig. 1. As the

temperature increases, below 420 K, the GdFe and NiCo moments are manifested in the two phases, while the GdFe moments dominate; at 420 K, the NiCo and GdFe moments balance out, leading to the appearance of a compensation point (Fig. 1d); above 420 K, NiCo moments become dominant, as the GdFe approaches its T_C.

Depth profiling with PNR. To confirm that the uncompensated elemental NiCo is indeed at the interface, we have employed PNR to probe the nuclear and magnetization depth profiles[33–36]. The fitted reflectometry data and corresponding profiles for sample A after field cooling in 15 mT, measured at 450 and 50 mT in-plane fields applied parallel to the cooling field, are shown in Fig. 4a,b, respectively. These measurement fields correspond to stages f and g in Fig. 3, and thus the only difference in the sample

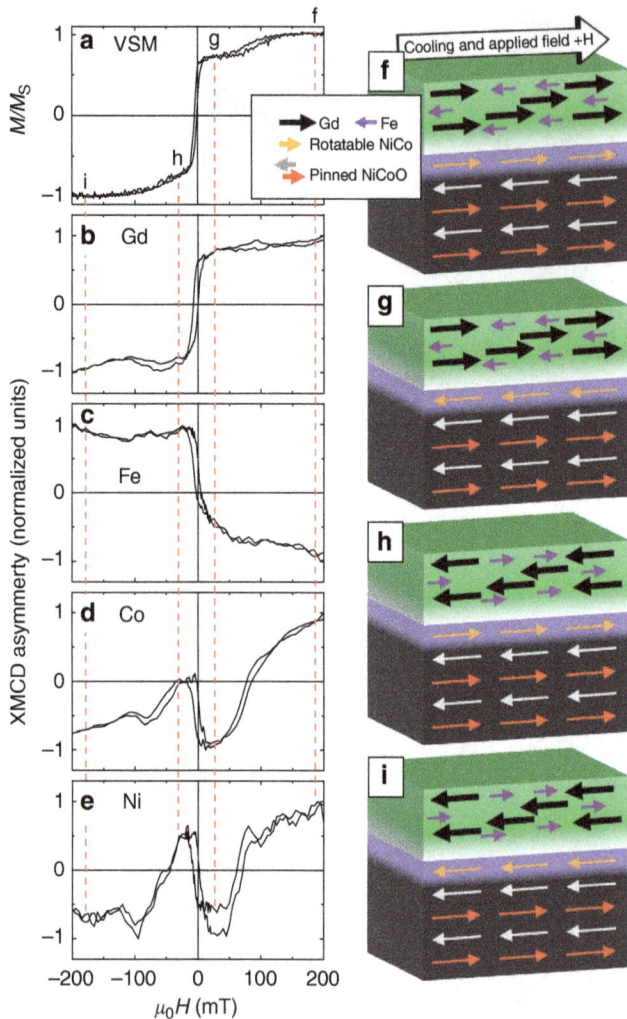

Figure 3 | Collective and element-specific magnetic hysteresis loops.
Room-temperature hysteresis loops of sample A ($x = 0.42$, with a Cu capping layer) measured by (**a**) VSM and element-specific XMCD for (**b**) Gd, (**c**) Fe, (**d**) Co and (**e**) Ni. Schematic illustrations of the sample magnetic configuration at various stages of the reversal are shown in **f–i**, respectively. In the layer structure, from top to bottom, the layers are GdFe (green), GdO$_y$ and GdFe mixture (white), NiCo (blue) and NiCoO (black), respectively. Arrows in the layers indicate the magnetization directions.

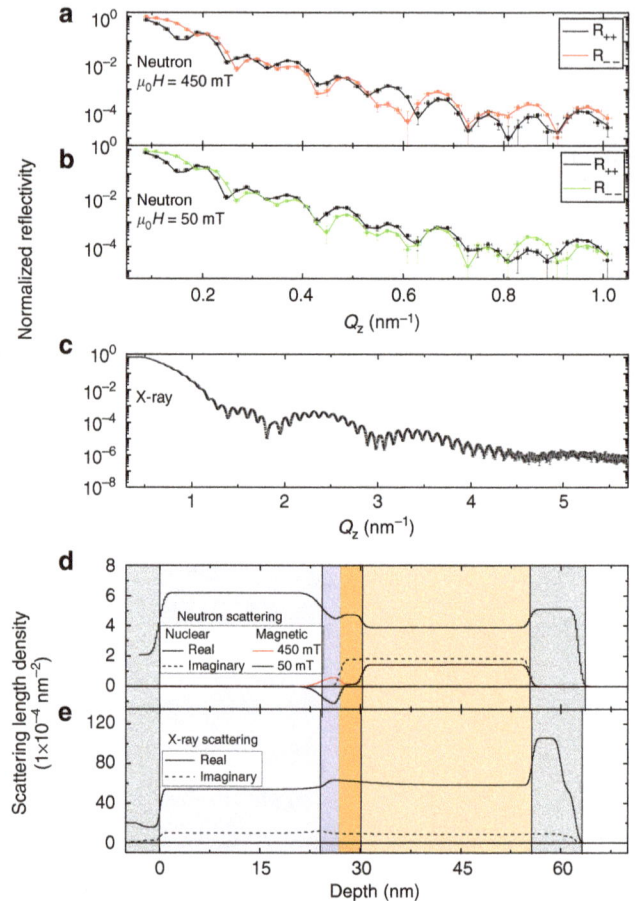

Figure 4 | Magnetic and structural depth profiles. PNR data (symbols) and fitted reflectometry (lines) for sample A ($x = 0.42$) in a (**a**) 450- and (**b**) 50-mT field applied parallel to the cooling field; (**c**) X-ray reflectometry data (symbols) and fitted reflectometry (line) of sample A. (**d**) The fitted profile determined from PNR with the real (solid) and imaginary (dashed) nuclear structure shown in black and magnetic depth profile in blue (50 mT) and red (450 mT); the structural profile from X-ray reflectivity (XRR) is shown in **e**. For **d** and **e**, the film structure from the left to right, at increasing depth, corresponds to the Si substrate (grey), NiCoO (light blue), interfacial NiCo (dark blue), interfacial GdFe/GdO$_y$ (peach), GdFe (brown), Ta cap (grey) and air (white). Error bars in Q_z identify machine precision; error bars in normalized reflectivity are defined by the s.d. and scales with the square root of the number of measurements.

magnetic configuration is the reversal of the rotatable NiCo. The plots show a clear, field-dependent difference in the reflectometry for $Q_Z > 0.4 \, \text{nm}^{-1}$.

The nuclear depth profile obtained from the neutron reflectivity (Fig. 4d) clearly differentiates the GdFe(O) and NiCo(O) layers by the imaginary component of the nuclear scattering. Moving from GdFe towards the substrate, at the GdFe/NiCoO interface there is a local maximum in the nuclear scattering length density (solid black line). We suggest that the bump can be attributed to GdFeO, where the oxygen increases the nuclear scattering length density relative to its neighbouring GdFe. The imaginary component of the nuclear scattering length density (corresponding to neutron absorbance, generally identifying Gd, as it is a strong neutron absorber) remains constant over the same region (dashed black line), indicating that the Gd remains localized, while it was infiltrated with additional non-absorbent elements (for example, oxygen). Underneath the GdFeO region the imaginary component drops to zero, whereas the real component exhibits a dip, indicating that this region does

not contain Gd and can be attributed to elemental NiCo. The magnetic depth profiles measured in 450 and 50 mT (Fig. 4d) show that the magnetic moment drops across the GdFe/NiCoO interface to zero in the NiCoO layer. For the 450 and 50 mT measurements, the interfacial magnetization is parallel and antiparallel to the GdFe layer magnetization, respectively. This difference corresponds to the transition between stages f and g in Fig. 3, which through XMCD was shown to originate from rotatable NiCo. This last piece of evidence shows that the NiCo responsible for phase 2 (Fig. 1) is indeed located at the interface and is probably the result of oxygen migration from NiCoO to GdFe. For comparison, the X-ray reflectivity spectrum—which is insensitive to magnetic ordering—is shown in Fig. 4c. The structural depth profile determined by X-ray reflectivity in Fig. 4e has features that track those in the PNR depth profile, especially the thicknesses, and shows a slight peak in the scattering length density at the GdFe/NiCoO interface, serving as another confirmation of the structural profile.

Comparing the integrated magnetization change of phase 2 in sample A to the saturation magnetization of the NiCo binary alloy, the interfacial layer thickness can be calculated to be 2.3 nm, consistent with the NiCo thickness obtained from PNR (≈ 2.5 nm). With increased Gd content, the magnetization associated with phase 2 increases, for example, increasing to 4 nm in sample D. It is worth noting that the FM moments of the interfacial layer are much larger than expected from the interfacial uncompensated AF moment in typical exchange bias systems[37,38].

Discussion

The mechanisms for the positive and controllable exchange bias can be understood by the AF interfacial exchange coupling between GdFe and uncompensated moments of NiCoO, with the latter existing as pinned and unpinned moments (including the metallic NiCo layer). For samples A and B where the GdFe T_C is above the NiCoO T_N, the pinned NiCoO moments give rise to the positive exchange bias in GdFe under increasing H_{FC} in phase 1. This is the common positive exchange bias due to the AF interfacial exchange coupling, as seen previously[25,39]. In the meantime, the unpinned moments, identified as the rotatable NiCo earlier, give rise to the subloops in phase 2, whose reversal are driven by the AF coupling to the Gd and FM coupling to the pinned uncompensated NiCoO moments. Cooling in small H_{FC} across the NiCoO T_N, the GdFe moments (dominated by the Gd lattice) are aligned with H_{FC}; as the NiCoO orders, its uncompensated pinned moments FM couple to the NiCo, which is AF coupled to GdFe, thus ordering opposite to H_{FC}. This establishes the negative exchange bias in GdFe (phase 1) by the pinned uncompensated NiCoO moments. During the hysteresis loop measurement, the rotatable NiCo moments, forced parallel to the field direction at saturation, switch first to be opposite to the Gd and into alignment with the pinned NiCo moments, leading to the subloop in the first quadrant (Fig. 1c); as the GdFe reverses, these rotatable NiCo moments switch again in negative fields, leading to the second subloop in the third quadrant. As H_{FC} becomes large enough to break the AF coupling, the uncompensated pinned NiCoO moments are forced to be parallel to both the GdFe and H_{FC}, and this configuration is frozen-in on cooling to room temperature, leading to a positive exchange bias in phase 1. As to the subloops in phase 2, the pinned NiCoO moments now act to stabilize the rotatable NiCo moments, whose reversal under the AF coupling with GdFe occurs at a more negative applied field, leading to the negative shift of the phase 2 loops. Thus, the interfacial NiCo moments and their AF exchange coupling with the GdFe are manifested in the complex exchange bias characteristics.

For the case of the higher Gd content samples C and D, the T_C in GdFe is tuned to be lower than the T_N of the NiCoO, an unusual situation[40]. Thus, during field cooling approaching the NiCoO T_N, only the rotatable NiCo moments remain FM and their orientation is set exclusively by the cooling field. Consequently, the pinned uncompensated NiCoO moments, through FM coupling to these NiCo, is also aligned with H_{FC} and this configuration is frozen-in at room temperature. The GdFe, through its AF coupling with the pinned uncompensated NiCoO moments, is always positively biased, unlike the cases in samples A and B where the positive bias is controllable under different H_{FC}. An illustration of the different reversal stages during the hysteresis loop measurement (Fig. 5a) at room temperature is shown in Fig. 5. Reversing from positive saturation (Fig. 5b), the GdFe layer is the first to reverse in a positive applied field (Fig. 5c). The rotatable NiCo is stabilized by both the pinned NiCoO moments and the GdFe, until a sufficiently large negative

field is applied (Fig. 5d). Thus, phase 2 only exhibits a single subloop, which is always negatively biased.

The interfacial rotatable NiCo layer, with a relatively significant thickness of a few nanometres, plays a prominent role in the complex bias phenomena exhibited in the GdFe/NiCoO system. Control experiments on single-layer films of GdFe measured up to 2 T do not show any two-phase behaviour, consistent with previous results[32,41,42], suggesting that the 2 T field is insufficient to break the Gd–Fe AF coupling. Similarly, single-layer films of NiCoO show only a linear magnetic field dependence in their hysteresis loops, characteristic of an AF. These control experiments indicate that the second phase is not intrinsic to each of the as-grown GdFe and NiCoO films, but rather a derived effect of the bilayer construction. It has been previously suggested that Gd is a strong reducing agent for some metal oxides[43], and that oxygen can be readily moved between GdO_y and Co[10]. Thus, we suggest that the mechanism at work is a redox reaction between the NiCoO and Gd, forming an interfacial region of elemental NiCo and GdO_y. Furthermore, Gd is also expected to reduce any iron oxide (discussed below), resulting in an interface that is probably a mixture of GdO_y and residual GdFe (with varying Gd:Fe ratio, even Fe). Although the GdO_y locally impedes the interfacial coupling, it is not expected to be continuous enough to completely suppress the exchange bias, which is still mediated through residual GdFe. The oxidized interface thickness and continuity are limited by oxygen diffusion within the GdFe and NiCoO, and thus is likely to be very thin. This scenario is consistent with the PNR profile. The real part of the nuclear scattering length density is shown to increase at the base of the GdFe, presumably due to the incorporation of oxygen, a strong neutron scatterer, whereas the imaginary part (which depends almost entirely on the Gd volume concentration) remains constant. At the same time, the nuclear scattering length density below the GdFeO is lower than its neighbouring NiCoO due to the loss of its oxygen, consistent with the required balancing of the redox equation.

To further investigate the plausibility of this explanation, the net heat of formation and Gibbs free energy at room temperature were calculated for $3NiO (CoO) + 2Gd \rightarrow Gd_2O_3 + 3Ni (Co)$ and were found to be around -1.1 MJ mol^{-1} for both terms[44], indicating that the reaction will occur spontaneously. Similar calculations for the formation of iron oxide ($CoO + Fe \rightarrow FeO_x + Co$) yield Gibbs free energies in the range of -30 to -160 kJ mol^{-1} for FeO, Fe_2O_3 and Fe_3O_4 (ref. 44), again indicating a spontaneous reaction. However, iron oxide is strongly reduced by Gd ($FeO_x + Gd \rightarrow Gd_2O_3 + Fe$, Gibbs free energy of about -1.0 MJ mol^{-1}). Thus, the calculated results support the spontaneous oxidation of Gd at the interface and reduction of the NiCoO. This type of interfacial redox behaviour has been seen previously in NiO/Co bilayer films[45], for which similar calculations also support spontaneous oxygen migration to the interfacial Co.

For samples A and B, the bias of the phase 2 subloops reflect the AF exchange interaction between the NiCo and GdFe, as well as the FM exchange interaction between the NiCo and the NiCoO. As the orientation of the pinned uncompensated NiCoO moments does not change with applied field after field cooling to room temperature, and the GdFe orientation does, the bias from the GdFe ($H_E^{NiCo/GdFe}$) changes on field cycling and the bias from the NiCoO ($H_E^{NiCo/NiCoO}$) remains constant. Thus, the bias field for the $+H$ subloop is determined to be $H_E^{+H} = H_E^{NiCo/NiCoO} + H_E^{NiCo/GdFe}$, whereas the $-H$ subloop is biased by $H_E^{-H} = H_E^{NiCo/NiCoO} - H_E^{NiCo/GdFe}$. Therefore, the bias fields can be separated: $H_E^{NiCo/NiCoO} = (H_E^{+H} + H_E^{-H})/2$ and $H_E^{NiCo/GdFe} = (H_E^{+H} - H_E^{-H})/2$. These values are determined to be 56 and -169 mT for sample A, and 36 and -184 mT for sample B,

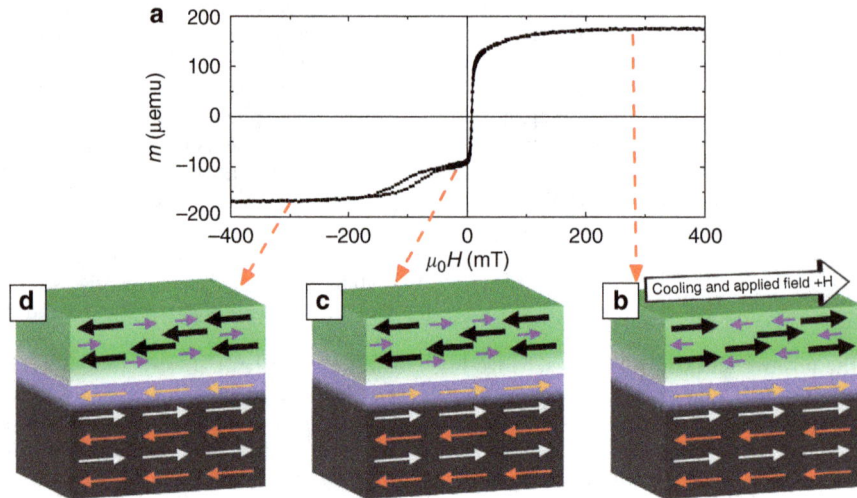

Figure 5 | Schematic illustrations of sample C magnetic configurations. Configurations during (**a**) the magnetic hysteresis loop measurement at room temperature are given at (**b**) saturation, (**c**) GdFe reversal and (**d**) NiCo reversal. In the layer structure, from top to bottom, the layers are GdFe (green), GdO_y and GdFe mixture (white), NiCo (blue) and NiCoO (black), respectively. Arrows in the layers indicate the magnetization directions.

respectively, after field cooling in 1.5 T. Using the generalized Meiklejohn–Bean approach[46], the exchange energy density is calculated to be $|J^{NiCo/NiCoO} S_{NiCo} S_{NiCoO}| = 1.3 \times 10^{-4} \, \mathrm{J\,m}^{-2}$ and $|J^{NiCo/GdFe} S_{NiCo} S_{GdFe}| = -3.8 \times 10^{-4} \, \mathrm{J\,m}^{-2}$ for sample A, and 1.1×10^{-4} and $-5.7 \times 10^{-4} \, \mathrm{J\,m}^{-2}$ for sample B, respectively. These values are much smaller than the bulk exchange values (scaled by the interface number density and atomic spin moment) for Co–Co ($7 \times 10^{-3} \, \mathrm{J\,m}^{-2}$) and Gd–Co ($-2 \times 10^{-3} \, \mathrm{J\,m}^{-2}$) (ref. 32), probably due to the interface details, as is typical in exchange-biased systems.

For samples C and D, the second phase is always negatively biased (there is no H_E^{+H} feature or any H_{FC} dependence); thus, the contributions of NiCoO and GdFe exchange coupling cannot be separated as above. However, the loss of the two subloop features does imply that the NiCo remains robust against the AF exchange coupling with the GdFe, indicating that $H_E^{NiCo/GdFe} < H_E^{NiCo/NiCoO}$. As $H_E^{NiCo/NiCoO}$ is not expected to change with GdFe stoichiometry, we can conclude that the NiCo/GdFe exchange coupling decreases with increased Gd in GdFe. The net exchange fields were extracted from the major loops to be 143 and 78 mT for samples C and D, respectively, under 1.5 T cooling field. As there is only one subloop in samples C and D, $H_E^{NiCo/NiCoO}$ and $H_E^{NiCo/GdFe}$ cannot be separated; the net coupling energy density is 5.4×10^{-4} and $3.5 \times 10^{-4} \, \mathrm{J\,m}^{-2}$, respectively.

In summary, we have demonstrated effective magneto-ionic manipulation of the $Gd_xFe_{1-x}/NiCoO$ interfaces and directly observed oxygen migration across buried interfaces and the impacts on the controlled positive exchange bias. The complex magnetization reversal is manifested in the hysteresis loops as multiple phases in bilayer samples with 42 and 48 at.% Gd: phase 1, identified as a single low anisotropy loop, was shown by XMCD spectroscopy to originate primarily from reversal of the GdFe, whereas phase 2, consisting of a pair of asymmetrically biased subloops, was shown to originate from reversal of rotatable NiCo moments. By varying the cooling field, the bias of phase 1 and asymmetry of phase 2 were shown to shift, with opposite trends. This controllability was suppressed and eventually destroyed by increasing the Gd content to 53 and 57 at. %, which lowered the GdFe Curie temperature below the NiCoO Néel temperature. The AF exchange coupling between the interfacial NiCo and GdFe causes the NiCo moments to be parallel to the GdFe at high fields and to be antiparallel at low fields. The field-dependent orientation of the interfacial moments controls

the AF orientation and the corresponding bias field direction. The interfacial NiCo was attributed to a redox reaction between the NiCoO and GdFe, leading to the formation of NiCo and GdO_y. These results demonstrate an effective way to tailor the interfacial characteristics and interlayer exchange coupling in metal/oxide heterostructures. Reversible control of the oxygen migration in such systems, for example, using an electric field, may enable concepts for energy-efficient spintronic devices.

Methods

Sample fabrication. Bilayer films of $Ni_{0.47}Co_{0.53}O$ (20 nm)/Gd_xFe_{1-x} (30 nm) ($x = 0.42$–0.57) were magnetron-sputtered on naturally oxidized Si (100) wafers at ambient temperature in 0.33 Pa Ar in a high-vacuum chamber (base pressure $< 6.7 \times 10^{-6}$ Pa). The NiCoO layer was radio frequency sputtered from a pressed composite target of CoO and NiO powders, while the GdFe was direct current co-sputtered from elemental targets. The samples were capped with 6 nm of Ta (or Cu for the XMCD samples).

Characterizations. X-ray diffraction revealed polycrystalline NiCoO and amorphous/nanocrystalline GdFe. Stoichiometry of the NiCoO was determined by energy dispersive X-ray spectroscopy, along with analysis of the nuclear scattering length density and bulk number density, to be $Ni_{0.47}Co_{0.53}O$ and the GdFe to be $Gd_{0.42}Fe_{0.58}$, $Gd_{0.48}Fe_{0.52}$, $Gd_{0.53}Fe_{0.47}$ and $Gd_{0.57}Fe_{0.43}$ (identified as samples A–D, respectively). Magnetic measurements were performed using a VSM with the field parallel to the cooling field axis, unless otherwise noted. Element-specific hysteresis loops were measured by XMCD at the Advanced Light Source Beamline 6.3.1. Loops for Fe, Ni, Co and Cu were determined by tuning to their respective $L_{2,3}$ edges, whereas Gd loops were determined by tuning to the $M_{4,5}$ edge, following previously outlined procedures[29,37,47]. Magnetic contrast was achieved by measuring the fluorescence yield signal with the left and right circularly polarized X-rays at 30° grazing incidence. XMCD asymmetry is achieved by calculating the difference of the left and right circularly polarized signals.

Polarized neutron reflectivity was used to probe depth-dependent nuclear and magnetic profiles of the films, performed on the polarized beam reflectometer and the multi-angle grazing-incidence k-vector reflectometer at the NIST Center for Neutron Research using wavelength $\lambda = 0.475$ nm neutrons. The applied magnetic field and corresponding neutron spin direction are in-plane and parallel to the field cooling direction. The reflectometry data are presented for the non-spin flip cases, with incident and scattered neutrons having the same spin, identified for the case of spin-up (down) by R_{++} (R_{--}). This configuration is sensitive to in-plane magnetization along the neutron spin direction. The spin flip reflectometry, which is sensitive to a net in-plane magnetization orthogonal to the applied field, showed no appreciable signal. Profile fitting was performed using the Refl1D software package[48]. Each fitted model consisted of the GdFe, NiCoO and Ta capping layers, as well as interfacial NiCo and GdO_y layers; all the measurements were fitted simultaneously, with the structural parameters between different models constrained to be the same. Alternative PNR fitting were also carried out for comparison, without the interfacial layer, and the resultant fits were significantly worse (Supplementary Fig. 4 and Supplementary Note 1). X-ray and neutron reflectometry data are presented with respect to the momentum transfer vector, Q.

References

1. Maier, J. Nanoionics: ion transport and electrochemical storage in confined systems. *Nat. Mater.* **4**, 805–815 (2005).
2. Waser, R., Dittmann, R., Staikov, G. & Szot, K. Redox-based resistive switching memories – nanoionic mechanisms, prospects, and challenges. *Adv. Mater.* **21**, 2632–2663 (2009).
3. Yang, J. J., Strukov, D. B. & Stewart, D. R. Memristive devices for computing. *Nat. Nanotechnol.* **8**, 13–24 (2013).
4. Jeong, J. *et al.* Suppression of metal-insulator transition in VO2 by electric field–induced oxygen vacancy formation. *Science* **339**, 1402–1405 (2013).
5. Bauer, U., Przybylski, M., Kirschner, J. S. D. & Beach, G. Magnetoelectric charge trap memory. *Nano Lett.* **12**, 1437–1442 (2012).
6. Bauer, U., Emori, S. & Beach, G. S. D. Voltage-controlled domain wall traps in ferromagnetic nanowires. *Nat. Nanotechnol.* **8**, 411–416 (2013).
7. Manchon, A. *et al.* X-ray analysis of the magnetic influence of oxygen in Pt/Co/AlOx trilayers. *J. Appl. Phys.* **103**, 07A912 (2008).
8. Bonell, F. *et al.* Reversible change in the oxidation state and magnetic circular dichroism of Fe driven by an electric field at the FeCo/MgO interface. *Appl. Phys. Lett.* **102**, 152401 (2013).
9. Rajanikanth, A., Hauet, T., Montaigne, F., Mangin, S. & Andrieu, S. Magnetic anisotropy modified by electric field in V/Fe/MgO(001)/Fe epitaxial magnetic tunnel junction. *Appl. Phys. Lett.* **103**, 062402 (2013).
10. Bi, C. *et al.* Reversible control of Co Magnetism by voltage-induced oxidation. *Phys. Rev. Lett.* **113**, 267202 (2014).
11. Bauer, U. *et al.* Magneto-ionic control of interfacial magnetism. *Nat. Mater.* **14**, 174–181 (2015).
12. Ikeda, S. *et al.* A perpendicular-anisotropy CoFeB–MgO magnetic tunnel junction. *Nat. Mater.* **9**, 721–724 (2010).
13. Wang, W. G., Li, M. G., Hageman, S. & Chien, C. L. Electric-field-assisted switching in magnetic tunnel junctions. *Nat. Mater.* **11**, 64–68 (2012).
14. Salazar-Alvarez, G. *et al.* Two-, three-, and four component magnetic multilayer onion nanoparticles based on iron oxides and manganese oxides. *J. Am. Chem. Soc.* **133**, 16738–16741 (2011).
15. Nogués, J. & Schuller, I. K. Exchange bias. *J. Magn. Magn. Mater.* **192**, 203–232 (1999).
16. Stamps, R. L. Mechanisms for exchange bias. *J. Phys. D Appl. Phys.* **33**, R247–R268 (2000).
17. Nogues, J. *et al.* Exchange bias in nanostructures. *Phys. Rep.* **422**, 65–117 (2005).
18. Tokunaga, Y., Taguchi, Y., Arima, T. & Tokura, T. Magnetic biasing of a ferroelectric hysteresis loop in a multiferroic orthoferrite. *Phys. Rev. Lett.* **112**, 037203 (2014).
19. Echtenkamp, W. & Binek, C. Electric control of exchange bias training. *Phys. Rev. Lett.* **111**, 187204 (2013).
20. Li, J. *et al.* Chirality switching and winding or unwinding of the antiferromagnetic NiO domain walls in Fe/NiO/Fe/CoO/Ag(001). *Phys. Rev. Lett.* **113**, 147207 (2014).
21. Yanes, R., Jackson, J., Udvardi, L., Szunyogh, L. & Nowak, U. Exchange bias driven Dzyaloshinskii-Moriya interactions. *Phys. Rev. Lett.* **111**, 217202 (2013).
22. Yan, S., Choi, D.-J., Burgess, J. A. J., Rolf-Pissarczyk, S. & Loth, S. Control of quantum magnets by atomic exchange bias. *Nat. Nanotechnol.* **10**, 40–45 (2015).
23. Fina, I. *et al.* Anisotropic magnetoresistance in an antiferromagnetic semiconductor. *Nat. Commun.* **5**, 4671 (2014).
24. Ciudad, D. *et al.* Sign control of magnetoresistance through chemically engineered interfaces. *Adv. Mater.* **26**, 7561–7567 (2014).
25. Nogués, J., Lederman, D., Moran, T. J. & Schuller, I. K. Positive exchange bias in FeF2-Fe bilayers. *Phys. Rev. Lett.* **76**, 4624 (1996).
26. Ambrose, T., Liu, K. & Chien, C. L. Doubly exchange-biased NiCoO/NiFe/Cu/NiFe/NiCoO spin valves. *J. Appl. Phys.* **85**, 6124–6126 (1999).
27. Yang, D. Z. *et al.* Positive exchange biasing in GdFe/NiCoO bilayers with antiferromagnetic coupling. *Phys. Rev. B* **71**, 144417 (2005).
28. Nogues, J. *et al.* Simultaneous in-plane and out-of-plane exchange bias using a single antiferromagnetic layer resolved by x-ray magnetic circular dichroism. *Appl. Phys. Lett.* **95**, 152515 (2009).
29. Arenholz, E., Liu, K., Li, Z. P. & Schuller, I. K. Magnetization reversal of uncompensated Fe moments in exchange biased Ni/FeF2 bilayers. *Appl. Phys. Lett.* **88**, 072503 (2006).
30. Olamit, J. *et al.* Loop bifurcation and magnetization rotation in exchange biased Ni/FeF2. *Phys. Rev. B* **72**, 012408 (2005).
31. Zhou, S. M., Liu, K. & Chien, C. L. Exchange coupling and macroscopic domain structure in a wedged permalloy/FeMn bilayer. *Phys. Rev. B* **58**, R14717–R14720 (1998).
32. Hansen, P., Clausen, C., Much, G., Rosenkranz, M. & Witter, K. Magnetic and magneto-optical properties of rare-earth transition-metal alloys containing Gd, Tb, Fe, Co. *J. Appl. Phys.* **66**, 756–767 (1989).
33. Majkrzak, C. F. Polarized neutron reflectometry. *Phys. B* **173**, 75–88 (1991).
34. Fitzsimmons, M. R. *et al.* Asymmetric magnetization reversal in exchange-biased hysteresis loops. *Phys. Rev. Lett.* **84**, 3986 (2000).
35. Kirby, B. J. *et al.* Vertically graded anisotropy in Co/Pd multilayers. *Phys. Rev. B* **81**, 100405 (2010).
36. Gilbert, D. A. *et al.* Realization of ground state artificial skyrmion lattices at room temperature. *Nat. Commun.* **6**, 8462 (2015).
37. Ohldag, H. *et al.* Correlation between exchange bias and pinned interfacial spins. *Phys. Rev. Lett.* **91**, 017203 (2003).
38. Fitzsimmons, M. R. *et al.* Pinned magnetization in the antiferromagnet and ferromagnet of an exchange bias system. *Phys. Rev. B* **75**, 214412 (2007).
39. Mangin, S., Montaigne, F. & Schuhl, A. Interface domain wall and exchange bias phenomena in ferrimagnetic/ferrimagnetic bilayers. *Phys. Rev. B* **68**, 140404 (2003).
40. Cai, J. W., Liu, K. & Chien, C. L. Exchange coupling in the paramagnetic state. *Phys. Rev. B* **60**, 72–75 (1999).
41. Cerdeira, M. A. *et al.* Magnetic properties and anisotropy of GdFe amorphous thin films. *J. Optoelectron. Adv. Mater.* **6**, 599–602 (2004).
42. Orehotsky, J. & Schroder, K. Magnetic properties of amorphous Fe$_x$Gd$_y$ alloy thin-films. *J. Appl. Phys.* **43**, 2413–2418 (1972).
43. Patnaik, P. *Handbook of Inorganic Chemicals* 302–306 (McGraw-Hill, 2002).
44. Wagman, D. D. *et al.* The NBS table of chemical thermodynamic properties. *J. Phys. Chem. Ref. Data* **11**, 166–230 (1982).
45. Tusche, C., Meyerheim, H. L., Hillebrecht, F. U. & Kirschner, J. Evidence for a mixed CoNiO layer at the Co/NiO (001) interface from surface x-ray diffraction. *Phys. Rev. B* **73**, 125401 (2006).
46. Binek, C., Hochstrat, A. & Kleemann, W. Exchange bias in a generalized Meiklejohn-Bean approach. *J. Magn. Magn. Mater.* **234**, 353–358 (2001).
47. Arenholz, E., Navas, E., Starke, K., Baumgarten, L. & Kaindl, G. Magnetic circular dichroism in core-level photoemission from Gd, Tb, and Dy in ferromagnetic materials. *Phys. Rev. B* **51**, 8211–8220 (1995).
48. Kirby, B. J. *et al.* Phase-sensitive specular neutron reflectometry for imaging the nanometer scale composition depth profile of thin-film materials. *Curr. Opin. Colloid Interface Sci.* **17**, 44–53 (2012).

Acknowledgements

This work has been supported by the NSF (DMR-1008791, ECCS-1232275 and DMR-1543582). D.A.G. and A.J.G. acknowledges the support of the NRC Research Associateship programme. R.K.D. acknowledges support from the Swedish Research Council (VR). The work at the Advanced Light Source was supported by the Director, Office of Science, Office of Basic Energy Sciences of the U.S. Department of Energy (DEAC02-05CH11231).

Author contributions

J.O., D.A.G. and K.L. conceived and designed the experiments. J.O. and D.A.G. synthesized the samples. J.O. performed magnetometry measurements. R.K.D., J.O. and E.A. performed XMCD studies. D.A.G., B.J.K., B.B.M. and J.A.B. carried out PNR investigations. D.A.G., J.O., A.J.G. and K.L. led the rest of the data analysis. D.A.G., J.O. and K.L. wrote the manuscript. K.L. coordinated the project. All authors contributed to discussions and manuscript revision.

Additional information

17

Ion selectivity of graphene nanopores

Ryan C. Rollings[1,*], Aaron T. Kuan[2,*] & Jene A. Golovchenko[1,2]

As population growth continues to outpace development of water infrastructure in many countries, desalination (the removal of salts from seawater) at high energy efficiency will likely become a vital source of fresh water. Due to its atomic thinness combined with its mechanical strength, porous graphene may be particularly well-suited for electrodialysis desalination, in which ions are removed under an electric field via ion-selective pores. Here, we show that single graphene nanopores preferentially permit the passage of K^+ cations over Cl^- anions with selectivity ratios of over 100 and conduct monovalent cations up to 5 times more rapidly than divalent cations. Surprisingly, the observed K^+/Cl^- selectivity persists in pores even as large as about 20 nm in diameter, suggesting that high throughput, highly selective graphene electrodialysis membranes can be fabricated without the need for subnanometer control over pore size.

[1] Department of Physics, Harvard University, Cambridge, Massachusetts 02138, USA. [2] School of Engineering and Applied Sciences, Harvard University, Cambridge, Massachusetts 02138, USA. * These authors contributed equally to this work. Correspondence and requests for materials should be addressed to J.A.G. (email: golovchenko@physics.harvard.edu).

Atomically thin graphene membranes have generated considerable interest for use as filtration membranes because their atomic thickness presents minimal resistance to fluid or ion flow while retaining high structural integrity. Recent investigations have suggested that porous graphene membranes can attain orders of magnitude higher flow rates than commercial reverse osmosis (RO) membranes, while still providing excellent salt rejection[1-7]. Unfortunately, RO salt rejection depends on a very tight distribution of subnanometer pores. A few large pores in a membrane can contribute large unselective water fluxes, impairing salt rejection. Thus, viable RO membranes depend on the complete elimination of pores larger than a nanometre or so, which remains a difficult fabrication challenge[6,8,9]. However, recent theoretical predictions suggest that graphene nanopores that are too large for RO may still be suitable for electrodialysis if they are electrostatically charged, allowing them to separate anions from cations[10-12]. Yet, experimental investigations of ion selectivity of graphene nanopores have been limited to subnanometer pores[7,13,14].

Here, we examine ion selectivity (cations versus anions, as well as among different cations) of single graphene nanopores with an emphasis on the relationship between ion selectivity and pore size. These experiments not only allow us to evaluate porous graphene membranes as electrodialysis membranes, but also shed light on the chemical structure of graphene pore edges and ion-specific interactions with graphene membranes.

Results

Nanopore fabrication. Graphene nanopores were fabricated using a recently reported electrical pulse method that enables rapid fabrication of very small single nanopores, as well as controllable, *in situ* enlargement of the nanopore[15]. This method allows measurements to be performed for multiple pore sizes with a single sample, which would be extremely difficult and time consuming using electron-beam drilling fabrication methods[16-20]. To create a pore, freestanding graphene membranes were placed in a flow-cell between two fluid reservoirs filled with 1 M KCl as schematically depicted in Fig. 1a. Ultra-short, high voltage pulses were applied across the membrane to nucleate and enlarge single nanopores. A transmission electron microscope (TEM) image of a graphene membrane after electrical pulse fabrication is shown in Fig. 1b. The outline of the pore can be clearly seen in a close-up of the image shown in Fig. 1c. Although TEM imaging can be used to measure pore sizes as was done here, preparing and imaging the samples after solution-based experiments is labour-intensive and low-yield (see the Methods section). Therefore, for the bulk of our experiments, we estimated the pore size based on the measured conductance of the nanopore in 1 M KCl solution using an analytical model of pore conductance[15,16,21,22], given by

$$G = \sigma \left(\frac{4t}{\pi D^2} + \frac{1}{D} \right)^{-1} \qquad (1)$$

where, G is the pore conductance, σ is the solution conductivity ($105\,mS\,cm^{-1}$), t is the effective thickness of the graphene membrane (0.6 nm, see ref. 16), and D is the pore diameter. Figure 1d shows the outline of the pore obtained via TEM imaging (grey) compared with the estimated size of the pore, based on the conductance of the pore in 1 M KCl at pH 2 (black circle). The close agreement suggests that equation 1 does an adequate job of estimating the pore size, precluding the need to image every sample with TEM. While the exact mechanism that produces electrically pulsed nanopores is not yet fully understood, it likely involves the oxidation of carbon at the pore edge[15], which results in carboxyl or other protonatable edge groups[23] that bestow a negative charge on the edge of the pore at neutral and higher pH (Fig. 1a, inset). Previous research on comparatively thick, insulating solid-state nanopores has demonstrated that negative charge at the periphery of the pore repels anions and attracts cations, which conduct the bulk of the ionic current[24-27]. However, such electrostatically controlled ion selectivity was thought to be negligible for pores in which the diameter is significantly larger than the membrane thickness[28]. Indeed, recent measurements have shown that sub-2 nm intrinsic defects in chemical vapour deposition (CVD) graphene membranes can distinguish between mono and divalent cations[14], but comparable measurements for larger pores have not been performed.

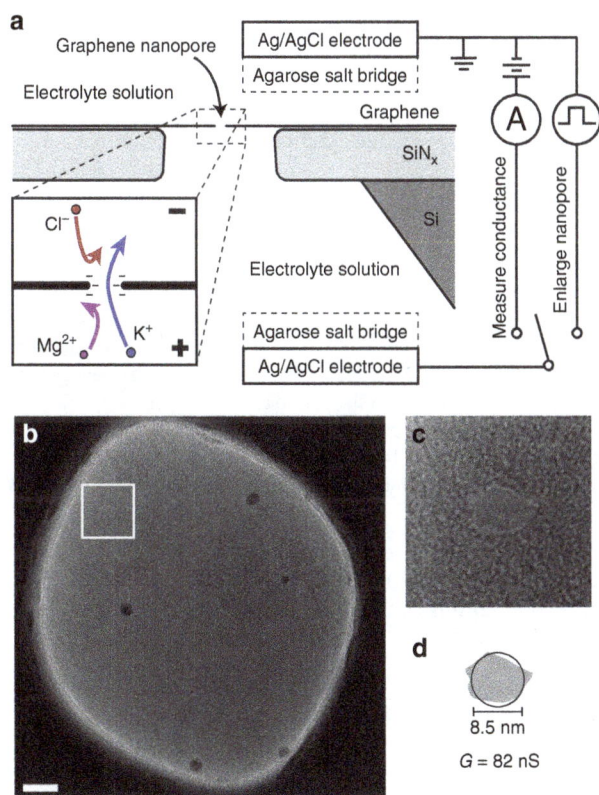

Figure 1 | Experimental setup. (**a**) Cross-sectional diagram of a suspended graphene nanopore sample immersed in electrolyte solution. Ag/AgCl electrodes that contact the solution via agarose salt bridges are used to measure ionic current through the nanopore or enlarge the nanopore using electrical pulses. Inset shows an illustration of anion and divalent cation rejection in a negatively charged nanopore. (**b**) TEM image of a suspended graphene membrane after a pore has been created via electrical pulse fabrication. Scale bar, 20 nm. White box indicates the location of the nanopore. (**c**) Close-up of area containing a nanopore. The sides of the image are 30 nm in length. (**d**) Comparison of the size of the pore with estimation of the pore size calculated from the pore conductance via equation 1. The grey outline is traced from the TEM image and the black circle is calculated from the conductance G.

Cation/anion selectivity. To measure cation/anion selectivity, current-voltage (I-V) curves were performed with a variety of KCl concentration gradients across the pore (Fig. 2a). The voltage bias was applied via Ag/AgCl electrodes contacting the fluidic cell via agarose salt bridges, which are used to eliminate the potential generated from redox reactions on electrodes in different salt concentrations. K^+ and Cl^- ions were selected as a representative cation/anion pair because they have very similar bulk mobilities (see Supplementary Table 1), and therefore exhibit

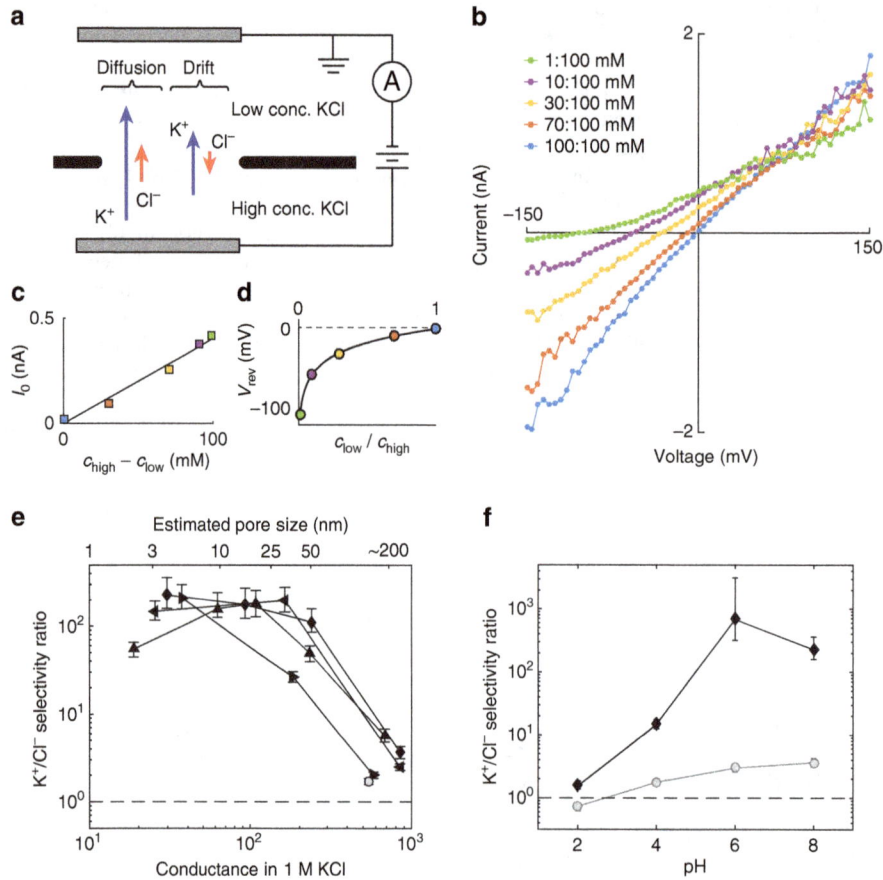

Figure 2 | K$^+$/Cl$^-$ selectivity. (a) Schematic of experimental setup: a concentration gradient and electric potential are simultaneously imposed across the nanopore and the net ionic current is measured. **(b)** Measured I-V curves for several concentration ratios. Each coloured curve indicates a different concentration ratio as indicated in the legend. **(c)** Zero-bias current indicates that K$^+$ ions pass more easily than Cl$^-$. Markers are coloured the same as in **b**. Solid line is a linear fit. **(d)** Reversal voltage as a function of concentration ratio, along with fit to the GHK voltage equation (solid line, equation 1), which is used to calculate selectivity. Markers are coloured the same as in **b**. **(e)** K$^+$/Cl$^-$ selectivity ratio as a function of pore size for several nanopores; different markers indicate different samples. The conductance in 1 M KCl (lower x-axis) is used to calculate the pore diameter (upper x-axis). Error bars indicate the 5th and 95th percentile estimates. The open circle indicates a control aperture with no suspended graphene. Dotted line indicates no selectivity. **(f)** K$^+$/Cl$^-$ selectivity ratio as a function of pH for a 3 nm pore (black diamonds), showing that selectivity increases with pH. In contrast, a sample with most of the graphene removed (grey circles) shows little selectivity at any pH. Error bars indicate the 5th and 95th percentile estimates. Dotted line indicates no selectivity.

negligible liquid junction potentials, about 1 mV for our measurements (as calculated by the Henderson equation[29]). Therefore, within experimental error (5 mV), the measured voltages were equal to the voltage drop across the graphene membrane. An example set of I-V curves for a 3 nm pore is shown in Fig. 2b. When there is no applied voltage (V = 0) both K$^+$ and Cl$^-$ ions diffuse from high to low concentration, and a net current (short-circuit current, I_0) is produced only if one ion diffuses at a higher rate than the other through the pore (Fig. 2c). The direction of this short-circuit current is consistent with the net flow of positive charges from high to low concentration, immediately indicating that the pore is cation selective. While the short-circuit current can identify selectivity, it is not a direct quantitative measure of selectivity because it also depends strongly on the conductance of the nanopore. A better choice is the reversal potential V_{rev}, the applied potential at which the net current is zero (Fig. 2d), which in the well-known Goldman–Hodgkin–Katz (GHK) model does not explicitly depend on pore size[30,31]. By assuming that each ion species contributes a current given by the Nernst–Planck equation, which is parametrized by an effective diffusion constant D_i^* that is different for each ion species i, the GHK model presents a quantitative measure of

selectivity that is useful for comparing selectivity among different pores[24,26,30]. In this context, the selectivity ratio S_{GHK} is defined as $D_{K^+}^*/D_{Cl^-}^*$, which is equal to the ratio of drift currents from each ion when there is no concentration gradient. The reversal potential is related to the selectivity of the pore via the GHK voltage equation

$$V_{rev} = \frac{k_B T}{e} \ln\left(\frac{S_{GHK} c_{high} + c_{low}}{S_{GHK} c_{low} + c_{high}}\right) \qquad (2)$$

where S_{GHK} is the selectivity ratio, c_{high} and c_{low} are the solution concentrations in the fluid reservoirs, e is the electron charge, k_B is the Boltzmann constant and T is the solution temperature.

The K$^+$/Cl$^-$ selectivity ratio S_{GHK} was calculated for each pore by fitting the reversal potentials to the GHK voltage equation (equation 2, Fig. 2d). Figure 2e shows the selectivity ratio S_{GHK} plotted as a function of pore size for four samples at pH 8 (see the Methods section for error estimation). The lower horizontal axis is the measured conductance in 1 M KCl, while the upper axis shows estimated pore diameter based on equation 1. The K$^+$/Cl$^-$ selectivity ratios at pH 8 were generally above 100, values comparable to the biological ion channels and polymer membranes many microns thick[32]. Surprisingly, the selectivity ratios were not significantly reduced until the pores were larger than about 20 nm. These high selectivities contrast with previous measurements by O'Hern et al.[13] on graphene membranes with many subnanometer pores, which reached a maximum selectivity ratio of $S_{GHK} = 1.3$. It is possible that the high K^+/Cl^- selectivities measured here are unique to electrically pulsed pores, or that a small number of large pores or tears in the centimeter-scale membranes of O'Hern et al.[13] drastically reduced selectivity by introducing parallel paths of non-selective ion flow. Indeed, pores larger than 100 nm (where the graphene pore is almost as large as the supporting silicon nitride aperture) showed minimal selectivity ($S_{GHK} < 5$, Fig. 2e). A control aperture with no graphene yielded $S_{GHK} \cong 2$ (Fig. 2e, open circle), indicating that the aperture itself contributes non-zero but comparatively minimal selectivity. S_{GHK} was also measured at different solution pHs to examine whether or not K^+/Cl^- selectivity is influenced by protonation/deprotonation of chemical groups (Fig. 2f). The measured selectivity for a 3 nm pore drops significantly between pH 6 and 4, and is negligible by pH 2. This pH dependence has a similar progression to deprotonation of edge groups expected at graphene edges (such as carboxyls)[33], suggesting that the deprotonation of chemical edge groups is necessary for cation/anion selectivity.

While the presence of negatively charged groups on the pore edge can explain strong K^+/Cl^- selectivity in subnanometer sized pores, it cannot account for larger pores (radius > Debye length $\cong 1$ nm in 100 mM KCl), where the edge charge is screened out by mobile counterions[25,28]. Moreover, in ultra-thin nanopores larger than a few nanometres in diameter, the pore conduction is mostly determined by access resistance extending out into the fluid[16,21,22], which is largely unaffected by the charge at the edge of the pore. Therefore, a different mechanism is needed to explain the high selectivities for large nanopores. In the Discussion section, we will present a new model of pore selectivity that is consistent with our K^+/Cl^- selectivity measurements.

Inter-cation selectivity. To measure selectivity among different cations, including divalent cations, we directly compared pore conductance in different cation-chloride solutions without a concentration gradient (Fig. 3a). This direct comparison can be made because the current due to Cl^- anions at pH 8 is negligible for small pores (as shown in the previous section). It is important to note, however, that different cations have significantly different electrophoretic mobilities, which result in differences in conductivity in bulk solution (see Supplementary Table 1). To account for these differences we introduce a normalized conductance g_i for each cation

$$g_i = \frac{G_i}{\mu_i/\mu_{K^+}} \tag{3}$$

where, G_i is the measured nanopore conductance in cation-chloride solution, μ_i is the bulk electrophoretic mobility of the cation and μ_{K^+} is the mobility of K^+ ions. Any observed differences in normalized conductance indicate inter-cation selectivity of the pore. Figure 3b shows normalized conductances for a variety of mono and divalent cations for several small nanopores 2–4 nm in diameter. It is immediately evident that divalent cations show much lower normalized conductance than monovalent cations, which agrees with the recent results on sub-2 nm defects in CVD graphene membranes[14]. Even among monovalent cations, the normalized conductances appear to follow the general trend $K^+ > Na^+ > Cs^+ > Li^+$. These results indicate that graphene nanopores can distinguish

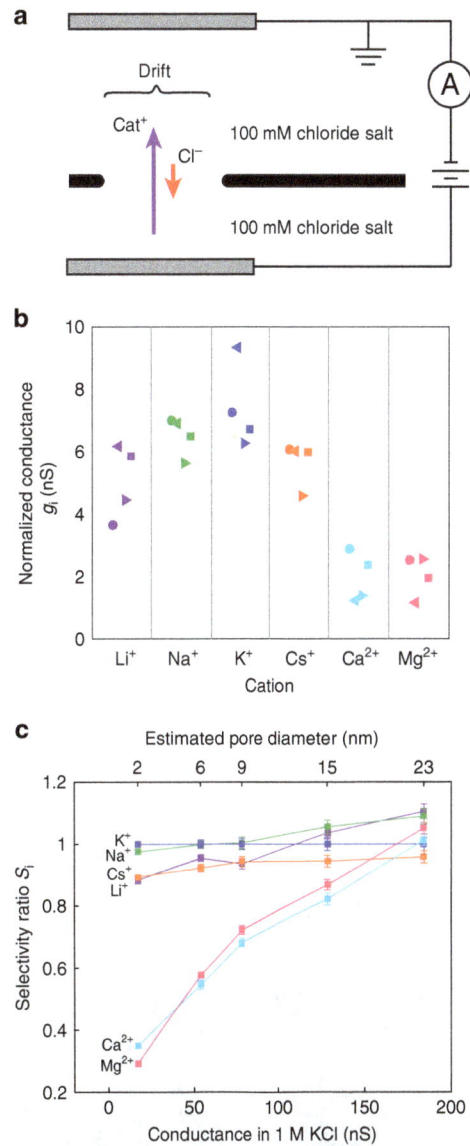

Figure 3 | Inter-cation selectivity. (**a**) Schematic of experimental setup: pore conductance is measured in a variety of 100 mM cation-chloride solutions. (**b**) Normalized conductance g_i (see equation 3 for definition) for four different nanopores 2–4 nm in diameter; different shaped markers indicate different samples. Data is sorted by cation. Monovalent cations pass more easily than divalent cations. (**c**) Inter-cation selectivity ratio S_i (see equation 4 for definition) of a graphene nanopore as a function of pore size. Error bars indicate the standard deviation. The conductance in 1 M KCl (lower x-axis) is used to estimate pore size using equation 1. Inter-cation selectivity decreases as pore size increases and is no longer significant above 20 nm.

strongly between mono and divalent cations, and weakly among monovalent cations.

To examine how inter-cation selectivity depends on pore size, we measured normalized cation conductances for a pore during the sequential stages of electrical pulse enlargement (Fig. 3c, see the Methods section for error estimation). To characterize the relative selectivities for different cations, we define the inter-cation selectivity ratio (relative to K^+) as

$$S_i = g_i/g_{K^+} \tag{4}$$

This definition of selectivity gives us $S_i = 1$ for a nanopore that does not distinguish between a cation i and K^+. As the pore is enlarged, the same ordering $K^+ > Na^+ > Cs^+ > Li^+ \gg Ca^{2+} > Mg^{2+}$ is preserved, but the selectivity of all cations is reduced as the pore size increases. For pores larger than 20 nm, no significant inter-cation selectivity remains. The deviations of g_i from 1 for such large pores are likely due to chlorine flux, which begin to contribute to the current for pores larger than 20 nm in diameter.

Discussion

The persistence of strong K^+/Cl^- selectivities in pores as large as 20 nm is surprising, and suggests that previously uncharacterized mechanisms may be responsible. We propose that the graphene surface (not just the pore edge) carries a pH-dependent surface charge due to deprotonatable oxygen-containing chemical groups on the graphene surface (oxidized graphene), or attached to the hydrocarbon contaminants on the graphene surface[34,35]. This surface charge, which would be negative at neutral pH values and neutralized in acidic solutions, would attract a screening cloud of positive counterions while also repelling anions in

solution. Because these mobile cations near the surface would be concentrated relative to anions, they would contribute a large cation-selective ion current, causing the total ionic current to be cation-selective.

To test the plausibility of this hypothesis, we measured pore conductance with various KCl concentrations at pH 2 and 8 (Fig. 4a). These measurements were taken on an 8.5 nm diameter nanopore, the same pore for which TEM imaging was shown in Fig. 1b,c. The presence of charged, deprotonatable surface groups on the entire graphene surface would cause the conductance at pH 8 to be significantly higher than at pH 2, even at salt concentrations high enough that the surface charge is screened out (Debye length < pore radius). The conductance data shown in Fig. 4a clearly shows this effect, with conductance at pH 8 considerably higher than at pH 2 for all concentrations below 3 M.

To quantitatively test the surface charge hypothesis, we numerically solved the Poisson–Nernst–Planck (PNP) equations for a 2D axisymmetric pore geometry with a variable surface charge on the suspended graphene surface (see Supplementary Fig. 1 for details). The conductance data agrees with a model of an 8.5 nm pore with surface charge $\sigma = -0.6 \, C \, m^{-2}$ (Fig. 4a). Initially, this surface charge density seems surprisingly high, considering that pristine graphene should not have any deprotonatable surface groups. However, it is possible that the high voltage pulses used to fabricate the nanopore oxidize the graphene surface, as well as hydrocarbon contaminants on the graphene surface[34,35]. As a comparison, graphene lightly treated with oxygen plasma was measured to have a surface charge density of $\sigma = -0.24 \, C \, m^{-2}$ at pH 7 (ref. 36). To illustrate how this surface charge density causes high K^+/Cl^- selectivity in this system, we examined two example paths in the numerical simulation (Fig. 4b): along the centre of the pore (path 1, Fig. 4b dotted line) and along the surface of the graphene (path 2, Fig. 4b, solid line). A plot of ion concentrations along the paths (Fig. 4c) shows increased K^+ concentration and decreased Cl^- concentration along the entirety of the surface path (path 2). As a result, the K^+ current density is elevated and the Cl^- current density reduced over the surface path (Fig. 4d).

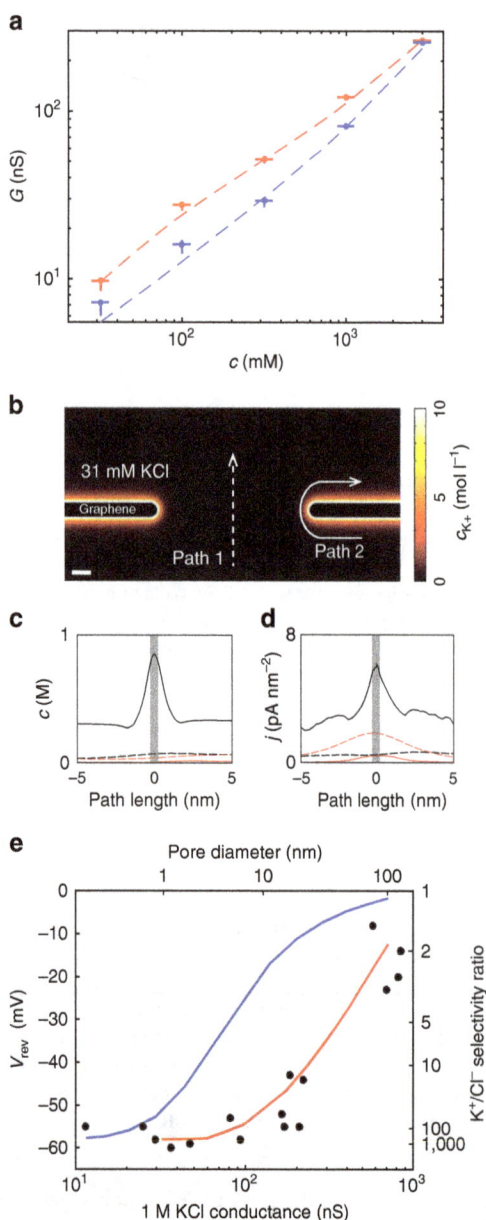

Figure 4 | Surface charge model of ion selectivity. (a) Conductance measurements as a function of KCl concentration for an 8.5 nm pore (the same pore shown in Fig. 1b–d). Blue and red markers show experimental data at pH 2 and pH 8, respectively. Vertical and horizontal error bars indicate standard deviation. Dotted lines show numerical predictions from a numerical PNP model (see Supplementary Fig. 1 for details) with surface charge density of $0 \, C \, m^{-2}$ (blue) and $-0.6 \, C \, m^{-2}$ (red). (b) Diagram of the numerical PNP system with negative surface charge ($\sigma = -0.6 \, C \, m^{-2}$). Scale bar, 1 nm. The K^+ concentration is plotted in a 31 mM KCl environment with 100 mV applied across the pore. Two illustrative current paths are indicated, path 1 (dotted line) through the centre of the pore and path 2 (solid line) along the membrane surface. (c) K^+ (blue) and Cl^- (red) concentrations plotted along the two illustrative paths shown in b. The grey bar indicates the thickness of the nanopore. Path 2 along the surface of the graphene has elevated K^+ and decreased Cl^- concentration. (d) K^+ (blue) and Cl^- current densities plotted along the two illustrative paths shown in b. Path 2 shows much greater K^+ current than Cl^- current, which results in K^+/Cl^- selectivity. The overall pore selectivity results from these highly selective, highly conductive current paths. (e) Comparison of measured reversal potentials (black dots) and numerical predictions from the model with surface charge density of $0 \, C \, m^{-2}$ (blue line) and $-0.6 \, C \, m^{-2}$ (red line). Both experimental data and numerical predictions are for a 10:100 mM concentration gradient. Right side, y-axis shows selectivity ratios calculated from the reversal potentials.

K^+/Cl^- selectivity is therefore a result of both increased K^+ current and reduced Cl^- current. Since these highly selective surface current paths are also highly conductive due to the elevated cation concentrations, they can cause even large pores to be very selective.

To quantitatively evaluate whether this model can account for the large measured selectivities, we simulated reversal potential measurements for a 10:100 mM concentration gradient and directly compared them with the experimental results shown in Fig. 2. Figure 4e shows the experimentally measured and numerically simulated reversal potentials as a function of pore size. Without the surface charge on the graphene membrane, the reversal potential and selectivity drop rapidly for pores larger than 1 nm, but the simulation including the surface charge models the measured data much better, predicting K^+/Cl^- selectivity ratios above 100 for pores as large as 5 nm. The spread in experimental data most likely indicates that the surface charge and pore shape vary from sample-to-sample.

With this surface charge mechanism for ion selectivity in mind, we must be careful in the interpretation of reversal potential data using the GHK equation (equation 2). In the GHK model, the ion selectivity does not depend on salt concentration. However, the surface charge model implies that selectivity would decrease with increasing salt concentration, because the surface charge is screened out more strongly in high salt solutions (Supplementary Fig. 2). The reduction of selectivity at high salt concentrations, an effect called salting-out, has previously been observed and modelled in biological porins[37]. To determine if the surface charge model accurately predicts this effect, we measured the reversal potential for a 10:1 concentration ratio ($c_{high}/c_{low} = 10$) as a function of salt concentration (c_{high}) and compared the results with predictions from the PNP model (Supplementary Fig. 3). Indeed, the reversal potential (and therefore selectivity) is lower at higher salt concentrations, although there is significant sample-to-sample variability on how much the selectivity drops off at salt concentrations of 1 M or higher. This trend agrees with predictions from the numerical PNP model (black line), which includes charge screening effects. Therefore, while the GHK model is useful for estimating selectivities at a given salt concentration, it cannot be interpreted as a full physical model because it does not encapsulate the vital electrostatic effects of surface charge. In comparison, the numerical PNP model is a more complete model but does not offer the convenience of analytical solutions.

In summary, we have shown that graphene nanopores up to about 20 nm in diameter show K^+/Cl^- selectivity ratios over 100 and monovalent/divalent cation selectivities up to 5. The K^+/Cl^- selectivities can be explained by elevated concentrations of mobile cations near the graphene surface. Future work studying the source of these increased surface concentrations (which may include mechanisms other than the fixed surface charge examined here) is still needed to complete our understanding of the mechanism responsible for ion selectivity. Modifying the surface or using different thin materials may also allow nanopores to select for anions instead of cations. Although we have limited this study to single pores, we expect that large-area porous membranes containing many nanopores, 20 nm or smaller, will retain high selectivity while supporting orders of magnitude larger ionic currents. Previous investigations of the desalination potential of graphene have focused on subnanometer diameter pores for RO, but the results shown here suggest that such strict fabrication limitations are not necessary. The loosening of the pore size upper boundary from around 1 to 20 nm means that existing techniques for creating porous graphene membranes can likely be used to create highly effective cation exchange membranes for electrodialysis. Furthermore, these surprising observations indicate that atomically thin nanopores can behave quite differently than their thicker counterparts, and should continue to be a rich platform for studying nanoscale mechanisms of ion transport.

Methods

Graphene membrane preparation. Single-layer graphene was grown on Cu foil (Alpha Aesar) at 1,000 °C with a flow of 10 sccm H_2 and 4 sccm CH_4 for 40 min. Graphene was transferred to an approximately 150-nm diameter aperture in a 300-nm thick low-stress LPCVD silicon nitride (SiN_x) membrane using established wet transfer techniques[15]. This SiN_x membrane was prepared using standard techniques, including photolithography and anisotropic KOH etching of the silicon substrate. The 150-nm diameter apertures were milled using an FIB (FEI/Micrion Vectra 980, 50 kV Ga^+). Once transferred, samples were annealed at 250 °C for at least 2 h under 200 sccm H_2 and 400 sccm Ar flow to remove surface contamination and were stored in a nitrogen-flushed drybox. Before wetting, the graphene membrane samples were annealed at 700 °C for 2 h under 200 sccm H_2 and 400 sccm Ar to further remove surface contamination.

Pore fabrication and fluidic cell measurements. Samples were loaded into an airtight PEEK fluidic cell with PDMS gaskets, and 99.999% pure CO_2 was flowed through the flow-cell for 3 min to purge out any air. The fluidic cell was then flushed on both sides of the membrane with 1 M KCl, pH 10, which reacts with CO_2 gas to form soluble carbonate anions, removing all gas in the fluidic cell and fully wetting the graphene membrane. The fluid reservoirs on each side of the fluidic cell were contacted via Ag/AgCl electrodes in 1 M KCl via agarose salt bridges to eliminate the potential generated from redox reactions on electrodes in asymmetric salt conditions. An Axopatch 200B patch clamp amplifier was used to apply d.c. voltage biases and measure ionic current. Pores were nucleated and enlarged using a pulse generator (HP8110A) as previously described[15]. Electrolyte solutions were prepared and buffered with 10 mM Tris, with the exception of 1 mM salt solutions, which were unbuffered. Pore diameters were estimated based on the conductance measurements using equation 2. It is worth noting that the additional conductance due to surface charge on the graphene surface is not taken into account in equation 2, so estimations of pore size based on the conductance measurements taken with neutral or basic pH solution may overestimate the pore size by as much as 50%.

Ion selectivity measurements. Reversal potentials were determined from interpolated I-V measurements. Selectivity S_{GHK} was estimated by nonlinear least squares regression to the measured reversal potential using equation 1. Replicate measurements of reversal potential produced values with a standard deviation of about 3 mV. S_{GHK} depends exponentially on reversal potential and for the extremely high selectivities measured here ($S_{GHK} > 100$), small deviations in reversal potential produce large, nonlinear changes in S_{GHK}. Error estimates in Fig. 2c,d were determined using Monte Carlo regression analysis on synthetic data sets, assuming a normal distribution of error in reversal potential measurements with a standard deviation of 3 mV. Reported error bars are the 5th and the 95th percentile of the resulting distribution of estimates of S_{GHK}. Reported error bars in selectivity ratio S_i in Fig. 3c are the standard errors of the mean from the linear least-squares regression from two repeated measurements.

TEM imaging. After electrical pulse fabrication and solution-based experiments, samples were removed from the fluidic cell and stored in deionized water. To avoid membrane damage due to surface tension, samples were critical-point dried before imaging. The nanopores were imaged in a JEOL 2010 F TEM operating at 200 kV. Efforts were made to minimize the beam exposure during imaging, because 200 kV electrons at high doses are capable of creating defects in graphene membranes. The yield for TEM imaging of electrical pulse-fabricated nanopores was very low (about 1 in 10) due to contamination of the membrane during drying and imaging. Contaminants from the solutions and from the air can easily cover the nanopores, especially under the electron beam, which can cause further deposition of contaminants. It seems that membranes that have been exposed to solution are more contamination-prone than membranes that have never been wet.

Numerical modelling. Numerical solutions to the PNP equations were calculated using the COMSOL Multiphysics software (COMSOL, Inc.). The model solves for concentrations of K^+ and Cl^- ions and the electric field using the steady-state Poisson equation and the Nernst–Planck equation for each ion individually (see Supplementary Fig. 1 and Supplementary Note 1). The negatively charged edge groups on the pore edge and the surface attraction to cations were modelled by imposing surface charge boundary conditions at the pore edge and on the graphene surface, respectively.

References

1. Cohen-Tanugi, D. & Grossman, J. C. Nanoporous graphene as a reverse osmosis membrane: recent insights from theory and simulation. *Desalination* **366**, 59–70 (2015).
2. Suk, M. E. & Aluru, N. R. Ion transport in sub-5-nm graphene nanopores. *J. Chem. Phys.* **140**, 084707 (2014).
3. Cohen-Tanugi, D. & Grossman, J. C. Water permeability of nanoporous graphene at realistic pressures for reverse osmosis desalination. *J. Chem. Phys.* **141**, 074704 (2014).
4. Cohen-Tanugi, D. & Grossman, J. C. Water desalination across nanoporous graphene. *Nano Lett.* **12**, 3602–3608 (2012).
5. Suk, M. E. & Aluru, N. R. Water transport through ultrathin graphene. *J. Phys. Chem. Lett.* **1**, 1590–1594 (2010).
6. O'Hern, S. C. *et al.* Nanofiltration across defect-sealed nanoporous monolayer graphene. *Nano Lett.* **15**, 3254–3260 (2015).
7. Surwade, S. P. *et al.* Water desalination using nanoporous single-layer graphene. *Nat. Nanotechnol.* **10**, 459–464 (2015).
8. Boutilier, M. S. H. *et al.* Implications of permeation through intrinsic defects in graphene on the design of defect-tolerant membranes for gas separation. *ACS Nano* **8**, 841–849 (2014).
9. O'Hern, S. C. *et al.* Selective molecular transport through intrinsic defects in a single layer of CVD graphene. *ACS Nano* **61**, 10130–10138 (2012).
10. Zhao, S., Xue, J. & Kang, W. Ion selection of charge-modified large nanopores in a graphene sheet. *J. Chem. Phys.* **139**, 114702 (2013).
11. He, Z., Zhou, J., Lu, X. & Corry, B. Bioinspired graphene nanopores with voltage-tunable ion selectivity for Na^+ and K^+. *ACS Nano* **7**, 10148–10157 (2013).
12. Sint, K., Wang, B. & Král, P. Selective ion passage through functionalized graphene nanopores. *J. Am. Chem. Soc.* **130**, 16448–16449 (2008).
13. O'Hern, S. C. *et al.* Selective ionic transport through tunable subnanometer pores in single-layer graphene membranes. *Nano Lett.* **14**, 1234–1241 (2014).
14. Jain, T. *et al.* Heterogeneous sub-continuum ionic transport in statistically isolated graphene nanopores. *Nat. Nanotechnol.* **10**, 1053–1057 (2015).
15. Kuan, A. T., Lu, B., Xie, P., Szalay, T. & Golovchenko, J. A. Electrical pulse fabrication of graphene nanopores in electrolyte solution. *Appl. Phys. Lett.* **106**, 203109 (2015).
16. Garaj, S. *et al.* Graphene as a subnanometre trans-electrode membrane. *Nature* **467**, 190–193 (2010).
17. Merchant, C. A. *et al.* DNA translocation through graphene nanopores. *Nano Lett.* **10**, 2915–2921 (2010).
18. Schneider, G. F. *et al.* DNA translocation through graphene nanopores. *Nano Lett.* **10**, 3163–3167 (2010).
19. Garaj, S., Liu, S., Golovchenko, J. A. & Branton, D. Molecule-hugging graphene nanopores. *Proc. Natl Acad. Sci. USA* **110**, 12192–12196 (2013).
20. Schneider, G. F. *et al.* Tailoring the hydrophobicity of graphene for its use as nanopores for DNA translocation. *Nat. Commun.* **4**, 2619 (2013).
21. Kowalczyk, S. W., Grosberg, A. Y., Rabin, Y. & Dekker, C. Modeling the conductance and DNA blockade of solid-state nanopores. *Nanotechnology* **22**, 315101 (2011).
22. Hall, J. E. Access resistance of a small circular pore. *J. Gen. Physiol.* **66**, 531–532 (1975).
23. Dreyer, D. R., Park, S., Bielawski, C. W. & Ruoff, R. S. The chemistry of graphene oxide. *Chem. Soc. Rev.* **39**, 228–240 (2009).
24. Chen, P. *et al.* Atomic layer deposition to fine-tune the surface properties and diameters of fabricated nanopores. *Nano Lett.* **4**, 1333–1337 (2004).
25. Kim, D.-K., Duan, C., Chen, Y.-F. & Majumdar, A. Power generation from concentration gradient by reverse electrodialysis in ion-selective nanochannels. *Microfluid Nanofluid.* **9**, 1215–1224 (2010).

26. Gillespie, D., Boda, D., He, Y., Apel, P. & Siwy, Z. S. Synthetic nanopores as a test case for ion channel theories: the anomalous mole fraction effect without single filing. *Biophys. J.* **95**, 609–619 (2008).
27. Siria, A. *et al.* Giant osmotic energy conversion measured in a single transmembrane boron nitride nanotube. *Nature* **494**, 455–458 (2013).
28. Vlassiouk, I., Smirnov, S. & Siwy, Z. Ionic selectivity of single nanochannels. *Nano Lett.* **8**, 1978–1985 (2008).
29. Bockris, J. & Reddy, A. *Modern Electrochemistry 1: Ionics* (Plenum, 1998).
30. Hille, B. *Ion Channels of Excitable Membranes* 3rd edn (Sinauer Associates, 2001).
31. Schultz, S. G. *Basic Principles of Membrane Transport* (Cambridge University Press, 1980).
32. Xu, T. Ion exchange membranes: state of their development and perspective. *J. Memb. Sci.* **263**, 1–29 (2005).
33. Konkena, B. & Vasudevan, S. Understanding aqueous dispersibility of graphene oxide and reduced graphene oxide through pKa measurements. *J. Phys. Chem. Lett.* **3**, 867–872 (2012).
34. Li, Z. *et al.* Effect of airborne contaminants on the wettability of supported graphene and graphite. *Nat. Mater.* **12**, 925–931 (2013).
35. Meyer, J. C., Girit, C. O., Crommie, M. F. & Zettl, A. Imaging and dynamics of light atoms and molecules on graphene. *Nature* **454**, 319–322 (2008).
36. Shan, Y. P. *et al.* Surface modification of graphene nanopores for protein translocation. *Nanotechnology* **24**, 495102 (2013).
37. Alcaraz, A., Nestorovich, E. M., Aguilella-Arzo, M., Aguilella, V. M. & Bezrukov, S. M. Salting out the ionic selectivity of a wide channel: the asymmetry of OmpF. *Biophys. J.* **87**, 943–957 (2004).

Acknowledgements

We acknowledge P. Frisella and T. Szalay for assistance in sample preparation, D. Branton for experimental advice, B. Lu for assistance with numerical modelling, and X. Chen and E. Levine for advice regarding the manuscript. This research was supported by the National Institutes of Health Award R01HG003703 to J.A.G. and the Department of Energy Office of Science Graduate Fellowship (DOE SCGF) to A.T.K., made possible in part by the American Recovery and Reinvestment Act of 2009, administered by ORISE-ORAU under contract no. DE-AC05-06OR23100.

Author contributions

All the authors conceived and designed the experiments, and wrote the paper. R.C.R. and A.T.K. performed the experiments and analysed the data, contributing equally to this work.

Additional information

18

Quantum teleportation from light beams to vibrational states of a macroscopic diamond

P.-Y. Hou[1], Y.-Y. Huang[1], X.-X. Yuan[1], X.-Y. Chang[1], C. Zu[1], L. He[1] & L.-M. Duan[1,2]

With the recent development of optomechanics, the vibration in solids, involving collective motion of trillions of atoms, gradually enters into the realm of quantum control. Here, building on the recent remarkable progress in optical control of motional states of diamonds, we report an experimental demonstration of quantum teleportation from light beams to vibrational states of a macroscopic diamond under ambient conditions. Through quantum process tomography, we demonstrate average teleportation fidelity (90.6 ± 1.0)%, clearly exceeding the classical limit of 2/3. The experiment pushes the target of quantum teleportation to the biggest object so far, with interesting implications for optomechanical quantum control and quantum information science.

[1] Center for Quantum Information, Institute for Interdisciplinary Information Sciences, Tsinghua University, Beijing 100084, China. [2] Department of Physics, University of Michigan, Ann Arbor, Michigan 48109, USA. Correspondence and requests for materials should be addressed to L.-M.D. (email: lmduan@umich.edu).

Quantum teleportation has found important applications for realization of various quantum technologies[1-4]. Teleportation of quantum states has been demonstrated between light beams[5-8], trapped atoms[9-12], superconducting qubits[13], defect spins in solids[14] and from light beams to atoms[15,16] or solid-state spin qubits[17,18]. It is of both fundamental interest and practical importance to push quantum teleportation towards more macroscopic objects. Observing quantum phenomenon in macroscopic objects is a big challenge, as their strong coupling to the environment causes fast decoherence that quickly pushes them to the classical world. For example, quantum coherence is hard to survive in mechanical vibration of macroscopic solids, which involves collective motion of a large number of strongly interacting atoms. Despite this challenge, achieving quantum control for the optomechanical systems becomes a recent focus of interest with remarkable progress[19-30]. This is driven in part by the fundamental interest and in part by the potential applications of these systems for quantum signal transduction[25-27], sensing[19] and quantum information processing[19-21]. There are typically two routes to achieve quantum control for the optomechanical systems: one needs to either identify some isolated degrees of freedom in mechanical vibrations and cool them to very low temperature to minimize their environmental coupling[19,28-30], or use the ultrafast laser technology to fast process and detect quantum coherence in such systems[20-24]. A remarkable example for the latter approach is provided by the optomechanical control in macroscopic diamond samples[20,21], where the motions of two separated diamonds have been cast into a quantum-entangled state[20].

In this paper, we report an experimental demonstration of quantum teleportation from light beams to the vibrational states of a macroscopic diamond sample of $3 \times 3 \times 0.3\,\mathrm{mm}^3$ in size under ambient conditions. The vibration states are carried by two optical phonon modes, representing collective oscillation of over 10^{16} carbon atoms. To facilitate convenient qubit operations, we use the dual-rail representation of qubits instead of the single-rail encoding used in the previous experiments[20-23] and generate entanglement between the paths of a photon and different oscillation patterns of the diamond represented by two phononic modes. Using quantum state tomography, we demonstrate entanglement fidelity of $(81.0 \pm 1.8)\%$ with the raw data and of $(89.7 \pm 1.2)\%$ after the background noise subtraction. Using this entanglement, we prepare arbitrary polarization states for the photon and teleport these polarization states to the phonon modes, with the Bell measurements on the polarization and

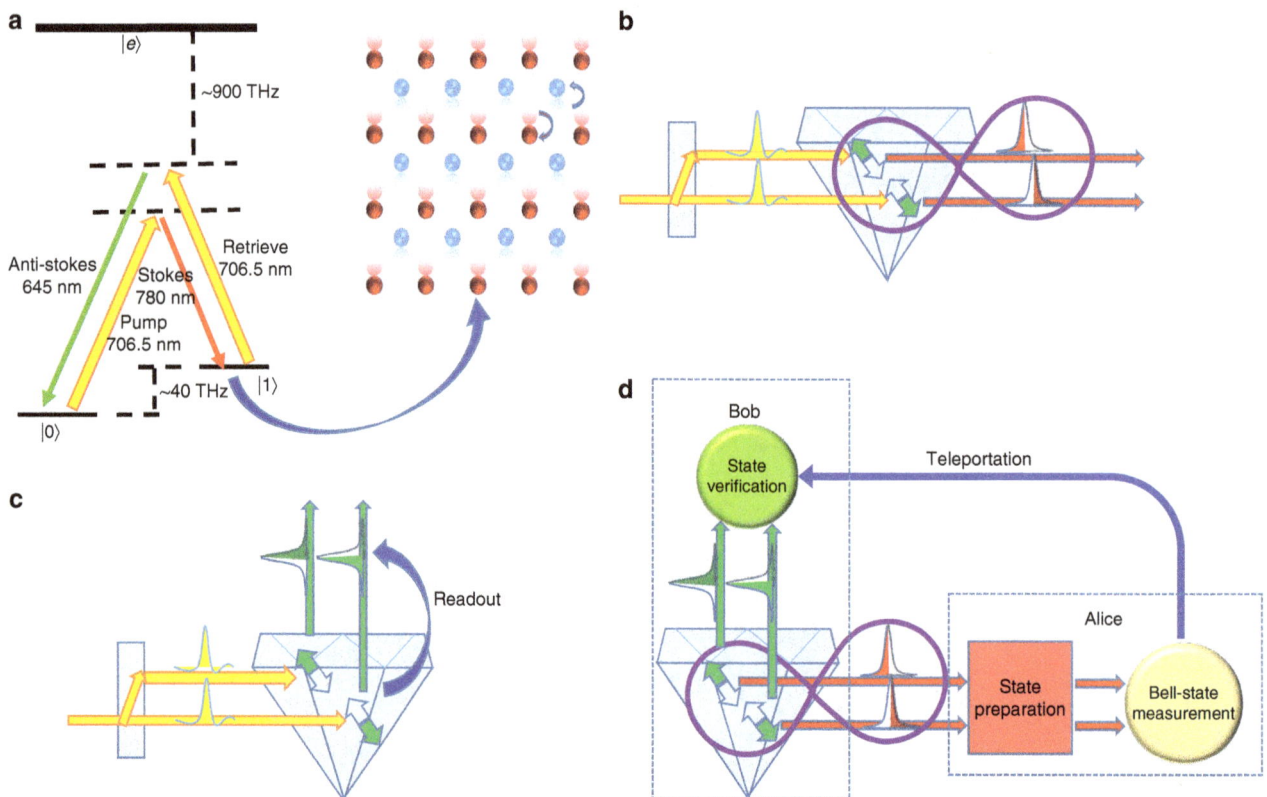

Figure 1 | Scheme for quantum teleportation with a diamond. (**a**) Illustration of the relevant level structure in the diamond. A write beam pumps the diamond in the ground state $|0\rangle$ and generates a Stokes photon in the forward direction and an excitation in the optical phonon mode of the diamond (denoted by the state $|1\rangle$). The optical phonon mode corresponds to the relative oscillation of the atoms in each unit cell of the diamond lattice, as illustrated by the figure on the right side. A read beam after a controllable delay converts the phonon excitation to an anti-Stokes photon that can be used for state readout. The corresponding wavelengths and frequencies are shown in the figure. The state $|e\rangle$ denotes the electron conduction band that is far detuned from the optical excitation. (**b**) A scheme for generation of the entanglement between a phonon in the diamond and a propagating photon. The phonon state is represented by a superposition of different oscillation modes of the diamond, while the photon state is represented by its spatial modes. (**c**) Readout of the phonon state with the read beams by coherently converting the phonon modes into the corresponding anti-Stoke photon modes. (**d**) A teleportation scheme using the photon–phonon entanglement. An input state is prepared by the message sender, Alice, on the photon's polarization degree of freedom. The photon thus carries two qubits, one by its polarization and one by its spatial modes. Alice performs Bell measurements on these two qubits. Conditional on certain measurement outcomes, the phonon state is projected to the same state input on Alice's side, which is read out and verified by Bob, the message receiver.

the path qubits carried by the same photon. The teleportation is verified by quantum process tomography (QPT), and we achieve a high average teleportation fidelity, $\sim (90.6 \pm 1.0)\%$ (or $(82.9 \pm 0.8)\%$) after (or before) subtraction of the background noise. To verify the phonon's state before its fast decay, our implementation of teleportation adopted the technique of reversed time ordering introduced in ref. 20 where the phonon's state is read out before the teleportation is completed. Similar to the pioneering teleportation experiment of photons[5], our implementation of teleportation is conditional, as the Bell measurements are not deterministic and require postselecting of successful measurement outcomes.

Results

Photon-to-phonon teleportation scheme. We illustrate our entanglement generation and quantum teleportation scheme in Fig. 1, using a type IIa single-crystal synthetic diamond sample cut along the 100 face from the Element Six company. Due to the strong interaction of atoms in the diamond, the optical phonon mode, which represents relative oscillation of the two sublattices in the stiff diamond lattice (Fig. 1a), has a very high excitation frequency $\sim 40\,THz$ near the momentum zero point in the Brillouin zone. The corresponding energy scale for this excitation is significantly higher than the room temperature thermal energy ($\sim 6\,THz$), and thus the optical phonon mode naturally stays at the vacuum state under ambient conditions, which simplifies its quantum control[20,21]. The coherence life time of the optical phonon mode is $\sim 7\,ps$ at room temperature, which is short, but accessible with the ultrafast laser technology for which the operational speed can be up to $\sim 10\,THz$ (refs 20,21).

We excite the optical phonon modes through ultrafast laser pulses of duration $\sim 150\,fs$ from the Ti–sapphire laser, with the carrier wavelength at 706.5 nm. The diamond has a large bandgap

of 5.5 ev, so the laser pulses are far detuned from the conduction band with a large gap $\sim 900\,THz$. Each laser pulse generates, with a small probability p_s, an excitation in the optical phonon mode and a Stokes photon of wavelength 780 nm in the forward direction (Fig. 1a). The relevant output state has the form

$$|\Psi\rangle = \left[1 + \sqrt{p_s}\, b_n^\dagger a_t^\dagger + o(p_s)\right]|vac\rangle, \qquad (1)$$

where $b_n^\dagger$ and $a_t^\dagger$ represent, respectively, the creation operators for an optical phonon and a Stokes photon, and $|vac\rangle$ denotes the common vacuum state for both the photon and the phonon modes.

To generate entanglement, we split the laser pulse into two coherent paths as shown in Fig. 1b, and the pulse in each path generates the corresponding phonon–photon correlated state described by equation 1. When there is an output photon, in one of the two paths, it is in the following maximally entangled state with the phonon excitation

$$|\Psi_{nt}\rangle = (|U\rangle_n|U\rangle_t + |L\rangle_n|L\rangle_t)/\sqrt{2}. \qquad (2)$$

Here $|U\rangle$ or $|L\rangle$ represents an excitation in the upper or lower path, and its subscript denotes the nature of the excitation, 'n' for a phonon and 't' for a photon. We drop the vacuum term in equation (1), as it is eliminated if we detect a photon emerging from one of the two paths. After entanglement generation, the photon state can be directly measured through single-photon detectors. To read out the phonon state, we apply another ultrafast laser pulse after a controllable delay within the coherence time of the optical phonon mode, and convert the phononic state to the same photonic state in the forward anti-Stokes mode at the wavelength of 645 nm (Fig. 1c). The state of the anti-Stokes photon is then measured through single-photon detectors together with linear optics devices. Note that the retrieval laser pulse could have a carrier frequency ω_r different from that of the

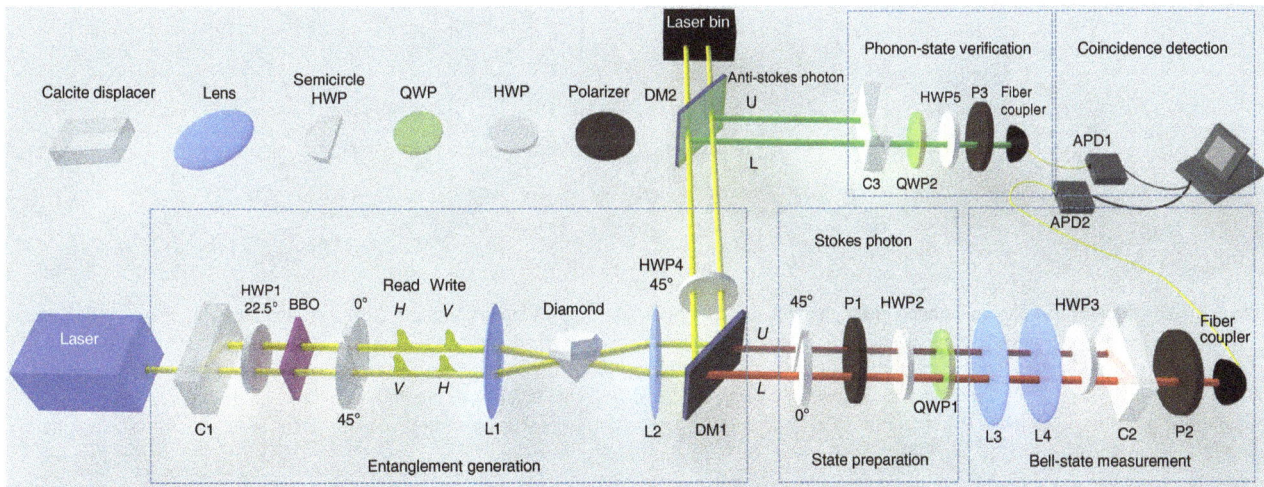

Figure 2 | Experimental set-up for the entanglement verification and quantum teleportation. Femtosecond laser pulses from a Ti-sapphire laser (coherent), with a repetition frequency of 76 MHz, a carrier wavelength of 706.5 nm and a polarization along the $|H\rangle + |V\rangle$ direction, are split by a birefringent calcite into two coherent paths with equal amplitudes. After rotation of the pulse polarization to equal superposition of $|H\rangle$ and $|V\rangle$ with a half-wave plate (HWP1) set at 22.5°, we introduce a time delay of 388 fs to the two polarization components H and V, with a birefringent beta barium borate (BBO) crystal. We use the lead pulse of H polarization as the write beam and the lagged pulse of V polarization as the read beam. After semicircle HWPs set at 0° and 45°, respectively, at the upper and lower paths, the polarization states of the pump beams are shown in the figure before the diamond sample. The write beam is focused by the lens L1 on the diamond sample and generates a Stokes photon in one of the paths, and an excitation in the corresponding optical phonon modes of the diamond. The Stokes photon, at the wavelength of 780 nm, is transmitted by the dichromatic mirror DM1 after the collection lens L2, with its two paths recombined by the calcite C2. The lens L3 and L4 are used to adjust the distance between the two optical paths, so that they can be combined at the calcite C2. The single-photon detector APD2, together with rotation of the polarizer P2, detects the two path (or polarization) components of the Stokes photon in different bases. To read out the state of the phonon modes, the read pulse converts the phonon to the anti-Stokes photon in the corresponding paths. The anti-Stokes photon, at a shorter wavelength of 645 nm, is reflected by both of the dichromatic mirrors DM1 and DM2, with its two paths recombined through the calcite C3. The photon coincidence counts are registered through a FPGA (Field-Programmable Gate Array) board with a 5 nm coincidence window.

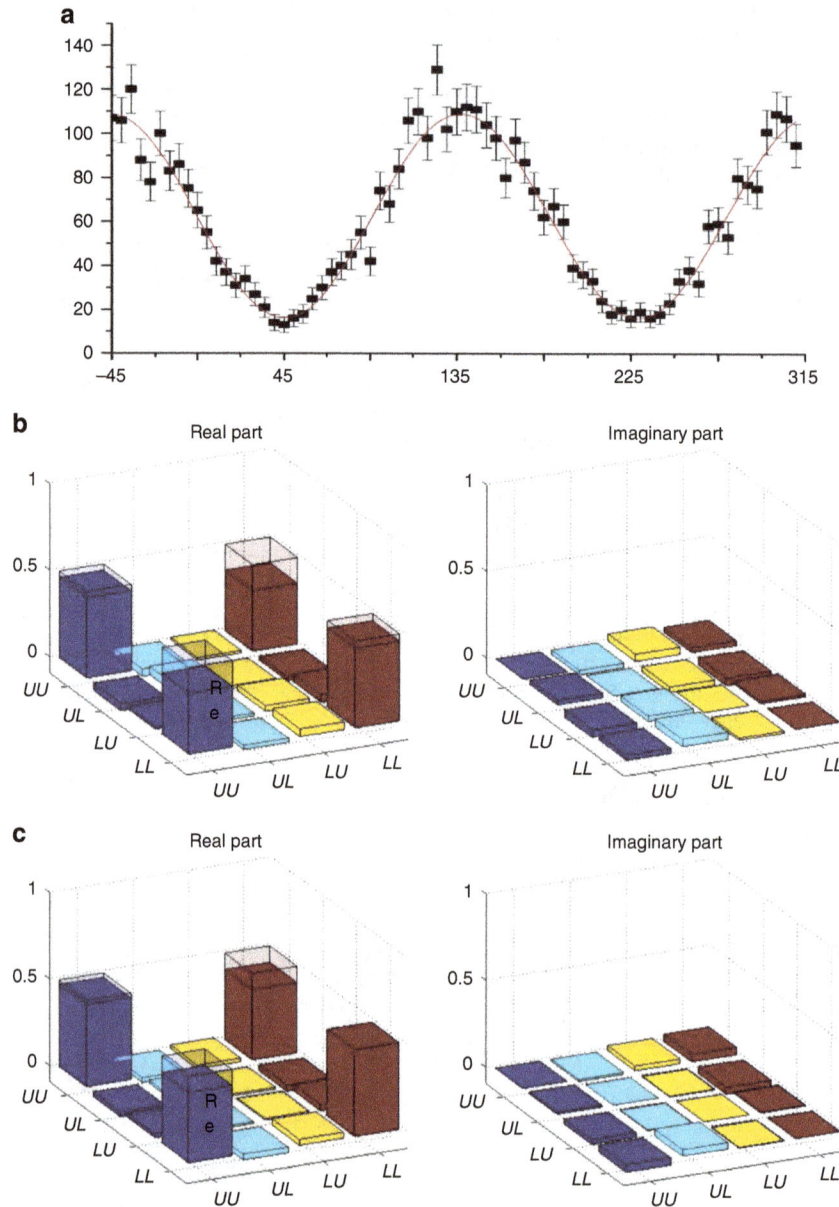

Figure 3 | Verification of the photon–phonon entanglement. (a) Coincidence counts of Stokes and anti-Stokes photons, as a function of the rotation angle (in degree) of the polarizer (P2 in Fig. 2) for the Stokes photon when the measurement basis of the anti-Stokes photon is fixed at $|U\rangle - |L\rangle$. The error bars denote the s.d. **(b)** Real and imaginary parts of the density matrix elements for the phonon–photon-entangled state reconstructed through the quantum state tomography. The hollow caps correspond to the values of matrix elements for a perfect maximally entangled state. **(c)** Same as **b**, but we subtract the background noise due to the accidental coincidences of the photon detectors. The coincidence count rate for Stokes and anti-Stokes photons is 8 per s for measurements in the UU and LL bases.

pump laser. For instance, with ω_r near the telecom band, our teleportation protocol would naturally realize a quantum-frequency transducer that transfers the photon's frequency to a desired band, without changing its quantum state. A quantum-frequency transducer is widely recognized as an important component for realization of long-distance quantum networks[25-27].

To realize teleportation, we need to prepare another qubit, whose state will be teleported to the phonon modes in the diamond. Similar to the teleportation experiments in refs 6,16, we use the polarization state of the photon to represent the input qubit, which can be independently prepared into an arbitrary state $c_0|H\rangle_t + c_1|V\rangle_t$, where $|H\rangle_t$ and $|V\rangle_t$ denote the horizontal and the vertical polarization states and c_0, c_1 are arbitrary

coefficients. The Bell measurements on the polarization and the path qubits carried by the same photon can be implemented through linear optic devices together with single-photon detection (Fig. 1d), and the teleported state to the phononic modes is retrieved and detected through its conversion to the anti-Stokes photon. Same as ref. 20, the short life time of the diamond's vibration modes requires us to retrieve and detect the phonon's state before applying detection on the Stokes photon, thus the phonon's state is measured before the teleportation protocol is completed. The reversed time ordering in this demonstration of quantum teleportation makes it unsuitable for application in quantum repeaters that requires a much longer memory time; however, it does not affect application of our teleportation experiment for realization of a

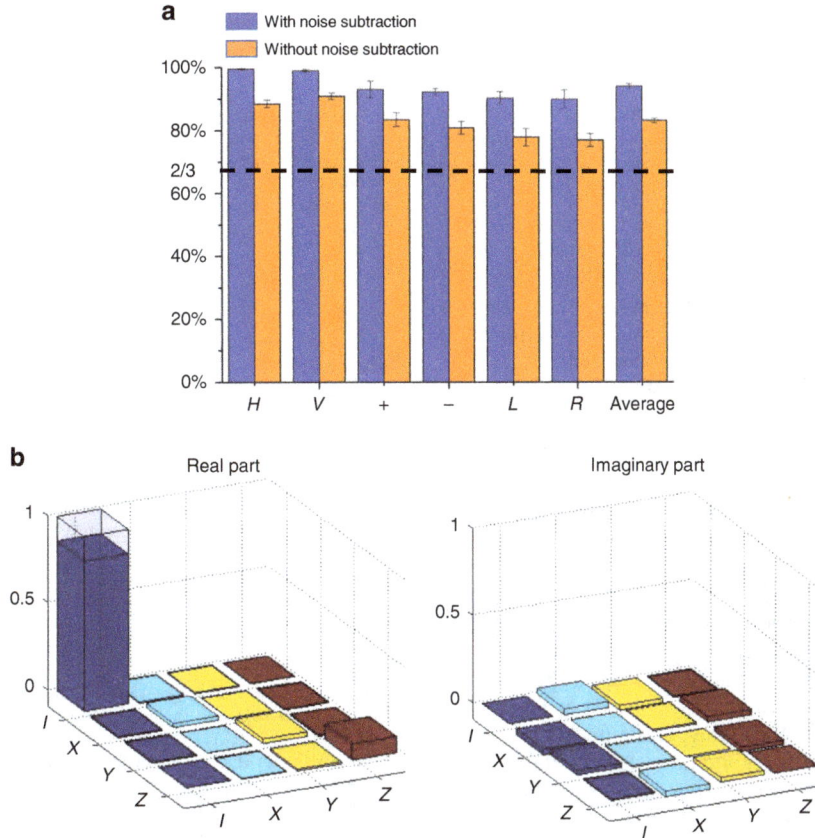

Figure 4 | Experimental results for photon-to-phonon quantum teleportation. (**a**) The teleportation fidelities for the six complementary bases states. The last two columns show the teleportation fidelity averaged over these six input states. The results are shown for both cases with or without subtraction of the background noise. The error bar denotes one s.d. The dashed line at fidelity 2/3 corresponds to the classical-quantum boundary for teleportation. (**b**) Real and imaginary parts of the process matrix elements for quantum teleportation reconstructed through the quantum process tomography (Methods). The hollow caps correspond to the values of process matrix elements for a perfect teleportation operation.

quantum-frequency transducer or a new source of entangled photons as discussed above.

Experimental realization of teleportation. Our experimental set-up is shown in Fig. 2. First, we verify entanglement generated between the Stokes photon and the optical phonon modes in the diamond. For this step, we remove the optical elements in the state preparation box shown in Fig. 2 and set the angle of half-wave plate (HWP3) to 0°. Different from the scheme illustrated in Fig. 1, we insert semicircle HWPs set at 0° and 45°, respectively, at the upper and the lower paths of the pump beam, so that both the Stokes photon and the anti-Stokes photon after the retrieval pulse have orthogonal polarizations along the two output paths, which can be combined together through the calcites C2 and C3. This facilitates the entanglement measurement through the detection in complementary local bases by rotating the polarizers P2 and P3, and the wave plates HWP5 and quarter-wave plate 2. Due to the different incident directions of the pump pulses at the upper and the lower paths, the corresponding phonon modes excited in the diamond have different momenta, so they represent independent modes even if they have partial spatial overlap. The phonon is converted to the anti-Stokes photon by the retrieval pulse, so we measure the photon–phonon state by detecting the coincidence counts between Stokes and anti-Stokes photons in different bases. In Fig. 3a, we show the registered coincidence counts, as we rotate the angle of the polarizer P2. The oscillation of the coincidence counts with a visibility of $(74.6 \pm 3.6)\%$ is an indicator of coherence of the underlying state. To verify

entanglement of the photon–phonon state, we use quantum state tomography to reconstruct the full density matrix from the measured coincidence counts[31], with the resulting matrix elements shown in Fig. 3b. From the reconstructed density matrix ρ_e, we find its entanglement fidelity, defined as the maximum overlap of ρ_e with a maximally entangled state, $F_e = (81.0 \pm 1.8)\%$, significantly higher than the criterion of $F_e = 0.5$ for verification of entanglement[32]. The error bars are determined by assuming a Poissonian distribution for the photon counts and propagated from the raw data to the calculated quantities through exact numerical simulation. The dominant noise in this system comes from the accidental coincidence between the detected Stokes and the anti-Stokes photons[20,21]. To measure the contribution of this accidental coincidence, we introduce an extra time delay of 13 ns, the repetition period of our pump pulses, to one of the detectors when we record the coincidence. When we subtract the background noise due to this accidental coincidence, the resulting matrix elements reconstructed from the quantum state tomography are shown in Fig. 3c. We find the entanglement fidelity is improved to $F_e = (89.7 \pm 1.2)\%$ after subtraction of the accidental coincidence.

To perform quantum teleportation using the photon–phonon entanglement, we first transform the effective photon–phonon-entangled state to the standard form of equation 2, by the semicircle HWPs in the state preparation box of Fig. 2. The polarizer P1, and the wave plates HWP2 and quarter-wave plate 1 then prepare the to-be-teleported photon polarization to arbitrary superposition states $|\Phi_{in}\rangle = c_0 |H\rangle_t + c_1 |V\rangle_t$. We perform Bell measurement through the calcite C2, the HWP3, the polarizer P2 and the detector APD2. For instance, with the HWP3 set at $0°$ and the polarizer P2 set along the direction $|H\rangle+|V\rangle$, a photon count in the detector APD2 corresponds to a projection to the Bell state $(|H\rangle_t|U\rangle_t+|V\rangle_t|L\rangle_t)/\sqrt{2}$ for the polarization and the path qubits of the photon before the measurement box. By rotating the angles of HWP3 and P2, we can also perform projection to any other Bell states.

The experimental result for teleportation is shown in Fig. 4. The teleportation fidelity is defined as $F=\langle\Phi_{in}|\rho_{out}|\Phi_{in}\rangle$, where $|\Phi_{in}\rangle$ is the input state at Alice's side and ρ_{out} denotes the output density matrix at Bob's side, reconstructed through quantum state tomography measurements. In Fig. 4a, we show the teleportation fidelity under six complementary bases states with $|\Phi_{in}\rangle=|H\rangle_t$, $|V\rangle_t$, $|\pm\rangle_t=(|H\rangle_t\pm|V\rangle_t)/\sqrt{2}$, $|L\rangle_t=(|H\rangle_t+i|V\rangle_t)/\sqrt{2}$, $|R\rangle_t=(|H\rangle_t-i|V\rangle_t)/\sqrt{2}$ in cases with and without subtraction of the background noise. The average fidelity over these six bases states is $(93.9\pm0.8)\%$ (or $(83.0\pm0.8)\%$) with (or without) background noise subtraction. This average fidelity is significantly $>2/3$, the boundary value for the fidelity that separates quantum teleportation from classical operations. For more complete characterization, we also perform QPT for the teleportation operation. In the ideal case, teleportation should be characterized by an identity transformation, meaning that Alice's input state is teleported perfectly to Bob's side. The experimentally reconstructed process matrix elements are shown in Fig. 4b (see Methods for explanation of QPT). The process fidelity is given by $F_P=(85.9\pm1.6)\%$, which corresponds to a teleportation fidelity $\bar{F}=(90.6\pm1.0)\%$ averaged over all possible input states, with equal weight in the qubit space.

Discussion

Teleportation of the quantum states from a photon to the vibration modes of a millimeter-sized diamond under ambient conditions generates a quantum link between the microscopic particle and the macroscopic world around us, usually under the law of classical physics. In our experiment, the ultrafast laser technology provides the key tool for the fast processing and detection of quantum states within its short life time in macroscopic objects, consisting of many strongly interacting atoms that are coupled to the environment. Combined with the tunability of the wavelength for the retrieval laser pulse[23], the technique introduced in our experiment would be useful for the realization of a new source of entangled photons based on the diamond optomechanical coupling with the dual-rail encoding. Such a source could generate entangled photons at wavelengths inconvenient to produce by other methods. For instance, we may generate entanglement between the ultraviolet and infrared photons, with the infrared photon good for quantum communication and the ultraviolet photon convenient to be interfaced with other qubits, such as the ion matter qubits. Such a photon source is hard to generate by the conventional spontaneous parametric down conversion method. In future, the tools based on the ultrafast pump and probe could be combined with the powerful laser cooling or low-temperature technology to provide more efficient ways for quantum control of the optomechanical systems, with important applications for realization of transduction of quantum signals[25,26], processing of quantum information or single-photon signals[19,20,23] and sensing of small mechanical vibrations[19].

Methods

Quantum process tomography. QPT[31] is defined by a completely positive map ε: $\rho_f\equiv\varepsilon(\rho_i)$ that transfers an arbitrary input state ρ_i to the output ρ_f. It can be characterized by a unique process matrix χ_{mn} through the map

$\rho_f=\sum_{mn}E_m\rho_iE_n^\dagger\chi_{mn}$, by choosing a fixed set of basis operator E_m. In our experiment, we set the basis operators E_m to be the identity operator I and the three Pauli matrices $X=\sigma_x$, $Y=-i\sigma_y$, $Z=\sigma_z$. This corresponds to a choice of six complementary input states $|H\rangle$, $|V\rangle$, $|+\rangle$, $|-\rangle$, $|L\rangle$ and $|R\rangle$ for the teleportation. We reconstruct the output state from teleportation by quantum state tomography and use them to calculate the process matrix χ through the maximally likelihood estimation[31]. The process fidelity is determined by $F_p=Tr(\chi\chi_{id})$, where χ_{id} is the identity process matrix corresponding to the perfect case. The process fidelity F_p determines the average teleportation fidelity $\bar{F}$ by the formula $\bar{F}=(2F_P+1)/3$ (ref. 31), where $\bar{F}$ is defined as the fidelity averaged over all possible states of the input qubit with equal weight.

References

1. Bennett, C. H. *et al.* Teleporting an unknown quantum state via dual classical and Einstein-Podolsky-Rosen channels. *Phys. Rev. Lett.* **70**, 1895–1899 (1993).
2. Briegel, H.-J., Dur, W., Cirac, J. I. & Zoller, P. Quantum repeaters: the role of imperfect local operations in quantum communication. *Phys. Rev. Lett.* **81**, 5932–5935 (1998).
3. Gottesman, D. & Chuang, I. L. Demonstrating the viability of universal quantum computation using teleportation and single-qubit operations. *Nature* **402**, 390–393 (1999).
4. Duan, L. M., Lukin, M. D., Cirac, J. I. & Zoller, P. Long-distance quantum communication with atomic ensembles and linear optics. *Nature* **414**, 413–418 (2001).
5. Bouwmeester, D. *et al.* Experimental quantum teleportation. *Nature* **390**, 575–579 (1997).
6. Boschi, D. *et al.* Experimental realization of teleporting an unknown pure quantum state via dual classical and Einstein-Podolsky-Rosen channels. *Phys. Rev. Lett.* **80**, 1121–1125 (1998).
7. Furusawa, A. *et al.* Unconditional quantum teleportation. *Science* **282**, 706–709 (1998).
8. Takeda, S. *et al.* Deterministic quantum teleportation of photonic quantum bits by a hybrid technique. *Nature* **500**, 315–318 (2013).
9. Riebe, M. *et al.* Deterministic quantum teleportation of atomic qubits. *Nature* **429**, 734–737 (2004).
10. Barrett, M. D. *et al.* Deterministic quantum teleportation with atoms. *Nature* **429**, 737–739 (2004).
11. Olmschenk, S. *et al.* Quantum teleportation between distant matter qubits. *Science* **323**, 486–489 (2009).
12. Krauter, H. *et al.* Deterministic quantum teleportation between distant atomic objects. *Nature Phys.* **9**, 400–404 (2013).
13. Steffen, L. *et al.* Deterministic quantum teleportation with feed-forward in a solid state system. *Nature* **500**, 319–322 (2013).
14. Pfaff, W. *et al.* Unconditional quantum teleportation between distant solid-state quantum bits. *Science* **345**, 532–535 (2014).
15. Sherson, J. F. *et al.* Quantum teleportation between light and matter. *Nature* **443**, 557–560 (2006).
16. Chen, Y.-A. *et al.* Memory-built-in quantum teleportation with photonic and atomic qubits. *Nature Phys.* **4**, 103–107 (2008).
17. Gao, W. B. *et al.* Quantum teleportation from a propagating photon to a solid-state spin qubit. *Nature Commun.* **4**, 3744 (2013).
18. Bussieres, F. *et al.* Quantum teleportation from a telecom-wavelength photon to a solid-state quantum memory. *Nature Photon.* **8**, 775–778 (2014).
19. Aspelmeyer, M., Kippenberg, T. J. & Marquardt, F. Cavity optomechanics. *Rev. Mod. Phys.* **86**, 1391–1452 (2014).
20. Lee, K. C. *et al.* Entangling macroscopic diamonds at room temperature. *Science* **334**, 1253–1256 (2011).
21. Lee, K. C. *et al.* Macroscopic non-classical states and terahertz quantum processing in room-temperature diamond. *Nature Photon.* **6**, 41–44 (2012).
22. England, D. G. *et al.* Storage and retrieval of THz-bandwidth single photons using a room-temperature diamond quantum memory. *Phys. Rev. Lett.* **114**, 053602 (2015).
23. Fisher, K. A. G. *et al.* Frequency and bandwidth conversion of single photons in a room-temperature diamond quantum memory. *Nature Commun.* 7, 11200 (2016).
24. Bustard, P. J. *et al.* Raman-induced slow-light delay of THz-bandwidth pulses. *Phys. Rev. A* **93**, 043810 (2016).
25. Stannigel, K., Rabl, P., Soensen, A. S., Zoller, P. & Lukin, M. D. Optomechanical transducers for long-distance quantum communication. *Phys. Rev. Lett.* **105**, 220501 (2010).
26. Rabl, P. *et al.* A quantum spin transducer based on nanoelectromechanical resonator arrays. *Nature Phys.* **6**, 602–608 (2010).
27. Dong, C.-H., Fiore, V., Kuzyk, M. C. & Wang, H.-L. Optomechanical dark mode. *Science* **338**, 1609–1613 (2012).
28. Gigan, S. *et al.* Self-cooling of a micromirror by radiation pressure. *Nature* **444**, 67–70 (2006).

29. O'Connell, A. D. *et al.* Quantum ground state and single-phonon control of a mechanical resonator. *Nature* **464,** 697–703 (2010).
30. Li, T.-C., Kheifets, S. & Raizen, M. G. Millikelvin cooling of an optically trapped microsphere in vacuum. *Nature Phys.* **7,** 527–530 (2011).
31. White, A. G. *et al.* Measuring two-qubit gates. *J. Opt. Soc. Am. B* **24,** 172–183 (2007).
32. Blinov, B. B. *et al.* Observation of entanglement between a single trapped atom and a single ion. *Nature* **428,** 153–157 (2004).

Acknowledgements

This work was supported by the Ministry of Education of China through its grant to Tsinghua University. L.M.D. acknowledges in addition support from the Intelligence Advanced Research Projects Activity (IARPA) quantum computing program, the Army Research Lab (ARL) quantum network program, and the Air Force Office of Scientific Research (AFOSR) Multidisciplinary University Research Initiative (MURI) program.

Author contributions

L.M.D. designed the experiment and supervised the project. P.Y.H., Y.Y.H., X.X.Y., X.Y.C., C.Z. and L.H. carried out the experiment. L.M.D. and P.Y.H. wrote the manuscript.

Additional information

Competing financial interests: The authors declare no competing financial interests.

19

Plasmonic piezoelectric nanomechanical resonator for spectrally selective infrared sensing

Yu Hui[1], Juan Sebastian Gomez-Diaz[2], Zhenyun Qian[1], Andrea Alù[2] & Matteo Rinaldi[1]

Ultrathin plasmonic metasurfaces have proven their ability to control and manipulate light at unprecedented levels, leading to exciting optical functionalities and applications. Although to date metasurfaces have mainly been investigated from an electromagnetic perspective, their ultrathin nature may also provide novel and useful mechanical properties. Here we propose a thin piezoelectric plasmonic metasurface forming the resonant body of a nanomechanical resonator with simultaneously tailored optical and electromechanical properties. We experimentally demonstrate that it is possible to achieve high thermomechanical coupling between electromagnetic and mechanical resonances in a single ultrathin piezoelectric nanoplate. The combination of nanoplasmonic and piezoelectric resonances allows the proposed device to selectively detect long-wavelength infrared radiation with unprecedented electromechanical performance and thermal capabilities. These attributes lead to the demonstration of a fast, high-resolution, uncooled infrared detector with ∼80% absorption for an optimized spectral bandwidth centered around 8.8 μm.

[1] Department of Electrical & Computer Engineering at Northeastern University, 360 Huntington Avenue, Boston, Massachusetts 02115, USA. [2] Department of Electrical & Computer Engineering at The University of Texas at Austin, 1616 Guadalupe St., UTA 7.215, Austin, Texas 78701, USA. Correspondence and requests for materials should be addressed to M.R. (email: rinaldi@ece.neu.edu) or to A.A. (email: alu@mail.utexas.edu).

Infrared detector technologies were originally developed primarily for military demands, such as night vision, missile tracking, target acquisition and surveillance. In the past few decades, the use of infrared technologies for civilian applications has been steadily growing. Nowadays, infrared detectors can be found in a wide variety of applications, including medical diagnostics, biological and chemical threat detection, electrical power system inspection, infrared spectroscopy, and thermal imaging. Photon detection and thermal sensing are the two main approaches used for the implementation of infrared detectors. Photonic detectors exploit the interaction between photons and electrons in a semiconductor material to produce an electrical output signal upon exposure to infrared radiation[1,2]. They have the advantages of high signal-to-noise ratio (hence high resolution) and fast response time. However, to achieve such a high performance, they typically need cryogenic cooling to prevent thermally generated carriers[3], making them bulky, expensive and power inefficient. On the other hand, thermal detectors rely on the temperature-induced change of material physical properties upon exposure to infrared radiation. They are generally less expensive, more compact and power efficient than semiconductor photon detectors, given their intrinsic capability to operate at room temperature, but they exhibit relatively worse resolution and slower response time. Several uncooled thermal detector technologies have been demonstrated including bolometers[4], thermopiles[5], pyroelectric detectors[6] and more recently, resonant detectors (electromechanical[7] and optical[8]).

Recently, the development of miniaturized, ultra-low-power and low-cost sensor technologies (including uncooled thermal detectors) have attracted great attention for the implementation of highly distributed wireless sensor networks, such as the internet of things, in which physical and virtual object are connected together through the exploitation of sensing and wireless communication functionalities. In this context, micro- and nanoelectromechanical systems (MEMS/NEMS) can have a tremendous impact, since they can simultaneously provide multiple sensing and wireless communication functionalities integrated in a small footprint. The use of MEMS/NEMS has been explored in a large number of applications, spanning from semiconductor-based technology[9] to fundamental science[10]. In particular, MEMS technology has also been employed successfully for miniaturized and ultra-sensitive uncooled infrared detectors such as microbolometers[11] and micromachined thermopile[12]. Among different MEMS/NEMS sensor technologies, the one based on a resonant-sensing mechanism offers significant advantages over other non-resonant approaches. In general, micro–nanoresonant sensors are characterized by a unique combination of high sensitivity to external perturbations, due to the greatly reduced dimensions of sensing element, and ultra-low-noise performance, due to the intrinsically high quality factor (Q) of such resonant systems. Furthermore, resonant sensors use frequency as the output variable, which is one of the physical quantities that can be monitored with the highest accuracy and converted to digital form by simply measuring zero crossings. Among all types of resonant sensors, the one based on MEMS/NEMS resonators can typically deliver the most compact and power efficient sensing solutions thanks to the use of on-chip transduction techniques (such as piezoelectric, electrostatic or thermal) in contrast with the bulky and off-chip optical actuation and readout techniques (requiring the use of power hungry lasers and other bulky optical components and interconnects) typically employed in optical resonators[8,13] or optomechanical resonant systems[14], which are not suitable for many low-power portable applications. The current bottleneck in the development of high-performance MEMS/NEMS resonant infrared detectors is the lack of deeply subwavelength and highly absorbing materials

compatible with standard microfabrication processes and efficient transduction techniques. Indeed, the conventional approach to enable infrared absorptance in MEMS/NEMS resonant structures involves the integration of an infrared absorber (that is, a thin layer of lossy dielectric[8,15,16] or a metal–insulator–metal grating[17]) on top of the vibrating body of the resonant transducer. Although a relatively weak infrared absorptance (<50% and polarization dependent) is typically achieved using this conventional approach, the electromechanical and thermal properties of the resonator (hence detection capability and power efficiency of the infrared sensor) are severely deteriorated due to the electrical and mechanical loading effects of the relatively bulky infrared absorbing material stack attached to the vibrating body of the micro/nanostructure[17].

The fundamental challenge associated with efficient light concentration in planar structures with deeply subwavelength thickness has recently been addressed in the field of nanoplasmonics[18–20]. Metasurfaces with optical properties not found in nature have been synthesized by tailoring plasmonic resonances sustained by arrays of nanostructures with subwavelength dimensions. These effects have been utilized in a wide range of applications, including beam steering[21,22], ultrathin focusing or diverging lenses[23,24], reflectors[25] and absorbers[26,27]. Thanks to these findings, plasmonically enhanced MEMS/NEMS has emerged as a promising research direction towards the development of miniaturized transducers able to convert electromagnetic energy into electric signals by simultaneously exploiting plasmonic, thermal and electromechanical properties[28–31]. In particular, uncooled infrared sensors consisting of a plasmonic absorber attached to a conventional MEMS thermal detector (that is, thermopile or microbolometer) have been demonstrated[32,33], showing enhanced responsivity at specific wavelengths of interest in the infrared range. More recently, the integration of bulk metamaterials and nano-plasmonic infrared/THz absorbers in beam-type nanomechanical structures has also been demonstrated[34]. It has been shown that the plasmonically enhanced light absorption can induce a substantial beam deflection through the intrinsically high thermomechanical coupling of such free-standing nanomechanical structures. Even though these devices exhibit improved detection capabilities compared with some conventional MEMS-based thermal detectors, they require the use of relatively cumbersome and complex off-chip optical readouts to monitor the thermally induced deformation of the nanobeam. In this perspective, the achievement of efficient actuation and sensing of mechanical vibration in a plasmonic nanomechanical structure with intrinsically enhanced light absorption capability (without the need of integrating additional infrared absorbing materials) is highly desirable, being able to combine high sensitivity with power efficient on-chip transduction and readout.

Here we propose an ultrathin (650 nm) piezoelectric plasmonic metasurface forming the vibrating body of a nanomechanical resonator with unprecedented optical and electromechanical performance. By combining plasmonic and piezoelectric electromechanical resonances, we demonstrate efficient transduction of vibration in a nanomechanical structure with a strong and polarization-independent absorption coefficient over an ultrathin thickness, addressing all fundamental challenges associated with the development of performing resonant infrared detectors.

Results

Device design. The proposed plasmonic piezoelectric NEMS resonator, illustrated in Fig. 1, is composed of an aluminum nitride (AlN) piezoelectric nanoplate (500-nm-thick) sandwiched between the two metal layers (Supplementary Fig. 1). The bottom

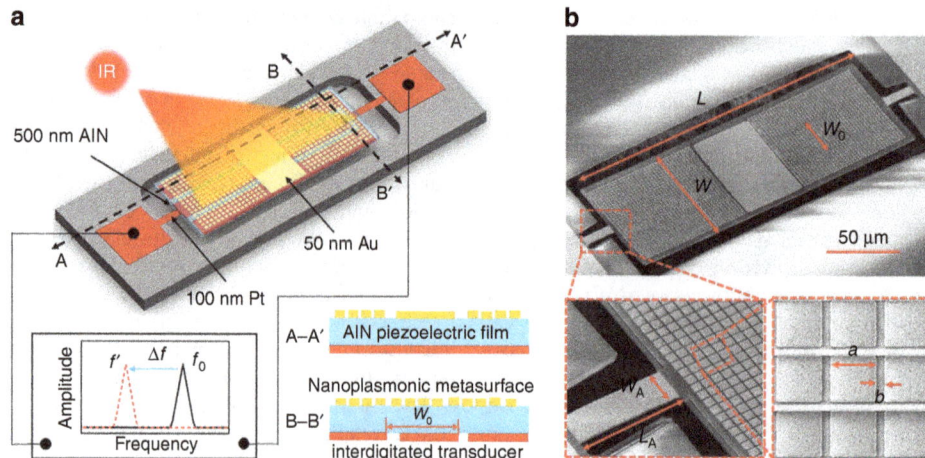

Figure 1 | Overview of the plasmonic piezoelectric nanomechanical resonant infrared detector. (a) Mock-up view: an aluminum nitride nanoplate is sandwiched between a bottom metallic interdigitated electrode and a top nanoplasmonic metasurface. The incident IR radiation is selectively absorbed by the plasmonic metasurface and heats up the resonator, shifting its resonance frequency from f_0 to f' due to the temperature dependence of its resonance frequency. (b) Scanning electron microscopy images of the fabricated resonator, metallic anchors and nanoplasmonic metasurface. The dimensions of the resonator are as follows: $L = 200\,\mu m$; $W = 75\,\mu m$; $W_0 = 25\,\mu m$ ($19 + 6\,\mu m$); $L_A = 20\,\mu m$; $W_A = 6.5\,\mu m$. The dimensions of the unit cell of the plasmonic metasurface are as follows: $a = 1635\,nm$; $b = 310\,nm$. IR, infrared.

layer (100-nm-thick platinum (Pt)) is patterned to form an interdigitated transducer (IDT) used to actuate and sense a high-order lateral-extensional mode of vibration in the nanoplate[35]; the top electrically floating layer (50-nm-thick gold) is patterned with the goal of confining the electric field induced by the bottom IDT across the piezoelectric nanoplate (Supplementary Fig. 2), while simultaneously enabling absorption of infrared radiation in the ultrathin piezoelectric nanoplate thanks to suitably tailored plasmonic resonances (Supplementary Fig. 3). The nanoplate is released from the silicon (Si) substrate to vibrate freely, and it is mechanically supported by two ultrathin Pt tethers (100-nm-thick, 6.5-μm-wide and 20-μm-long), which also provide electrical contact[36]. Such Pt tethers greatly improve the thermal isolation between the nanoplate and the Si substrate compared with conventional ones composed of an AlN–Pt stack[37]. The mechanical resonance frequency of the plasmonic piezoelectric resonator is defined by the equivalent Young's modulus E_{eq} and density ρ_{eq} of the resonant material stack, and the pitch of the interdigitated electrode W_0 (electrode width plus spacing, see Fig. 1a), given by $f_0 = \frac{1}{2W_0}\sqrt{\frac{E_{eq}}{\rho_{eq}}}$. When an a.c. signal is applied to the bottom IDT of the device, the top electrically floating electrode acts to confine the electric field across the device thickness, and a high-order contour-extensional vibration mode is excited through the equivalent d_{31} piezoelectric coefficient of AlN[35] when the frequency of the a.c. signal coincides with the natural resonance frequency, f_0, of the resonator (Supplementary Note 1). If an infrared beam impinges on the device from the top (Fig. 1a), it is selectively absorbed by the metasurface, leading to a large and fast increase of the device temperature ΔT, due to the excellent thermal isolation and extremely low-thermal mass of the free-standing nanomechanical structure. Such infrared-induced temperature rise results in a shift in the mechanical resonance frequency of the resonator (from f_0 to $f_0 - \Delta f$) due to the intrinsically large temperature coefficient of frequency (TCF) of the device[38]. Therefore, the incident IR power can be readily detected by monitoring the resonance frequency of the device.

As we show in the following, the proposed plasmonic nanomechanical resonant structure provides all the fundamental features necessary for the implementation of uncooled infrared detectors with unprecedented performance (Supplementary

Note 2). First, we need to maximize its absorption within the spectrum of interest: to this end, an array of subwavelength patches (Fig. 1b) is patterned within the top metal electrode of the device. Proper patterning of such plasmonic nanostructures in the top metal layer allows the whole device to behave as a spectrally selective and polarization-independent infrared ultrathin absorber[26], significantly enhancing the electromagnetic field concentration within the AlN dielectric (Supplementary Fig. 3). While a single array of subwavelength patches is fundamentally bound to absorb no more than half of the impinging radiation for symmetry constraints[39], the presence of the piezoelectric nanoplate allows us to go beyond this limit and absorb a large portion of infrared energy at resonance. At the same time, we need to achieve maximum thermal isolation of the resonant body from the heat sink, which is ensured by minimizing the thickness of the tethers used to support the nanoplate. The anchors of piezoelectric MEMS/NEMS resonators are conventionally composed of a thick and thermally conductive piezoelectric layer, directly patterned on the same layer forming the vibrating body of the resonator[35], and a thin metal layer employed to route the electrical signal to the actuation electrode integrated in the body of the resonator. On the contrary, here the relatively thick piezoelectric material is completely removed from the anchors, minimizing their thicknesses (ultimately limited by the need of a thin metal layer for electrical routing), resulting in a resonant thermal detector with markedly enhanced responsivity (0.68 Hz nW^{-1}). Third, we need to achieve also a low-thermal time constant (440 μs), which is obtained by exploiting the unique properties of high-quality ultrathin AlN films deposited on a Si substrate with a low-temperature sputtering process, enabling low-volume piezoelectric nanoplate resonant structures with a significantly reduced thermal mass. Despite the resonator volume scaling, we are also able to obtain low-noise performance, thanks to the piezoelectric transduction properties of the ultrathin AlN film, which enable efficient on-chip piezoelectric actuation and sensing of a high $Q > 1,000$ and high-frequency bulk vibration mode in a free-standing nanoplate, leading to a resonator with very low noise spectral density ($\sim 1.46\,\text{Hz}\,\text{Hz}^{-1/2}$ at 100 Hz measurement bandwidth).

The plasmonic metasurface was designed using a transmission line approach, assuming a continuous 100-nm-thick Pt layer beneath the piezoelectric thin film (see Methods). The patch and unit cell dimensions were chosen to provide a Fabry–Perot-like resonance at ~8.8 μm. While a conventional longitudinal resonance would lead to a significant thickness, severely affecting the mechanical and thermal response of the resonator, in our design we tailored the plasmonic metasurface patterned on top of the grounded AlN nanoplate to have a large capacitive surface reactance $X_s = -1/(\omega C_s)$, under a $e^{-i\omega t}$ time convention. Stacked on top of a grounded slab, the dominant resonance is achieved when $X_s = -Z_0 \tan(\beta d)$, where Z_0 and β are the characteristic impedance and propagation constant of the AlN substrate. It confirms that, by tailoring the surface reactance of the plasmonic metasurface to be largely capacitive, it is possible to induce an ultrathin Fabry–Perot resonance in the substrate. It is worth noting that, despite the presence of a small perturbation in the bottom metal layer (interdigitated configuration with two 6-μm-wide gaps rather than a perfectly continuous ground plane), the induced Fabry–Perot-like resonance is mainly determined by the physical dimensions of the gold patches patterned on the top surface of the AlN nanoplate, which are polarization independent[26].

In our device, the plasmonic nanostructures cover 80% of the top metal layer, as a trade-off between large absorption, achieved by coating the entire layer, and high electromechanical transduction efficiency, achieved by removing a portion of the metasurface and replacing it with continuous metal (Supplementary Note 3). We achieved an electromechanical coupling coefficient, $k_t^2 \sim 1\%$ (Supplementary Fig. 4). Figure 2a,b

presents the predicted absorption with and without top subwavelength patches, highlighting how the metasurface can largely increase the absorption at resonance, despite the deeply subwavelength thickness of the device. We also theoretically and experimentally demonstrate that a strong and spectrally selective absorption of long-wavelength infrared (LWIR) radiation, with lithographically determined centre frequency and peak values >85% (over the device area covered by the plasmonic metasurface), can be readily achieved (Fig. 2c,d). Our measurements match very well with the theoretical simulations within the entire band of interest.

Device fabrication and characterization. On the basis of this design, the proposed infrared detector (Fig. 1) was fabricated using a post-complementary metal-oxide-semiconductor (CMOS) compatible microfabrication process involving a combination of photolithography (four masks) and electron-beam lithography (one step). The Fourier transform infrared (FTIR) absorption spectrum of the device was first measured (see Methods) showing that ~80% of the impinging optical power (normal incidence to the device surface) is absorbed at the desired spectral wavelength (Fig. 2b), which validates the performance of the designed piezoelectric metasurface. By removing the plasmonic pattern from the device top electrode would lead to negligible absorption, confirming the uniqueness of the proposed design (Fig. 2a).

We also show that, by altering the lateral size of the nanoplasmonic structure (gold patch), it is possible to accurately

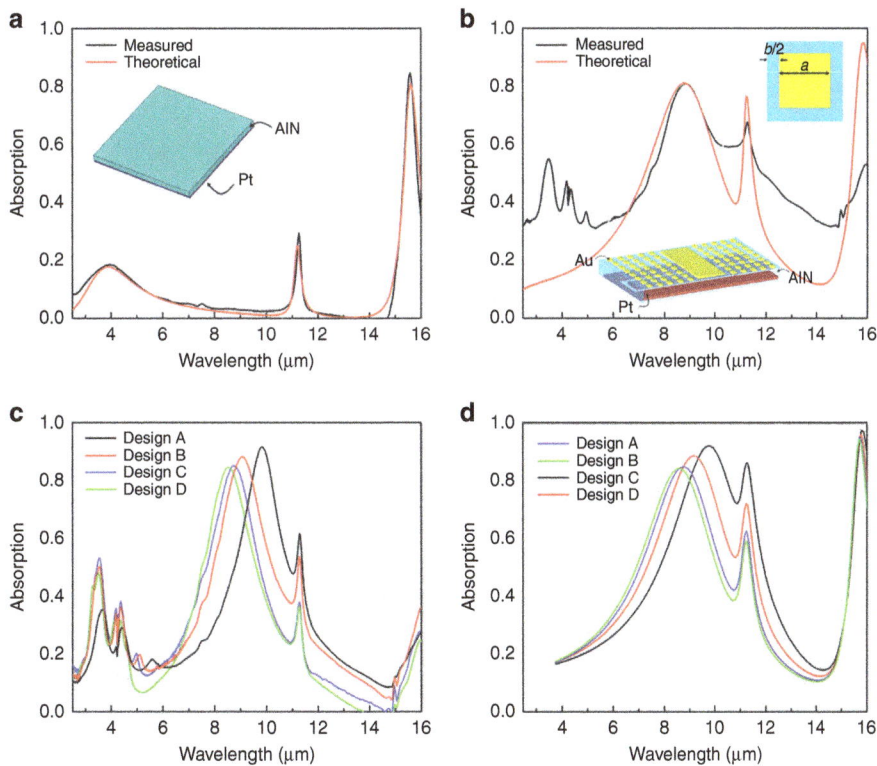

Figure 2 | Absorption properties of the proposed plasmonic piezoelectric nanomechanical resonator. (**a**) Simulated (transmission line theory) and measured (FTIR) absorption spectra of a 500-nm-thick AlN slab grounded by a Pt layer (without plasmonic nanostructures). It shows two intrinsic absorption peaks, associated with AlN at 11.3 μm (888 cm^{-1}) and 15.5 μm (647 cm^{-1})[40], and one at 4 μm associated with the resonant structure. (**b**) Simulated and measured absorption spectra of the fabricated plasmonic piezoelectric nanomechanical resonator. The dimensions of the Au patches that compose the metasurface are $a = 1,635$ nm, $b = 310$ nm, and the thickness of the Au, AlN and Pt layers are 50, 500 and 100 nm, respectively. (**c,d**) Measured and simulated absorption properties of the piezoelectric plasmonic resonant structure with varied Au patch sizes, demonstrating its functionality of spectrally elective detection of infrared radiation in the LWIR range. The unit cell sizes are as follows: design A: $a = 1,780$ nm, $b = 128$ nm; design B: $a = 1,680$ nm, $b = 253$ nm; design C: $a = 1,640$ nm, $b = 313$ nm; design D: $a = 1,620$ nm, $b = 331$ nm.

tailor the absorption peak over a wide range for spectrally selective infrared detection (Fig. 2c,d). We do note the presence of absorption peaks at around 3–4 μm in Fig. 2b,c, which we attribute to an in-plane resonance arising between the fingers of the interdigitated bottom Pt electrode.

The electromechanical performance of the resonator was characterized by measuring its admittance versus frequency (Fig. 3a). A high $Q = 1,116$ and electromechanical coupling coefficient $k_t^2 = 0.86\%$ were extracted by equivalent model fitting (Fig. 3a; Methods), demonstrating the unique advantages of the proposed design in terms of high electromechanical transduction efficiency and low loss. The thermal properties (thermal resistance, temperature distribution and TCF) of the infrared detector were characterized by both finite element analysis and experimental verification (Supplementary Figs 5–8; Supplementary Table 1; Supplementary Note 4). The response of the fabricated infrared detector in the LWIR band was characterized using a 1,500-K globar (2–16-μm emission) as an infrared source. For the sake of comparison, the incoming infrared radiation was also detected using a conventional AlN MEMS resonator with same frequency sensitivity to absorbed heat but without plasmonic pattern (hence non-enhanced infrared absorptance). Thanks to its properly engineered optical properties, the piezoelectric plasmonic resonator showed fourfold enhanced responsivity (Fig. 3b), despite its absorption band (full width at half maximum of 1.5 μm) being much narrower than the emission band of the source. With a narrowband source at the frequency of interest, the responsivity would be much larger. The smallest impinging optical power that can be detected was experimentally estimated by measuring the device responsivity, R_s, and noise spectral density (Supplementary Note 5), demonstrating a low noise equivalent power (NEP) ~ 2.1 nW Hz$^{-1/2}$ at the designed spectral wavelength (for which infrared absorptance $\sim 80\%$). The NEP is arguably considered the most important performance metric for an infrared detector (Supplementary Note 5) and the value measured for the realized proof-of-concept detector proposed here is already comparable to the best commercially available uncooled broadband thermal detectors, while providing unique spectral selectivity in the LWIR band. The response time of the detector was also evaluated by measuring the attenuation of the device response when exposed to infrared radiation modulated at increasingly faster rates (Fig. 3c; Methods), showing a low-thermal time constant, $\tau \sim 440$ μs.

Discussion

Differently from more conventional approaches involving the integration of an infrared absorbing material (such as Si_3N_4, SiO_2 or a metal–insulator–metal grating) on top of an optical/mechanical resonant thermal detector[8], our approach uses an individual plasmonic piezoelectric nanostructure acting simultaneously as absorber, resonator and transducer. It is also worth noting that our proposed plasmonic NEMS resonant sensor exploits a high-frequency on-chip piezoelectric

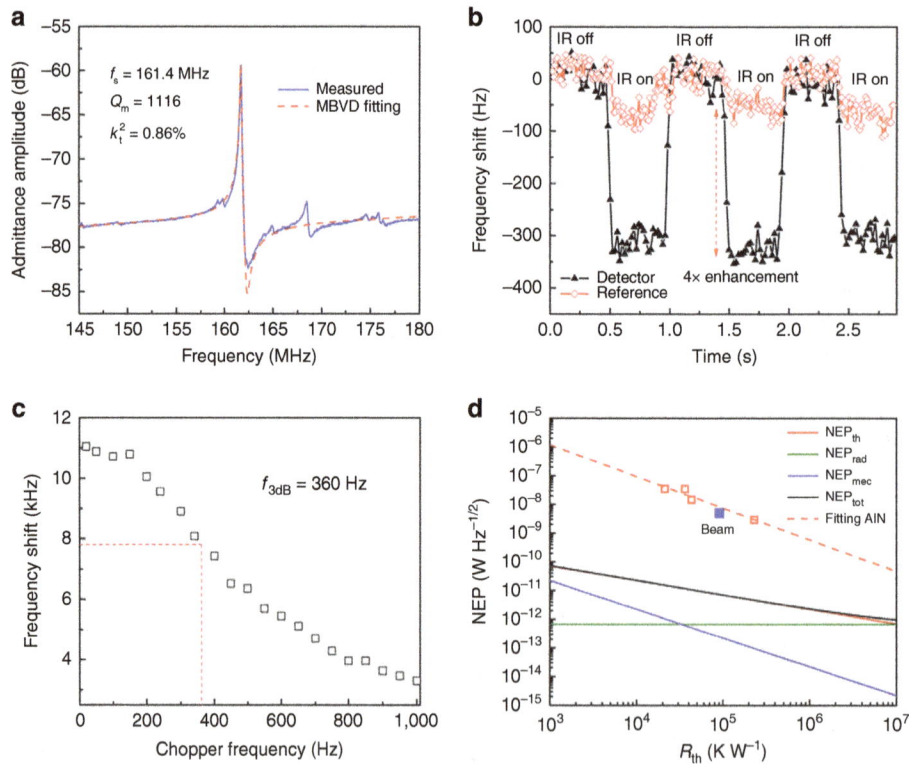

Figure 3 | Device performance. (a) Measured admittance curve versus frequency and MBVD model fitting of the resonator for $Q_{IR} = 0$. The extracted values of the MBVD parameters (see Methods) are as follows: $R_s = 80\,\Omega$; $R_m = 880\,\Omega$; $L_m = 1$ mH; $C_m = 1$ fF; $R_0 = 2$ kΩ; $C_0 = 145$ fF. (b) Measured response of the plasmonic piezoelectric resonator and a conventional AlN MEMS resonator to a modulated IR radiation emitted by a 1,500-K globar (2–16 μm broadband spectral range). (c) Measured frequency response of the detector. The 3dB cutoff frequency, f_{3dB}, was found to be 360 Hz, resulting in a time constant $\tau = 1/(2\pi f_{3dB})$ of 440 μs. (d) NEP for different values of thermal resistance (R_{th}). The solid lines indicate the calculated NEP values associated with each of the three fundamental noise contributions (as expressed in Supplementary Equations 5–6), assuming: resonator area $= 200 \times 75$ μm^2, $\varepsilon = 1$, $T_0 = 300$ K, $P_c = 0$ dBm, $|TCF| = 30$ p.p.m. K^{-1}, $Q = 2,000$. The individual data points indicate the measured NEP values of four fabricated AlN resonant plasmonic IR detectors using four different anchor designs (hence four different R_{th} values) and a Si_3N_4 nanobeam (spectrally selective, static measurement using off-chip optical readout)[34]. IR, infrared.

transduction mechanism that has been a key enabler for MEMS technologies. In particular, AlN-based piezoelectric micro-acoustic devices are nowadays the commercial standard used in the radio frequency (RF) front ends of modern smart phones (that is, see Avago Technologies thin-film bulk acoustic resonator (FBAR)). Therefore, the demonstration of a resonant infrared detector based on a bulk-extensional mode plasmonically enhanced piezoelectric resonator, instead of more conventional MEMS, optical resonators or optomechanical structures, is a key advancement over earlier works that really enables the use of these emerging MEMS plasmonic technologies in low-power and miniaturized wireless communication devices. The unique capability of such AlN-based MEMS technology to deliver high-performance and CMOS compatible sensors and RF components makes it the best candidate for the realization of the next-generation miniaturized, low-power, multi-functional and reconfigurable wireless sensing platforms that will be crucial for the development of the internet of things.

Importantly, even though the proposed resonator possesses a unique combination of efficient electromechanical transduction, strong and spectrally selective infrared absorption capability and very low NEP, there is still plenty of room to improve the performance of these piezoelectric plasmonic NEMS devices. First, novel designs could be employed to further improve the absorption of the piezoelectric metasurface to near unity[41]. Second, to reach the thermal fluctuation noise limit (Fig. 3d), all the noise sources contributing to the generation of frequency fluctuations, such as the resonator flicker noise, random walk and drifts[42], need to be carefully investigated and mitigated. Moreover, the volume of the plasmonic piezoelectric resonator can be further reduced (for instance, by scaling the thickness of the piezoelectric nanoplate, see Supplementary Fig. 9) and its design can be optimized, investigating optimal materials and

innovative geometries for the device anchors to increase the thermal resistance up to $\sim 10^7 \, K \, W^{-1}$, which is typical of conventional microbolometers. As a result, we expect this technology to achieve NEP in the order of $\sim 1 \, pW \, Hz^{-1/2}$ (Fig. 3d), thus enabling the implementation of multi-spectral thermal imagers with noise equivalent temperature difference as low as $\sim 1 \, mK$.

In conclusion, we have demonstrated an uncooled NEMS resonant infrared detector with unique spectrally selective infrared detection capability. Sensing and actuation of a high-frequency (162 MHz) bulk acoustic mode of vibration in a free-standing ultrathin piezoelectric plasmonic metasurface have been demonstrated and exploited for the implementation of a NEMS resonator with unique combined optical and electro-mechanical properties. By exploiting the piezoelectric properties of AlN thin films, efficient on-chip transduction of the NEMS plasmonic resonant structure has been achieved, eliminating the need for the cumbersome and complex off-chip optical readouts employed in previous devices. Thanks to properly tailored absorption properties, strong and spectrally selective detection of infrared radiation, for an optimized spectral bandwidth centered around 8.8 µm and high thermomechanical coupling between electromagnetic and mechanical resonances in a single ultrathin piezoelectric nanoplate have been experimentally verified. This work sets a milestone towards the development of a new technology platform based on the combination of nanoplasmonics and piezoelectric nanoelectromechanical systems, which can potentially deliver fast (hundreds of µs), high resolution (NEP as low as $\sim 1 \, pW \, Hz^{-1/2}$) and spectrally selective uncooled infrared detectors suitable for the implementation of high-performance, miniaturized and power efficient infrared/THz spectrometer and multi-spectral imaging systems.

Methods

Fabrication. The plasmonic piezoelectric resonant infrared detector was fabricated using a post-CMOS compatible microfabrication process involving a combination of photolithography (four masks) and electron-beam lithography (one step), as illustrated in Fig. 4. The fabrication started with a high-resistivity (resistivity > 20,000 Ω m) 4-inch silicon wafer: (a) 100-nm-thick Pt was sputter-deposited and patterned by lift-off process to define the bottom IDT; (b) 500-nm-thick high-quality c axis orientated AlN film was sputter-deposited on top of the Pt IDT, and wet etched in H_3PO_4 to open the vias to get access to the bottom electrode and dry etched by inductively coupled plasma in Cl_2-based chemistry to define the shape of the AlN resonator; (c) 100-nm-thick Au was deposited by electron-beam deposition and patterned by lift-off process to define the probing pad; (d) 50-nm-thick Au was deposited by electron-beam deposition and patterned by electron-beam lithography and lift-off process to define the nanoplasmonic metasurface; (e) the Si substrate underneath the resonator was etched by xenon difluoride (XeF_2) to completely release the device.

Modelling. *Nanoplasmonic piezoelectric resonant infrared detector—electrothermal equivalent circuit.* The proposed plasmonic piezoelectric NEMS resonant infrared detector is modelled by a two-port network with both electrical (voltage, V, used to drive the electromechanical resonance) and thermal inputs (infrared power, Q_{IR}, absorbed in the piezoelectric resonant structure), as shown in Fig. 5a. At the

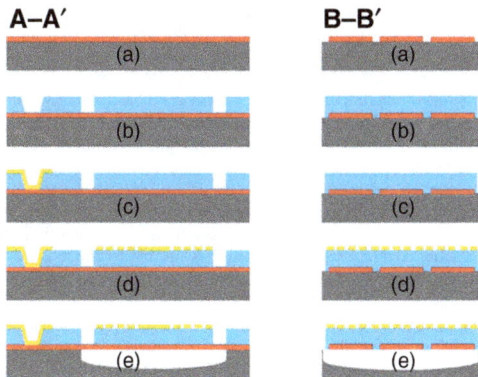

Figure 4 | Microfabrication process for the plasmonic piezoelectric NEMS resonant infrared detector. A–A′ and B–B′ denote longitudinal and transversal axis of the device, as illustrated in Fig. 1.

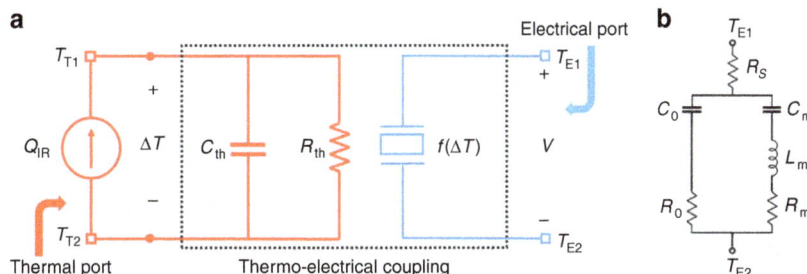

Figure 5 | Electrothermal equivalent circuit of the device. (**a**) Equivalent thermoelectrical circuit of the nanoplasmonic piezoelectric NEMS resonator. (**b**) Modified Butterworth–Van Dyke (MBVD) equivalent circuit.

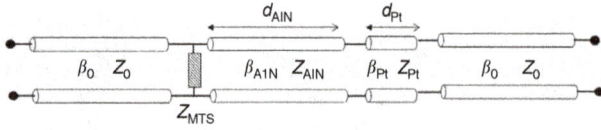

Figure 6 | Electromagnetic circuit model of the piezoelectric resonator.
A normally incident Transverse electromagnetic (TEM) wave impinges on
the structure shown in Fig. 1. The outer transmission line sections represent
the free space, Z_{MTS}, is the surface impedance of the array of gold patches
(metasurface), and the inner transmission line section takes into account
the AlN dielectric and the ground platinum layer, respectively. Each
section is characterized by its characteristic impedance Z and propagation
constant β.

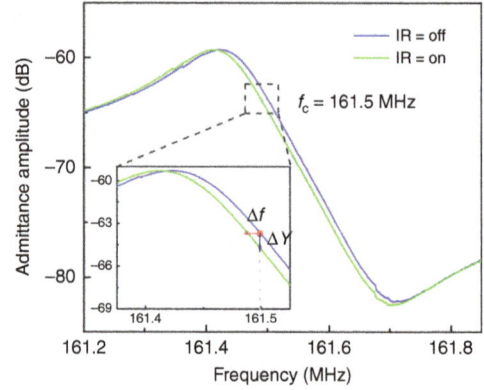

**Figure 7 | Measured admittance curves versus frequency of the
resonator for IR on and off.** The inset shows the zoomed in admittance
shift around the resonance frequency. The exciting frequency of 161.5 MHz
is also marked at which the frequency response of the device was
measured. IR, infrared.

thermal port, the free-standing resonant structure is simply modelled as a thermal
mass, with thermal capacitance C_{th}, coupled with the heat sink at a constant
temperature T_0 via the thermal conductance G_{th} (thermal resistance, $R_{th} = 1/G_{th}$).
The thermal capacitance is intrinsically related to the properties and overall volume
of the material stack forming the vibrating structure ($C_{th} = c \cdot v \cdot \rho$; where c is the
specific heat capacity, v is the volume and ρ is the material density). The thermal
conductance is instead mainly determined by the geometry and material properties
of the tethers connecting the free-standing vibrating body of the device to the
substrate (Supplementary Note 2). At the electrical port, the resonator is modelled
with a modified Butterworth–Van Dyke (MBVD) equivalent circuit model[43],
shown in Figure 5b. The MBVD circuit consists of two branches in parallel: an
acoustic branch composed by the series combination of the motional resistance, R_m
(quantifying dissipative losses), motional capacitance, C_m (inversely proportional
to the stiffness), and motional inductance, L_m (proportional to the mass); and an
electrical branch composed by the series combination of the capacitance, C_0
(capacitance between the device terminals), and resistance, R_0 (representing the
dielectric loss). A series resistance, R_s, is also included in the circuit to represent the
electrical loss associated with the metal electrodes and routing (Fig. 5b).

Nanoplasmonic piezoelectric resonant infrared detector—electromagnetic model.
This section briefly presents the analysis and design of the proposed plasmonic
nanomechanical resonator from an electromagnetic point of view. Specifically, this
analysis permits obtaining the frequency-dependent absorption coefficient of the
resonator, η, which determines the resonator increase of temperature (ΔT) when
infrared radiation is impinging onto the detector (Supplementary Note 2).

Thanks to the structure symmetry and assuming a continuous Pt layer beneath
the AlN, the analysis of a single unit cell of the metasurface suffices to investigate
the electromagnetic behaviour of the whole device. This analysis is carried out
using the equivalent transmission line of Fig. 6. This circuit is totally rigorous[44,45],
that is, exactly equivalent to solving Maxwell's equations, assuming that (i) a
transverse electromagnetic (TEM) wave is normally impinging on the resonator
and (ii) the operation frequency is well below the cutoff frequency of the
higher-order modes excited in the metasurface. Note that these two conditions
are indeed fulfilled in this case.

The characteristic impedance and propagation constant of the different
transmission line sections involved in the model are

$$Z_0 = \sqrt{\frac{\mu_0}{\varepsilon_0}}; \quad Z_{AlN} = \sqrt{\frac{\mu_0}{\varepsilon_0 \varepsilon_{AlN}}}; \quad Z_{Pt} = \sqrt{\frac{\mu_0}{\varepsilon_0 \varepsilon_{Pt}}}$$
$$\beta_0 = \frac{\omega}{c}; \quad \beta_{AlN} = \frac{\omega}{c}\sqrt{\varepsilon_{AlN}}; \quad \beta_{Pt} = \frac{\omega}{c}\sqrt{\varepsilon_{Pt}} \tag{1}$$

where the subscripts '0', 'AlN' and 'Pt' denote free space, aluminum nitrate
dielectric and Pt, respectively, 'c' is the speed of light, ω is the radial frequency and
ε is permittivity, respectively. In addition, the Pt layer is modelled following a
simple Drude model

$$\varepsilon_{Pt} \approx 1 - \frac{\omega_p^2}{\omega^2 + j\omega\tau^{-1}} \tag{2}$$

with plasma frequency $\omega_p = 7.88 \times 10^{15}$ rad s^{-1} and relaxation time $\tau = 7.8$ fs
(ref. 46). We characterize the aluminum nitrate substrate using a Lorentzian
permittivity model as

$$\varepsilon_{AlN} \approx 4 + \frac{\omega_{p1}^2}{\omega_{t1}^2 - \omega^2 + j\omega\tau_1^{-1}} + \frac{\omega_{p2}^2}{\omega_{t2}^2 - \omega^2 + j\omega\tau_2^{-1}} \tag{3}$$

with $\omega_{p1} = 8.663 \times 10^{13}$ rad s^{-1}, $\omega_{t1} = 1.6894 \times 10^{14}$ rad s^{-1}, $\tau_1 = 0.2367$ ps,
$\omega_{p2} = 2.5115 \times 10^{14}$ rad s^{-1}, $\omega_{t2} = 1.2558 \times 10^{14}$ rad s^{-1}, $\tau_2 = 0.3185$ ps. The
electromagnetic model of the array of gold patches is very challenging and there are
no closed-form expressions available for a general case. However, in case of
electrically small unit cells (that is, $a + b \ll \lambda_0$, being λ_0 the operating wavelength)
and densely packed planar patches ($b \ll a$), the surface impedance of the array of

gold patches can be approximated by[44]

$$Z_{MTS} = \frac{a+b}{a\sigma_{Au}} - j\frac{\pi}{2\omega\varepsilon_0\varepsilon_{eff}(a+b)\log\left(\csc\left(\frac{\pi b}{2a+b}\right)\right)} \tag{4}$$

where σ_{Au} is the gold conductivity and $\varepsilon_{eff} = (1 + \varepsilon_{AlN})/2$ is the effective
permittivity surrounding the metasurface. We have numerically verified that
equation 4 provides accurate results in the specific metasurfaces under
consideration. However, note that the near field created by the actual thickness of
the gold patches (~ 50 nm) slightly decreases the effective distance between the
patches. This phenomenon, common in microwave and optics[47], is easily taken
into account using an effective separation distance b slightly lower than the
physical one.

The proposed circuit is analysed using an analytical transfer-matrix approach,
thus providing the absorption coefficient of the resonator versus frequency. The
metasurface employed to cover the AlN resonator is designed in two steps: (i) the
input impedance and absorption of the AlN slab grounded by the Pt layer are
computed using the equivalent circuit of Fig. 6 in the absence of the metasurface
(that is, $Z_{MTS} \approx \infty$), and (ii) the dimensions of the unit cell and patches of the
metasurfaces are chosen with the help of (equations 2–4) aiming to match the
input impedance of the structure to the free-space impedance at the operation
frequency, therefore maximizing the absorption coefficient. Finally, the
electromagnetic behaviour of the designed structure is confirmed using the
full-wave commercial software Computer Simulation Technology (CST)
mimcrowave studio.

It is important to point out the good agreement between theory and
measurement (Fig. 2a,b), especially taking into account that the fabricated
device does not have a continuous Pt layer, but a patterned one (patterned
IDT with a partial coverage of $\sim 70\%$, which is needed to excite a 160-MHz
contour-extensional mode of vibration in the plasmonic piezoelectric nanoplate).
However, the influence of this pattern is limited because (i) most of the unit cells
(around 70%) are grounded by a continuous Pt layer and (ii) it basically introduces
a small capacitance to the ultrathin Pt layer electromagnetic response. We have
further verified with CST full-wave simulations that the influence of this pattern is
negligible.

Measurements. The electromechanical performance of the plasmonic
piezoelectric resonator was characterized by measuring its equivalent electrical
admittance with an Agilent E5071C vector network analyzer after performing an
open-short-load calibration on a standard substrate. The scattering parameter, S_{11},
of this one-port electrical network was first measured and then converted into the
admittance, Y, according to: $Y = (1/50)(1 + S_{11})/(1 - S_{11})$. The values of the
equivalent circuit elements were extracted from the measurement by MBVD model
fitting. The mechanical quality factor, Q (inversely proportional to the mechanical
damping affecting the resonator), and electromechanical coupling coefficient,
k_t^2 (a numerical measure of the conversion efficiency between electrical and
mechanical energy in an electromechanical resonator), were extracted according to:
$Q = (1/R_m)(L_m/C_m)^{0.5}$, and $k_t^2 = (\pi^2/8)(C_m/C_0)$, respectively[48]. The device
admittance amplitude–frequency nonlinearity was measured by monitoring the
admittance amplitude versus frequency shift for different levels of RF power.
Further details about the measurement and analysis can be found in the
Supplementary Note 5.

The reflectance, R, spectra of the fabricated nanoplasmonic metasurface and
plasmonic piezoelectric resonator were measured using a Bruker V70 Fourier
transform infrared coupled with a Hyperion 1000 microscope. A reflective gold

mirror was used as a background for calibration. The absorption spectrum, A, of each sample was then directly obtained as $1 - R$, assuming minimum transmission through the sample, which is a reasonable assumption given the large metal coverage of both the top Au plasmonic structures and bottom Pt electrode.

The response of the fabricated infrared detector in the LWIR band (Fig. 3b) was characterized using a 1,500-K globar (2–16-μm emission) as an infrared source coupled with a Bruker Hyperion 1000 microscope for focusing the infrared beam onto the device. The incident infrared radiation was modulated by an optical chopper at 1 Hz. Measurement bandwidth (intermediate frequency (IF) bandwidth) of the network analyzer was set to 100 Hz to obtain the best noise performance. All experiments were performed in an open environment in ambient temperature and pressure.

The thermal time constant of the detector was evaluated by measuring the attenuation of the device response when exposed to infrared radiation modulated at increasingly faster rates (Fig. 3c). An EOS Photonics $\sim$5-μm continuous-wave quantum cascade laser with a relatively high output power ($\sim$50 mW) was employed as a heat source. The infrared radiation emitted by the quantum cascade laser was modulated using an optical chopper (modulation frequency was varied from 1 Hz to 1 kHz) and focused onto the infrared detector using a zinc selenide (ZnSe) lens (with 70% transmission for infrared radiation in the 0.6–16 μm spectral range). The 3dB cutoff frequency, $f_{3\,dB}$, was found to be 360 Hz, resulting in a time constant $\tau = 1/(2\pi f_{3\,dB})$ of 440 μs (Fig. 3c).

The transient responses of the device were measured by exciting the resonator at a single frequency, $f_c = 161.5$ MHz, for which the slope of admittance amplitude curve versus frequency is maximum (121.3 dB MHz^{-1})[7], and by monitoring the variations over time of the device admittance amplitude (ΔY) using our network analyzer. The admittance amplitude change was then converted to frequency change (Δf) by multiplying the slope of the admittance versus frequency (Fig. 7). The measurement bandwidth (IF bandwidth) of the network analyzer was set to 10 kHz (sampling time of 0.1 ms), which is small enough compared with the thermal time constant of the detector.

References

1. Reine, M. B. HgCdTe photodiodes for IR detector: a review. *Proc. SPIE* **4288**, 266–277 (2001).
2. Kozlowski, Lester J. *et al.* Recent advances in staring hybrid focal plane arrays: comparison of HgCdTe, InGaAs, and GaAs/AlGaAs detector technologies. *Proc. SPIE* **2274**, 93–116 (1994).
3. Rogalski, A. Infrared detectors: status and trends. *Prog. Quant. Electron.* **27**, 59–210 (2003).
4. Chen, C., Yi, X., Zhao, X. & Xiong, B. Characterization of VO$_2$ based uncooled microbolometer linear array. *Sens. Actuators A* **90**, 212–214 (2001).
5. Schaufelbuhl, A. *et al.* Uncooled low-cost thermal imager based on micromachined CMOS integrated sensor array. *J. Microelectromech. Syst.* **10**, 503–510 (2001).
6. Kang, D. H., Kim, K. W., Lee, S. Y., Kim, Y. H. & Keun Gil, S. Influencing factors on the pyroelectric properties of Pb (Zr, Ti) O$_3$ thin film for uncooled infrared detector. *Mater. Chem. Phys.* **90**, 411–416 (2005).
7. Hui, Y. & Rinaldi, M. Fast and high resolution thermal detector based on an aluminum nitride piezoelectric microelectromechanical resonator with an integrated suspended heat absorbing element. *Appl. Phys. Lett.* **102**, 093501 (2013).
8. Watts, M. R., Shaw, M. J. & Nielson, G. N. Optical resonators: Microphotonic thermal imaging. *Nat. Photon.* **1**, 632–634 (2007).
9. Masmanidis, S. C. *et al.* Multifunctional nanomechanical systems via tunably coupled piezoelectric actuation. *Science* **317**, 780–783 (2007).
10. Li, M., Tang, H. X. & Roukes, M. L. Ultra-sensitive NEMS-based cantilevers for sensing, scanned probe and very high-frequency applications. *Nat. Nanotechnol.* **2**, 114–120 (2007).
11. Niklaus, F., Vieider, C. & Jakobsen, H. in Photonics Asia 2007 68360D (International Society for Optics and Photonics, 2007).
12. Graf, A., Arndt, M., Sauer, M. & Gerlach, G. Review of micromachined thermopiles for infrared detection. *Meas. Sci. Technol.* **18**, R59 (2007).
13. Exner, A. T., Pavlichenko, I., Lotsch, B. V., Scarpa, G. & Lugli, P. Low-cost thermo-optic imaging sensors: a detection principle based on tunable one-dimensional photonic crystals. *ACS Appl. Mater. Interfaces* **5**, 1575–1582 (2013).
14. Tallur, S. & Bhave, S. A. A silicon electromechanical photodetector. *Nano Lett.* **13**, 2760–2765 (2013).
15. Hui, Y. & Rinaldi, M. in *The 17th International Conference on Solid-State Sensors, Actuators and Microsystems* (TRANSDUCERS & EUROSENSORS XXVII), 2013 Transducers & Eurosensors XXVII: 968–971 (Barcelona, Spain, 2013).
16. Qian, Z., Hui, Y., Liu, F., Kai, S. & Rinaldi, M. in *18th International Conference on Solid-State Sensors, Actuators and Microsystems* (TRANSDUCERS), 2015 Transducers 1429–1432 (Anchorage, AK, USA, 2015).
17. Gokhale, V. J., Myers, P. D. & Rais-Zadeh, M. in *IEEE Sensors*, 982–985 (Valencia, Spain, 2014).
18. Kildishev, A. V., Boltasseva, A. & Shalaev, V. M. Planar photonics with metasurfaces. *Science* **339**, 1232009 (2013).
19. Yu, N. & Capasso, F. Flat optics with designer metasurfaces. *Nat. Mater.* **13**, 139–150 (2014).
20. Zhao, Y., Belkin, M. & Alù, A. Twisted optical metamaterials for planarized ultrathin broadband circular polarizers. *Nat. Commun.* **3**, 870 (2012).
21. Pfeiffer, C. & Grbic, A. Metamaterial Huygens' surfaces: tailoring wave fronts with reflectionless sheets. *Phys. Rev. Lett.* **110**, 197401 (2013).
22. Esquius-Morote, M., Gomez-Diaz, J. S. & Perruisseau-Carrier, J. Sinusoidally modulated graphene leaky-wave antenna for electronic beamscanning at THz. *IEEE Trans. THz Sci. Technol.* **4**, 116–122 (2014).
23. Pozar, D. Flat lens antenna concept using aperture coupled microstrip patches. *Electron. Lett.* **32**, 2109–2111 (1996).
24. Monticone, F., Estakhri, N. M. & Alù, A. Full control of nanoscale optical transmission with a composite metascreen. *Phys. Rev. Lett.* **110**, 203903 (2013).
25. Yu, N. *et al.* Light propagation with phase discontinuities: generalized laws of reflection and refraction. *Science* **334**, 333–337 (2011).
26. Mosallaei, H. & Sarabandi, K. in *Antennas and Propagation Society International Symposium, 2005 IEEE*, 615–618 (Washington, DC, USA, 2005).
27. Argyropoulos, C., Le, K. Q., Mattiucci, N., D'Aguanno, G. & Alu, A. Broadband absorbers and selective emitters based on plasmonic Brewster metasurfaces. *Phys. Rev. B* **87**, 205112 (2013).
28. Ou, J.-Y., Plum, E., Zhang, J. & Zheludev, N. I. An electromechanically reconfigurable plasmonic metamaterial operating in the near-infrared. *Nat. Nanotechnol.* **8**, 252–255 (2013).
29. Yamaguchi, K., Fujii, M., Okamoto, T. & Haraguchi, M. Electrically driven plasmon chip: active plasmon filter. *Appl. Phys. Express* **7**, 012201 (2014).
30. Dennis, B. *et al.* Compact nanomechanical plasmonic phase modulators. *Nat. Photon.* **9**, 267–273 (2015).
31. Valente, J., Ou, J.-Y., Plum, E., Youngs, I. J. & Zheludev, N. I. A magneto-electro-optical effect in a plasmonic nanowire material. *Nat. Commun.* **6** (2015).
32. Ogawa, S., Okada, K., Fukushima, N. & Kimata, M. Wavelength selective uncooled infrared sensor by plasmonics. *Appl. Phys. Lett.* **100**, 021111 (2012).
33. Talghader, J. J., Gawarikar, A. S. & Shea, R. P. Spectral selectivity in infrared thermal detection. *Light Sci. Appl.* **1**, e24 (2012).
34. Yi, F., Zhu, H., Reed, J. C. & Cubukcu, E. Plasmonically enhanced thermomechanical detection of infrared radiation. *Nano Lett.* **13**, 1638–1643 (2013).
35. Rinaldi, M. & Piazza, G. in *2011 Joint Conference of the IEEE International on Frequency Control and the European Frequency and Time Forum (FCS)*, 1–5 (San Fransisco, CA, USA, 2011).
36. Hui, Y., Qian, Z., Hummel, G. & Rinaldi, M. in *Proceedings of the 2014 Solid-State Sensors, Actuators and Microsystems Workshop (Hilton Head 2014)* 387–391 (Hilton Head Island, SC, USA, 2014) 387–390 (2014).
37. Hui, Y. & Rinaldi, M. in *28th IEEE International Conference on Micro Electro Mechanical Systems (MEMS)* 984–987 (Estoril, Portugal, 2015).
38. Kuypers, J. H., Lin, C.-M., Vigevani, G. & Pisano, A. P. in *2008 IEEE International Frequency Control Symposium*, 240–249 (Honolulu, HI, USA, 2008).
39. Thongrattanasiri, S., Koppens, F. H. & de Abajo, F. J. G Complete optical absorption in periodically patterned graphene. *Phys. Rev. Lett.* **108**, 047401 (2012).
40. Ibáñez, J. *et al.* Far-infrared transmission in GaN, AlN, and AlGaN thin films grown by molecular beam epitaxy. *J. Appl. Phys.* **104**, 033544 (2008).
41. Hendrickson, J., Guo, J., Zhang, B., Buchwald, W. & Soref, R. Wideband perfect light absorber at midwave infrared using multiplexed metal structures. *Optics Lett.* **37**, 371–373 (2012).
42. Rubiola, E. *Phase Noise and Freqency Stability in Oscillators* (Cambridge Univ. Press, 2008).
43. Larson, III J. D., Bradley, R., Wartenberg, S. & Ruby, R. C. in *IEEE Ultrasonics Symposium*, 863–868 (San Juan, 2000).
44. Tretyakov, S. *Analytical Modeling in Applied Electromagnetics* (Artech House, 2003).
45. Padooru, Y. R., Yakovlev, A. B., Kaipa, C. S., Medina, F. & Mesa, F. Circuit modeling of multiband high-impedance surface absorbers in the microwave regime. *Phys. Rev. B* **84**, 035108 (2011).
46. Rakic, A. D., Djurišic, A. B., Elazar, J. M. & Majewski, M. L. Optical properties of metallic films for vertical-cavity optoelectronic devices. *Appl. Opt.* **37**, 5271–5283 (1998).
47. Pozar, D. M. *Microwave Engineering* (John Wiley & Sons, 2009).
48. Rinaldi, M., Zuniga, C., Zuo, C. & Piazza, G. in *IEEE Transactions on Ultrasonics, Ferroelectrics and Frequency Control*, Vol. 57, 38–45 (2010).

Acknowledgements

We thank the staff of the George J. Kostas Nanotechnology and Manufacturing Facility, at Northeastern University, and of the Center for Nanoscale Systems, at Harvard University, for their support in device fabrication. This work was supported by the Defense Advanced Research Projects Agency (DARPA) Young Faculty Award N66001-12-1-4221, the National Science Foundation (NSF) Career Award ECCS-1350114, the Awareness and Localization of Explosives-Related Threats (ALERT) Department of

Homeland Security Center of Excellence, the Air Force Office of Scientific Research and the Welch Foundation with grant No. F-1802.

Author contributions

M.R. conceived the concept and initiated the research. Y.H. and M.R. designed the device and the experiments. Y.H. fabricated the devices, performed the experiments and developed the electrothermal model of the device. Y.H. and J.S.G.-D. analysed and processed the data. J.S.G.-D. and A.A. developed the electromagnetic model of the device, performed the analytical, and numerical electromagnetic simulations and fitted the measured infrared absorptance data. Z.Q. contributed to the fabrication of devices and data analysis. M.R. and A.A. coordinated and supervised the research. All authors contributed to the preparation of the manuscript.

Additional information

Competing financial interests: The authors declare no competing financial interests. US Patent Application 14/969,948 has been filed.

20

Experimental perfect state transfer of an entangled photonic qubit

Robert J. Chapman[1,2], Matteo Santandrea[3,4], Zixin Huang[1,2], Giacomo Corrielli[3,4], Andrea Crespi[3,4], Man-Hong Yung[5], Roberto Osellame[3,4] & Alberto Peruzzo[1,2]

The transfer of data is a fundamental task in information systems. Microprocessors contain dedicated data buses that transmit bits across different locations and implement sophisticated routing protocols. Transferring quantum information with high fidelity is a challenging task, due to the intrinsic fragility of quantum states. Here we report on the implementation of the perfect state transfer protocol applied to a photonic qubit entangled with another qubit at a different location. On a single device we perform three routing procedures on entangled states, preserving the encoded quantum state with an average fidelity of 97.1%, measuring in the coincidence basis. Our protocol extends the regular perfect state transfer by maintaining quantum information encoded in the polarization state of the photonic qubit. Our results demonstrate the key principle of perfect state transfer, opening a route towards data transfer for quantum computing systems.

[1] Quantum Photonics Laboratory, School of Engineering, RMIT University, Melbourne, Victoria 3000, Australia. [2] School of Physics, The University of Sydney, Sydney, New South Wales 2006, Australia. [3] Istituto di Fotonica e Nanotecnologie, Consiglio Nazionale delle Ricerche, Piazza Leonardo da Vinci 32, Milano I-20133, Italy. [4] Dipartimento di Fisica, Politecnico di Milano, Piazza Leonardo da Vinci 32, Milano I-20133, Italy. [5] Department of Physics, South University of Science and Technology of China, Shenzhen 518055, China. Correspondence and requests for materials should be addressed to A.P. (email: alberto.peruzzo@rmit.edu.au).

Transferring quantum information between locations without disrupting the encoded information *en route* is crucial for future quantum technologies[1-8]. Routing quantum information is necessary for communication between quantum processors, addressing single qubits in topological surface architectures, and for quantum memories as well as many other applications.

Coupling between stationary qubits and mobile qubits via cavity and circuit quantum electrodynamics has been an active area of research with promise for long-distance quantum communication[9-12]; however, coupling between different quantum information platforms is challenging as unwanted degrees of freedom lead to increased decoherence[13]. Quantum teleportation between distant qubits allows long-distance quantum communication via shared entangled states[14-17]; however, in most quantum information platforms this would again require coupling between stationary and mobile qubits. Physically relocating trapped ion qubits has also been demonstrated[18,19], however, with additional decoherence incurred during transport.

By taking advantage of coupling between neighbouring qubits, it is possible to transport quantum information across a stationary lattice[2]. This has the benefits that one physical platform is being used and the lattice sites remain at fixed locations. The most basic method is to apply a series of SWAP operations between neighbouring sites such that, with enough iterations, the state of the first qubit is relocated to the last. This method requires a high level of active control on the coupling and is inherently weak as individual errors accumulate after each operation, leading to an exponential decay in fidelity as the number of operations increases[20].

The perfect state transfer (PST) protocol utilizes an engineered but fixed coupled lattice. Quantum states are transferred between sites through Hamiltonian evolution for a specified time[2-7]. For a one-dimensional system with N sites, the state intially at site n is transferred to site $N-n+1$ with 100% probability without need for active control on the coupling[21]. PST can be performed on any quantum computing architecture where coupling between sites can be engineered, such as ion traps[18] and quantum dots[22]. Figure 1 presents an illustration of the PST protocol. The encoded quantum state, initially at the first site, is recovered at the final site after a specific time. In the intermediate stages, the qubit is in a superposition across the lattice. Aside from qubit relocation, the PST framework can be applied to entangled W-state preparation[23], state amplification[24] and even quantum computation[25-29].

To date, most research on PST has been theoretical[2-7,20,21,23,24,28,30-42], with experiments[43,44] being limited to demonstrations where no quantum information is transferred, and do not incorporate entanglement, often considered the defining feature of quantum mechanics[45]. Here, we present the implementation of a protocol that extends PST for relocating a polarization-encoded photonic qubit across a one-dimensional lattice, realized as an array of 11 evanescently coupled waveguides[46-48]. We show that the entanglement between a photon propagating through the PST waveguide array and another photon at a different location is preserved.

Results

PST Hamiltonian. The Hamiltonian for our system in the nearest-neighbour approximation is given by the tight-binding formalism

$$\hat{H} = \sum_{\sigma \in \{H,V\}} \sum_{n=1}^{N-1} C_{n,n+1} \left(\hat{a}_{n+1,\sigma}^\dagger \hat{a}_{n,\sigma} + \hat{a}_{n,\sigma}^\dagger \hat{a}_{n+1,\sigma} \right), \quad (1)$$

where $C_{n,n+1}$ is the coupling coefficient between waveguides n and $n+1$, and $\hat{a}_{n,\sigma}$ ($\hat{a}_{n,\sigma}^\dagger$) is the annihilation (creation) operator

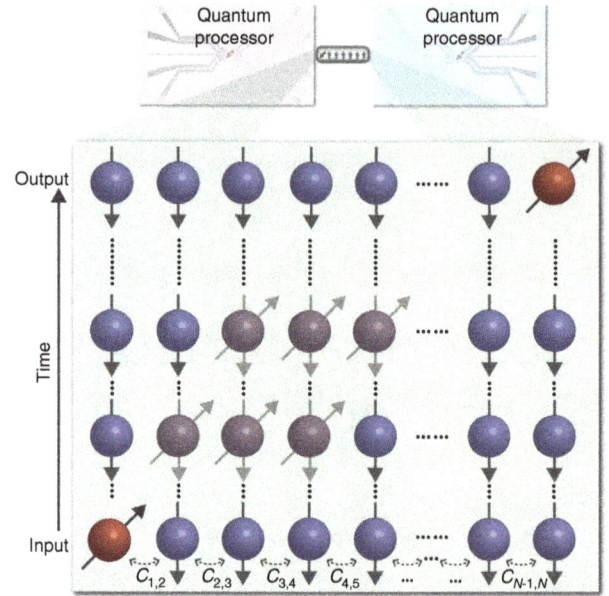

Figure 1 | Illustration of a one-dimensional perfect state transfer lattice connecting two quantum processors. By engineering the Hamiltonian of a lattice, the state at the first site is transferred to the last site after a specific time. This Hamiltonian defines the perfect state transfer protocol[3], which can be used for routing quantum information inside a quantum processor.

applied to waveguide n and polarization σ (horizontal or vertical). Hamiltonian evolution of a state $|\psi_0\rangle$ for a time t is calculated via the Schrödinger equation, giving the final state $|\psi(t)\rangle = \exp(\frac{-i\hat{H}t}{\hbar})|\psi(0)\rangle$ (ref. 49). Equation (1) is constructed of independent tight-binding Hamiltonians acting on each orthogonal polarization. This requires there to be no cross-talk terms $\hat{a}_{n,H}^\dagger \hat{a}_{m,V}$ or $\hat{a}_{n,V}^\dagger \hat{a}_{m,H}$ $\forall m, n$. The spectrum of coupling coefficients $C_{n,n+1}$ is crucial for successful PST. Evolution of this Hamiltonian with a uniform coupling coefficient spectrum, equivalent to equally spaced waveguides, is not sufficient for PST with over three lattice sites as simulated in Fig. 2a. PST requires the coupling coefficient spectrum to follow the function

$$C_{n,n+1} = C_0 \sqrt{n(N-n)}, \quad (2)$$

where C_0 is a constant, N is the total number of lattice sites and evolution is for a specific time $t_{PST} = \frac{\pi}{2C_0}$ (refs 3, 4). This enables arbitrary-length PST as simulated in Fig. 2b for 11 sites. The coupling coefficient spectrum for each polarization must be equal and follow equation (2) for the qubit to be faithfully relocated and the polarization-encoded quantum information to be preserved. The distance between waveguides dictates the coupling coefficient; however, for planar systems, the coupling coefficient of each polarization will in general be unequal due to the waveguide birefringence. To achieve equal coupling between polarizations, the waveguide array is fabricated along a tilted plane in the substrate[50]. This is made possible by the unique three-dimensional capabilities of the femtosecond laser-writing technique (see Supplementary Note 1 for further fabrication and device details). We measure a total propagation loss of 1.8 ± 0.2 dB; however, our figure of merit is how well preserved the polarization quantum state is after the transfer protocol. Therefore we calculate fidelity without loss. Ideally the PST protocol exhibits unit fidelity and efficiency, where the quantum state is reliably transferred and the encoded state is preserved. Due to loss in our experiment, we have less than unit efficiency; however, this loss is largely unrelated to the PST Hamiltonian in equation (1). Further optimizing the fabrication process could

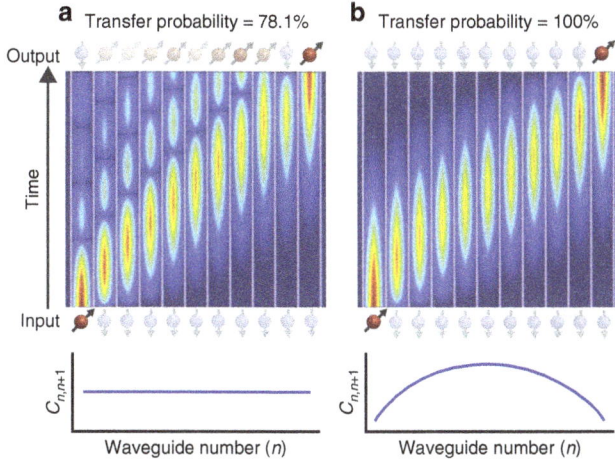

Figure 2 | Propagation simulations with different coupling coefficient spectra. (a) A photon is injected into the first waveguide of an array of eleven coupled waveguides with the Hamiltonian in equation (1) and a uniform coupling coefficient spectrum. With the constraint that reflections off boundaries are not allowed, we calculate a maximum probability of transferring the photon to waveguide 11 of 78.1% (ref. 2). **(b)** A photon is injected into the first waveguide of an array of eleven coupled waveguides, this time with the coupling coefficient spectrum of equation (2). After evolution for a pre-determined time, the photon is received at waveguide 11 with 100% probability[3-7].

reduce the level of propagation loss (see Methods and Supplementary Note 2 for further details on loss).

We inject photons into waveguides 1, 6 and 10 of the array, which after time t_{PST} transfer to waveguides 11, 6 and 2, respectively. Figure 3a–c presents propagation simulations for each transfer. Input waveguides extend to the end of the device to allow selective injection.

Transfer characterization. To characterize the coupling coefficient spectra, we inject horizontally and vertically polarized laser light at 808 nm into each input waveguide. Laser light is more robust to noise than single photons and we can monitor the output with a CCD camera to fast gather results. Using laser light at the same wavelength as our single photons will give an output intensity distribution equivalent to the output probability distribution for detecting single photons[46]. Ideally light injected into waveguide n will output the device only in waveguide $N - n + 1$; however, this assumes an approximate model of nearest-neighbour coupling only. Taking into account coupling between further separated waveguides reduces the transfer probability. This decrease is greater for light injected closer to the centre of the array (see Supplementary Note 1). Figure 3d–f presents our measured output probability distribution for horizontally $\left(P_n^H\right)$ and vertically $\left(P_n^V\right)$ polarized laser light injected into each input waveguide, where n is the output waveguide number. Fidelity between the probability distributions for each polarization is given by $F_{distribution} = \sum_n \sqrt{P_n^H P_n^V}$. This fidelity is closely related to how similar the two coupling coefficient spectra are. We measure an average probability distribution fidelity for all transfers of 0.976 ± 0.006 (see Supplementary Table 1 for all fidelity values). We encode quantum information in the polarization state of the photon and are interested in reliably relocating this qubit. We use a single optical fibre to capture photons from the designed output waveguide, which, in all cases, is the waveguide with the greatest output probability.

Quantum process tomography. We perform quantum process tomography to understand the operation performed on the single-photon polarization state during each PST transfer. We inject single-photon states $|\psi_{in}\rangle = (\alpha \hat{a}_{S,H}^\dagger + \beta \hat{a}_{S,V}^\dagger)|0\rangle$ into each input waveguide $S \in \{1,6,10\}$, where α (β) is the probability amplitude of the horizontal (vertical) component of the photon and $|\alpha|^2 + |\beta|^2 = 1$. From quantum process tomography on the output polarization states, we can generate a process matrix χ_{pol} for each transfer[1,51]. We aim to perform the identity operation so that the quantum information encoded in the polarization can be recovered after relocation. We measure a polarization phase shift associated with each transfer. This phase shift can be compensated for with a local polarization rotation applied before injection. Figure 3g–i presents our measured process matrix for each transfer. Across all transfers we demonstrate an average fidelity of the polarization process including compensation to an identity of 0.982 ± 0.003 (see Supplementary Note 3 for details of the compensation scheme and Supplementary Table 2 for all fidelities). Process fidelity is calculated as $F_{process} = \text{Tr}\{\chi_1 \chi_{pol + comp}\}$ (ref. 52), where χ_1 is the process matrix for the identity operation and $\chi_{pol + comp}$ is the combined polarization operation and compensation process matrix.

Ideally the output state for each transfer is $|\psi_{out}\rangle = (\alpha \hat{a}_{T,H}^\dagger + \beta \hat{a}_{T,V}^\dagger)|0\rangle$, where $T \in \{11,6,2\}$ and the probability amplitude of each polarization component remains equal to the input state. Our high-fidelity measurements on single-photon relocation demonstrate that we can route a polarization-encoded photonic qubit across our device and faithfully recover the encoded quantum information.

Entangled state transfer. Entanglement is likely to be a defining feature of quantum computing, and preserving entanglement is therefore critical to the success of any qubit relocation protocol. We prepare the Bell state $\frac{1}{\sqrt{2}}(|H_1 V_2\rangle + |V_1 H_2\rangle)$ using the spontaneous parametric downconversion process. The polarization is controlled using rotatable half and quarter waveplates (HWPs and QWPs), and polarizing beam splitters (PBSs) as shown in Fig. 4 (ref. 53) (see Methods for details). This set-up prepares a general state $\alpha|H_1 V_2\rangle + \beta|V_1 H_2\rangle$ when measuring in coincidence, where $|\alpha|^2 + |\beta|^2 = 1$. Photon 1 is injected into the waveguide array, while photon 2 propagates through polarization-maintaining fibre (PMF). In terms of waveguide occupancy, our input state is $|\psi_{in}\rangle = \frac{1}{\sqrt{2}}(\hat{a}_{S,H}^\dagger \hat{a}_{0,V}^\dagger + \hat{a}_{S,V}^\dagger \hat{a}_{0,H}^\dagger)|00\rangle$ for each input waveguide $S \in \{1,6,10\}$, where $\hat{a}_{0,\sigma}^\dagger$ denotes the creation operator acting on polarization σ in PMF. Full two-qubit polarization tomography[54] is performed on the output and the fidelity calculated as

$$F_{quantum} = \left(\text{Tr} \left\{ \sqrt{ \sqrt{\rho_{input}} \rho_{output} \sqrt{\rho_{input}} } \right\} \right)^2, \quad (3)$$

where ρ_{output} is the density matrix after the PST protocol has been applied and ρ_{input} is the density matrix after propagation through a reference straight waveguide[55]. After all qubit relocations we measure an average polarization state fidelity of 0.971 ± 0.014. Fidelity is measured in the two-photon coincidence basis. This value is therefore the fidelity on the quantum state transferred without taking into account the loss (see Methods and Supplementary Note 2 for loss analysis). We can use the results from quantum process tomography to generate a characterized model of our device. We can now use this model to calculate the similarity between the predicted output state and our measured output state as

$$S_{quantum} = \left(\text{Tr} \left\{ \sqrt{ \sqrt{\rho_{predicted}} \rho_{output} \sqrt{\rho_{predicted}} } \right\} \right)^2. \quad (4)$$

We calculate an average similarity of 0.987 ± 0.014 across all transfers (see Supplementary Table 3 for all fidelities

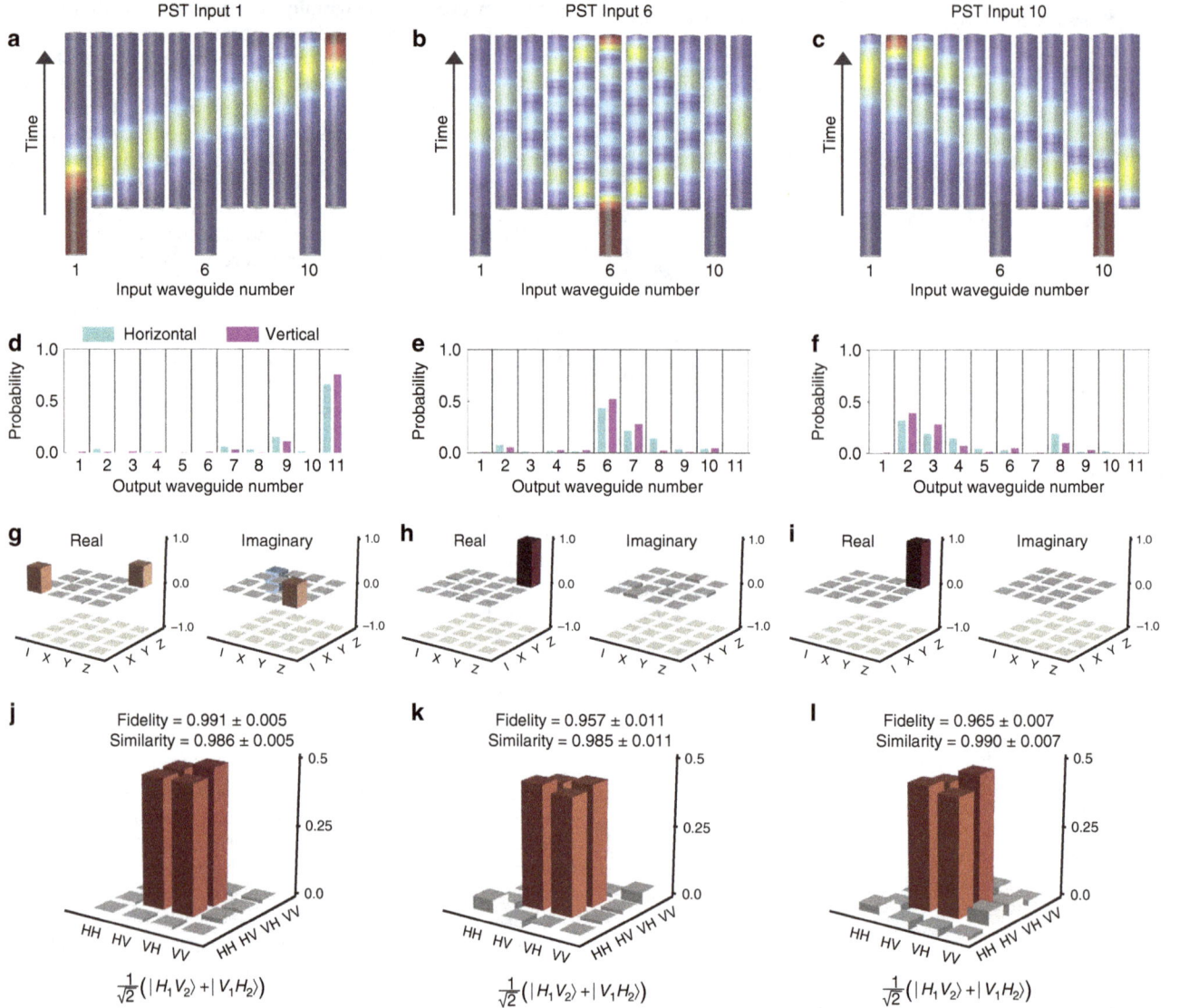

Figure 3 | Experimental data from the characterization and performance of perfect state transfer waveguide array. (**a–c**) Propagation simulations showing the device implementation to enable specific waveguide input. (**d–f**) Output probability distributions for each input of the PST array for horizontally and vertically polarized laser light. (**g–i**) Quantum process matrix for each transfer in the PST array measured with single-photon quantum process tomography. (**j–l**) Two-photon quantum state tomography is performed after photon 1 of the polarization entangled Bell state $\frac{1}{\sqrt{2}}(|H_1V_2\rangle + |V_1H_2\rangle)$ has been relocated. Results have had the small imaginary components removed for brevity.

and similarities). Figure 3j–l presents our measured density matrix after each entangled state transfer.

Ideally, the output state for each transfer is $|\Psi_{\text{out}}\rangle = \frac{1}{\sqrt{2}}(\hat{a}_{T,H}^\dagger \hat{a}_{0,V}^\dagger + \hat{a}_{T,V}^\dagger \hat{a}_{0,H}^\dagger)|00\rangle$, where $T \in \{11,6,2\}$. With high fidelity the probability amplitude of each component is preserved and the state remains almost pure. This result demonstrates that with our device we can relocate a polarization qubit between distant sites and preserve entanglement with another qubit at a different location. In principle our device could route qubits from any waveguide n to waveguide $N-n+1$. Quantum error correction protocols require sophisticated interconnection to access individual qubits for control and measurement within large, highly entangled surface code geometries[56]. PST is a clear gateway towards accessing qubits in such systems without disrupting quantum states and entanglement throughout the surface code.

Decohered state transfer. Decoherence has applications in quantum simulation to emulate systems in nature[57], and it is therefore important to note that this approach for relocating quantum information can be applied to states of any purity[3]. We prepare decohered states by introducing a time delay between the horizontal and vertical components of the polarization qubit. We implement this delay by extending one arm of the source, which reduces the overlap of the photons after they are both incident on the PBS, as shown in Fig. 4. This delay extends the state into a time-bin basis, which we trace over on measurement, leading to a mixed state. The purity of the state can be calculated as the convolution of the horizontal and vertical components with a time delay τ:

$$\text{Purity}(\tau) \equiv \int_{-\infty}^{\infty} H(t)V(\tau - t)\mathrm{d}t, \qquad (5)$$

where τ is controlled by altering the path length of the vertical

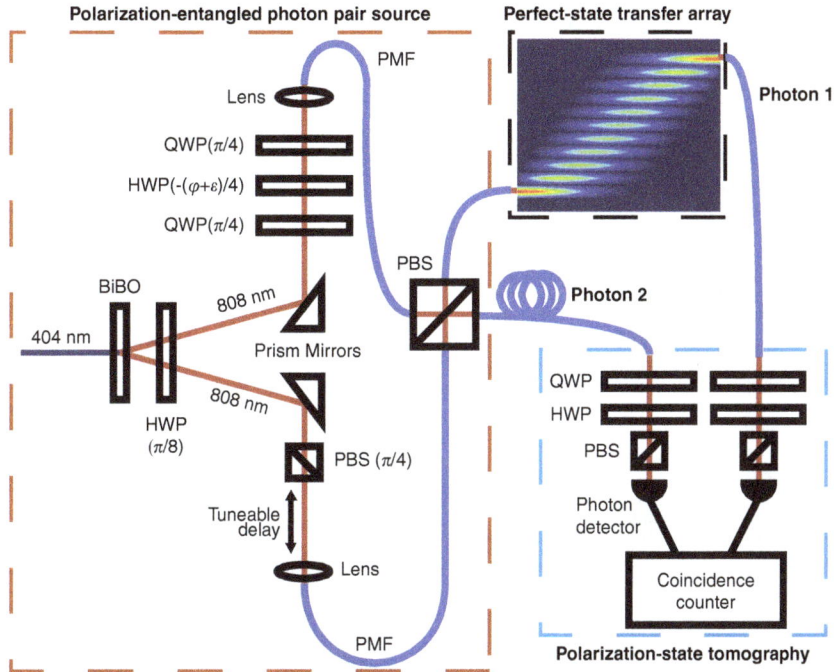

Figure 4 | Experimental set-up. Polarization entangled photons are generated in free space before coupling into PMF. Photon 1 is injected into the perfect state transfer array, while photon 2 travels through PMF. Full two-qubit polarization tomography is performed on the output. See Methods for experimental set-up details.

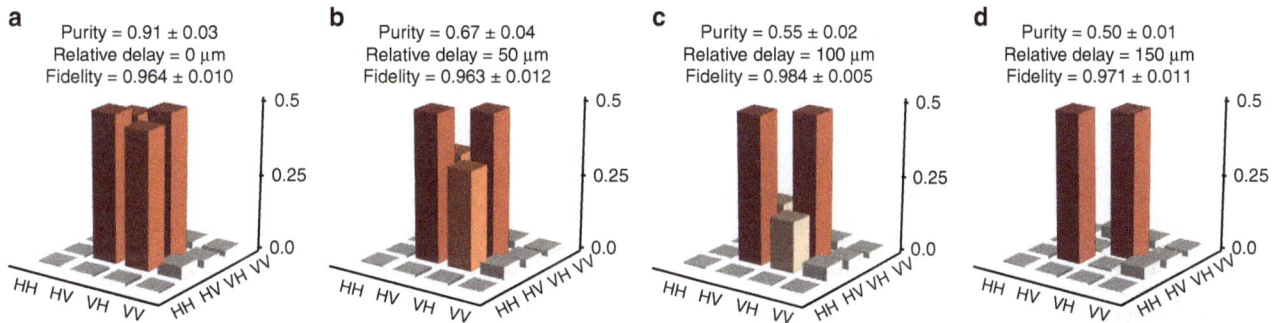

Figure 5 | Perfect state transfer of entangled states with varying purity. Photon 1 of the state $\frac{1}{\sqrt{2}}(|H_1V_2\rangle + |V_1H_2\rangle)$ is injected into waveguide 1 of the PST array. A delay is applied to the vertical component to control the purity of the state. (**a**) Relative delay of 0 μm, (**b**) 50 μm, (**c**) 100 μm and (**d**) 150 μm. Results have had the small imaginary components removed for brevity.

component of the state. $H(V)$ is the horizontal (vertical) component of the photon. Figure 5 presents density matrices for PST from waveguide 1 to waveguide 11 applied to entangled states of varying purity. The injected states are recovered with an average fidelity of 0.971 ± 0.019 and an average similarity of 0.978 ± 0.019 (see Supplementary Table 4 for all values).

Discussion

We have proposed and experimentally demonstrated a protocol for relocating a photonic qubit across eleven discrete sites, maintaining the quantum state with high fidelity and preserving entanglement with another qubit at a different location. We can aim to improve our fidelity by reducing next-nearest-neighbour coupling by further separating the waveguides and having a longer device. This would increase the contrast between nearest- and next-nearest-neighbour coupling to better fit the Hamiltonian in equation (1). A by-product of longer devices, however, is an increase in propagation loss. Depth-dependent spherical aberrations in the laser irradiation process may also affect the homogeneity of the three-dimensional waveguide array.

Additional optics in the laser writing set-up could be employed to reduce this effect. Protocols for relocating quantum information across discrete sites are essential for future quantum technologies. Our protocol builds on the PST with extension to include an additional degree of freedom for encoding quantum information. This demonstration opens pathways towards faithful quantum state relocation in quantum computing systems.

Methods

Experimental set-up. Horizontally polarized photon pairs at 807.5 nm are generated via type 1 spontaneous parametric downconversion in a 1-mm-thick BiBO crystal, pumped by an 80-mW, 403.75-nm CW diode laser. Both photons are rotated into a diagonal state $\frac{1}{\sqrt{2}}(|H\rangle + |V\rangle)$ by a half waveplate (HWP) with fast axis at 22.5° from vertical. One photon has a phase applied by two 45° quarter waveplates (QWP) on either side of a HWP at θ°. The second photon has its diagonal state optimized with a PBS at ~45°.

Each photon is collected in PMF and are incident on the two input faces of a fibre pigtailed PBS. When measuring in the coincidence basis, this post-selects the entangled state $\frac{1}{\sqrt{2}}(|H_1V_2\rangle + e^{i\phi}|V_1H_2\rangle)$, where $\phi = 4(\theta + \epsilon)$ and ϵ is the intrinsic phase applied by the whole system. The experimental set-up is illustrated in Fig. 4.

PMF is highly birefringent, resulting in full decoherence of the polarization state after ~1 m of fibre giving a mixed state. To maintain polarization superposition over several metres of fibre, we use $90°$ connections to ensure that both polarizations propagate through equal proportions of fast- and slow-axis fibre. Slight length differences between fibres and temperature variations mean the whole system applies a residual phase ϵ to the state, which can be compensated for in the source using the phase-controlling HWP.

Polarization state tomography combines statistics from projection measurements to generate the density matrix of a state. Single-photon rotations are applied by a QWP and HWP before a PBS. Single-qubit tomography requires four measurements and two-qubit tomography requires 16. Accidental counts are removed by taking each reading with and without an electronic delay. This helps reduce noise in our measurements.

Photon count rate. In our experiment, we prepare polarization Bell states with a count rate of $\sim 2 \times 10^3$ s^{-1}. After the PST array we measure a count rate of $\sim 10^2$ s^{-1}. The propagation loss of the array is only 1.8 dB. Most of the total loss (~ 13 dB) is indeed due to mode mismatch between the waveguides and fibres, imperfect coupling, reflections at interfaces, and non-unit relocation efficiency. We integrate our measurements for 30 s to reduce the statistical noise due to the Poisson distribution of the photon count rate.

References

1. Nielsen, M. A. & Chuang, I. L. *Quantum Computation and Quantum Information* 10th anniversary edition (Cambridge University Press, 2000).
2. Bose, S. Quantum communication through an unmodulated spin chain. *Phys. Rev. Lett.* **91**, 207901 (2003).
3. Christandl, M., Datta, N., Ekert, A. & Landahl, A. J. Perfect state transfer in quantum spin networks. *Phys. Rev. Lett.* **92**, 187902 (2004).
4. Plenio, M. B., Hartley, J. & Eisert, J. Dynamics and manipulation of entanglement in coupled harmonic systems with many degrees of freedom. *New J. Phys.* **6**, 36 (2004).
5. Gordon, R. Harmonic oscillation in a spatially finite array waveguide. *Opt. Lett.* **29**, 2752 (2004).
6. Nikolopoulos, G. M., Petrosyan, D. & Lambropoulos, P. Electron wavepacket propagation in a chain of coupled quantum dots. *J. Phys. Condens. Matter* **16**, 4991 (2004).
7. Nikolopoulos, G. M., Petrosyan, D. & Lambropoulos, P. Coherent electron wavepacket propagation and entanglement in array of coupled quantum dots. *Europhys. Lett.* **65**, 297–303 (2004).
8. DiVincenzo, D. P. The physical implementation of quantum computation. *Fortschr. Phys.* **48**, 771–783 (2000).
9. Wallraff, A. *et al.* Strong coupling of a single photon to a superconducting qubit using circuit quantum electrodynamics. *Nature* **431**, 162–167 (2004).
10. Majer, J. *et al.* Coupling superconducting qubits via a cavity bus. *Nature* **449**, 443–447 (2007).
11. Herskind, P. F., Dantan, A., Marler, J. P., Albert, M. & Drewsen, M. Realization of collective strong coupling with ion Coulomb crystals in an optical cavity. *Nat. Phys.* **5**, 494–498 (2009).
12. Paik, H. *et al.* Observation of high coherence in Josephson junction qubits measured in a three-dimensional circuit QED architecture. *Phys. Rev. Lett.* **107**, 240501 (2011).
13. Schoelkopf, R. J. & Girvin, S. M. Wiring up quantum systems. *Nature* **451**, 664–669 (2008).
14. Bennett, C. H. *et al.* Teleporting an unknown quantum state via dual classical and Einstein-Podolsky-Rosen channels. *Phys. Rev. Lett.* **70**, 1895–1899 (1993).
15. Bouwmeester, D. *et al.* Experimental quantum teleportation. *Nature* **390**, 575–579 (1997).
16. Furusawa, A. *et al.* Unconditional quantum teleportation. *Science* **282**, 706–709 (1998).
17. Braunstein, S. L. & Kimble, H. J. Teleportation of continuous quantum variables. *Phys. Rev. Lett.* **80**, 869–872 (1998).
18. Kielpinski, D., Monroe, C. & Wineland, D. J. Architecture for a large-scale ion-trap quantum computer. *Nature* **417**, 709–711 (2002).
19. Seidelin, S. *et al.* Microfabricated surface-electrode ion trap for scalable quantum information processing. *Phys. Rev. Lett.* **96**, 253003 (2006).
20. Yung, M.-H. Quantum speed limit for perfect state transfer in one dimension. *Phys. Rev. A* **74**, 030303 (2006).
21. Christandl, M. *et al.* Perfect transfer of arbitrary states in quantum spin networks. *Phys. Rev. A* **71**, 032312 (2005).
22. Loss, D. & DiVincenzo, D. P. Quantum computation with quantum dots. *Phys. Rev. A* **57**, 120–126 (1998).
23. Kay, A. Perfect efficient, state transfer and its application as a constructive tool. *Int. J. Quantum Inf.* **08**, 641–676 (2010).
24. Kay, A. Unifying quantum state transfer and state amplification. *Phys. Rev. Lett.* **98**, 010501 (2007).
25. Raussendorf, R. & Briegel, H. J. A one-way quantum computer. *Phys. Rev. Lett.* **86**, 5188–5191 (2001).
26. Zhou, X., Zhou, Z.-W., Guo, G.-C. & Feldman, M. J. Quantum computation with untunable couplings. *Phys. Rev. Lett.* **89**, 197903 (2002).
27. Benjamin, S. C. & Bose, S. Quantum computing in arrays coupled by "always-on" interactions. *Phys. Rev. A* **70**, 032314 (2004).
28. Kay, A. Computational power of symmetric Hamiltonians. *Phys. Rev. A* **78**, 012346 (2008).
29. Mkrtchian, G. F. Universal quantum logic gates in a scalable Ising spin quantum computer. *Phys. Lett. A* **372**, 5270–5273 (2008).
30. Cook, R. J. & Shore, B. W. Coherent dynamics of N-level atoms and molecules. III. An analytically soluble periodic case. *Phys. Rev. A* **20**, 539–544 (1979).
31. Burgarth, D. & Bose, S. Conclusive and arbitrarily perfect quantum-state transfer using parallel spin-chain channels. *Phys. Rev. A* **71**, 052315 (2005).
32. Burgarth, D., Giovannetti, V. & Bose, S. Efficient and perfect state transfer in quantum chains. *J. Phys. A Math. Gen.* **38**, 6793 (2005).
33. Yung, M.-H. & Bose, S. Perfect state transfer, effective gates, and entanglement generation in engineered bosonic and fermionic networks. *Phys. Rev. A* **71**, 032310 (2005).
34. Plenio, M. B. & Semio, F. L. High efficiency transfer of quantum information and multiparticle entanglement generation in translation-invariant quantum chains. *New J. Phys.* **7**, 73 (2005).
35. Zhang, J. *et al.* Simulation of Heisenberg XY interactions and realization of a perfect state transfer in spin chains using liquid nuclear magnetic resonance. *Phys. Rev. A* **72**, 012331 (2005).
36. Kay, A. Perfect state transfer: beyond nearest-neighbor couplings. *Phys. Rev. A* **73**, 032306 (2006).
37. Bose, S. Quantum communication through spin chain dynamics: an introductory overview. *Contemp. Phys.* **48**, 13–30 (2007).
38. Kostak, V., Nikolopoulos, G. M. & Jex, I. Perfect state transfer in networks of arbitrary topology and coupling configuration. *Phys. Rev. A* **75**, 042319 (2007).
39. Di Franco, C., Paternostro, M. & Kim, M. S. Perfect state transfer on a spin chain without state initialization. *Phys. Rev. Lett.* **101**, 230502 (2008).
40. Gualdi, G., Kostak, V., Marzoli, I. & Tombesi, P. Perfect state transfer in long-range interacting spin chains. *Phys. Rev. A* **78**, 022325 (2008).
41. Paz-Silva, G. A., Rebic, S., Twamley, J. & Duty, T. Perfect mirror transport protocol with higher dimensional quantum chains. *Phys. Rev. Lett.* **102**, 020503 (2009).
42. Perez-Leija, A., Keil, R., Moya-Cessa, H., Szameit, A. & Christodoulides, D. N. Perfect transfer of path-entangled photons in Jx photonic lattices. *Phys. Rev. A* **87**, 022303 (2013).
43. Bellec, M., Nikolopoulos, G. M. & Tzortzakis, S. Faithful communication Hamiltonian in photonic lattices. *Opt. Lett.* **37**, 4504 (2012).
44. Perez-Leija, A. *et al.* Coherent quantum transport in photonic lattices. *Phys. Rev. A* **87**, 012309 (2013).
45. Einstein, A., Podolsky, B. & Rosen, N. Can quantum-mechanical description of physical reality be considered complete? *Phys. Rev.* **47**, 777–780 (1935).
46. Perets, H. B. *et al.* Realization of quantum walks with negligible decoherence in waveguide lattices. *Phys. Rev. Lett.* **100**, 170506 (2008).
47. Rai, A., Agarwal, G. S. & Perk, J. H. H. Transport and quantum walk of nonclassical light in coupled waveguides. *Phys. Rev. A* **78**, 042304 (2008).
48. Peruzzo, A. *et al.* Quantum walks of correlated photons. *Science* **329**, 1500–1503 (2010).
49. Bromberg, Y., Lahini, Y., Morandotti, R. & Silberberg, Y. Quantum and classical correlations in waveguide lattices. *Phys. Rev. Lett.* **102**, 253904 (2009).
50. Sansoni, L. *et al.* Two-particle bosonic-fermionic quantum walk via integrated photonics. *Phys. Rev. Lett.* **108**, 010502 (2012).
51. O'Brien, J. L. *et al.* Quantum process tomography of a controlled-NOT gate. *Phys. Rev. Lett.* **93**, 080502 (2004).
52. Gilchrist, A., Langford, N. K. & Nielsen, M. A. Distance measures to compare real and ideal quantum processes. *Phys. Rev. A* **71**, 062310 (2005).
53. Matthews, J. C. F. *et al.* Observing fermionic statistics with photons in arbitrary processes. *Sci. Rep.* **3**, 1539 (2013).
54. James, D. F. V., Kwiat, P. G., Munro, W. J. & White, A. G. Measurement of qubits. *Phys. Rev. A* **64**, 052312 (2001).
55. Jozsa, R. Fidelity for mixed quantum states. *J. Mod. Opt.* **41**, 2315–2323 (1994).
56. Devitt, S. J., Munro, W. J. & Nemoto, K. Quantum error correction for beginners. *Rep. Prog. Phys.* **76**, 076001 (2013).
57. Lloyd, S. Universal quantum simulators. *Science* **273**, 1073–1078 (1996).

Acknowledgements

G.C., A.C. and R.O. acknowledge support from the European Union through the projects FP7-ICT-2011-9-600838 (QWAD-Quantum Waveguides Application and Development; www.qwad-project.eu) and H2020-FETPROACT-2014-641039 (QUCHIP-Quantum Simulation on a Photonic Chip; www.quchip.eu). M.-H.Y. acknowledges support by the National Natural Science Foundation of China under Grants No. 11405093. A.P. acknowledges an Australian Research Council Discovery Early Career Researcher Award under project number DE140101700 and an RMIT University Vice-Chancellor's Senior Research Fellowship.

Author contributions

All authors contributed to all aspects of this work.

Additional information

Competing financial interests: The authors declare no competing financial interests.

21

Microelectromechanical reprogrammable logic device

M.A.A. Hafiz[1], L. Kosuru[1] & M.I. Younis[1]

In modern computing, the Boolean logic operations are set by interconnect schemes between the transistors. As the miniaturization in the component level to enhance the computational power is rapidly approaching physical limits, alternative computing methods are vigorously pursued. One of the desired aspects in the future computing approaches is the provision for hardware reconfigurability at run time to allow enhanced functionality. Here we demonstrate a reprogrammable logic device based on the electrothermal frequency modulation scheme of a single microelectromechanical resonator, capable of performing all the fundamental 2-bit logic functions as well as *n*-bit logic operations. Logic functions are performed by actively tuning the linear resonance frequency of the resonator operated at room temperature and under modest vacuum conditions, reprogrammable by the a.c.-driving frequency. The device is fabricated using complementary metal oxide semiconductor compatible mass fabrication process, suitable for on-chip integration, and promises an alternative electromechanical computing scheme.

[1] Physical Sciences and Engineering Division, King Abdullah University of Science and Technology, Thuwal 23955-6900, Saudi Arabia. Correspondence and requests for materials should be addressed to M.I.Y. (email: mohammad.younis@kaust.edu.sa).

The quest for mechanical computation is a century old and can be traced back to at least 1822 when Babbage presented his concept of difference engine[1]. Although the interest remained within the research community, the subsequent development in the fields of electronic transistor[2] and magnetic storage[3,4] outperformed the mechanical approach in computation both in terms of speed of operation and data density. However, recent advancements in micro-/nano-fabrication and measurement techniques have renewed the interest in the field of mechanical computation in the last decade[5–21].

The key to any computing machine are logic elements. The first demonstrated dynamic mechanical XOR logic gate was based on a piezoelectric nanoelectromechanical system (NEMS) structure where the presence (absence) of high-amplitude vibration in the linear regime denotes a logical high (low) state[7]. Later, OR/NOR and AND/NAND logic gates have been demonstrated utilizing the bistability of a nonlinearly resonating NEMS resonator mediated by the noise floor[12]. A universal logic device capable of performing AND, OR and XOR logic gates as well as multibit logic circuits has been implemented by parametrically exciting a single electromechanical resonator[15]. Same research group also demonstrated XOR and OR logic gates in an electromechanical membrane resonator under high vacuum and at room temperature condition[16]. On the basis of feedback control, a memory and OR logic operation have been demonstrated on a single microelectromechanical system (MEMS) resonator working in the nonlinear regime[20]. Recently, an unconventional and reversible logic gate (Fredkin gate) has been presented based on four coupled linearly resonating NEMS resonators[21] where AND, OR, NOT and FANOUT gate operations have been demonstrated. Note that room temperature and atmospheric operations are desirable prerequisites for any practical device implementation.

Here we demonstrate a reprogrammable logic device, capable of performing 2-bit AND, NAND, OR, NOR, XOR, XNOR and NOT logic operations using a single microelectromechanical resonator operating in the linear regime. The logic operations are performed by electrothermal modulation of the linear resonance frequency of the resonator, where two separate d.c. voltage sources represent logic inputs. The device can be programmed to perform any of these logic operations by simply tuning the a.c.-driving frequency. Also, we use this scheme of electrothermal frequency tuning to demonstrate 3-bit AND, NAND, OR and NOR logic gates on a single MEMS resonator. This can be extended to n-bit logic operations by adding a single d.c. voltage source per bit. This device works under room temperature and modest vacuum conditions and is fabricated using standard complementary metal oxide semiconductor-based fabrication techniques suitable for mass fabrication and on-chip integrated system development.

Results

Device fabrication and experimental set-up. The resonator is fabricated on a highly conductive Si device layer of silicon on insulator wafer by a two-mask process using standard photolithography, electron beam evaporation for metal layer deposition for actuating pad, deep reactive ion etch for silicon device layer etching and vapour hydrofluoric acid etch to remove the oxide layer underneath the resonating structure. It consists of a clamped–clamped arch-shaped microbeam with two adjacent electrodes to electrostatically induce the vibration and detect the generated a.c. output current due to the in-plane motion of the microbeam. The dimensions of the curved beam are 500 μm in length, 3 μm in width and 30 μm in thickness. The gap between the actuating electrode and the resonating beam is 8 μm at the

fixed anchors and 11 μm at the midpoint of the microbeam due to its 3-μm initial curvature.

Figure 1a shows the schematic of the arch microbeam and the two-port electrical transmission measurement configuration for electrostatic actuation and sensing that includes the parasitic current compensation circuit for enhanced transmission signal measurements[22]. The drive electrode is provided with an a.c. actuation signal from one of the outputs of a single-to-differential driver (AD8131), and the beam electrode is biased with a d.c. voltage source. The output current induced at the sense electrode is coupled with the variable compensation capacitor, C_{comp}, and followed by a low-noise amplifier whose output is coupled to the network analyser input port. Two logic inputs are provided with two d.c. voltage sources, V_A and V_B, connected in parallel across the microbeam with series resistors, R_A and R_B, and switches, A and B, respectively. The electrical wiring scheme for the logic inputs is depicted in red to differentiate it from the rest of the electrical connections. The binary logic input 1(0) is represented by connecting (disconnecting) V_A and V_B from the electrical network by the two switches, A and B, respectively. Hereafter, switch ON (OFF) condition for switches A and B corresponds to

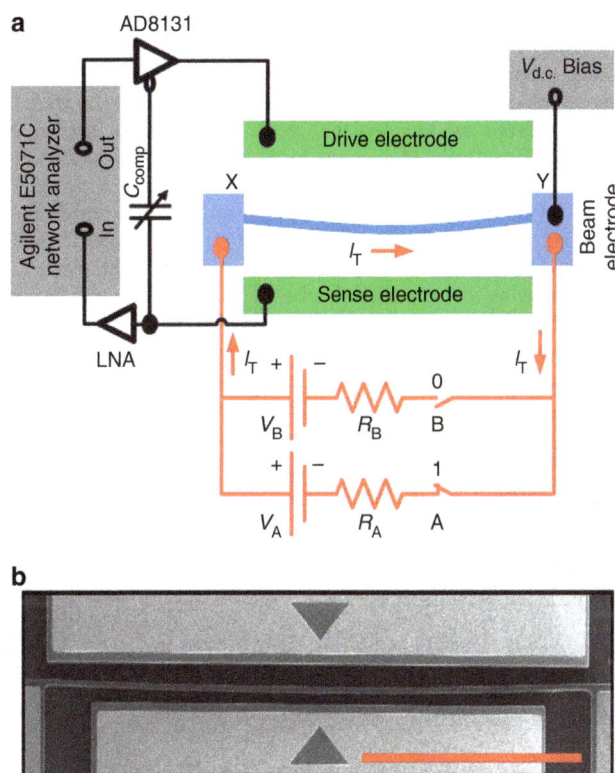

Figure 1 | Clamped–clamped arch resonator. (a) Schematic of the arch beam resonator and the two-port electrical transmission measurement configuration together with a parasitic current compensation circuitry using single-to-differential driver (AD8131) and a variable compensation capacitor, C_{comp}. The drive electrode is provided with an a.c. signal from one of the outputs from AD8131 and the beam electrode is biased with a d.c. voltage source. The output current induced at the sense-electrode is coupled with the compensation capacitor and followed by a low-noise amplifier (LNA) whose output is coupled to the network analyser input port. Two voltage sources, V_A and V_B and switches, A and B are connected in parallel across the beam to perform logic operations by electrothermal tuning of the resonance frequency. The arrow in the red represents the current flowing through the beam, responsible for electrothermal frequency modulation. **(b)** An SEM image of the microbeam resonator. Scale bar, 200 μm.

the binary logic input 1(0). The sensing electrode is used to obtain the logic output, where a relative high (low) S_{21} transmission signal corresponds to the logic output 1(0). Figure 1b shows an SEM image of the arch microbeam resonator.

Electrothermal frequency modulation. Electrothermal frequency modulation has an essential role in the execution of the logic functions in this architecture. Figure 2 shows four different electrical circuit configurations between nodes X and Y, shown in Fig. 1a. All the four logic input conditions, (0,0), (0,1), (1,0) and (1,1) are shown in Fig. 2a–d, respectively. For the case of (0,0) logic input condition, the total current flowing through the microbeam is $I_T = 0$ as depicted in the electrical circuit in Fig. 2a. In this case, the resonator exhibits series resonance peak and parallel resonance dip (anti-resonance) at 117.663 and 117.361 kHz, respectively, with an a.c. actuation voltage of 2 dBm (0.28 V_{rms}) and $V_{d.c.}$ of 45 V at 1 torr pressure and at room temperature (see Supplementary Note 1 and Supplementary Fig. 1a,b). The corresponding frequency response is plotted in black in Fig. 3. Note that due to over compensation of the feed through by the parallel variable compensation capacitance, C_{comp}, the parallel resonance appears earlier than the series resonance[22]. However, this does not put any limitation on the successful logic operation by the device. Moreover, we use both the series and parallel resonances for implementing the logic gates. For logic input (0,1) or (1,0) conditions, either V_B or V_A is connected to the microbeam as depicted in the electrical circuits shown in Fig. 2b,c, respectively. Hence, the total current that flows through the microbeam is either $I_T = I_B$ or $I_T = I_A$. We chose $V_A = 0.4$ V, $V_B = 0.7$ V, and $R_A = R_B = 50\,\Omega$ so that it satisfies the condition of the same current amount at each case; $I_A = I_B$. Note that we measured the microbeam resistance $R_{MB} = 114\,\Omega$. The electrical current flowing through the microbeam generates heat and causes

thermal expansion, which induces compressive axial force. This compressive force causes an increase in the microbeam curvature[23–25] and increases its stiffness. Hence, the series resonance frequency increases to 121.431 kHz for either (0,1) or (1,0) logic input conditions. The frequency responses due to the logic input (0,1) and (1,0) conditions are plotted as red and blue, respectively, in Fig. 3. For logic input condition (1,1), both the voltage sources V_A and V_B are connected to the microbeam as depicted in the electrical circuit shown in Fig. 2d. The total current generated in this case is $I_T = I'_A + I'_B > I_A$ or I_B. Hence, the series resonance frequency further increases to 128.969 kHz as depicted in green in Fig. 3. Thus, one can modulate the resonance frequencies (series and parallel) of the microbeam through the electrothermal effect by controlling the amount of current flow in the microbeam. Towards this, we build different logic gates by properly choosing the a.c.-driving frequency. We identify three regions in the frequency response plot of Fig. 3 to build all the six logic gates. Region I corresponds to frequency of operation for logic gates OR/NOR, region II corresponds to logic gates XOR/XNOR and finally, region III corresponds to logic gates AND/NAND. NOT logic operation can be built on any of these frequencies by proper conditioning of one of the inputs. The detail execution of the logic gates will be discussed in the following sections.

NOR/OR. The frequency responses of the resonator for different logic input conditions are shown in Fig. 4a, which lies in the region I of Fig. 3. To demonstrate NOR gate operation, the frequency of 117.663 kHz is chosen as it shows high S_{21} transmission signal denoted as the logic output 1 (in black) for (0,0) logic input

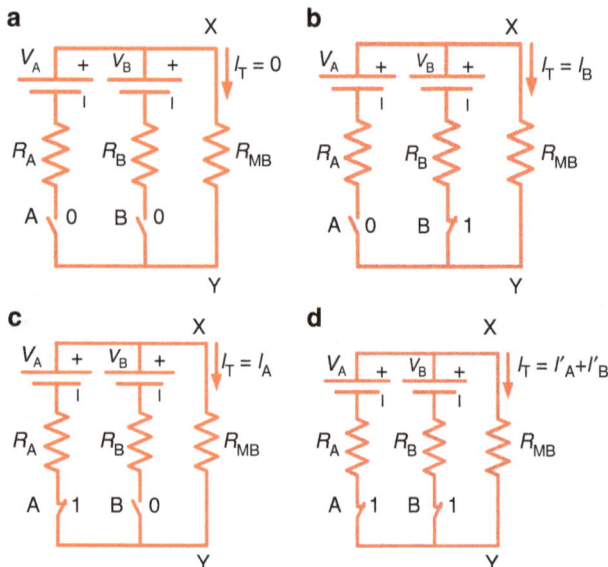

Figure 2 | Electrical circuit configuration of the logic input conditions.
(**a**) The electrical circuit represents the (0,0) logic input condition where the total current I_T through the beam R_{MB} is zero. (**b**) The circuit represents the (0,1) logic input condition corresponds to switch A, OFF and switch B, ON where the total current I_T flowing through the beam R_{MB} is I_B.
(**c**) The circuit represents the (1,0) logic input condition corresponds to switch A, ON and switch B, OFF where the total current I_T flowing through the beam R_{MB} is I_A. (**d**) The circuit represents the (1,1) logic input condition corresponds to switch A, ON and switch B, ON where the total current I_T flowing through the beam R_{MB} is $I'_A + I'_B$.

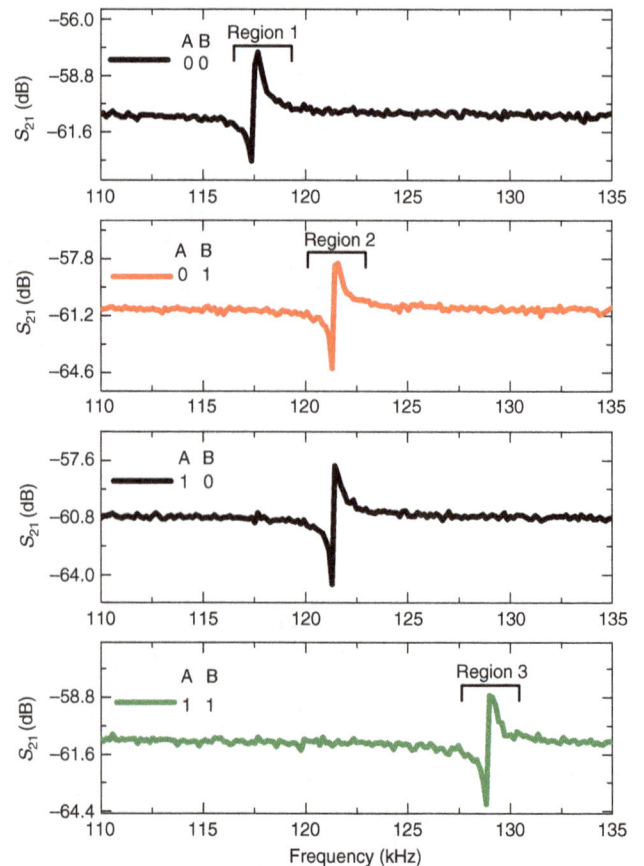

Figure 3 | Electrothermal frequency modulation. Frequency responses of the resonator for different logic input conditions, (0,0), (0,1), (1,0) and (1,1), shown in black, red, blue and green, respectively.

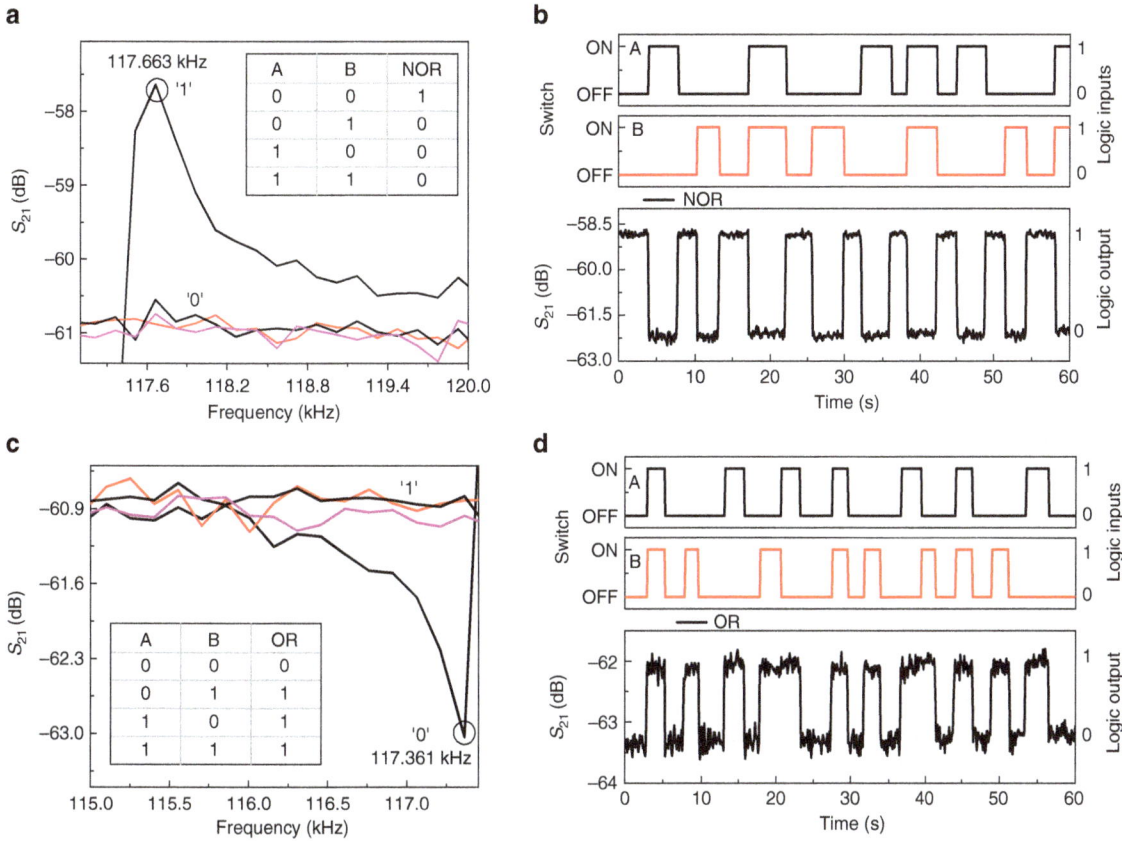

Figure 4 | Demonstration of 2-bit NOR and OR logic gates. (**a**) Frequency responses of the resonator for different logic input conditions where (0,0) logic input condition, shown in black has high S_{21} transmission signal at 117.663 kHz and others have low S_{21} transmission signal represented by 1 and 0, respectively. Truth table of NOR logic output is shown in the inset. (**b**) Demonstration of NOR logic operation when the frequency of the a.c. input signal is chosen as 117.663 kHz. Two input signals A and B are shown in black and red, respectively, where the switch OFF/ON corresponds to 0/1 logic input conditions. The S_{21} transmission signal in blue corresponds to the logic output and fulfills the NOR truth table. (**c**) Frequency responses of the resonator for different logic input conditions, where (0,0) logic input condition shown in black has low S_{21} transmission signal at 117.361 kHz and others have high S_{21} transmission signal, represented by 0 and 1, respectively. Truth table for OR logic output shown in the inset. (**d**) Demonstration of OR logic operation when the a.c. input signal frequency is chosen as 117.361 kHz. Two input signals, A and B are shown in black and red, respectively, and the switch OFF/ON corresponds to 0/1 logic input conditions. The S_{21} transmission signal in blue corresponds to the logic output that fulfills the OR truth table.

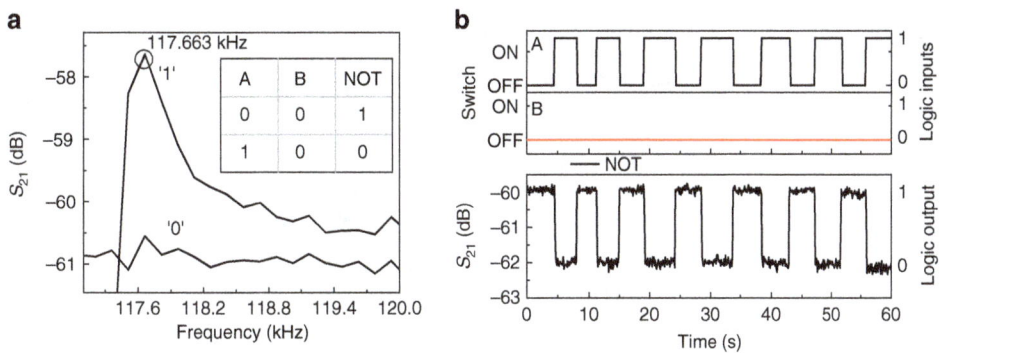

Figure 5 | Demonstration of NOT gate. (**a**) Frequency responses of the resonator for different logic input conditions, where (0,0) logic input condition shown in black has high S_{21} transmission signal at 117.663 kHz and others have low S_{21} transmission signal represented by 1 and 0, respectively. Truth table of NOT logic gate is shown in the inset. (**b**) Demonstration of NOT logic operation when the frequency of the a.c. input signal is chosen as 117.663 kHz. Two input signals, A and B are shown in black and red, respectively, where the switch OFF/ON corresponds to 0/1 logic input conditions. S_{21} transmission signal in blue corresponds to the logic output and fulfills the NOT truth table.

condition only. The resonator is tuned away from its series resonance frequency of 117.663 kHz by other logic input conditions, (0,1), (1,0) and (1,1), respectively. Hence, shows low S_{21} transmission signal denoted as logic output 0 (in black) at the

frequency of 117.663 kHz. The NOR gate truth table is shown in the inset of Fig. 4a. The time response of the resonator showing binary inputs A and B and the corresponding logic output is depicted in Fig. 4b. It clearly shows NOR logic operation as the output is 1 (high) only when both the inputs A and B are 0 (switch OFF), and the output is 0 (low) for all the other conditions, (0,1), (1,0) and (1,1).

To demonstrate OR logic gate, we exploit the parallel resonance dip at 117.361 kHz, shown in black circle in Fig. 4c. Here the low level of S_{21} transmission signal is considered as the logic output 0 (in green), and otherwise as the logic output 1 (in green). The OR gate truth table is shown in the inset of Fig. 4c. Figure 4d shows the time response of the resonator output for OR logic gate operation with the corresponding binary inputs A and B. It clearly shows OR logic operation as the logic output is 0 (low) when both the inputs A and B are 0, and logic output is 1 (high) for all the other conditions.

NOT. To perform NOT operation on the input A, the a.c.-driving frequency is set to be at 117.663 kHz and the input B is set to 0 (switch OFF). For this set condition, a high S_{21} transmission signal (logic output 1) is achieved for the logic input A set at 0 (switch OFF) and vice versa as shown in Fig. 5a. We note that NOT operation can also be built on input B by properly setting input A (switch OFF/ON) and a.c.-driving frequency. The time response for the NOT operation is shown in Fig. 5b. It is evident from the output signal that when the input A is 0, the output is 1 and vice versa.

XOR/XNOR. Frequency responses of the resonator for different logic input conditions are shown in Fig. 6a, which lies in the region II of Fig. 3. To implement XOR gate, the frequency of operation is chosen as 121.431 kHz, shown in black circle in Fig. 6a. At this operating frequency, it shows low S_{21} transmission signal denoted as the logic output 0 (in black) for the logic input conditions (0,0) and (1,1). For other logic input conditions, (0,1) and (1,0), it shows high S_{21} transmission signal denoted as the logic output 1 (in black). The truth table for XOR logic gate is shown in the inset of Fig. 6a. Figure 6b shows the time response of the resonator output for XOR logic gate operation with the corresponding binary inputs A and B. It clearly shows XOR logic gate operation as the logic output is 1 (high) when the inputs A and B are complementary to each other. On the other hand, the logic output is 0 (low) for the same logic input conditions, (0,0) and (1,1).

To demonstrate XNOR logic gate, we exploit the parallel resonance dip at 121.281 kHz, shown in black circle in Fig. 6c. Here the low level of S_{21} transmission signal is considered as the logic output 0 (in green), and otherwise as the logic output 1 (in green). XNOR truth table is shown in the inset of Fig. 6c. Figure 6d shows the time response of XNOR logic gate output and the corresponding binary logic inputs A and B. It clearly shows XNOR logic gate operation as the logic output is 1 (high) when both the inputs A and B are same, (0,0) and (1,1), and otherwise the logic output is 0 (low). Note that occasional spikes observed in the S_{21} transmission signal (in blue) in Fig. 6b,d are due to the switching between (0,1) and (1,0) logic input conditions. However, the resonator still performs the desired logic operations successfully.

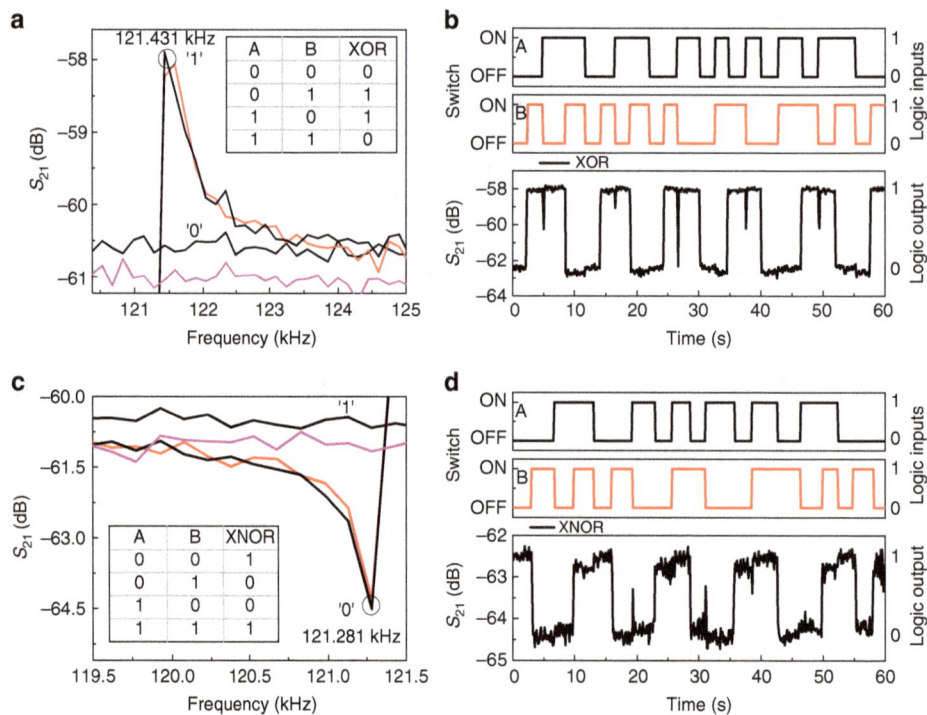

Figure 6 | Demonstration of 2-bit XOR and XNOR logic gates. (a) Frequency responses of the resonator for different logic input conditions, where (0,1) and (1,0) logic input condition shown in red and blue has high S_{21} transmission signal at 121.43 kHz and others have low S_{21} transmission signal represented by 1 and 0, respectively. Truth table of XOR logic gate is shown in the inset. **(b)** Demonstration of XOR logic operation when the operation frequency is chosen as 121.43 kHz. Two input signals, A and B are shown in black and red, respectively, where the switch OFF/ON corresponds to 0/1 logic input conditions. S_{21} transmission signal in blue corresponds to the logic output that fulfills the XOR truth table. **(c)** Frequency responses of the resonator for different logic input conditions, where (0,0) and (1,1) logic input conditions shown in red and blue, respectively, has low S_{21} transmission signal at 121.281 kHz and others have high S_{21} transmission signal represented by 0 and 1, respectively. Truth table of XNOR logic output is shown in the inset. **(d)** Demonstration of XNOR logic operation when the operating frequency is fixed at 121.281 kHz. Two input signals, A and B are shown in black and red, respectively, where the switch OFF/ON corresponds to 0/1 logic input conditions. S_{21} transmission signal in blue corresponds to the logic output and fulfills the XNOR truth table.

AND/NAND. Frequency responses of the resonator for different logic input conditions are shown in Fig. 7a, which falls in the region III of Fig. 3. To demonstrate AND gate operation, the frequency of 128.969 kHz is chosen, which is shown in black circle in Fig. 7a. When both the inputs A and B are 1 (switch ON), the high S_{21} transmission signal is observed at this operating frequency and denoted as the logic output 1 (in black). For other logic input conditions, (0,1), (1,0) and (0,0), it shows the low S_{21} transmission signal, which is denoted as the logic output 0 (in black). This is expressed in a truth table in the inset of Fig. 7a. The time response of the resonator for AND gate operation and the corresponding binary logic inputs A and B are shown in Fig. 7b. It clearly shows AND gate operation as the output is 1 (high) only when both the inputs A and B are 1, otherwise 0 (low).

To demonstrate NAND gate, the frequency of operation is chosen at 128.819 kHz, shown in black circle in Fig. 7c. Here the low level of S_{21} transmission signal of the parallel resonance dip is considered as the logic output 0 (in green), and otherwise as the logic output 1 (in green). NAND gate truth table is shown in inset of Fig. 7c. Figure 7d shows the time response of NAND logic gate output and the corresponding binary logic inputs A and B. It shows NAND logic operation as the logic output is 0 (low) only when both the inputs A and B are 1 (switch ON).

Three-bit logic gates. We also implemented 3-bit logic gates by adding a third voltage source V_C (0.44 V) with series resistor R_C (50 Ω) and switch C, connected in parallel with the other two voltage sources, V_A (0.4 V) and V_B (0.7 V), in the electrical circuit

shown in Fig. 1a. Figure 8 shows the frequency responses of the resonator for different logic input conditions with an a.c. actuation voltage of 2 dBm (0.28 V_{rms}) and $V_{d.c.}$ of 40 V at 1 torr pressure and at room temperature. Three-bit NOR gate is realized by choosing the a.c.-driving frequency at 119.022 kHz marked in black circle as shown in Fig. 8. For (0,0,0) logic input condition, the frequency response shows high S_{21} transmission signal corresponds to the logic output (1). For all the other logic input conditions, the response shows low S_{21} transmission signal at this frequency, which corresponds to the logic output (0). Similar to the 2-bit OR logic operation, a 3-bit OR logic function can be realized by selecting the frequency of the anti-resonance dip as the a.c.-driving frequency. Next, a 3-bit AND gate is realized by choosing the frequency of operation at 132.105 kHz marked in black circle in Fig. 8, where only (1,1,1) logic input condition shows high S_{21} transmission signal corresponds to the logic output (1). For all the other logic input conditions shows low S_{21} transmission signal corresponds to the logic output (0). By selecting the corresponding anti-resonance frequency as the a.c.-driving frequency, a 3-bit NAND gate can be realized. Figure 9a shows the time response of the 3-bit NOR logic gate at the operation frequency of 119.022 kHz. Three logic input signals, A, B and C are shown in black, red and blue, respectively, where the switch OFF/ON corresponds to 0/1 logic input conditions. S_{21} transmission signal in green corresponds to the logic output, and fulfills the NOR truth table. Figure 9b shows the demonstration of a 3-bit AND logic function at the a.c.-driving frequency of 132.105 kHz. The output response shown in green fulfills the AND truth table.

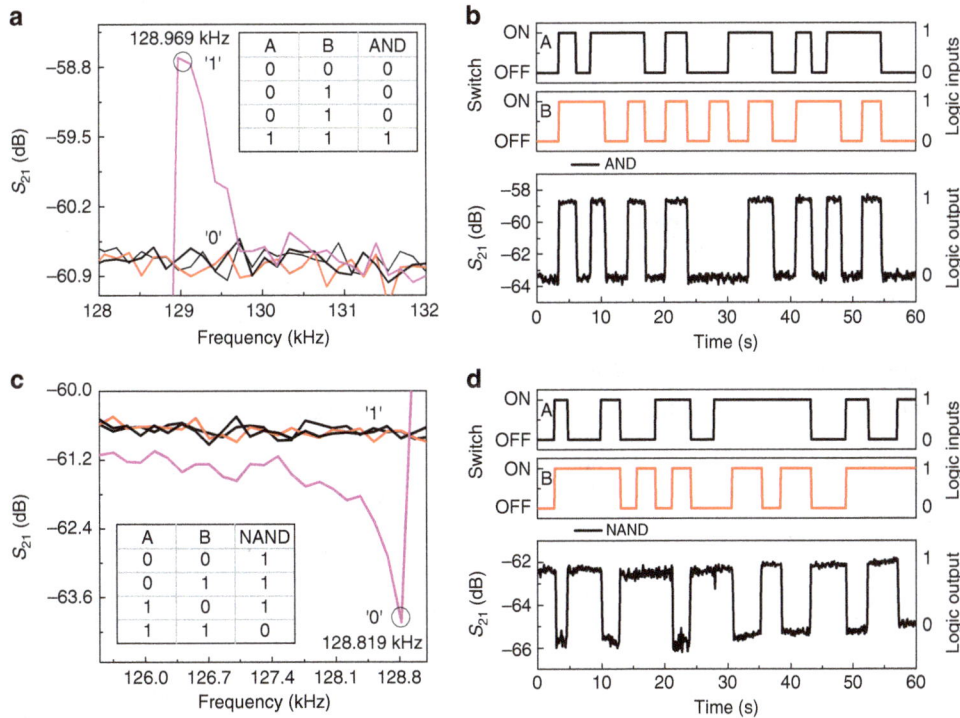

Figure 7 | Demonstration of 2-bit AND and NAND logic gates. (**a**) Frequency responses of the resonator for different logic input conditions, where (1,1) logic input condition shown in magenta has high S_{21} transmission signal at 128.969 kHz and others have low signal represented by 1 and 0, respectively. Truth table of AND logic output is shown in the inset. (**b**) Demonstration of AND logic operation when the operation of frequency is chosen as 128.969 kHz. Two input signals, A and B are shown in black and red, respectively, where the switch OFF/ON corresponds to 0/1 logic input conditions. S_{21} transmission signal in blue corresponds to the logic output and fulfills the AND truth table. (**c**) Frequency responses of the resonator for different logic input conditions, where (1,1) logic input condition has low S_{21} transmission signal at 128.819 kHz and others have high S_{21} transmission signal represented by 0 and 1, respectively. Truth table of NAND logic output is shown in the inset. (**d**) Demonstration of NAND logic operation when the operation of frequency is chosen as 128.819 kHz. Two input signals, A and B are shown in black and red, respectively, where the switch OFF/ON corresponds to 0/1 logic input conditions. S_{21} transmission signal in blue corresponds to the logic output and fulfills the NAND truth table.

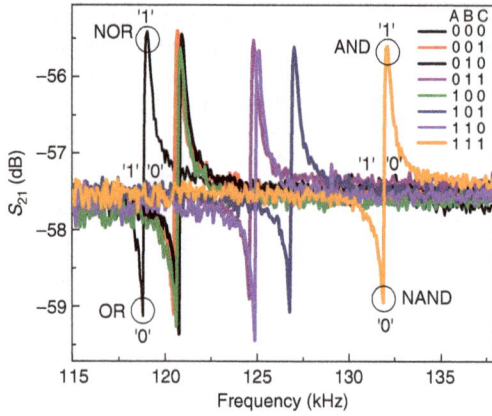

Figure 8 | Realization of 3-bit logic gates. Frequency responses of the resonator for three different input logic conditions. NOR gate is realized by choosing the frequency of operation at 119.022 kHz, where (0,0,0) logic input condition has high S_{21} transmission signal and all the others have low S_{21} transmission signal. By choosing the corresponding anti resonance dip frequency, 3-bit OR gate can be realized. A 3-bit AND gate is realized by choosing the frequency of operation at 132.105 kHz, where (1,1,1) logic input condition has high S_{21} transmission signal and all others have low S_{21} transmission signal. By choosing the corresponding anti resonance dip frequency, a 3-bit NAND gate can be realized.

One final remark is regarding the chosen d.c. bias voltage of this study. The demonstrated logic gates can be also operated at lower d.c. bias voltage. For example, we demonstrated a 2-bit NOR logic gate with a 20 V d.c. bias voltage (see Supplementary Note 2 and Supplementary Fig. 2a,b).

Discussion

An important feature of a logic gate is the operation speed. The speed of operation of the proposed logic device is governed by the speed of the electrothermal frequency modulation and the resonator switching speed. The characteristic time associated with electrothermal heating and cooling is typically much longer than the period of free vibrations of MEMS/NEMS structures[26,27]. Hence, electrothermal actuators have been mainly explored for static or low-frequency operations[26,27]. It is possible to calculate the thermal time constant of the microbeam[28,29] using the equation $\tau = \left[\frac{\pi^2 K_{Si}}{c\rho l^2} + \frac{F_s K_{air}}{gtc\rho}\right]^{-1}$, where l is the length of the microbeam, g is the gap between the beam and the substrate, t is the thickness of the beam, ρ is the density of silicon, c is the heat capacity of silicon, K_{Si} and K_{air} are the thermal conductivity of silicon and air, respectively. The beam shape factor, F_s, is a correction term that depends on the geometry of the beam. This correction term is necessary because the heat is conducted to the substrate not only through the bottom surface of the beam but also from the sides and the top surface. The formula for the shape factor[30] is given by $F_s = \frac{t}{w}\left(\frac{2g}{t}+1\right)+1$, where w is the width of the beam. F_s for the studied microbeam is calculated to be 12.33. The calculated thermal time constant for the microbeam used in this study is 152 µs, which indicates an electrothermal switching speed of 6.5 kHz. The theoretical open-loop switching speed of the MEMS resonator is estimated to be, $f/Q \sim 238$ Hz. Thus, it can be inferred that the maximum operating speed of the proposed logic device is limited by the ring-up or ring-down time of the resonator rather than the thermal time constant. It is worth to note that by scaling the device dimensions to nanoscale, both the mechanical response time and the thermal time constant will be improved significantly. As an example, we have estimated the thermal time constant[28,29] to be in the order of 10^{-6} s for a

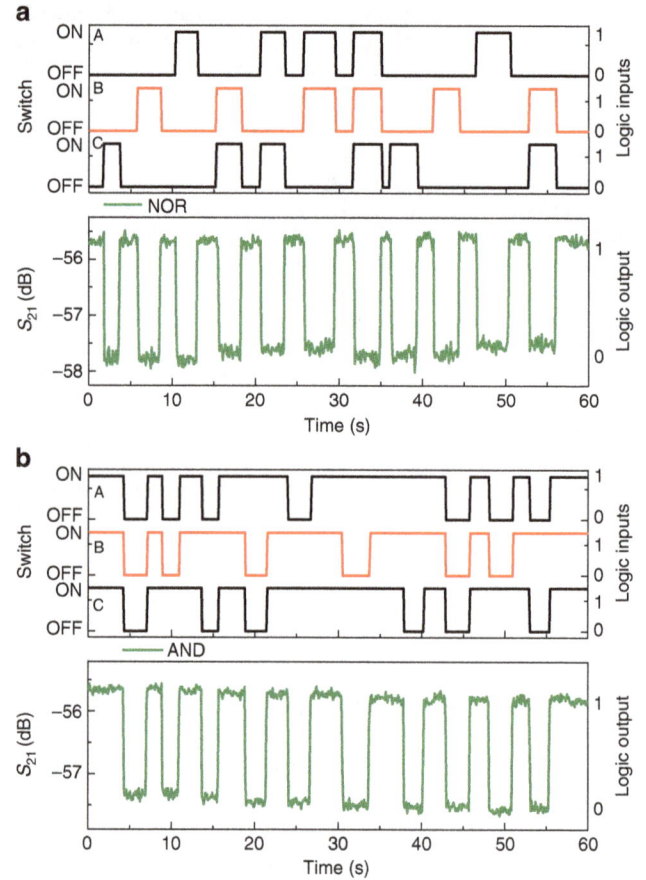

Figure 9 | Demonstration of 3-bit logic gates. (**a**) Demonstration of 3-bit NOR logic operation when the operating frequency is chosen at 119.022 kHz. Three input signals, A, B, and C are shown in black, red, and blue, respectively, where the switch OFF/ON corresponds to 0/1 logic input conditions. S_{21} transmission signal in green corresponds to the logic output and fulfills the NOR truth table. (**b**) Demonstration of 3-bit AND logic operation when the operation frequency is chosen at 132.105 kHz. Three input signals, A, B, and C are shown in black, red, and blue, respectively, where the switch OFF/ON corresponds to 0/1 logic input conditions. S_{21} transmission signal in green corresponds to the logic output and fulfills the AND truth table.

clamped–clamped beam resonator with a length of 20 µm, width of 300 nm and thickness of 500nm (ref. 12). This translates into a maximum electrothermal modulation speed in the order of 10^6 Hz. For the same resonator, the reported open-loop operation speed was around 48 kHz (ref. 12). It implies that the operation speed of logic devices built in these dimensions will be defined by the mechanical response time rather than the thermal response time. By considering a length of 600 µm and width of 50 µm for our device (includes electrodes and anchors), an integration density in the order of 10^4 per cm^2 can be achieved. Moreover, we note that the use of nanomechanical resonators would significantly increase the integration density. For a resonator with a length and width of 1 µm and 100 nm (resonance frequency around 1 GHz)[10], respectively, an integration density as high as 10^8 devices per cm^2 is plausible.

Another important aspect of a logic gate is the switching energy necessary to perform the desired logic operation. In this proposed scheme, the energy provided for the necessary switching events for the logic operation is in the form of resistive heating of the microbeam using the electrothermal circuit consisting of R_A, R_B and R_{MB}. While only a fraction of the total energy provided to the system is used by the microbeam for the state change during the logic operations, most of the energy is lost in the form of heat dissipation to the environment through R_A, R_B and R_{MB}. We estimated the maximum power cost for performing a single-logic operation as $P_{diss} \approx \frac{V_{RA}^2}{R_A} + \frac{V_{RB}^2}{R_B} + \frac{V_{RMB}^2}{R_{MB}} \approx 10^{-2}$ W. One can note that this energy cost is relatively high compared with other reported energy cost in performing a single-logic operation on nanomechanical resonator-based systems, such as in the work of Guerra et al.[12] and Wenzler et al.[21], which is based on electrostatic actuation. As traditionally well-known, thermal actuation, which is the base of this work, is considered less-energy efficient compared with other actuation methods. Nevertheless, the same principle demonstrated in this work applies when using other actuation techniques, as long as they can actively tune the stiffness of the resonating structure. It is also expected that the energy cost can be further reduced by orders of magnitude by optimizing device geometry.

The sensitivity of the proposed device to temperature variation is another important factor that needs to be addressed. The bandwidth, Δf, of the resonator of this study is estimated to be around 240 Hz ($\Delta f = f/Q$) at 1 torr. It implies that for the resonance frequency chosen as the operating frequency, the device will perform the desired logic operation successfully as long as the frequency shift due to the change in the ambient temperature lies within ± 120 Hz. We estimated the frequency shift due to temperature change according to ref. 31 $f(T) = f_0(1 + TC_f(T - T_0))$, where $TC_f = -30$ p.p.m. per °C, is the temperature coefficient of frequency for silicon resonators[32]. For the ambient temperature change between $-10\,°C$ and $+60\,°C$ from room temperature at 25 °C, the frequency shift is estimated to be $f_{shift} = \pm 120$ Hz, which is within the bandwidth of the resonator. Hence, the device will perform the desired logic operations successfully by selecting the resonance frequency as the driving frequency within this range of temperature variations. Additional temperature compensation scheme would be necessary to perform successful logic operation beyond this temperature range for the current device. Apart from this, the variation of resonance frequency due to phase noise is estimated to be around 105 Hz (see Supplementary Note 3 and Supplementary Fig. 3). Hence the device can still perform the desired logic operation successfully at a given operating conditions since the bandwidth is larger than the noise related frequency shift.

With regards to the potential interference between series and parallel resonances while selecting the a.c. operating frequency, it is noted that lowering down the compensation capacitance will broaden the separation between the series and parallel resonances. Also, improving the bandwidth will help to choose proper operating frequencies with lower margin of error.

A note is worth to be mentioned regarding the survivability of the resonators to mechanical shock. As was demonstrated[33–36] theoretically and experimentally, microstructures similar to the studied resonator shows excellent shock resilience up to 30,000–50,000 g. Downscaling the dimensions of the resonators will further improve the shock resilience.

The flexibility to cascade multiple gates is of paramount importance for realizing complex logic circuits. For the proposed scheme it is limited by two current challenges that warrant more future research. First, the strength of the output a.c. signal, which requires a transimpedence amplifier. Second is the fact that the signal waveforms as logic inputs and logic outputs are of different form. The output signal, a.c., needs to be converted into a d.c. signal. The d.c. output signal can be then used as an input to the next logic element, and hence enables sequencing. Also, the d.c. current can be split into various branches or pass through multiple in-series resonators. If a single operating frequency is desired to be used throughout the grid of logic resonators, then one possibility is to fabricate several devices to have slightly different resonance frequencies, such that all can be driven at the same frequency. Also, the devices can be individually tuned by a separate d.c. biasing mechanism for each.

In summary, we demonstrated a reprogrammable logic device based on electrothermal tuning of the resonanance frequency, capable of performing all the fundamental 2-bit logic operations; AND, NAND, OR, NOR, XOR, XNOR and NOT, at room temperature and at modest vacuum conditions. We also demonstrated a single MEMS resonator-based reprogrammable 3-bit AND, NAND, OR and NOR logic gates. This device can be easily modified to perform n-bit OR/NOR and AND/NAND logic operations by simply adding one voltage source per bit in parallel in the electrical network responsible for the electrothermal frequency modulation. We program the device to perform a desired logic operation by simply choosing appropriate a.c.-driving frequency. This logic device operates in the linear regime of the resonator, and hence, may further reduce the voltage load if operated under low damping conditions. Although we have used an arch-shaped microbeam resonator, the same principle of electrothermal frequency modulation is equally applicable for a straight clamped–clamped MEMS/NEMS resonator. In fact, the demonstrated principle applies on any MEMS/NEMS resonator devices working in the linear frequency regime with a proper frequency tuning mechanism that can alter the stiffness property of the resonator, and hence, its linear resonance frequency. Future directions in this research can be targeted to simplify the bulky S_{21} parameter measurement set-up used in this paper. This complexity can be minimized by integrating necessary complementary metal oxide semiconductor devices, such as transimpedance amplifier, on-chip. This practical demonstration of essential elements of computation using MEMS resonators provide fundamental building blocks for alternative computing scheme in the electromechanical domain.

References

1. Babbage, H. P. Babbage's Calculating Engines Vol. 2 (The MIT Press, 1984).
2. Davis, M. The Universal Computer (Norton, 2000).
3. Theis, T. N. & Horn, P. M. Basic research in the information technology industry. Phys. Today 56, 44–49 (2003).
4. Swade, D. D. Redeeming Charles Babbage's mechanical computer. Sci. Am. 268, 86–91 (1993).
5. Halg, B. On a micro-electro-mechanical nonvolatile memory cell. IEEE Trans. Electron Devices 37, 2230–2236 (1990).
6. Badzey, R. L., Zolfagharkhani, G., Gaidarzhy, A. & Mohanty, P. A controllable nanomechanical memory element. Appl. Phys. Lett. 85, 3587–3589 (2004).
7. Masmanidis, S. C. et al. Multifunctional nanomechanical systems via tunably coupled piezoelectric actuation. Science 317, 780–783 (2007).
8. Charlot, B., Sun, W., Yamashita, K., Fujita, H. & Toshiyo, H. Bistable nanowire for micromechanical memory. J. Micromech. Microeng. 18, 045005 (2008).
9. Guerra, D. N., Imboden, M. & Mohanty, P. Electrostatically actuated silicon-based nanomechanical switch at room temperature. Appl. Phys. Lett. 93, 033515 (2008).
10. Mahboob, I. & Yamaguchi, H. Bit storage and bit flip operations in an electromechanical oscillator. Nat. Nanotechnol. 3, 275–279 (2008).
11. Roodenburg, D., Spronck, J. W., Van der Zant, H. S. J. & Venstra, W. J. Buckling beam micromechanical memory with on-chip readout. Appl. Phys. Lett. 94, 183501 (2009).
12. Guerra, D. N. et al. A noise assisted reprogrammable nanomechanical logic gate. Nano Lett. 10, 1168–1171 (2010).
13. Noh, H., Shim, S. B., Jung, M., Khim, Z. G. & Kim, J. A mechanical memory with a dc modulation of nonlinear resonance. Appl. Phys. Lett. 97, 033116 (2010).
14. Venstra, W. J., Westra, H. J. R. & van der Zant, H. S. J. Mechanical stiffening, bistability, and bit operations in a microcantilever. Appl. Phys. Lett. 97, 193107 (2010).
15. Mahboob, I., Flurin, E., Nishiguchi, K., Fujiwara, A. & Yamaguchi, H. Interconnect-free parallel logic circuits in a single mechanical resonator. Nat. Commun. 2, 198 (2011).

16. Hatanaka, D., Mahboob, I., Okamoto, H., Onomitsu, K. & Yamaguchi, H. An electromechanical membrane resonator. *Appl. Phys. Lett.* **101**, 063102 (2012).

17. Uranga, A. *et al.* Exploitation of non-linearities in CMOS-NEMS electrostatic resonators for mechanical memories. *Sen. Actuators A Phys.* **197**, 88–95 (2013).

18. Mahboob, I., Mounaix, M., Nishiguchi, K., Fujiwara, A. & Yamaguchi, H. A multimode electromechanical parametric resonator array. *Sci. Rep.* **4**, 4448 (2014).

19. Yao, A. & Hikihara, T. Counter operation in nonlinear micro-electro-mechanical resonators. *Phys. Lett. A* **377**, 2551–2555 (2013).

20. Yao, A. & Hikihara, T. Logic-memory device of a mechanical resonator. *Appl. Phys. Lett.* **105**, 123104 (2014).

21. Wenzler, J. S., Dunn, T., Toffoli, T. & Mohanty, P. A nanomechanical Fredkin gate. *Nano Lett.* **14**, 89 (2014).

22. Lee, J. E. Y. & Seshia, A. A. Parasitic feedthrough cancellation techniques for enhanced electrical characterization of electrostatic microresonators. *Sens. Actuators A Phys.* **156**, 36 (2009).

23. Younis, M. I., Ouakad, H., Alsaleem, F. M., Miles, R. & Cui, W. Nonlinear dynamics of MEMS arches under harmonic electrostatic actuation. *J. Microelectromech. Syst.* **19**, 647–656 (2010).

24. Alkharabsheh, S. & Younis, M. I. Dynamics of MEMS arches of flexible supports. *J. Microelectromech. Syst.* **22**, 216–224 (2013).

25. Alkharabsheh, S. & Younis, M. I. Statics and dynamics of MEMS arches under axial forces. *J. Vib. Acoust.* **135**, 021007 (2013).

26. Pelesko, J. A. & Bernstein, D. H. *Modeling of MEMS and NEMS* (CRC Press, 2002).

27. Kaajakari, V. *Practical MEMS* (Small Gear Publishing, 2009).

28. Schreiber, D. S., Cheng, W. J., Maloney, J. M. & DeVoe, D. L. in *Proceedings of 2001 ASME International Mechanical Engineering Congress and Exposition* 141–147 (New York, NY, USA, 2001).

29. Wang, Y., Zhihong, L., Daniel, M. T. & Norman, T. C. A low-voltage lateral MEMS switch with high RF performance. *J. Microelectromech. Syst.* **13**, 902 (2004).

30. Lin, L. & Chiao, M. Electrothermal responses of line shape microstructures. *Sens. Actuators A Phys.* **55**, 35–41 (1996).

31. Zhua, H., Tua, C., Shanb, G. & Lee, J. E. Y. Dependence of temperature coefficient of frequency (TCf) on crystallography and eigenmode in N-doped silicon contour mode micromechanical resonators. *Sen. Actuators A Phys.* **215**, 189–196 (2014).

32. Jeong, J. H., Chung, S. H., Lee, Se-Ho. & Kwon, D. Evaluation of elastic properties and temperature effects in Si thin films using an electrostatic microresonator. *J. Microelectromech. Syst.* **12**, 524–530 (2003).

33. Younis, M. I. *MEMS Linear and Nonlinear Statics and Dynamics* (Springer, 2011).

34. Ouakad, H., Younis, M. I. & Alsaleem, F. Dynamic response of an electrostatically actuated microbeam to drop-table test. *J. Micromech. Microeng.* **22**, 095003 (2012).

35. Younis, M. I., Jordy, D. & Pitarresi, J. Computationally efficient approaches to characterize the dynamic response of microstructures under mechanical shock. *J. Microelectromech. Syst.* **16**, 628–638 (2007).

36. Younis, M. I., Miles, R. & Jordy, D. Investigation of the response of microstructures under the combined effect of mechanical shock and electrostatic forces. *J. Micromech. Microeng.* **16**, 2463–2474 (2006).

Acknowledgements

We acknowledge Ulrich Buttner, EMPIRe Lab at KAUST for helping with laser cutting the chips. This research has been funded by KAUST.

Author contributions

M.A.H. conceived the idea, designed and fabricated the MEMS resonator. M.A.H. and L.K. performed the measurements and analysed the data. All authors discussed the results and wrote the paper. M.I.Y supervised the project.

Additional information

Competing financial interests: The authors declare no competing financial interests.

22

Generating giant and tunable nonlinearity in a macroscopic mechanical resonator from a single chemical bond

Pu Huang[1,2,3,*], Jingwei Zhou[1,2,*], Liang Zhang[1,2,*], Dong Hou[1,2,3], Shaochun Lin[1,2,3], Wen Deng[1,2,4], Chao Meng[1,2], Changkui Duan[1,2], Chenyong Ju[1,2,3], Xiao Zheng[1,2,3], Fei Xue[4] & Jiangfeng Du[1,2,3]

Nonlinearity in macroscopic mechanical systems may lead to abundant phenomena for fundamental studies and potential applications. However, it is difficult to generate nonlinearity due to the fact that macroscopic mechanical systems follow Hooke's law and respond linearly to external force, unless strong drive is used. Here we propose and experimentally realize high cubic nonlinear response in a macroscopic mechanical system by exploring the anharmonicity in chemical bonding interactions. We demonstrate the high tunability of nonlinear response by precisely controlling the chemical bonding interaction, and realize, at the single-bond limit, a cubic elastic constant of $1 \times 10^{20}\,\mathrm{N\,m^{-3}}$. This enables us to observe the resonator's vibrational bi-states transitions driven by the weak Brownian thermal noise at 6 K. This method can be flexibly applied to a variety of mechanical systems to improve nonlinear responses, and can be used, with further improvements, to explore macroscopic quantum mechanics.

[1] National Laboratory for Physics Sciences at the Microscale, University of Science and Technology of China, Hefei 230026, China. [2] Department of Modern Physics, University of Science and Technology of China, Hefei 230026, China. [3] Synergetic Innovation Center of Quantum Information and Quantum Physics, University of Science and Technology of China, Hefei 230026, China. [4] High Magnetic Field Laboratory, Chinese Academy of Science, Hefei 230026, China. * These authors contributed equally to this work. Correspondence and requests for materials should be addressed to J.D. (email: djf@ustc.edu.cn).

onlinearity in micro- and nano-mechanical systems has been used to study processes such as fluctuation-enhanced dynamics[1-5], synchronization[6,7], mode mixing[8], noise control[9,10], signal amplification[11,12] and logic devices[13-15]. Its dynamics can be modelled by a driven Duffing oscillator equation as follows:

$$m\ddot{x} + \frac{m\omega_0}{Q}\dot{x} + kx + \alpha x^3 = F_{drive}\cos(\omega t). \quad (1)$$

Here m, ω_0, Q and $k = m\omega_0^2$ are the mass, resonance frequency, quality factor and linear spring constant, respectively. And αx^3 is the Duffing nonlinearity with α the Duffing constant, that is, the cubic elastic constant. Under weak drive F_{drive}, the nonlinear response is negligible due to its cubic dependence on the amplitude x of the resonator, and so the resonator behaves like a simple harmonic oscillator. This is the well-known Hooke's law of elasticity[16]. On the other hand, nontrivial dynamics of the resonator emerges when the drive is strong enough. A famous example is the occurrence of the driven Duffing biastability when the drive strength reaches a certain threshold F_c (ref. 17), with the corresponding threshold power P_c.

Since F_c is inversely proportional to $\sqrt{\alpha}$ as $F_c \sim \omega_0^3\sqrt{m^3/(\alpha Q^3)}$, the larger is α, the weaker driving force is required for the system to reach nonlinear regime. One benefit of low driving force is low driving noise, while in those micro- and nano-mechanical systems reported in the literature, driving noise is far beyond the system's intrinsic Brownian thermal noise even at room temperature[1-5,9,11]. On the other hand, $P_c \sim m^2\omega_0^5/(\alpha Q^2)$, so increasing α can lower the threshold power, which is favourable in some scenarios[13-15]. In practice, since the system's fundamental parameters m, ω_0 and Q are limited by various factors, such as working bandwidth, material or fabrication, a universal way to increase and to tune α independently is significant.

What is more demanding is from quantum science, where strong nonlinearity can make quantum effects emerge from a classical harmonic resonator[18]. Such quantum nonlinearity is still elusive in macro-scale mechanical systems due to the naturally weak nonlinear response. Generally speaking, the emergence of quantum behaviour requires quantum nonlinear strength $g \propto \alpha x_{zpf}^4$ to overcome the system's decoherence[19-24], and since the quantum fluctuation x_{zpf} is usually extremely tiny for a macro-scale resonator, generating ultra-strong nonlinear response α is of paramount importance.

In this work we demonstrate a system with a macroscopic mechanical resonator coupled to a single chemical bond, where the anharmonicity of the chemical bond deformation potential induces a giant nonlinear response of the resonator and can be tuned using external force. When driving to the nonlinear bi-states regime, stochastic transitions between bi-states are observed, which are demonstrated to be induced by the intrinsic Brownian thermal noise of the resonator.

Results

Theoretical model and density functional calculations. Our system consists of a macroscopic mechanical resonator tightened to an anchor via a chemical bonding structure shown schematically in Fig. 1a. Strong nonlinearity is achieved when the resonator moves along x-direction by deforming the chemical bond. This is because, although the resonator alone follows the elasticity theory with linear dynamics, the response of the chemical bond is highly anharmonic. This is illustrated by the chemical bond's energy curve $U_{chem}(x)$ (Fig. 1b) obtained by density functional theory calculations (see Methods for details). For simplicity, $x = 0$ is set at the minimum of $U_{chem}(x)$ (Fig. 1b). By applying on the chemical bond an external control force F, the resonator's equilibrium position x_{eq} can be tuned, and along with this the spring constant for sufficient small vibration is modified by $\Delta k = \partial^2 U_{chem}/\partial x^2$. As shown in Fig. 1c, when the resonator's equilibrium position is far away from the atom contact, the chemical bonding interaction is weak. By tuning the control force, that is, shifting the resonator's equilibrium position towards $x = 0$, the strength of Δk first reaches a local maximum and then

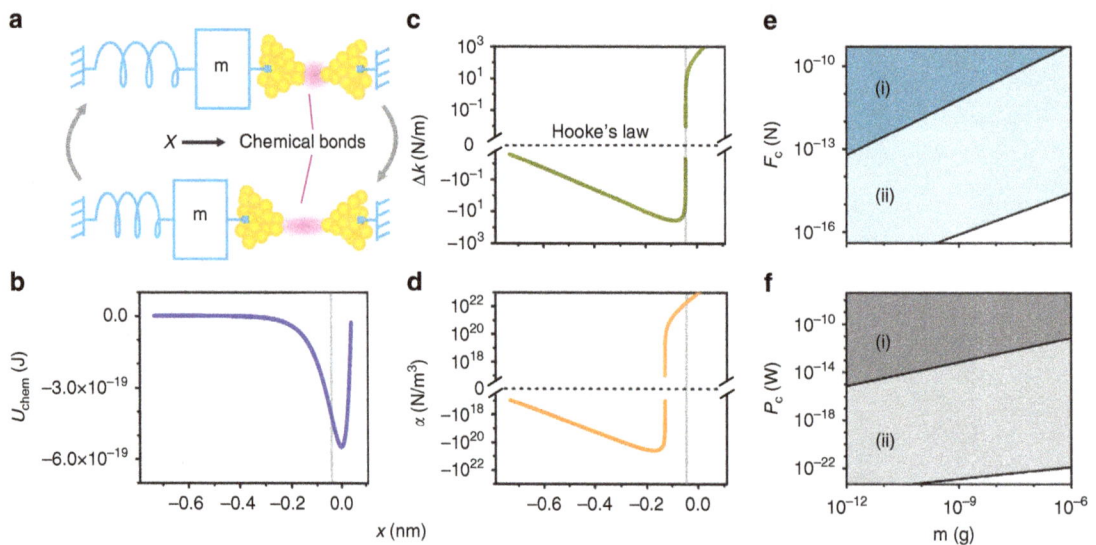

Figure 1 | Concept of the system and theoretical results. (**a**) A macroscopic resonator tightened to an anchor via chemical bonds. The displaced resonator can compress (top of panel) and stretch (bottom of panel) the chemical bond of gold-atom contact. (**b-d**) Density functional theory calculation results of (**b**) the chemical bonding interaction energy $U_{chem}(x)$, (**c**) the modified spring constant $\Delta k = \partial^2 U_{chem}/\partial x^2$ and (**d**) the enhanced nonlinearity coefficient $\alpha = (1/6)\partial^4 U_{chem}/\partial x^4$ as a function of the resonator displacement x. (**e,f**) Estimated threshold drive force F_c and corresponding threshold power P_c as a function of the resonator mass, m, with: (i) intrinsic nonlinear response of the resonator (dark blue, dark grey); and (ii) enhanced nonlinear response of the resonator (light blue, light grey) by chemical bonding interaction, at the point indicated as grey vertical line in **b-d**, where chemical bonding-induced linear response $\Delta k = 0$ (see Methods for details of the model).

drops to zero at the point where the chemical bonding attraction reaches maximum. The Duffing constant from chemical bonding, that is, $\alpha = (1/6)\partial^4 U_{chem}/\partial x^4$, changes differently (Fig. 1d): its strength first reaches a local maximum, then drops to zero before approaching the maximum attraction point, and finally changes its sign and increases in strength. At the maximum attraction point, $\alpha = 2 \times 10^{22}\,N\,m^{-3}$ is reached, while Δk is zero.

Figure 1e,f shows the threshold driving force F_c and the corresponding power P_c as a function of the macroscopicity of the resonator (here characterized by the resonator's mass) with/without using the chemical bonding force-induced nonlinearity. With specific basic mechanical parameters, that is, resonance frequency ω_0, quality factor Q and mass m, introducing chemical bonding force can reduce F_c as well as P_c significantly. The term proportional to x^3 in the expansion of chemical bonding potential U_{chem} and higher-order terms beyond the Duffing nonlinearity are also presented in our system, but are insignificant for bi-states dynamics studied in current experiments.

Experimental realization of the system. Now we demonstrate the above idea by using a macroscopic doubly clamped beam whose fundamental vibrational mode couples to a gold-atom contact, as shown in Fig. 2a. The atom contact is made by electric current migrations (see Methods) on a nano-bridge, which anchors the beam to a stiff electrode (Fig. 2b). The dimensions of beams is $l \times w \times t = 50 \times 1.5 \times 0.51\,\mu m$. The mass of the beam is $\sim 0.2\,ng$. The vibrations of the beam deform the gold–gold bonds of the contact along x direction. The beam has a typical intrinsic frequency around $\omega_0/2\pi = 1.6\,MHz$, with measured quality factor Q ranging from 1,000 to over 3,000, depending on specific device. The device is placed in an ultrahigh vacuum chamber and has an environment temperature of 6 K.

We carried out the experiments (see Methods) with several devices. The measured beam's resonance frequency and the corresponding spring constant are plotted as functions of control force F in Fig. 3a–c for devices A to C correspondingly. Device A is in non-contact regime while B and C are in contact regime but with different atom structure in the contact. In typical contact regime devices, the frequency varies by about 1 MHz when the beam is pulled out to non-contact regime. Figure 3d–f shows the corresponding electron conductance results measured simultaneously as those for Fig. 3a–c. In the non-contact regime, the conductance due to quantum tunnelling shows a strong dependence on the control force and a value as low as $\sim 300\,k\Omega^{-1}$ is reached, which confirms that a short-range chemical bonding force has been significantly involved[25], while in the contact regime, quantized conductance is observed and is insensitive to control force. Such a ballistic conductance is resulted from small number of metallic bonds[26].

From the dependence of the spring constant k on the applied static force F, we can obtain the Duffing constant as

$$\alpha(F) = \frac{1}{6\xi^2}\left(\frac{\partial^2 k}{\partial F^2}k^2 + \left(\frac{\partial k}{\partial F}\right)^2 k\right), \quad (2)$$

where ξ is the shape factor with value the order of 1 and depends on the definition of the mode shape normalization of the device (Supplementary Note 1). To reliably obtain the Duffing constant from the data with noise, we have smoothed the α using a running average of five data points. Figure 3g–i plots the measured Duffing constant α as a function of control force F. In the single-bond case, the maximum strength of the Duffing constant achieved in our experiments is $(1.1 \pm 0.2) \times 10^{20}\,N\,m^{-3}$, with an enhancement of six orders relative to the estimated intrinsic nonlinearity due to the elongation of the beam ($\alpha_0 \approx 1 \times 10^{14}\,N\,m^{-3}$), and is many orders of magnitude larger in strength than those reported previously[11,12,27–31] (see Supplementary Table 1 for comparison to other systems). High tunability is easily reached by control force. In practical, both non-contact and contact regimes can be used depending on detailed applications. In the non-contact regime, jump-to-contact leads to system's instability (see Supplementary Note 1 for detail), while in contact regime such instability is naturally avoided. The frequency response of the device was studied by increasing the drive force, and the results are plotted in Fig. 3j–m, with Fig. 3j from device A, Fig. 3k,l from device B corresponding to regions I and II in Fig. 3h, respectively, and Fig. 3m from device C. The hysteresis responses are due to Duffing induced bi-states[17], with their directions agreeing with the signs of the nonlinear coefficients measured in Fig. 3g–i.

Observation of thermally activated bi-states dynamics. In the following we present a demonstration of a dynamical effect of the strong nonlinear response. Thermal fluctuation effects in nonlinear regime can lead to various complex dynamics, and have been studied in various systems, with an example being the centre of mass motion of optical trapped particles[32]. However, such a thermal nonlinear regime is exclusive in widely studied micro- and nano-mechanical systems. A well-known phenomenon is noise-activated bi-states transitions in a driven Duffing oscillator[33]. As a consequence of lacking strong enough nonlinear response, this phenomenon has only been previously observed by introducing strong artificial noise that far above the thermal fluctuation[1–5,9]. With the successful realization of strong nonlinear response here, we have observed in our system the bi-states transitions activated by the Brownian thermal noise even at cryogenic temperatures.

In the process, we chose a device (device D) of a relative high quality factor $Q = 3,100$, and tuned the control force to a point with the Duffing constant approximately $-1 \times 10^{17}\,N\,m^{-3}$. We avoided using the maximum absolute Duffing constant, in

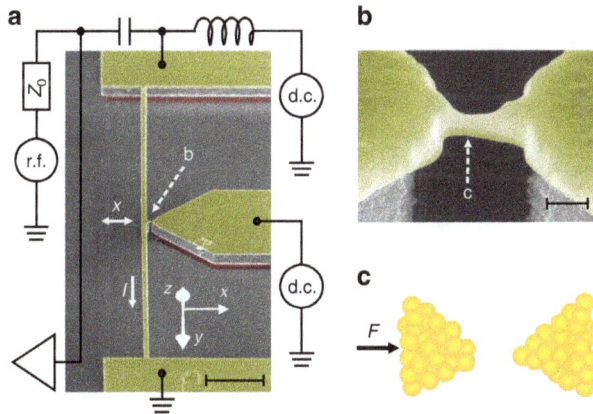

Figure 2 | Experimental set-up. (a) Scanning electron microscopy of a representative device in false colour. The macroscopic resonator is a doubly clamped silicon beam with thin layer of gold deposited on it, with dimension $l \times w \times t = 50 \times 1.5 \times 0.51\,\mu m$ and total mass $\sim 0.2\,ng$. The centre of the beam has horizontal displacement x. In the presence of a 6-T external magnetic field along the z direction, the electric current (I) can excite and detect the motion of the beam, with the schematic circuits shown. Scale bar, 5 μm. **(b)** Nano-bridge connecting the beam to a stiff electrode, before experiments. Scale bar, 100 nm. **(c)** Cartoon plot of the atom contact generated on the nano-bridge indicated by 'c' in **(b)**. The gold–gold bonding interaction is then tuned by force F, which is controlled by a d.c. current through the beam, and the electrostatic interaction of the contact is minimized applying a d.c. bias on the tip.

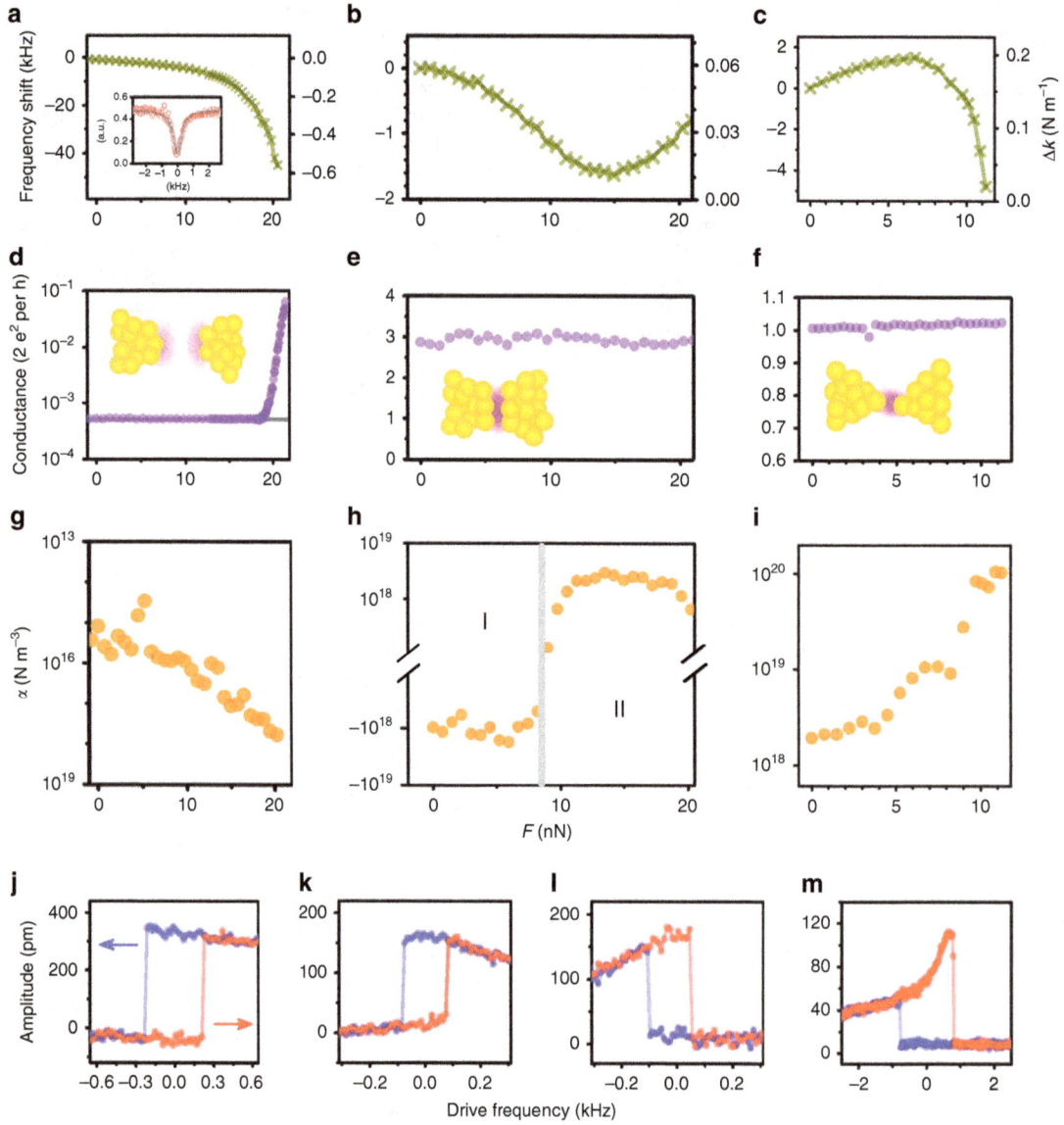

Figure 3 | Tuning the nonlinear response by chemical bonding force. (**a–c**) Frequency shift (left axis) and the corresponding effective spring constant change Δk (right axis) as a function of the external control force F for three different devices (A, B and C) which work in non-contact (device A) and contact regimes (devices B and C). Typical data to obtain the resonant frequency under weak drive with the beam in near linear regime are plotted in inset. (**d–f**) Conductance in unit of quantized conductance ($2e^2/h$) between atom contact measured simultaneously with **a–c** with inset cartoons the corresponding atom structures (schematic). The grey line in **d** indicates the noise level of measurement circuit. (**g–i**) Duffing constant α estimated from **a–c** correspondingly. (**j–m**) Typical hysteresis response under drive frequency sweeping corresponding to device A (**j**) device B ((**k**) for region I and (**l**) for region II) and device C (**m**). Note that the frequency shift, effective string constant change (Δk), control force F and driving frequency are all relative with large constants being subtracted for the ease of displaying.

which case the corresponding bi-states amplitude would be too small and that would make the motion sensing the noise. We drove the system to the bistable state regime and observed the switching behaviour by recording the amplitude of vibrations of the beam, as shown by the green trajectory in Fig. 4a.

To verify that such switchings are indeed from the Brownian thermal noise of the beam, we introduced an amplitude modulation to the driving signal of the form $F_{drive}(t) = F_{drive} + \delta F_{drive} \cos(\Omega t)$, with $\Omega = 0.5\,Hz$ and $\delta F_{drive} = 0.18\,pN$. The amplitudes of the vibration of the beam show periodic switchings (Fig. 4a, purple trajectory) instead of random switchings. Figure 4b shows the corresponding power density, $S_{mod}(\Omega)$ with modulations and $S_{noise}(\Omega)$ without modulations, from which the signal to noise ratio (SNR) is calculated with

$SNR(\Omega) = S_{mod}(\Omega)/S_{noise}(\Omega)$. By using the standard stochastic resonance theory (Methods), we estimated the total force noise as $\sqrt{S^F_{total}} = (3.6 \pm 0.6) \times 10^{-16}\,N\,Hz^{-1/2}$. It agrees nicely with the resonator's Brownian thermal noise, estimated[34] by $\sqrt{S^F_{th}} = 4m\omega_0 k_B T/Q$ as $3.3 \times 10^{-16}\,N\,Hz^{-1/2}$. In the limit of Ω approaching zero, the modulation becomes a perturbative force signal $\delta F_{drive} e\cos(\omega t)$ added to the driving force with the same phase. In the rotating frame of the driving signal, the bi-states dynamics is described by an over-damped double well whose shape is tuned by δF_{drive} (ref. 33), and becomes sensitive to the tuning near the bifurcation point. By applying a weak force $\delta F_{drive} = 2\,fN$ that is only several times larger than the resonator's

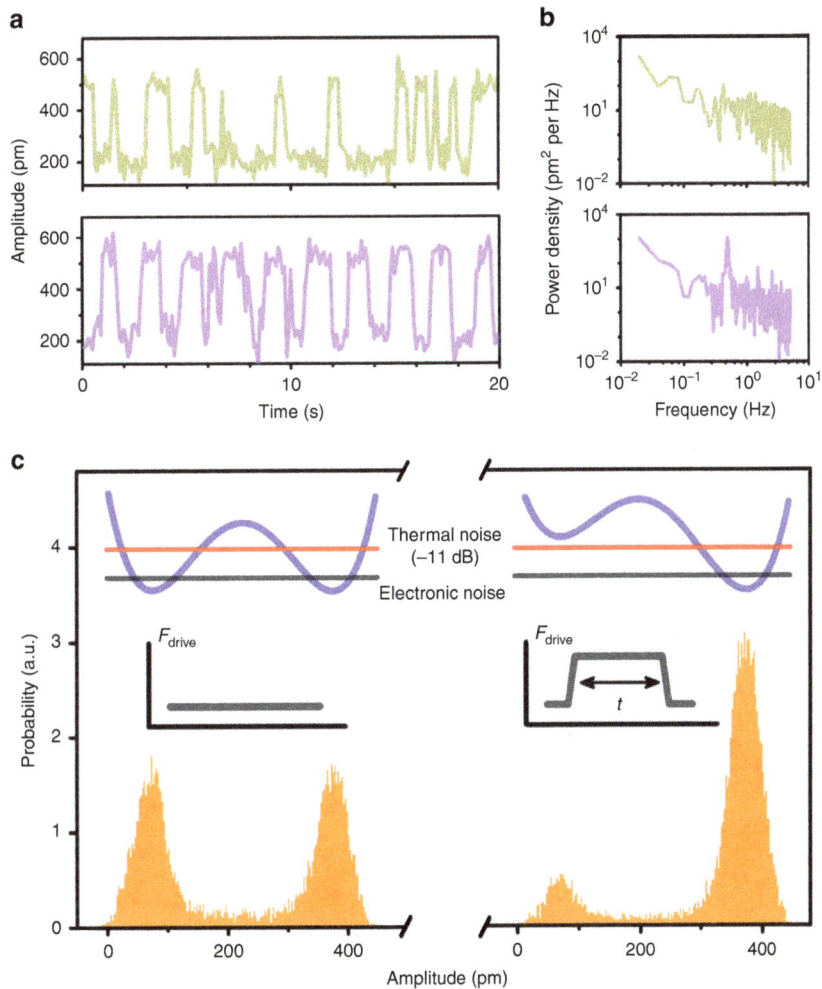

Figure 4 | Brownian thermal noise induced bi-states transitions. (**a**) Switching between bistable states with constant-amplitude drive (upper panel) and modulated-amplitude drive (lower panel). To observe bi-states transitions, the driving frequency is fixed in the middle point of hysteresis response and the drive amplitude is $F_{drive} = 4$ pN, and the modulation frequency and amplitude are $\Omega = 0.5$ Hz and $\delta F_{drive} = 0.18$ pN, respectively. (**b**) Power spectrum density. Compared with the constant-amplitude drive (upper panel), the amplitude-modulated drive (lower panel) induces an additional peak in the power spectrum density at the modulation frequency $\Omega = 0.5$ Hz. (**c**) The amplitude distribution of the bistable resonator depends on the driving amplitude. Very different histograms for constant drive with $F_{drive} = 4.000$ pN (left panel) and $F_{drive} = 4.002$ pN (right panel) are observed. The change in amplitude distribution can be understood using an effective double-well potential (sketched in the inset). The total electronic noise is 11 dB below the Brownian thermal noise in our experiment.

Brownian thermal noise force, significant changes in the statistic distribution are observed, as shown in Fig. 4c. The sensitivity of such a response to the external force is limited by the total force noise $\sqrt{S_{total}^F}$. It is noted that, in our device, in addition to the thermal noise, the electronic noise from the circuit also contributes to the total force noise, but with a weak power density of $S_{elec}^F = S_{total}^F - S_{th}^F$. We estimated that S_{elec}^F is 11 dB below the S_{th}^F level, corresponding to an electronic noise temperature of 500 mK, and is mainly limited by our room-temperature detection circuit (see Supplementary Table 2 for detailed data).

Discussion

In conclusion, we have demonstrated a highly controllable nonlinearity in a macroscopic mechanical system, with its fluctuation dynamics dominated by Brownian thermal noise. The universal existence and the small scale of the chemical bonding force make our method applicable to current widely used micro- and nano-mechanical systems in improving their

nonlinear responses. With further improvements, nonlinearity-induced quantum behaviours in macroscopic mechanics[19–24] are foreseeable in the type of systems described here.

Methods

Theoretical description of the system. Density functional theory calculations were used to estimate qualitatively the chemical bonding force (see Supplementary Note 1 for detailed description). Owing to stability considerations, a small structure of two gold-atom clusters was adopted in the calculation (see Supplementary Fig. 1 for data). By changing only the relative position x of the left cluster relative to the right one while otherwise keeping the relative positions of all the gold atoms fixed, we obtained a chemical bonding energy function $U_{chem}(x)$, from which the nonlinearity is calculated as the fourth-order derivative. Other structures of clusters are also considered (see Supplementary Fig. 2 for data). Owing to the existing systematic error in performing density functional theory calculations, the second-order derivative of the total energy shows slightly an oscillatory behaviour at large distance range. We smoothed the calculated data by fitting it to the function $\frac{A}{x^a} + \frac{B}{x^b}$. In reality, however, the relaxation of gold atoms can occur, which may modify the energy–displacement curve $U_{chem}(x)$. We have considered this effect and find out that the results do not change significantly. To estimate the long-range forces, we considered analytically the van der Waals attraction and the electrostatic force[25] for the geometric model of two gold cylinders close to each other, similar to the case of our nano-bridge structure and in modelling nonlinear dynamics, we have dropped the quadratic nonlinearity in the chemical interaction (see Supplementary Fig. 3 for data). To estimate the intrinsic macroscopic mechanical property, we modelled the macroscopic resonator as a doubly clamped beam. The lowest vibrational mode has a weak intrinsic nonlinearity due to elongation. In calculating the threshold drive force F_c and the corresponding power P_c in Fig. 1e,f, we took the mechanical quality factor 3,000 and the ratio between the beam's thickness and length $t/l \geq 0.005$. We also estimated the electrostatic force-induced nonlinear response based on the same doubly clamped beam (see Supplementary Fig. 4 for data).

Fabrication of the device. The device was fabricated on a commercial silicon-on-insulator wafer using nanolithography (see Supplementary Note 2 for detailed descriptions). The nano-bridge connecting the beam and the stiff electrode is about 200 nm in length and 80 nm in diameter. Focus-ion beam was adopted to narrow it down to <50 nm. Once such a device, that is, the doubly clamped beam with a suspended nano-bridge, was successfully fabricated, it was placed in the ultrahigh vacuum environment of the cryogenic system, and then the bridge was electro-migrated to form an atomic point contact[35]. After the atom contact was produced, the beam was disconnected from the stiff electrode and its lowest vibrational mode was measured. The position of the equilibrium of the beam is controlled by applying a d.c. current through the beam so that the atom interaction can be tuned.

Device characterization and electronic noise. The resonant frequency $(2\pi)^{-1}\omega$ of the device under a control force F is measured directly when the device is working in the linear regime under a weak drive. The spring constant k is calculated as $k = m\omega^2$. Here the effective mass m for the first vibrational mode of the beam is estimated by using finite element simulation, which is widely used for similar systems[36]. The electric conductance is measured by using $G = dI(dV)^{-1}$.

The Brownian thermal noise on the beam is calculated to be 3.3×10^{-16} N Hz$^{-1/2}$. On the basis of this and the calculated mass, the measured resonator frequency and quality factor, we estimate the thermal motion amplitude noise on resonance as 1.0×10^{-13} m Hz$^{-1/2}$ (refs 34,36), which cannot be measured directly due to the sensitivity limitation of our room-temperature measurement electrics. We modelled the mechanical resonator as LCR elements following standard procedure[37] (see Supplementary Note 3 for detailed descriptions), and analysed the noise in the equivalence electric circuit (see Supplementary Fig. 5 for circuit diagram). The separation of the atom contact can be controlled with a precision of about 1 pm without any feedback control for thousands of seconds (see Supplementary Fig. 6 for the data), and once jump-to-contact occurred, a strong enough control force is used to pulled the tip out again and device's main nonlinear character is commonly preserved (see Supplementary Fig. 7 for data). The intrinsic nonlinearity is measured from the frequency response to the driving strength in a standard way based on the Duffing nonlinear model[17] (see Supplementary Fig. 8 for data). To do this, we have pulled the beam far apart from the stiff electrode with contact separation larger than 30 nm, so the interaction due to the atom contact is negligible. A 6-T magnetic field is used in our experiments so that control force can be generated using a small current. This decreases the system total quality factor from its intrinsic value (see Supplementary Fig. 9 for data). We also applied a voltage bias <200 mV between the beam and the stiff electrode to compensate the contact potential of the atom contact so that the contribution of electrostatic force in our experiments is minimized (see Supplementary Fig. 10 for data).

In our experiment, there are mainly three sources of electronic noise. The first one is the Johnson–Nyquist noise $S_R^V = 4Rk_BT$, with T room temperature and R the circuit's resistant, which produces a force noise with power density S_R^F. The second one is the current leakage from the input of the voltage preamplifier S_{ba}^I. We model the preamplifier back-action by a current source similar to that described in ref. 38. In doing this, we assume that there is no correlation between voltage imprecision S_{im}^V and the back-action S_{ba}^I (ref 39), so S_{ba}^I leads to a force noise with power density S_{ba}^F. Another one is from the radio frequency driving signal, which generates a phase noise and works as an equivalent force noise on the beam with power density S_{pha}^F. We estimate the total electronic-induced force noise as $S_{elec}^F = S_R^F + S_{ba}^F + S_{pha}^F$, and in the experiments, $\sqrt{S_{elec}^F} = 0.9 \times 10^{-16}$ N Hz$^{-1/2}$, with power density about 11 dB below the S_{th}^F.

Measurement of the total force noise on the beam. The dynamics of our system is modelled by the standard Duffing oscillator of equation of motion

$$m\ddot{x} + \gamma\dot{x} + m\omega_0^2 x + \alpha x^3 = F_{\text{drive}}(t)\cos(\omega t) + F_{\text{noise}}(t). \qquad (3)$$

where the dissipation rate $\gamma = m\omega_0/Q$, $F_{\text{drive}}(t) = F_{\text{drive}} + \delta F_{\text{drive}}\cos(\Omega t)$ is the amplitude-modulated driving force and $F_{\text{noise}}(t)$ is the total noise with power density S_{total}^F. The minimum force that drives the system into the nonlinear regime where bifurcation occurs is F_c, with the corresponding vibration frequency being $(2\pi)^{-1}\omega_c$, and the amplitude x_c can be calculated from equation (3)[17]. Near the nonlinear bifurcation point, we transform this equation to an over-damped one by following the standard procedure[33]. For the case of modulation frequency Ω being much smaller than the decay rate ω_0/Q of the system, the SNR is related to the

system's noise power as[40]

$$\text{SNR} = \pi\frac{\gamma_k}{\delta\omega}x_m^2\left(\frac{\delta F_{\text{drive}}m\omega}{S_{\text{total}}^F}\right)^2, \qquad (4)$$

with x_m the half of the vibration amplitude of the bistable states, $\delta\omega$ the nonlinearity frequency-induced shift from linear resonance peak and γ_k the measured random switching rate without modulation. So from the measured SNR we obtained S_{total}^F.

References

1. Aldridge, J. S. & Cleland, A. N. Noise-enabled precision measurements of a duffing nanomechanical resonator. *Phys. Rev. Lett.* **94**, 156403 (2005).
2. Badzey, R. L. & Mohanty, P. Coherent signal amplification in bistable nanomechanical oscillators by stochastic resonance. *Nature* **437**, 995–998 (2005).
3. Stambaugh, C. & Chan, H. B. Noise-activated switching in a driven nonlinear micromechanical oscillator. *Phys. Rev. B* **73**, 172302 (2006).
4. Ono, T., Yoshida, Y., Jiang, Y. G. & Esashi, M. Noise-enhanced sensing of light and magnetic force based on a nonlinear silicon microresonator. *Appl. Phys. Express* **1**, 123001 (2008).
5. Venstra, W. J., Westra, H. J. & van der Zant, H. S. Stochastic switching of cantilever motion. *Nat. Commun.* **4**, 2624 (2013).
6. Cross, M. C., Zumdieck, A., Lifshitz, R. & Rogers, J. L. Synchronization by nonlinear frequency pulling. *Phys. Rev. Lett.* **93**, 224101 (2004).
7. Matheny, M. H. *et al.* Phase synchronization of two anharmonic nanomechanical oscillators. *Phys. Rev. Lett.* **112**, 014101 (2014).
8. Matheny, M. H., Villanueva, L. G., Karabalin, R. B., Sader, J. E. & Roukes, M. L. Nonlinear mode-coupling in nanomechanical systems. *Nano Lett.* **13**, 1622–1626 (2013).
9. Almog, R., Zaitsev, S., Shtempluck, O. & Buks, E. Noise squeezing in a nanomechanical duffing resonator. *Phys. Rev. Lett.* **98**, 078103 (2007).
10. Villanueva, L. G. *et al.* Surpassing fundamental limits of oscillators using nonlinear resonators. *Phys. Rev. Lett.* **110**, 177208 (2013).
11. Karabalin, R. B. *et al.* Signal amplification by sensitive control of bifurcation topology. *Phys. Rev. Lett.* **106**, 094102 (2011).
12. Suh, J., LaHaye, M. D., Echternach, P. M., Schwab, K. C. & Roukes, M. L. Parametric amplification and back-action noise squeezing by a qubit-coupled nanoresonator. *Nano Lett.* **10**, 3990–3994 (2010).
13. Mahboob, I. & Yamaguchi, H. Bit storage and bit flip operations in an electromechanical oscillator. *Nat. Nanotechnol.* **3**, 275–279 (2008).
14. Venstra, W. J., Westra, H. J. R. & van der Zant, H. S. J. Mechanical stiffening, bistability, and bit operations in a microcantilever. *Appl. Phys. Lett.* **97**, 193107 (2010).
15. Bagheri, M., Poot, M., Li, M., Pernice, W. P. H. & Tang, H. X. Dynamic manipulation of nanomechanical resonators in the high-amplitude regime and non-volatile mechanical memory operation. *Nat. Nanotechnol.* **6**, 726–732 (2011).
16. Landau, L. D. & Lifshitz, E. M. *Theory of Elasticity* (Pergamon, 1986).
17. Lifshitz, R. & Cross, M. C. *Review of Nonlinear Dynamics and Complexity* (Wiley-VCH, 2009).
18. Yurke, B. & Stoler, D. Generating quantum-mechanical superpositions of macroscopically distinguishable states via amplitude dispersion. *Phys. Rev. Lett.* **57**, 13 (1986).
19. Carr, S. M., Lawrence, W. E. & Wybourne, M. N. Accessibility of quantum effects in mesomechanical systems. *Phys. Rev. B* **64**, 220101 (2001).
20. Peano, V. & Thorwart, M. Macroscopic quantum effects in a strongly driven nanomechanical resonator. *Phys. Rev. B* **70**, 235401 (2004).
21. Savel'ev, S., Hu, X. D. & Nori, F. Quantum electromechanics: qubits from buckling nanobars. *New J. Phys.* **8**, 105 (2006).
22. Katz, I., Retzker, A., Straub, R. & Lifshitz, R. Signatures for a classical to quantum transition of a driven nonlinear nanomechanical resonator. *Phys. Rev. Lett.* **99**, 040404 (2007).
23. Serban, I. & Wilhelm, F. K. Dynamical tunneling in macroscopic systems. *Phys. Rev. Lett.* **99**, 137001 (2007).
24. Sillanpaa, M. A., Khan, R., Heikkila, T. T. & Hakonen, P. J. Macroscopic quantum tunneling in nanoelectromechanical systems. *Phys. Rev. B* **84**, 195433 (2011).
25. Giessibl, F. J. Advances in atomic force microscopy. *Rev. Mod. Phys.* **75**, 949–983 (2003).
26. Ohnishi, H., Kondo, Y. & Takayanagi, K. Quantized conductance through individual rows of suspended gold atoms. *Nature* **395**, 780–783 (1998).
27. Eichler, A. *et al.* Nonlinear damping in mechanical resonators made from carbon nanotubes and graphene. *Nat. Nanotechnol.* **6**, 339–342 (2011).
28. Chan, H. B., Aksyuk, V. A., Kleiman, R. N., Bishop, D. J. & Capasso, Federico Nonlinear micromechanical Casimir oscillator. *Phys. Rev. Lett.* **87**, 211801 (2001).

29. Lee, S. I., Howell, S. W., Raman, A. & Reifenberger, R. Nonlinear dynamics of microcantilevers in tapping mode atomic force microscopy: a comparison between theory and experiment. *Phys. Rev. B.* **66**, 115409 (2002).

30. Kozinsky, I., Postma, H. W., Ch., Bargatin, I. & Roukes, M. L. Tuning nonlinearity, dynamic range, and frequency of nanomechanical resonators. *Appl. Phys. Lett.* **88**, 253101 (2006).

31. Sankey, J. C., Yang, C., Zwickl, B. M., Jayich, A. M. & Harris, J. G. E. Strong and tunable nonlinear optomechanical coupling in a low-loss system. *Nat. Phys.* **6**, 707–712 (2010).

32. Gieseler, J., Novotny, L. & Quidant, R. Thermal nonlinearities in a nanomechanical oscillator. *Nat. Phys.* **9**, 806–810 (2013).

33. Dykman, M. & Krivoglaz, M. Fluctuations in nonlinear systems near bifurcations corresponding to the appearance of new stable states. *Physica A* **104**, 480–494 (1980).

34. Huang, P. *et al.* Demonstration of motion transduction based on parametrically coupled mechanical resonators. *Phys. Rev. Lett.* **110**, 227202 (2013).

35. Park, H., Lim, A. K. L., Alivisatos, A. P., Park, J. & McEuen, P. L. Fabrication of metallic electrodes with nanometer separation by electromigration. *Appl. Phys. Lett.* **75**, 301–303 (1999).

36. Knobel, G. & Cleland, A. Nanometre-scale displacement sensing using a single electron transistor. *Nature* **424**, 291–293 (2003).

37. Cleland, A. N. & Roukes, M. L. External control of dissipation in a nanometer-scale radiofrequency mechanical resonator. *Sens. Actuators A* **72**, 256–261 (1999).

38. Devoret, M. H. & Schoelkopf, R. J. Amplifying quantum signals with the single-electron transistor. *Nature* **406**, 1039–1046 (2000).

39. Clerk, A. A., Devoret, M. H., Girvin, S. M., Marquardt, F. & Schoelkopf, R. J. Introduction to quantum noise, measurement, and amplification. *Rev. Mod. Phys.* **82**, 1155–1208 (2010).

40. Gammaitoni, L., Hänggi, P., Jung, P. & Marchesoni, F. Stochastic resonance. *Rev. Mod. Phys.* **70**, 223 (1998).

Acknowledgements

We thank L. Jiang for many stimulating discussions and comments and for his help in improving the manuscript. We thank Y.X. Liu, Z.Q. Yin, L. Tian C.P. Sun, and M. Lukin for helpful discussions. We thank Yiqun Wang from Suzhou Institute of Nano-Tech and Nano-Bionics for nano-fabricating supports and Guizhou Provincial Key laboratory of Computational Nano-Material Science for calculation supports. This work was supported by the 973 Program (Grant No. 2013CB921800), the NNSFC (Grant Nos. 11227901, 91021005, 11104262, 31470835, 21233007, 21303175, 21322305, 11374305 and 11274299), and the 'Strategic Priority Research Program (B)' of the CAS (Grant Nos. XDB01030400 and 01020000).

Author contributions

J.D. supervised the experiments; J.D. and P.H. proposed the idea and designed the experimental proposal; P.H., J.Z., L.Z., W.D. and F.X. prepared the experimental set-up; P.H., J.Z., L.Z., S.L. and C.J. performed the experiments; J.Z. fabricated the device; D.H. and X.Z. carried out the theoretical calculation; M.C. carried out the finite element simulation; P.H., C.D., F.X. and J.D. wrote the paper; all authors analysed the data, discussed the results and commented on the manuscript.

Additional information

23

Functionalization mediates heat transport in graphene nanoflakes

Haoxue Han[1,*], Yong Zhang[2,3,*], Nan Wang[3,*], Majid Kabiri Samani[3], Yuxiang Ni[4], Zainelabideen Y. Mijbil[5,6], Michael Edwards[3], Shiyun Xiong[7], Kimmo Sääskilahti[8], Murali Murugesan[3], Yifeng Fu[3,9], Lilei Ye[9], Hatef Sadeghi[5], Steven Bailey[5], Yuriy A. Kosevich[1,10], Colin J. Lambert[5], Johan Liu[2,3] & Sebastian Volz[1]

The high thermal conductivity of graphene and few-layer graphene undergoes severe degradations through contact with the substrate. Here we show experimentally that the thermal management of a micro heater is substantially improved by introducing alternative heat-escaping channels into a graphene-based film bonded to functionalized graphene oxide through amino-silane molecules. Using a resistance temperature probe for *in situ* monitoring we demonstrate that the hotspot temperature was lowered by $\sim 28\,°C$ for a chip operating at $1{,}300\,W\,cm^{-2}$. Thermal resistance probed by pulsed photothermal reflectance measurements demonstrated an improved thermal coupling due to functionalization on the graphene–graphene oxide interface. Three functionalization molecules manifest distinct interfacial thermal transport behaviour, corroborating our atomistic calculations in unveiling the role of molecular chain length and functional groups. Molecular dynamics simulations reveal that the functionalization constrains the cross-plane phonon scattering, which in turn enhances in-plane heat conduction of the bonded graphene film by recovering the long flexural phonon lifetime.

[1] Laboratoire EM2C, CNRS, CentraleSupélec, Université Paris-Saclay, Grande Voie des Vignes, 92295 Châtenay-Malabry, France. [2] SMIT Center, School of Automation and Mechanical Engineering and Institute of NanomicroEnergy, Shanghai University, 20 Chengzhong Road, Shanghai 201800, China. [3] Electronics Materials and Systems Laboratory, Department of Microtechnology and Nanoscience, Chalmers University of Technology, Kemivägen 9, SE-412 96 Gothenburg, Sweden. [4] Department of Mechanical Engineering, University of Minnesota, 111 Church Street SE, Minneapolis, Minnesota 55455, USA. [5] Quantum Technology Center, Physics Department, Lancaster University, Lancaster LA1 4YB, UK. [6] Science Department, Veterinary Medicine College, Al-Qasim Green University, Babylon, Iraq. [7] Max Planck Institute for Polymer Research, Ackermannweg 10, D-55128 Mainz, Germany. [8] Department of Biomedical Engineering and Computational Science, Aalto University, FI-00076 Aalto, Finland. [9] SHT Smart High Tech AB, Ascherbergsgatan 46, SE-411 33 Gothenburg, Sweden. [10] Department of Polymers and Composite Materials, Semenov Institute of Chemical Physics, Russian Academy of Sciences, Kosygin Street 4, 119991 Moscow, Russia. * These authors contributed equally to this work. Correspondence and requests for materials should be addressed to C.J.L. (email: c.lambert@lancaster.ac.uk) or to J.L. (email: johan.liu@chalmers.se) or to S.V. (email: sebastian.volz@ecp.fr).

nisotropic properties of two-dimensional (2D) layered materials make them promising in the application of next-generation electronic devices, among which graphene and few-layer graphene (FLG) have been most intensively studied for thermal management, due to their extraordinarily high in-plane thermal conductivity $(\kappa)^{1-5}$. For instance, Yan et al.[6] reported that the maximum hotspot temperature can be lowered by $\sim 20\,°C$ in transistors operating at $\sim 13\,W\,mm^{-1}$ using FLG as a heat spreader for a gallium nitride (GaN) transistor. Gao et al.[7] reported that the maximum hotspot temperature decreased from 121 to $108\,°C$ $(\Delta T = 13\,°C)$ for a heat flux of $430\,W\,cm^{-2}$ after the introduction of a single-layer graphene heat spreader. Moreover, the simulations of graphene heat spreaders were also reported for silicon-on-insulator integrated circuits[8] and three-dimensional (3D) integrated circuits[9]. The thermal conductivity of a graphene laminate film supported on substrate was also investigated and found to remain rather large[10]. However, in most practical applications, graphene/FLG will be supported by and integrated with insulators, both in electronic circuitry and heat-spreader applications[11]. Therefore, thermal energy flow will be limited both by the in-plane thermal conductivity (κ) of the supported graphene/FLG and by the thermal boundary resistance (R) at the graphene/FLG–substrate interface[12].

Owing to their exceedingly large surface-to-volume ratio, the properties of 2D layered materials are very sensitive to the interactions with external bodies. Indeed, when supported on an amorphous substrate, κ of suspended graphene decreased by almost one order of magnitude, from $\sim 4{,}000$ (ref. 13) to $\sim 600\,W\,m^{-1}\,K^{-1}$ (ref. 14). Such a striking discrepancy in κ significantly limits the thermal performance of graphene/FLG in real applications. It is reported that the different behaviours are due to the strong scattering of the important heat-carrying flexural acoustic (ZA) modes[15] to the substrate[16]. More specifically, it was identified that the phonon relaxation times of graphene ZA modes are suppressed when supported on a SiO_2 substrate. These studies have improved our fundamental understanding in the physics behind the problem, and it was suggested that making rational choice of the substrate material[14,17] and modulating its coupling to graphene[12] may be useful to improve κ of the supported graphene/FLG.

The thermal boundary resistance (R) of a graphene/FLG–substrate interface is another limiting factor to their thermal performance in devices. Covalent functionalization has been proved to efficiently promote heat transfer between interfaces by introducing additional thermal pathways through the functionalizing molecules[18-33]. For example, self-assembled monolayers (SAMs) were used to functionalize metallic surfaces to enhance heat transport across metal–water[20,27], metal–gas[28], metal–semiconductor[21] and metal–polymer[31] interfaces. Functionalization was used in graphene and carbon nanotube nanocomposites to mitigate the high thermal boundary resistance between the graphene/carbon nanotube fillers and the polymer matrices[18,25,26,30]. Functionalized molecules also assist to align and densely pack multilayer graphene sheets and reduce the interlayer thermal resistance of graphene[25]. Recently, it was shown that plasma-functionalized graphene raised the cross-plane thermal conductance between aluminium and its substrate by a factor of two[19]. Nevertheless, the functionalization-introduced point defects will further decrease κ of the supported graphene/FLG, as they introduce phonon-scattering centres[25,32,33]. To correct this drawback, a robust solution that maintains the high thermal conductivity of graphene/FLG when supported, while effectively reducing the interface thermal resistance is needed.

Here we demonstrate that thermal management of a micro heater is considerably improved via introducing alternative heat-spreading channels implemented with graphene-based film (GBF) bonded to functionalized graphene oxide (FGO) through amino-silane molecules. We probed interface thermal resistance by photothermal reflectance measurements to demonstrate an improved thermal coupling due to functionalization on the graphene–graphene oxide interface and the graphene oxide–silica interface. Molecular dynamics simulations and ab initio calculations reveal that the functionalization constrains the cross-plane scattering of low-frequency phonons, which in turn enhances in-plane heat conduction of the bonded graphene film by recovering the long flexural phonon lifetime. Our results provide evidence that a graphene film deposited on a FGO substrate provides a very attractive platform for thermal management applications.

Results

Graphene-based film and graphene oxide and device. A GBF bonded to the FGO substrate through silane molecules is shown in Fig. 1a,b. To synthesize GBF and FGO experimentally (Fig. 1c), we first prepared a graphene oxide (GO) dispersion (see Experimental method). The FGO was obtained by functionalizing GO with a silane-based chemistry suitable for reactive oxide-forming surfaces including the basal plane of GO and SiO_2. 3-Amino-propyltriethoxysilane (APTES) has three –Si–O– groups and one $–NH_2$ end, as shown in Fig. 1a. Owing to the simple chemistry and unique multifunctional nature of APTES, it can easily bind two different substrates. In our case, the –Si–O end of APTES binds to the GO substrate. $–Si–OC_2H_5$ groups of APTES hydrolyse in water and form crosslink bonds with each other. The crosslinked Si–O structure acts as a strong bonding layer between the substrate and GBFs. On the other hand, the $–NH_2$ end of APTES binds onto carboxyl groups on the functionalized graphene film. The FGO layer has a thickness of $\sim 5\,nm$. The graphene film was fabricated from chemically reduced GO and can recover relatively high in-plane thermal conductivity after thermal annealing[34]. The graphene film was then spin-coated[35] with the FGO and the resulting bundle was transferred to a thermal evaluation device[7], resulting in the formation of molecular bridges between the graphene surface and the device's SiO_2 substrate. The thermal evaluation device was integrated with micro Pt-based heating resistors as the hotspot and temperature sensors[7], as shown in Fig. 1d, acting as a simulation platform of an electronic component to demonstrate the heat-spreading capability of the supported graphene film.

In situ **temperature measurement with resistance thermometry.** A direct current I was input into the circuit by applying an outer voltage V in Fig. 1e, and hence the generated power is calculated as $P = V \times I$. Since the lateral dimension of the hotspot $(A = 400 \times 400\,\mu m^2)$ is much larger than its thickness (260 nm), most of the heat is dissipated through the lateral direction of the hotspot. Hence the heat flux is defined as $Q = P/A$, and the direction is parallel to the substrate. The calibration relationship between the resistance $R\,(\Omega)$ and the temperature $T\,(°C)$ of the thermal evaluation chip is $R(T) = 0.21T + 112$. The temperature measurement uncertainty is $\epsilon = \pm 0.5\,°C$. Figure 1e shows the temperature measured *in situ* at the hotspot and compares the thermal performance of the graphene film with and without the functionalization. With a constant heat flux of $1{,}300 \pm 2.3\,W\,cm^{-2}$ at the Pt chip, the hotspot temperature decreased from 135 ± 1.2 to $118 \pm 1.1\,°C$ $(\Delta T = 17 \pm 2.3\,°C)$ with a GBF deposited on non-functionalized GO compared with the case of a bare chip. Such a remarkable temperature decrease is far beyond the measurement uncertainty $\Delta T \gg \epsilon$ (see Supplementary Fig. 1 and Supplementary Note 1 for an uncertainty analysis).

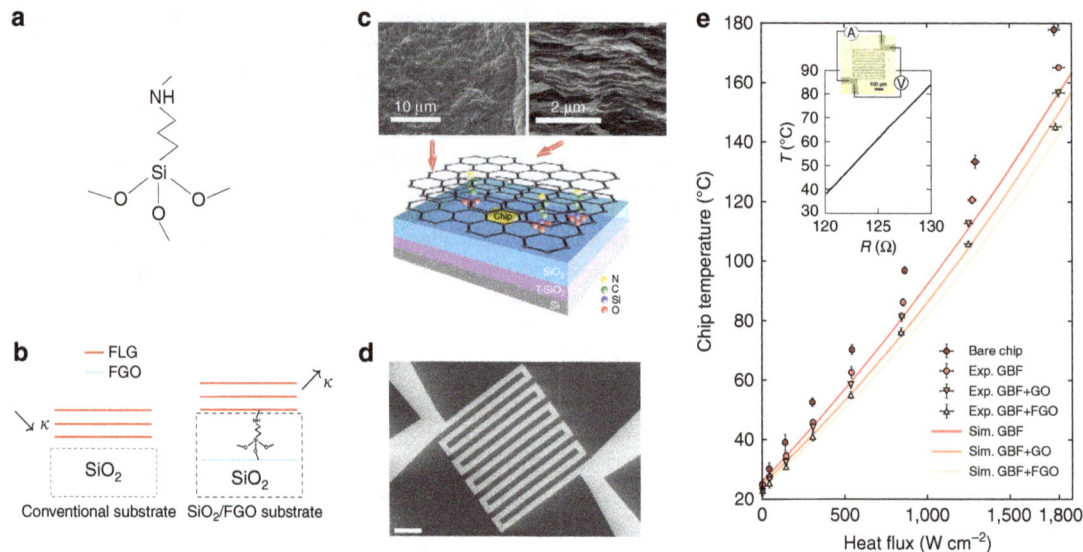

Figure 1 | Graphene-based film on FGO as heat spreader for hotspot. (**a**) Sketch of the chemical bonds of the silane molecule. (**b**) Schematic of a graphene film on different supports. Left: conventional silica substrate. Right: the proposed silica/FGO substrate. (**c**) Schematic of the measurement set-up. The chip is embedded in the SiO₂ substrate. T-SiO₂ stands for thermally grown SiO₂. Scanning electron microscopy (SEM) image of the in-plane and the cross-section of the GBF. (**d**) SEM image of the chip as the hotspot. Scale bar, 100 µm. (**e**) Measured (filled markers) and finite-element-simulated (lines) chip temperatures versus the in-plane heat fluxes dissipated in a bare hotspot (rectangles), a hotspot covered by a GBF (circles), a chip covered by a GBF with non-functionalized GO (up triangle) and a chip covered by a GBF with APTES FGO (down triangle). 'Exp.' and 'Sim.' stand for experiments and simulations, respectively. Inset: calibration relationship between the resistance R (Ω) and the temperature T (°C) of the thermal evaluation chip. Scale bar on the chip, 100 µm. The error bars correspond to the s.d.'s from the measurements on five samples.

Furthermore, with the same heat flux input, the hotspot temperature further decreased from 118 ± 1.1 to 107 ± 0.8 °C ($\Delta T = 11 \pm 1.9$ °C) thanks to the presence of the APTES functionalization. The heat-spreading performance is thus enhanced by ~57% via the functionalization. We have implemented a finite-element model (Supplementary Figs 2–4, Supplementary Table 1 and Supplementary Note 2) of the heat-spreading device by taking the results of atomistic simulation as input parameters. As shown in Fig. 1e, the heat-spreading performance of the equivalent macroscopic finite-element model agrees reasonably well with the one measured by experiments.

Thermal resistance measurement with photothermal reflectance. To further confirm the enhanced heat spreading assisted by molecular functionalization, we measured the interface thermal resistance by using the pulsed photothermal reflectance (PPR) method[36,37]. To enhance heat absorption, a gold layer was evaporated on the surface of the GO and FGO layers after drop coating. The sample was first excited by a Nd:YAG laser pulse. This caused a fast rise in the surface temperature followed by a relaxation. The change of surface temperature was monitored by a probe laser, which reflects off from the samples' surface. Since the relaxation time is governed by the thermal properties of the underlying layers and interfacial thermal resistance between the layers, by obtaining the temperature profile one can extract the thermal properties of the layers and interface thermal resistance between the layers through a heat conduction model. Four sets of samples were fabricated, as shown in Fig. 2 and the thermal resistance R_1 between the Au-Cr film and the (functionalized) GO layer, and the resistance R_2 between (functionalized) GO layer and GBF or SiO₂ were measured. The experimental set-up and the procedure of thermal resistance extraction by fitting the photothermal response to the model (Supplementary Figs 5 and 6 and Supplementary Note 3). The normalized surface temperatures of the four sample sets are

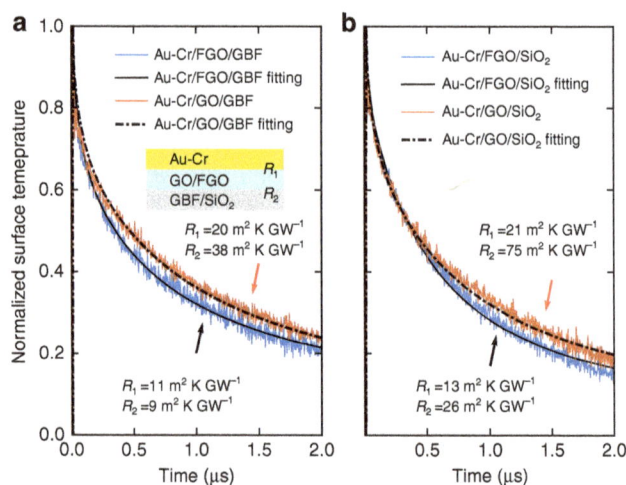

Figure 2 | Normalized surface temperature of PPR measurements of the thermal interface resistance. (**a**) Au-Cr/FGO/GBF and Au-Cr/GO/GBF samples and (**b**) Au-Cr/FGO/SiO₂ and Au-Cr/GO/SiO₂ samples. Inset: sample geometry for the PPR measurement. R_1 and R_2 refer to the thermal interface resistance between the Au-Cr film and the (APTES functionalized) GO layer, and that between (functionalized) GO layer and GBF or SiO₂. The thermal resistances are also reported in Table 1.

shown in Fig. 2 and the extracted thermal resistances are reported in Table 1. A fourfold reduction was achieved in the thermal resistance between GO and GBF from 3.8×10^{-8} to $0.9 \times 10^{-8} \mathrm{m}^2 \mathrm{K} \mathrm{W}^{-1}$. On the GO–SiO₂ interface, the functionalization remarkably reduced the thermal resistance by a factor of almost three, from 7.5×10^{-8} to $2.6 \times 10^{-8} \mathrm{m}^2 \mathrm{K} \mathrm{W}^{-1}$. We also observed a better thermal coupling on the metal–dielectric interface between Au-Cr and GO due to the surface chemistry.

Table 1 | Au/(F-)GO and (F-)GO/X interface thermal resistances.

Interface	R_1 ($m^2 K W^{-1}$)	R_2 ($m^2 K W^{-1}$)
Au-Cr/GO/GBF	2.0E-8	3.8E-8
Au-Cr/FGO/GBF	1.1E-8	0.9E-8
Au-Cr/GO/SiO$_2$	2.1E-8	7.5E-8
Au-Cr/FGO/SiO$_2$	1.3E-8	2.6E-8

X = GBF or SiO$_2$. The fitting uncertainty is 20% (Supplementary Fig. 7).

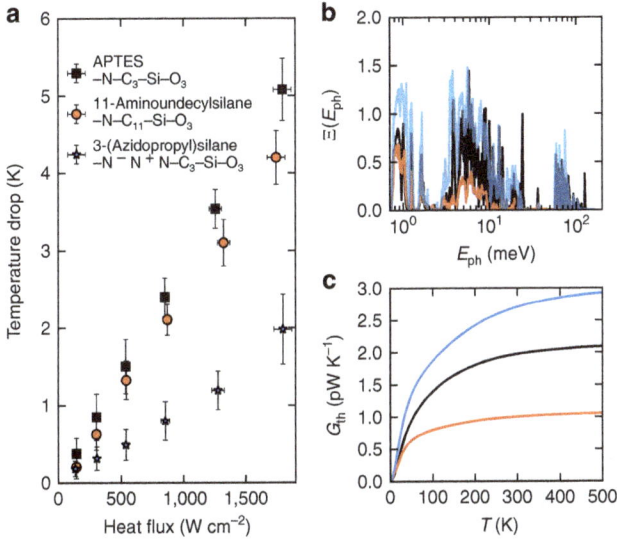

Figure 3 | Heat-spreading performance of GBF on FGO with different functional agents. The molecules are APTES, 11-Aminoundecyltriethoxysilane and 3-(Azidopropyl)triethoxysilane. (**a**) Measured temperature drop of heat spreaders with different functionalization molecules compared with that with non-functionalized GO. The error bars correspond to the s.d.'s of measurements on five samples for each molecule type. (**b**) Phonon transmission function $\Xi(E_{ph})$ for three different types of molecules used in the experiment. (**c**) Phonon thermal conductance through the molecules as a function of temperature. The colour code of the data curves in **b** and **c** is the same as in **a**.

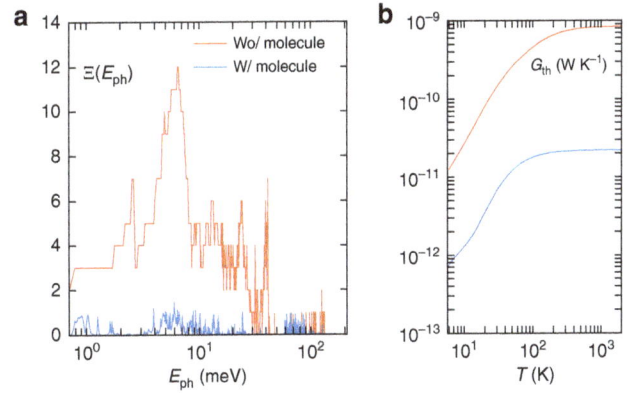

Figure 4 | Phonon transport through the APTES molecule. (**a**) Phonon transmission $\Xi(E_{ph})$ versus phonon energy $E_{ph} = \hbar\omega$ (red curve) between two adjacent graphene layers and (blue curve) through the APTES molecule bonding the two graphene layers. 'w/' and 'wo/' stand for with and without, respectively. (**b**) Thermal conductances G_{th} contributed by phonons versus temperature for the two cases.

Heat spreading through different functional molecules. To gain a deeper insight into the impact of molecular structure on the thermal transport along the molecules, we used APTES, 11-Aminoundecyltriethoxysilane and 3-(Azidopropyl)-triethoxysilane as different functional agent on the GO to evaluate their heat-spreading performance on the same thermal test platform. The same concentration of 0.1492 mol kg^{-1} was used for all three types of molecules. Figure 3 shows the temperature reduction of the hotspot covered by GBFs with FGO using three different molecules. The results show that the heat spreader of GO functionalized with APTES has the best cooling performance. To properly understand this difference, an exploration of the internal vibrational properties of the molecule is crucial[38]. We hence investigated how the differences in the phonon transmission impact the interfacial thermal transport. With this aim, we probed the phonon transmission $\Xi(E_{ph})$ by atomistic Green's function to characterize the local heat conduction with and without the presence of the molecule (Supplementary Note 4). $\Xi(E_{ph})$ enables a precise measurement of the atomic-scale molecule-graphene heat transport that the conventional models fail to provide. The phonon transmission functions through different molecules and the associated thermal

conductances versus temperature are shown in Fig. 3b,c. The phonon transmission at low phonon energies across the 11-Aminoundecyltriethoxysilane molecule is comparable to that across the APTES molecule, whereas at high phonon energies (>4 meV), the phonon transmission is considerably suppressed. Such a distinct phonon transport behaviour is determined by the molecule chain length. By comparing the chemical structures, we can see that 11-Aminoundecyltriethoxysilane ($-N-C_{11}-Si-O_3$) has the same functional groups as APTES ($-N-C_3-Si-O_3$) but has a longer carbon backbone. Such a long chain length has a rather weak impact on the low-frequency phonons due to their very long wavelength but can strongly suppress the phonon transport at high frequencies. On the other hand, when comparing the phonon transmission through the junction of APTES and 3-(Azidopropyl)triethoxysilane ($-N^-N^+N-C_3-Si-O_3$), it is evident to identify a stronger transmission at all frequencies. By comparing the chemical structures in Fig. 3a, we can see that 3-(Azidopropyl)triethoxysilane has the same carbon backbone as APTES but has an azido group instead of an amino group. The azido group has strong interaction with the sp^2-bonded carbon lattice of graphene to form three-membered heterocycle structures. Given the similar backbone structure, the phonon eigenmodes in the molecule have not been significantly altered, but the transmission is enhanced due to the stronger thermal coupling to the graphene reservoirs. However, the introduction of nitrogen atoms into the sp^2 carbon structure can strongly scatter phonons by distorting the structure of the graphene sheet, which leads to a poor thermal performance of the heat spreader.

Phonon transport analysis in APTES. We now investigate the detailed vibrational and electronic transport properties of the APTES molecule since it presents the best performance in heat spreading. We probed the phonon transmission $\Xi(E_{ph})$ to characterize the local heat conduction with and without the presence of the APTES molecule. As shown in Fig. 4, the transmission function $\Xi(E_{ph})$ for the two adjacent graphene layers without any molecule displays a clear stepwise structure that provides the number of phonon channels. Low-energy phonons ($E_{ph} \leq 10$ meV) dominate heat conduction since the adjacent graphene flakes interact only through weak van der Waals (vdW) forces that inhibit the transmission of high-frequency phonons[39]. When the graphene layers are bridged by a amino-silane molecule, the high-frequency phonons act as the major contributors in the heat conductance G_{ph}, creating more phonon channels through the covalent bond vibrations. This is in line with the transmission calculation of Segal et al.[40] who observed a contribution to heat conduction by the higher-frequency phonons within the molecule coupled to the low-frequency phonons responsible for heat transport in the thermal reservoirs. The oscillations in the transmission spectrum may originate from phonon interferences within the alkane chain[41–43]. Fabry–Pérot-like interference effect occurs in the frequency region of $E_{ph} = 20$–100 meV, as was previously observed in an alkane SAM interface[41]. Such Fabry–Pérot-like interferences originate from the multiple reflected phonons interfering constructively within the alkane chain, as the local maxima in the transmission (Fig. 4) through the molecule can attain the same intensity of that through pristine graphene films at given frequencies. Although destructive quantum interference was believed unlikely to occur in a linear alkane chain[42], we observe strong destructive interference patterns in the high-frequency range ($\omega/2\pi = 40 - 60$ THz), which may correspond to two-path destructive phonon interferences[43,44]. We also investigated the interlayer electron transport in the graphene and the effect of the silane intercalation in such a hybrid nanostructure through a ab initio calculation combined with Green's function (Supplementary Fig. 8 and Supplementary Note 5). The main heat carrier in this system is phonon, as the thermal conductance due to the electron contributes to the total thermal conductance by ~4% at room temperature with functionalization and by ~2% without molecules.

Thermal conductance and conductivity calculations. To explore the effect of functional APTES molecules on the in-plane thermal conductance of the graphene film, we first perform molecular dynamics simulations to study a nanoscale molecular junction between two stacks of multilayer graphene nanoflakes. For a weak oxidation, the thermal resistance of the graphene film and its oxidized substrate on one hand, and on the other hand its thermal conductivity are very close to those with a non-oxidized graphene support. The dependence of thermal resistance and conductivity of graphene film to the oxidation rate of the GO substrate is reported in Supplementary Fig. 9 and Supplementary Note 6. We consider defect-free and isotopically pure graphene flakes of 10×10 nm^2 thermalized at 300 K (Methods, Supplementary Fig. 10 and Supplementary Note 7). The FGO substrate has a 5-nm thickness as in the experiments. Conventional silica substrate results in a substantial decrease of the basal-plane thermal conductivity of graphene due to the non-conformality of the substrate–graphene interface[14]. To compare, the FGO substrate proposed herein minimize the perturbation of substrate on the morphology of graphene (Fig. 1c), thus maintaining its high thermal conductivity. Periodic boundary conditions were used in the in-plane directions so that the molecular dynamics system corresponds to two thin films connected through silane molecules with the number density ρ, defined as the molecule number per graphene unit area.

To illustrate the intriguing role of the functionalizing molecule, the in-plane thermal conductivity κ of the film and its interfacial thermal resistance R (Supplementary Fig. 10 and Supplementary Note 7) with the FGO substrate is plotted as a function of the graphene layer number l_G in the film in Fig. 5a,b, respectively. First, a supercell containing a single molecule is studied, which corresponds to $\rho = 0.081$ nm^{-2}. For $l_G \geq 2$, the presence of the APTES molecule results in an unexpected increase both in the graphene film thermal conductivity κ and in R. An overall decaying trend of the in-plane thermal conductivity κ of the

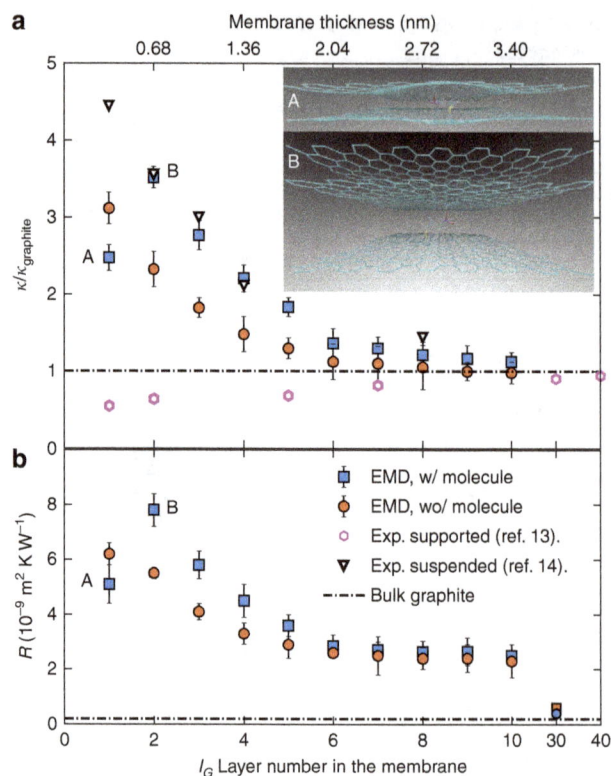

Figure 5 | Thermal resistance and in-plane thermal conductivity of the GBF versus the graphene layer number. (a) Molecular dynamics simulation results of in-plane thermal conductivity κ of the graphene film and (b) interfacial thermal resistance R between the FGO substrate and the graphene film versus the graphene layer number l_G in the film. The molecule density is $\rho = 0.081$ nm^{-2}. Cases with (red circles) and without (blue squares) the APTES molecule are compared. 'w/' and 'wo/' stand for with and without, respectively. The values of thermal conductivity are normalized to that of the single-layer graphene. Inset A illustrates a saddle-like surface generated by the APTES molecule for $l_G = 1$. Inset B shows that the saddle-like curvature disappears for $l_G \geq 1$. The measurements of suspended and supported graphene are from refs 13,14, respectively. EMD stand for equilibrium molecular dynamics. 'Exp.' and 'Sim.' stand for experiments and simulations, respectively. The simulated (respectively measured) thermal conductivity is normalized with respect to that of the simulated (respectively measured) graphite to allow a reasonable comparison. The error bar corresponds to the s.d. of the ensemble average on 50 independent trajectories.

graphene film and its resistance R with the substrate versus the layer number l_G is observed until approaching the value of bulk graphite. This is due to the increased cross-plane coupling of the low-energy phonons. A similar decay was found both in experimental measurements of κ (ref. 13) of suspended graphene and in simulation-based estimation of R (ref. 45). Unlike silica-supported graphene[14], the FLG on FGO support recovers the high thermal conductivity and follows the same decaying trend versus l_G as the suspended graphene[13]. Therefore, the proposed FGO substrate better conserves the high thermal conductivity of graphene compared with the silica substrate. Interestingly, for $l_G = 1$, the presence of the molecule reduces κ, which goes against the case where no molecule interconnects the graphene film and the substrate. This breakdown of the thermal conductivity enhancement is due to a saddle-like surface generated around the molecule's chemical bonds of amino and silano groups connecting the graphene, with the bond centre as the saddle point, as shown in the inset A of Fig. 5a. The saddle-like surface strongly scatters all phonon modes, thus decreasing κ of the graphene film. Such a curved surface was found quite common in defective graphene[46,47], resulting from the Jahn–Teller effect to lower the energy by geometrical distortion[48]. The thermal conductance of a single silane molecule is determined to be $82\,\mathrm{pW\,K^{-1}}$ through the molecular footprint (Supplementary Fig. 11 and Supplementary Note 8). Our simulated result is comparable to recent measurements of the thermal conductance of alkane thiols SAM at a silica–gold interface[38].

It is clearly seen from Fig. 5a that the critical layer number for the in-plane thermal conductivity switch is $l_c = 2$. This critical number depends on the level of conformation distortion of the functionalized graphene sheet. Several factors can play a role (Supplementary Figs 12–14 and Supplementary Note 9): (i) molecule density: a diluted molecule distribution creates segregated ripples on the functionalized graphene sheet and even on the sheets further above thus interrupting the phonon mean free path. In this case more graphene sheets are required to recover the flat conformation, that is, $l_c > 2$. For a high molecule density, the ripples tend to merge with themselves, recovering the flat surface in an extended area. Such merging hence recovers the interrupted phonon MFP and alleviates the detrimental influence of the separate ripples on the basal-plane thermal conductivity of graphene. In this case the two layers of graphene with functionalization have higher basal-plane thermal conductivity compared with that without functionalization, that is, $l_c = 2$. (ii) Functional group of the molecule: when the functional agent forms a stronger bond with its substrate than that of the amino-silane molecule, more severe static graphene distortion occurs thus increasing the critical number l_c. (iii) Substrate of the graphene film: a highly mismatched substrate yields remarkable surface distortion of the graphene sheets in such a way that two layers of graphene are not sufficient to recover the high basal-plane thermal conductivity.

Functionalization effects on phonon lifetime of graphene. We investigate the microscopic origin of the thermal conductivity κ enhancement in the graphene film by probing the mode-wise phonon relaxation time (Supplementary Fig. 15 and Supplementary Note 10). The phonon relaxation time τ measures the temporal response of a perturbed phonon mode to relax back to equilibrium due to the net effect of different phonon-scattering mechanisms. τ can be defined as[49] $\partial n/\partial \tau = (n - n_0)/\tau$ where n and n_0 are the phonon occupation numbers out of and at thermal equilibrium. Under the single-mode-relaxation-time approximation, the thermal conductivity is given by $\kappa = \sum C_i v_i^2 \tau_i$ where C_i and v_i are the specific heat per volume unit and the group velocity of the i-th phonon mode. The phonon dispersion of the supported graphene film for $\rho = 0.081\,\mathrm{nm}^{-2}$ and $l_G = 2$, and the extracted relaxation time τ for all the phonon modes are shown in Fig. 6. By inserting the APTES molecule, the relaxation time of the acoustic flexural modes τ_{ZA} largely increase at low frequencies ($\omega < 10\,\mathrm{THz}$), whereas the longitudinal and transverse modes undergo a slight decrease in $\tau_{LA,LO,TA,TO}$. The notable increase in τ_{ZA} accounts for the enhancement in κ of the graphene film bonded to the substrate since the ZA modes contribute considerably to κ as much as 77% at 300 K (ref. 15). We attribute the increase in τ_{ZA} to the weakened coupling between the graphene film and the substrate, which is reflected as the increased thermal resistance R for $\rho = 0.081\,\mathrm{nm}^{-2}$ and $l_G \geq 2$, as is shown in Fig. 5b. An approximate expression from the perturbation theory for the relaxation time due to phonon leakage towards the contact interface yields $\tau^{-1} \propto g(\omega)K^2/\omega^2$, where $g(\omega)$ depends on the phonon density of states and K is the average vdW coupling constant between the graphene film and its

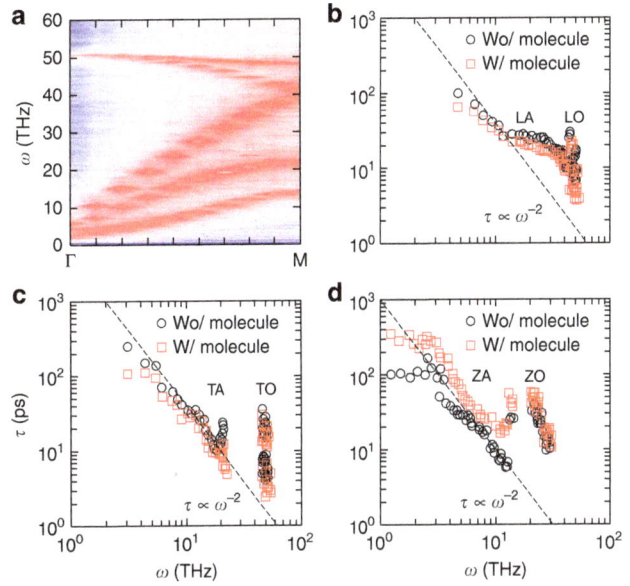

Figure 6 | Phonon dispersion and lifetimes of the graphene film.
(**a**) Phonon dispersion of the graphene film bonded to the APTES-FGO substrate from molecular dynamics simulations, for $\rho = 0.081\,\mathrm{nm}^{-2}$ and $l_G = 2$. (**b-d**) Mode-wise phonon relaxation time for longitudinal modes including longitudinal acoustic (LA) and optical (LO) branches, for transverse modes including transverse acoustic (TA) and optical (TO) branches, and for flexural modes including flexural acoustic (ZA) and optical (ZO) branches, respectively. 'w/' and 'wo/' stand for with and without, respectively. For low-frequency phonons, τ follows the ω^{-2} scaling rule of the Umklapp processes at low frequency and high-temperature limits.

substrate. A previous calculation[16] shows that K_{ZA} largely depends on the substrate morphology. The presence of the molecule strongly decreases the film–substrate coupling K_{ZA}, thus yielding a longer time τ_{ZA}. This demonstrates that the amino-silane molecules refrain the film–substrate phonon scattering, which in turn enhances in-plane heat conduction in the bonded graphene film. Note that ZA mode becomes 'massive' as shown in Fig. 6a, that is, the ZA branch does not reach all the way to zero frequency but shifts to higher frequencies. The reduced inter-plane scattering compensates the reduction in the phonon group velocity at Brillouin zone centre, so that the in-plane thermal conductivity increases after all (Supplementary Fig. 16 and Supplementary Note 11).

The cross-plane heat conduction between the top graphene films and the lower FGO layers is governed by the competing effects of (i) the intercalation of the molecules that tends to weaken the interlayer coupling of graphene therefore increasing the cross-plane thermal resistance and (ii) the additional heat channels introduced by the molecules covalently bonding the graphene sheets that tend to facilitate the cross-plane thermal coupling. We investigated the in-plane thermal conductivity κ of the graphene film and its thermal resistance R with the functionalized substrate versus the equivalent APTES molecule number density ρ in Fig. 7. At a low molecule density, the molecule population is too small in such a way that the additional thermal phonon transport through the molecules cannot compensate the weakened cross-plane thermal coupling through the interlayer vdW interactions due to the intercalation effect by the molecules (also see Supplementary Fig. 17 and Supplementary Note 12 for a detailed analysis on the κ–ρ relation at low molecular densities). Therefore, the net effect is an increase in the

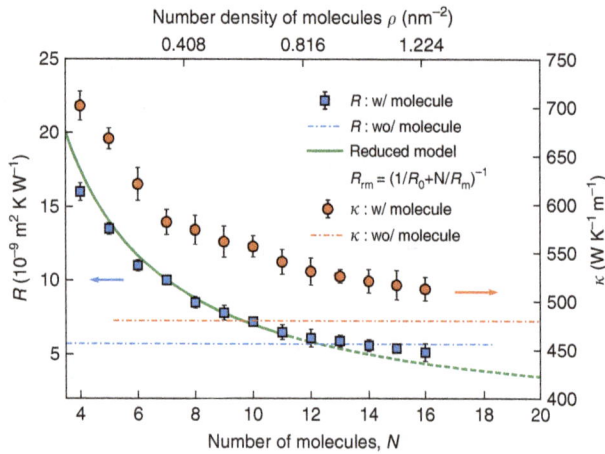

Number density of molecules ρ (nm⁻²)

Figure 7 | Thermal resistance and in-plane thermal conductivity of the graphene film versus the molecule density. Thermal conductivity κ (red circles) of the graphene film and its thermal resistance R (blue squares) with the APTES functionalized substrate versus the equivalent molecule number density ρ. Green line: resistance R_{rm} predicted by a reduced model of parallel thermal resistors is to compare with the molecular dynamics (MD) results for small ρ. $R_{rm} = (1/R_0 + N/R_m)^{-1}$, where R_0 is the thermal resistance between two adjacent graphene layers and R_m is the additional resistance induced by a single molecule. The values of R_0 and R_m are determined in MD simulations: $R_0 = 0.203$ mm² K W⁻¹ and $R_m = 0.069$ mm² K W⁻¹. Red and blue dashed lines represent, respectively, κ and R without the interconnecting molecule at the film–substrate interface. 'w/' and 'wo/' stand for with and without, respectively. The error bar corresponds to the s.d. of the ensemble average on 50 independent trajectories.

cross-plane thermal resistance between the top graphene films and the lower GO layers compared to the non-functionalization case. In the limit of small number density, we consider the molecules as independent heat conductors connecting the film and the substrate. R agrees well with a reduced model of parallel thermal resistors. Thus, R decreases with the number of molecules. κ also decreases with the molecule number since the molecules increase the cross-plane phonon scattering in the graphene film. For larger ρ, the interactions among the molecules gain importance and R starts to deviate from the prediction of the parallel-thermal-resistor model, as shown in Fig. 7. The molecule population is large enough so that the additional thermal phonon transport through the molecules becomes very strong and compensates the weakened cross-plane thermal coupling through the interlayer interactions due to the intercalation effect by the molecules. Therefore, the net effect is that the cross-plane thermal resistance between the top graphene films and the lower FGO layers falls lower than that of the case without functionalization in the GO. For large ρ, the thermal resistance R is lower than that without molecules. However κ remains enhanced by a factor of ~15% compared with its value for the non-functionalized graphene substrate. The molecule density ρ effectively tunes the thermal conductivity of the supported graphene.

Discussion

The correlation between the in- and cross-plane thermal transport of 2D materials has been recently studied from both theoretical[15,17] and experimental[13,14,16] points of view. Among them, Ghosh et al.[13] studied the dimensional crossover from a 2D graphene to a 3D graphite system. The in-plane thermal conductivity decreased with the increase of the layer number in

the FLG. Wei et al.[17] used a 2D material model to show a negative correlation between the in-plane thermal conductivity and the cross-plane interaction. These studies have considered different mechanisms of the correlation, yet no general conclusion is applicable to any quasi-2D system. We show that the functionalization molecules mediate the cross-plane phonon scattering and in turn effectively control the in-plane thermal transport in the graphene-based heat spreader.

Our work tackles the key technological challenge of efficient thermal management in the industry of the next-generation integrated circuit. The excessive heat from the hotspots accelerates the failure rate and slows down the operating speed of microelectronic devices[50,51]. Extremely high power density beyond 1,000 W cm⁻² can be observed in microprocessors[51]. Coolers are most commonly based on metallic materials due to their relatively high thermal conductivities, for example, 200–400 W m⁻¹ K⁻¹ for Al and Cu, and the low cost. However, for these technologically important metals, the thermal conductivity of a 100-nm-thick film constitutes only ~20% of the thermal conductivity of the bulk[52,53]. On the other hand, GBF used in the present work can have an in-plane thermal conductivity as high as 1,600 W m⁻¹ K⁻¹ (ref. 54), which is more than two orders of magnitude higher than the ones of the metal thin films. Such a superior thermal conductivity significantly facilitates the heat conduction when GBFs are used as heat spreaders together with a better cross-plane thermal coupling with FGO. The APTES functionalization yields an additional temperature decrease ΔT of the hotspot of 11 °C compared with the case of GBF with non-functionalized GO ($\Delta T = 17$ °C), resulting in a large total temperature decrease of 28 °C when operating at 1,300 W cm⁻². In comparison, the highest ΔT achieved so far at a similar high heat flux of 1,250 W cm⁻² is 19 °C by using a miniaturized thermoelectric cooler[51]. The contribution of functionalization to the overall ΔT is robust especially at higher heat flux beyond 1,000 W cm⁻², as shown in Fig. 1e. Therefore, we propose a significant package-level solution for the thermal management of hotspots in high-power electronics at the micro- and nanometre length scale.

Methods

Experimental method. (i) Sample synthesis: graphite (Sigma, 4 g); H_2SO_4 (92 ml, 98%); $NaNO_3$ (2 g); and $KMnO_4$ (12 g) were used to prepare GO dispersion by following Hummers' method[55]. The obtained GO dispersion was reduced by L-ascorbic acid, and polyvinyl alcohol was also added for better suspension. The GBF was prepared via vacuum filtration with polycarbonate filter paper with a pore size of 3 μm. The film thickness was controlled by the filtration volume and the graphene concentration in the suspension. After dissolving the filter paper in pure acetone, a freestanding GBF was obtained. The thickness of the GBF was measured as ~20 μm. Raman spectroscopy data of the GBF before and after the spin-coating with FGO are shown in Supplementary Fig. 18. (ii) Functionalization: GO powder (20 mg) and dicyclohexylcarbodiimide (5 mg) were mixed with APTES (30 ml) by ultrasonication for 2 h to produce a homogeneous suspension. Then the suspension was heated up to 100 °C for 3 h, with continuous stirring to realize the functionalization. Fourier transform infrared spectroscopy data provide evidence for the functionalization (Supplementary Fig. 19). (iii) Transfer process: the GBF was first transferred onto a thermal release tape, and then spin-coated[35] with the FGO layer at 4,000 r.p.m. for 2 min onto the film. The thermal release tape was removed by heating the device.

Simulation set-up. Classical molecular dynamics simulations were performed using LAMMPS[56]. Adaptive intermolecular reactive empirical bond order potential[57] was used to simulate the graphene's C–C interactions. The intramolecular forces are taken into account through the ReaxFF potential[58], which uses distance-dependent bond-order functions to represent the contributions of chemical bonding to the potential energy. Periodic boundary conditions are applied in the in-plane directions and free boundary condition in the cross-plane direction of the graphene system. First, each supercell was relaxed at the simulation temperature to achieve zero in-plane stress. Then the systems were thermalized by using a Langevin heat bath. The system reached thermal equilibrium after 2 ns computed in the microcanonical ensemble. Temperature and heat flux were sampled in the microcanonical ensemble in the in the following 1 ns.

Autocorrelation functions for the resistance and the thermal conductivity were calculated over this latter duration.

Thermal conductance. Thermal conductance $\kappa = \kappa_e + \kappa_{ph}$, where κ_{ph} (κ_e) is the thermal conductance due to the phonons (electrons). From the phonon transmission $\Xi(E_{ph})$ the thermal conductance due to the phonon could be calculated as $\kappa_{ph}(T) = \int_0^\infty d\omega \left(-\frac{\partial f_{BE}}{\partial E}\right) \hbar\omega \Xi(\omega)/2\pi$, where ω is the frequency, T refers to the mean temperature of the system, f_{BE} is the Bose–Einstein phonon statistics, and $\hbar$ represents the reduced Plancks constant. The thermal conductance due to the electrons[59] could be calculated from the electron transmission coefficient T_{el} as $\kappa_e(T) = \frac{L_0 L_2 - L_1^2}{hTL_0}$, where $L_n = \int_{-\infty}^\infty dE (E - E_F)^n \, T_{el}(E) \left(-\frac{\partial f_{FD}}{\partial E}\right)$ in which f_{FD} is the Fermi–Dirac electron statistics.

References

1. Yan, Z., Nika, D. L. & Balandin, A. A. Thermal properties of graphene and few-layer graphene: applications in electronics. *IET Circ. Device. Syst.* **9**, 4–12 (2015).
2. Geim, A. K. & Novoselov, K. S. The rise of graphene. *Nat. Mater.* **6**, 183–191 (2007).
3. Balandin, A. A. *et al.* Superior thermal conductivity of single-layer graphene. *Nano Lett.* **8**, 902–907 (2008).
4. Balandin, A. A. Thermal properties of graphene and nanostructured carbon materials. *Nat. Mater.* **10**, 569 (2011).
5. Kong, Q. Q. *et al.* Hierarchical graphenecarbon fiber composite paper as a flexible lateral heat spreader. *Adv. Funct. Mater.* **24**, 4222–4228 (2014).
6. Yan, Z., Liu, G., Khan, J. M. & Balandin, A. A. Graphene quilts for thermal management of high-power GaN transistors. *Nat. Commun.* **3**, 827 (2012).
7. Gao, Z., Zhang, Y., Fu, Y., Yuen, M. M. & Liu, J. Thermal chemical vapor deposition grown graphene heat spreader for thermal management of hot spots. *Carbon* **61**, 342–348 (2013).
8. Subrina, S., Kotchetkov, D. & Balandin, A. A. Heat removal in silicon-on-insulator integrated circuits with graphene lateral heat spreaders. *IEEE Electron Device Lett.* **30**, 1281–1283 (2009).
9. Subrina, S. Modeling based design of graphene heat spreaders and interconnects in 3-D integrated circuits. *J. Nanoelectron. Optoelectron.* **5**, 281–286 (2010).
10. Malekpour, H. *et al.* Thermal conductivity of graphene laminate. *Nano Lett.* **14**, 5155 (2014).
11. Prasher, R. Graphene spreads the heat. *Science* **328**, 185–186 (2010).
12. Ong, Z. Y. & Pop, E. Effect of substrate modes on thermal transport in supported graphene. *Phys. Rev. B* **84**, 075471 (2011).
13. Ghosh, S. *et al.* Dimensional crossover of thermal transport in few-layer graphene. *Nat. Mater.* **9**, 555 (2010).
14. Sadeghi, M. M., Jo, I. & Shi, L. Phonon-interface scattering in multilayer graphene on an amorphous support. *Proc. Natl Acad. Sci. USA* **110**, 16321 (2013).
15. Lindsay, L., Broido, D. A. & Mingo, N. Flexural phonons and thermal transport in graphene. *Phys. Rev. B* **82**, 115427 (2010).
16. Seol, J. H. *et al.* Two-dimensional phonon transport in supported graphene. *Science* **328**, 213 (2010).
17. Wei, Z., Chen, Y. & Dames, C. Negative correlation between in-plane bonding strength and cross-plane thermal conductivity in a model layered material. *Appl. Phys. Lett.* **102**, 011901 (2013).
18. Luo, T. & Lloyd, J. R. Enhancement of thermal energy transport across graphene/graphite and polymer interfaces: a molecular dynamics study. *Adv. Mater.* **22**, 2495 (2012).
19. Hopkins, P. E. *et al.* Manipulating thermal conductance at metal-graphene contacts via chemical functionalization. *Nano Lett.* **12**, 590 (2012).
20. Ge, Z., Cahill, D. G. & Braun, P. V. Thermal conductance of hydrophilic and hydrophobic interfaces. *Phys. Rev. Lett.* **96**, 186101 (2006).
21. Wang, R. Y., Segalman, R. A. & Majumdar, A. Room temperature thermal conductance of alkanedithiol self-assembled monolayers. *Appl. Phys. Lett.* **89**, 173113 (2006).
22. Ramanathan, T. *et al.* Functionalized graphene sheets for polymer nanocomposites. *Nat. Nanotechnol.* **3**, 327–331 (2008).
23. Konatham, D. & Striolo, A. Thermal boundary resistance at the graphene-oil interface. *Appl. Phys. Lett.* **95**, 163105 (2009).
24. Collins, K. C., Chen, S. & Chen, G. Effects of surface chemistry on thermal conductance at aluminum-diamond interfaces. *Appl. Phys. Lett.* **97**, 083102 (2010).
25. Liang, Q., Yao, X., Wang, W., Liu, Y. & Wong, C. P. A three-dimensional vertically aligned functionalized multilayer graphene architecture: an approach for graphene-based thermal interfacial materials. *ACS Nano* **5**, 2392–2401 (2011).
26. Ni, Y. *et al.* Highly efficient thermal glue for carbon nanotubes based on azide polymers. *Appl. Phys. Lett.* **100**, 193118 (2012).
27. Harikrishna, H., Ducker, W. A. & Huxtable, S. T. The influence of interface bonding on thermal transport through solid-liquid interfaces. *Appl. Phys. Lett.* **102**, 251606 (2013).
28. Liang, Z., Evans, W., Desai, T. & Keblinski, P. Improvement of heat transfer efficiency at solid-gas interfaces by self-assembled monolayers. *Appl. Phys. Lett.* **102**, 061907 (2013).
29. OBrien, P. J. *et al.* Bonding-induced thermal conductance enhancement at inorganic heterointerfaces using nanomolecular monolayers. *Nat. Mater.* **12**, 118 (2013).
30. Taphouse, J. H., Smith, O. N. L., Marder, S. R. & Cola, B. A. A pyrenylpropyl phosphonic acid surface modifier for mitigating the thermal resistance of carbon nanotube contacts. *Adv. Funct. Mater.* **24**, 465 (2014).
31. Sun, F. *et al.* Molecular bridge enables anomalous enhancement in thermal transport across hard-soft material interfaces. *Adv. Mater.* **26**, 6093 (2014).
32. Chien, S. K. & Yang, Y. T. Influence of hydrogen functionalization on thermal conductivity of graphene: Nonequilibrium molecular dynamics simulations. *Appl. Phys. Lett.* **98**, 033107 (2011).
33. Kim, J. Y., Lee, J.-H. & Grossman, J. C. Thermal transport in functionalized graphene. *ACS Nano* **6**, 9050–9057 (2012).
34. Xin, G. *et al.* Large-area freestanding graphene paper for superior thermal management. *Adv. Mater.* **26**, 4521–4526 (2014).
35. Pasternack, R. M., Amy, S. R. & Chabal, Y. J. *Langmuir* **24**, 12963–12971 (2008).
36. Chen, G. & Hui, P. Pulsed photothermal modeling of composite samples based on transmission-line theory of heat conduction. *Thin Solid Films* **339**, 58–67 (1999).
37. Zhao, Y. *et al.* Pulsed photothermal reflectance measurement of the thermal conductivity of sputtered aluminum nitride thin films. *J. Appl. Phys.* **96**, 4563 (2004).
38. Meier, T. *et al.* Length-dependent thermal transport along molecular chains. *Phys. Rev. Lett.* **113**, 060801 (2014).
39. Yang, J. *et al.* Phonon transport through point contacts between graphitic nanomaterials. *Phys. Rev. Lett.* **112**, 205901 (2014).
40. Segal, D., Nitzan, A. & Hänggi, P. Thermal conductance through molecular wires. *J. Chem. Phys.* **119**, 6840 (2003).
41. Hu, L. *et al.* Phonon interference at self-assembled monolayer interfaces: Molecular dynamics simulations. *Phys. Rev. B* **81**, 235427 (2010).
42. Markussen, T. Phonon interference effects in molecular junctions. *J. Chem. Phys.* **139**, 244101 (2013).
43. Han, H. *et al.* Phonon interference and thermal conductance reduction in atomic-scale metamaterials. *Phys. Rev. B* **89**, 180301(R) (2014).
44. Han, H., Li, B., Volz, S. & Kosevich, Y. A. Ultracompact interference phonon nanocapacitor for storage and lasing of coherent terahertz lattice waves. *Phys. Rev. Lett.* **114**, 145501 (2015).
45. Ni, Y., Chalopin, Y. & Volz, S. Significant thickness dependence of the thermal resistance between few-layer graphenes. *Appl. Phys. Lett.* **103**, 061906 (2013).
46. Lusk, M. T. & Carr, L. D. Nanoengineering defect structures on graphene. *Phys. Rev. Lett.* **100**, 175503 (2008).
47. Jiang, J.-W., Wang, B.-S. & Wang, J.-S. First principle study of the thermal conductance in graphene nanoribbon with vacancy and substitutional silicon defects. *Appl. Phys. Lett.* **98**, 113114 (2011).
48. Hughbanks, T. & Hoffmann, R. Chains of trans-edge-sharing molybdenum octahedra: metal-metal bonding in extended systems. *J. Am. Chem. Soc* **105**, 3528 (1983).
49. Ziman, J. M. *Electrons and Phonons* (Oxford Univ. Press, 1960).
50. Trew, R. J., Green, D. S. & Shealy, J. B. AlGaN/GaN HFET reliability. *IEEE Microw. Mag.* **10**, 116127 (2009).
51. Yang, B., Wang, P. & Bar-Cohen, A. 'Mini-contact enhanced thermoelectric cooling of hot spots in high power devices'. *IEEE Trans. Compon. Packag. Technol.* **30**, 432–438 (2007).
52. Nath, P. & Chopra, K. L. Thermal conductivity of copper films. *Thin Solid Films* **20**, 5362 (1974).
53. Langer, G., Hartmann, J. & Reichling, P. Thermal conductivity of thin metal films measured by photothermal profile analysis. *Rev. Sci. Instrum* **68**, 15101513 (1997).
54. Kumar, P. *et al.* Large-area reduced graphene oxide thin film with excellent thermal conductivity and electromagnetic interference shielding effectiveness. *Carbon* **94**, 494–500 (2015).
55. Hummers, W. S. & Offeman, R. E. Preparation of graphitic oxide. *J. Am. Chem. Soc.* **80**, 1339–1339 (1958).
56. Plimpton, S. Fast parallel algorithms for short-range molecular dynamics. *J. Comput. Phys.* **117**, 1 (1995).
57. Stuart, S. J., Tutein, A. B. & Harrison, J. A. A reactive potential for hydrocarbons with intermolecular interactions. *J. Chem. Phys.* **112**, 6472 (2000).
58. Chenoweth, K., van Duin, A. C. & Goddard, W. A. ReaxFF reactive force field for molecular dynamics simulations of hydrocarbon oxidation. *J. Phys. Chem. A* **112**, 1040–1053 (2008).
59. Sadeghi, H., Sangtarash, S. & Lambert, C. Enhancing the thermoelectric figure of merit in engineered graphene nanoribbons. *Beilstein J. Nanotechnol.* **6**, 1176 (2015).

Acknowledgements

The work is sponsored by EU FP7 programme 'Nanotherm' under the Grant Agreement No 318117. This work is also supported by the National Chinese Science Foundation Projects (51272153 and 61574088) and the Science and Technology Commission of Shanghai Municipality programme (12JC1403900), as well as Shanghai Education Commission programme (Shanghai University Peak Discipline Construction Project), by the Swedish Foundation for Strategic Research (SSF) Frame Project Contract No: SE13-0061 and SSF Project Contract No: EM11-0002. This work is supported by EPSRC project EP/N017188/1 and by the EU ITN MOLESCO project number 606728. We acknowledge the financial support from The Production Area of Advance programme, Chalmers University of Technology, Sweden.

Author contributions

S.V. and J.L. proposed and designed the project; H.H., Y.Z. and N.W. contributed equally to this work; H.H. performed phonon and electron Green's function calculations and molecular dynamics simulations; Y.Z. characterized the samples with Raman spectroscopy, Fourier transform infrared spectroscopy and RTD measurement; N.W. and M.M. prepared GBF and functional materials; M.K.S. conducted the PPR measurement with assistance from Y.Z.; M.E. and H.H. performed FE simulations; Y.F. designed the experiments and analysed the experimental data; L.Y. supervised the experiments on functional materials; H.H., Y.N., K.S., S.Y., Y.A.K. and S.V. analysed and interpreted the phonon transport calculations; H.H. and Z.Y.M. conducted first-principle and Green's function calculations with assistance from S.X., H.S., S.B. and C.L.; H.H. integrated results among different research units and wrote the manuscript.

Additional information

Competing financial interests: The authors declare no competing financial interests.

24

Microwave a.c. conductivity of domain walls in ferroelectric thin films

Alexander Tselev[1], Pu Yu[2,3], Ye Cao[1], Liv R. Dedon[4], Lane W. Martin[4,5], Sergei V. Kalinin[1] & Petro Maksymovych[1]

Ferroelectric domain walls are of great interest as elementary building blocks for future electronic devices due to their intrinsic few-nanometre width, multifunctional properties and field-controlled topology. To realize the electronic functions, domain walls are required to be electrically conducting and addressable non-destructively. However, these properties have been elusive because conducting walls have to be electrically charged, which makes them unstable and uncommon in ferroelectric materials. Here we reveal that spontaneous and recorded domain walls in thin films of lead zirconate and bismuth ferrite exhibit large conductance at microwave frequencies despite being insulating at d.c. We explain this effect by morphological roughening of the walls and local charges induced by disorder with the overall charge neutrality. a.c. conduction is immune to large contact resistance enabling completely non-destructive walls read-out. This demonstrates a technological potential for harnessing a.c. conduction for oxide electronics and other materials with poor d.c. conduction, particularly at the nanoscale.

[1] Center for Nanophase Materials Sciences, Oak Ridge National Laboratory, Oak Ridge, Tennessee 37831, USA. [2] State Key Laboratory for Low-Dimensional Quantum Physics, Department of Physics and Collaborative Innovation Center for Quantum Matter, Tsinghua University, Beijing 100084, China. [3] RIKEN Center for Emergent Matter Science (CEMS), Wako, Saitama 351-0198, Japan. [4] Department of Materials Science and Engineering and Department of Physics, University of California, Berkeley, Berkeley, California 94720, USA. [5] Materials Science Division, Lawrence Berkeley National Laboratory, Berkeley, California 94720, USA. Correspondence and requests for materials should be addressed to A.T. (email: atselev@utk.edu) or to P.M. (email: maksymovychp@ornl.gov).

n ferroelectrics, domains of uniform polarizations are separated by domain walls. Domain walls can be created and reconfigured by electric fields[1,2], and their lateral width is only a few nanometre due to strong coupling between lattice strain and ferroelectric polarization[3,4]. In the past few years, the concept of domain wall electronics pursued utilization of domain walls[5,6] and other topological defects in ferroelectrics in view of applications in electronic devices[7,8]. Electronic conductivity is a basic example of the desired functionality. Despite electrically insulating nature of ferroelectric materials, several types of ferroelectric domain walls showed d.c. conduction[9-15]. However, the elucidation of the mechanism of the wall conduction and the progress towards practical utilization of domain wall circuits have so far been impeded by low conduction of domain walls and a large ferroelectric-electrode contact resistance[9-13,16]. Domain wall conductance was attributed to electronic reconstructions at the domain wall[9], segregation of mobile donor or charge-trapping defects[11,13], or finite charge of non-equilibrated domain wall morphology[1,10,12,14,17]. Electrostatically charged domain walls were proposed to be more conducting[14,15], but they are unstable and cannot be generally made without special conditioning[15], nanoscale topologies[18] or composition control[14].

The problem of contact resistance is particularly important in domain wall electronics. The contact interface Schottky barrier conceals the intrinsic mechanisms of charge transport along the in-depth region of a domain wall. Overcoming contact resistance necessitates electric fields that can become comparable or exceed the threshold for domain wall motion[10,11]. As a result, electronic measurements and prospective read-out of domain walls are generally destructive, capable of displacing and erasing the domain wall. At the same time, domain walls can reconstruct in an applied field, creating partially charged and more conducting configurations[10,11]. Domain wall reconstruction by itself can become practically useful provided there is a clear way to control domain morphology and analyse various conducting entities non-destructively[10,11], in a broad analogy to memristor electronics[19].

Here we reveal an alternative regime of ferroelectric domain wall conduction at a.c. frequencies in the gigahertz range. Measurements at high frequencies are insensitive to the contact barriers at the electrode–ferroelectric interface, enabling quantitative non-destructive measurement of the domain wall conductance. We found that nominally uncharged domain walls in conventional ferroelectric thin films of lead zirconate and bismuth ferrite can be strongly conducting, on par with charged walls. The a.c. conductivity is at least 100 times higher than at d.c. with the same probing voltage. From the behaviour of the a.c. domain wall conductance as a function of bias voltage, temperature and time, we conclude that the cloud of free carriers forming around the transient charged domain walls during polarization switching incompletely dissipates after the domain stabilization and partially remains at the wall. The domain wall assumes a rough morphology assisted by a disorder potential in the film. Such a configuration hinders d.c. conduction, but remains many fold more conducting at gigahertz frequencies than the surrounding domains, paving a way to practical utilization of domain wall circuits.

Results

Microwave microscopy of pristine films.
The experiments were carried out on a 100-nm-thick epitaxial thin film of $Pb(Zr_{0.2}Ti_{0.8})O_3$ (100) (PZT) grown using pulsed laser deposition on a single-crystal TiO_2-terminated $SrTiO_3$ (001) substrate with 50 nm $SrRuO_3$ as a bottom electrode (Fig. 1a). Hysteresis loops of remnant piezoresponse of the as-grown PZT film indicate

good ferroelectric properties of the film (Supplementary Fig. 1). The PZT film has a tetragonal lattice with the polarization axis along the surface normal. The as-grown ferroelectric polarization points to the film surface. Ferroelectric domains were created and manipulated using scanning probes, while a.c. conductivity was imaged and measured using a near-field scanning microwave impedance microscope (sMIM) operating at a frequency $f \approx 3$ GHz. Microwave microscopy has previously been shown to be capable of detecting conductance of the domain walls in bulk crystals[20,21]. As illustrated in Fig. 1a, the sMIM probe terminates a microwave transmission line. From the standpoint of the measurement system, the sample can be viewed as a capacitor C_b and a resistor with conduction G_b connected in parallel; the complex admittance of the tip–sample system $Y = G + i2\pi fC$ is measured, and real and imaginary parts are displayed in two channels: sMIM-G and sMIM-C, respectively.

Figure 1b,c shows topography and piezoresponse force microscopy (PFM) images of a stripe domain structure formed by applying sequentially $V_{bias} = -7$ V and $+7$ V to the scanning probe. Ferroelectric polarization reverses between stripes separated by nominally 180° domain walls. Figure 1d shows a microwave conductance image (sMIM-G channel) of the structure taken with a zero d.c. bias at the probe and an a.c. voltage amplitude $\lesssim 300$ mV. Domain walls are clearly conducting unlike the bulk of the surrounding domains, while no contrast between the two is seen in the simultaneous permittivity image (sMIM-C channel; Supplementary Fig. 2). We have created domain structures of more complicated geometries that also showed microwave conductance of the domain walls. Erasing domain walls by reconfiguring ferroelectric domains leaves no observable traces in the sMIM-G images. A 10-fold reduction of the a.c. voltage amplitude did not change the domain wall contrast (Supplementary Fig. 3). Imaging with a small-d.c. bias of ~ 2 $V_{d.c.}$ revealed oppositely polarized domains in the sMIM-C channel (Supplementary Fig. 4a) due to dielectric tunability of PZT (which has opposite signs in oppositely polarized domains); however, the walls do not produce any contrast in the sMIM-C image. In the corresponding sMIM-G image (Supplementary Fig. 4b), domain walls showed approximately the same contrast as with the zero d.c. bias at the probe. Furthermore, the large a.c. conductance was detected with a film of $BiFeO_3$ (Supplementary Fig. 5) and several other PZT films, pointing to a general character of this phenomenon.

To gain insight into the mechanism of microwave conductance of the domain walls, we measured sMIM response as a function of d.c. bias at fixed locations on the sample surface. As displayed in Fig. 1e, d.c. bias sweeps between -8 and $+8$ V reveal dielectric tunability of PZT in the sMIM-C channel with a curve shape typical for a ferroelectric. Kinks and jumps in the permittivity correspond to polarization switching under the probe. Meanwhile, the sMIM-G channel (Fig. 1f) shows abrupt increase of conductivity at the switching events. The jumps can be identified as conductance due to tilted and charged domain walls of ferroelectric nanodomains, which were previously surmised to be electronically conducting due to accumulation of mobile carriers partially screening the bound charge along curved and tilted domain walls[18]. During domain growth, the a.c. conductance gradually decreases down to the bulk value, indicating that walls relax to the weakly charged state and eventually move out of the probed volume due to the growth of the domain in the d.c. electric field of the probe.

Effect of annealing.
An increased a.c. conductance in the PZT film subject to heat treatment also supports that the a.c.

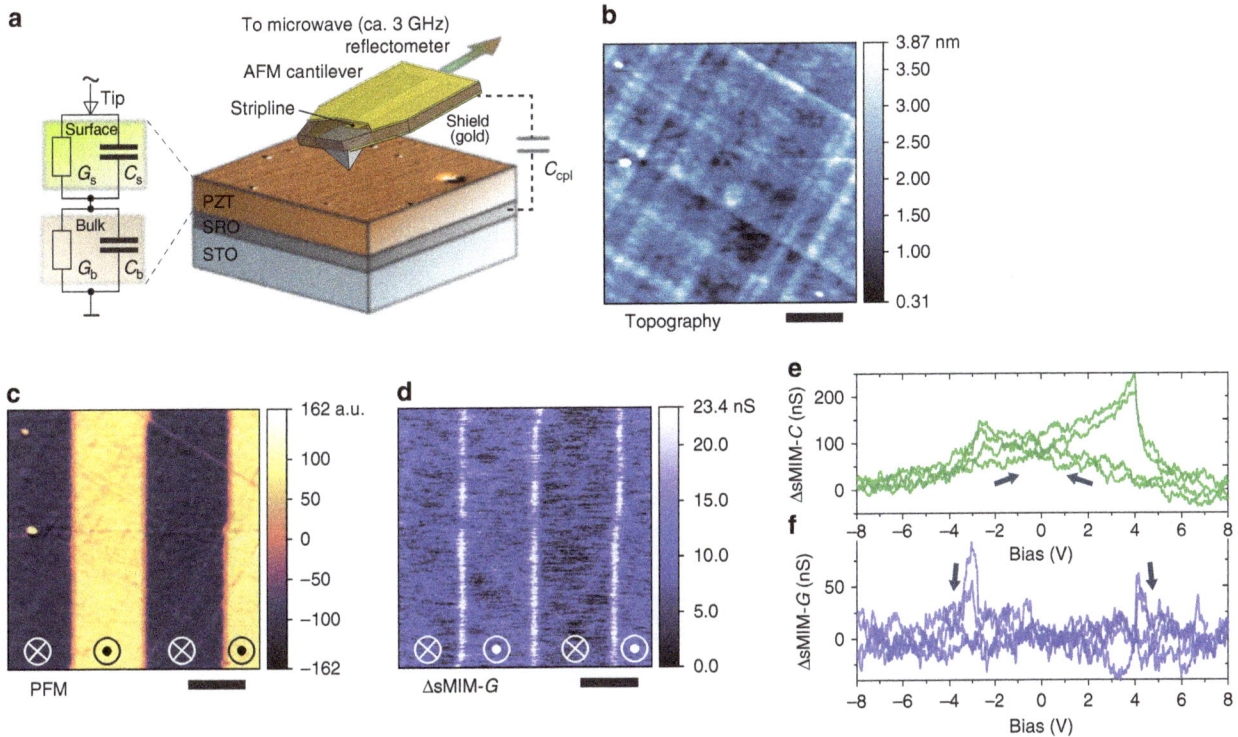

Figure 1 | Microwave imaging of conducting domain walls. (a) Schematic of the sample and sMIM microwave probe on an atomic force microscope (AFM) cantilever. An 100-nm-thick epitaxial thin film of Pb($Zr_{0.2}Ti_{0.8}$)O_3 (PZT) with a 50-nm-thick SrRuO$_3$ (SRO) bottom electrode is deposited on an SrTiO$_3$(001) (STO) substrate. Microwaves of a frequency $f \approx 3$ GHz are delivered to the sensing tip of the probe and to a sample through a stripline fabricated on a silicon nitride AFM cantilever. A reflectometer measures amplitude and phase of the wave reflected from the tip, and represents the results as a change of the admittance $Y = G + i2\pi fC$ of the tip–sample system through two channels sMIM-G and sMIM-C, corresponding to the conductance G and capacitance C, respectively. Due to the capacitance of the space charge layer or conduction-blocking layers on the surface, electrical behaviour of the film surface can be described by a parallel resistor–capacitor circuit $G_s || C_s$ shown in the lumped elements diagram. The intrinsic dielectric response and conduction of the material bulk are represented by the pair $G_b || C_b$. Capacitor C_{cpl} represents the large coupling capacitance between the SRO bottom electrode and the microwave probe shield. **(b)** Image of the surface topography. **(c)** Combined out-of-plane PFM image showing a stripe domain structure with polarization **P** orientated up ⊙ and down ⊗ in the ferroelectric domains as indicated in the image. **(d)** sMIM-G image (that is, a map of the variation ΔsMIM-G of the reflectometer sMIM-G signal) clearly reveals conductivity in the walls of the stripe domains seen in **b**. Images in **b**–**d** were obtained from the same area of the pristine PZT film. **(e)** Signal from the sMIM capacitance channel and **(f)** simultaneous signal from the sMIM conductance channel as functions of d.c. bias applied to the probe at a single point on the pristine film. Arrows indicate direction of the hysteresis. The signal in **f** is shown in respect to a reference signal approximately corresponding to the domain bulk. The images in **b**–**d** were acquired with a zero d.c. bias applied to the probe. Scale bars, 1 µm (**b**–**d**).

conduction of domain walls is associated with mobile charge carriers. Specifically, we annealed the PZT film under reducing conditions (in vacuum, 10^{-8} torr at 350 °C for 20 min). This depletes a small amount of oxygen and creates oxygen vacancies. The vacancies act as electron donors increasing the number of charge carriers in the film. The sMIM-G response of the domain walls in the annealed film was ~1.5–2 times larger than the value of the as-grown film (compare Figs 1d and 2a,b). Noteworthy, annealing of the film also produced spontaneous a.c.-conducting walls (Fig. 2a). The magnitude of the conduction is comparable in spontaneous and recorded walls (Fig. 2c), as is the local sMIM response versus bias (Fig. 2g,h).

d.c. conduction of the domain walls. Both of the PZT films show d.c. conductance of the walls when probed by conductive atomic force microscopy (c-AFM; Fig. 3a,b). A rough estimate at a high bias (at 10 V) from the d.c. *I–V* curves for the domain bulk (Supplementary Figs 6 and 7) corresponds to ca. 2×10^{-3} S m^{-1} for the pristine film and ca. 5×10^{-3} S m^{-1} for the annealed film. In turn, to quantify the a.c. conductivity, we performed

calibrated measurements and numerical modelling of the tip–sample admittance alteration in the presence of a conducting domain wall (see the Methods section and Supplementary Figs 8 and 9 for details). The a.c. conductivity of the pristine film bulk was estimated 0.4–0.7 S m^{-1} at 3 GHz. The domain wall a.c. conductivity fell in a range 4–8 S m^{-1} assuming a 3-nm wall thickness, that is, ca. 10 times higher than in the bulk. At the same time, the d.c. conductivity is 100–200 times smaller than a.c. conductivity measured at 3 GHz, while a similar estimate for the domain wall in the pristine film (data of Fig. 3e) yields d.c. conductivity of ca. 0.1 S m^{-1}, that is, ~50 times lower than the a.c. value. We note that generally conductivity at gigahertz frequencies can be larger by orders of magnitude than at d.c. and show a relatively week, power-law, temperature dependence[22–28]. A fundamental reason behind this enhancement is that charge carriers localized by energy barriers at d.c. can contribute to a.c. conduction by oscillating between the barriers at high frequencies[22–24,26,28,29].

Temperature dependence of the domain wall a.c. conduction. In the PZT film, the a.c. conduction also showed negligible

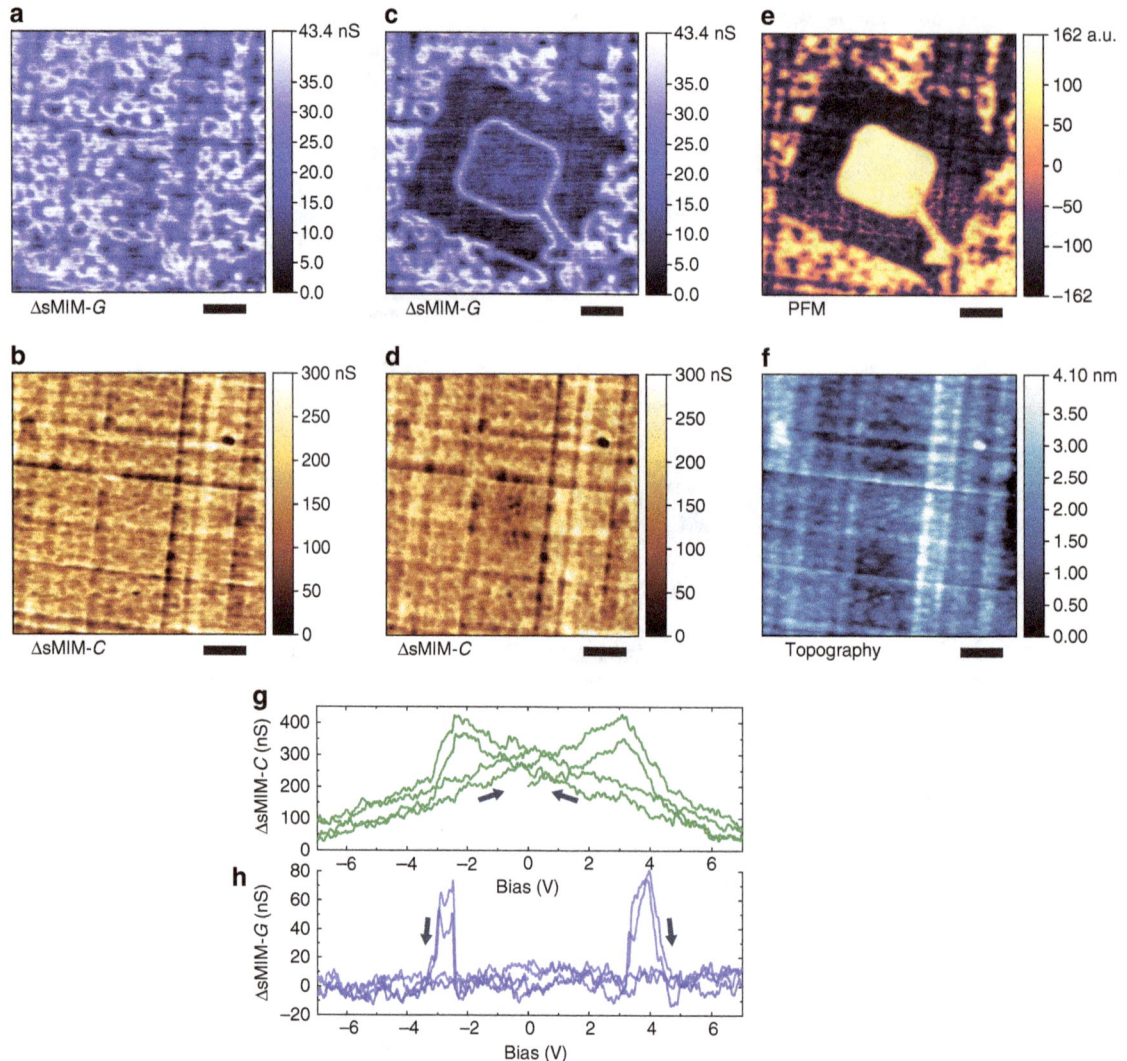

Figure 2 | a.c. conductivity and manipulation of conducting domain walls in the annealed PZT film. Images in **a–f** were obtained from the same area of the annealed film with a zero d.c. bias of the probe. (**a**) sMIM-*G* image, where conducting walls of spontaneous domains of a few-hundred-nm size are clearly seen. (**b**) sMIM-*C* image recorded simultaneously with the image in **a**. (**c**) sMIM-*G* image of the box-in-box domain structure written by the probe after the image in **a** was taken. Spontaneous domains and domain walls were erased inside the structure; the domain walls of the structure are conducting. (**d**) sMIM-*C* image recorded simultaneously with the image in **c**. (**e**) Combined out-of-plane PFM image acquired right after the image in **c**. (**f**) Image of the film surface topography recorded simultaneously with the image in **e**. Comparing images in **b-d** with the image in **f**, it is seen that the sMIM-*C* signal is dominated by a cross-talk with the surface topography. No traces of domain walls are seen in **d**. (**g**) Single-point signal from the sMIM capacitance channel and (**h**) simultaneous signal from the sMIM conductance channel as functions of d.c. bias applied to the probe for the annealed film. Arrows indicate the direction of the hysteresis. The signal in **h** is shown in respect to a reference signal approximately corresponding to the domain bulk. Scale bars, 1 μm (**a–f**).

temperature dependence up to 115 °C (Fig. 4a–c). This is unlike d.c. conduction of domain walls, which could be both linear[14] and exponential[18]. Noteworthy as well, microwave conduction displays negligible dependence on d.c. bias away from switching events (Figs 1f, 2h and 4e). This is in stark contrast to rectifying d.c. *I–V* curves (Fig. 3e; Supplementary Figs 6 and 7), which are strongly non-linear, hysteretic, with current seen only at positive bias above a threshold of ~ 1.3 V. Likewise, the images of the sMIM-*G* conductance weakly depend on the d.c. bias between -2 and 2 V. This behaviour of the a.c. response indicates no influence of the contact effects.

Temporal stability. The a.c. conductance of domain walls was stable over at least 48 h and showed no degradation. This time is sufficient for the 180° walls to equilibrate aligned along the polar

direction of the film. Such domain walls are nominally uncharged, and yet they show clear a.c. conduction, being minimally perturbed by the a.c. voltage (down to below 60 mV peak to peak at 3 GHz).

Discussion

Whereas the a.c. conduction associated with mobile charge carriers oscillating between energy barriers explains the observed microwave response of the ferroelectric domain walls, domain wall vibrations near equilibrium positions forced by the high-frequency electric field of the probe should be considered as an alternative origin of the observed sMIM-*G* response to complete the picture. Such vibrations would result in microwave energy dissipation, which can be indistinguishable from the mobile charge high-frequency conduction in the sMIM measurements.

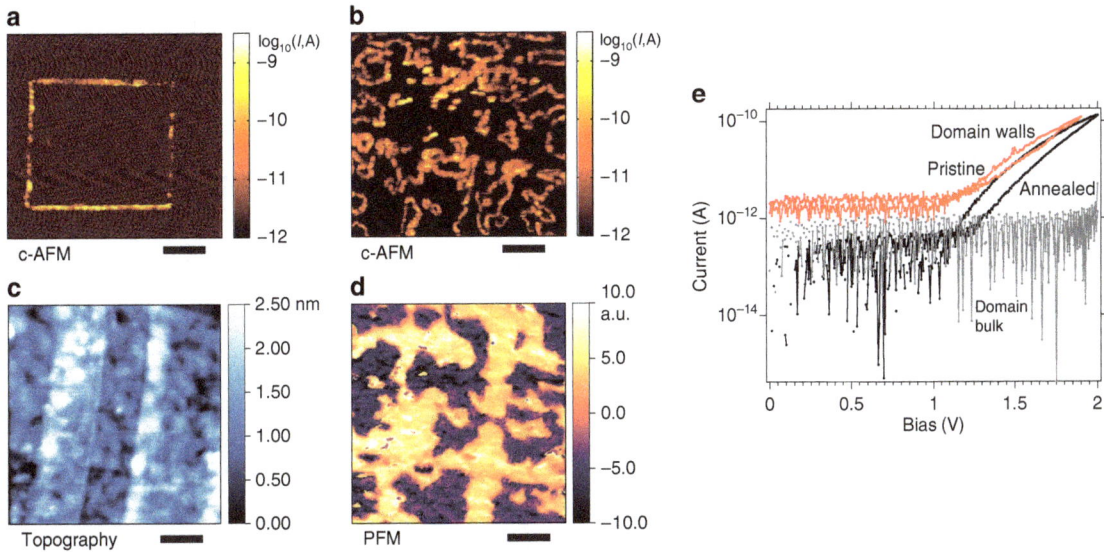

Figure 3 | d.c. conduction of domain walls in PZT. (a,b) Conductive AFM (c-AFM) images of the d.c. current, *I*, in the PZT film obtained with a metal-coated AFM probe in ultrahigh vacuum. Probe bias in **a** and **b** was fixed at +1.9 V. The image in **a** reveals a conducting domain wall written in the pristine film; the image in **b** shows a spontaneously formed network of conducting domain walls in the annealed film. **(c,d)** Image of surface topography and a combined out-of-plane PFM image, respectively, acquired simultaneously from the annealed film. The spots on the film, where images in **b–d** were recorded are different, but located close to each other. **(e)** Probe current versus bias for wall locations and for the domain bulk away from walls. Red, blue and grey correspond, respectively, to a domain wall in the pristine film, a domain wall in the annealed film and domain bulk (which does not show detectable conduction at d.c. both in the pristine and annealed films). Scale bars, 280 nm **(a)**; 400 nm **(b–d)**.

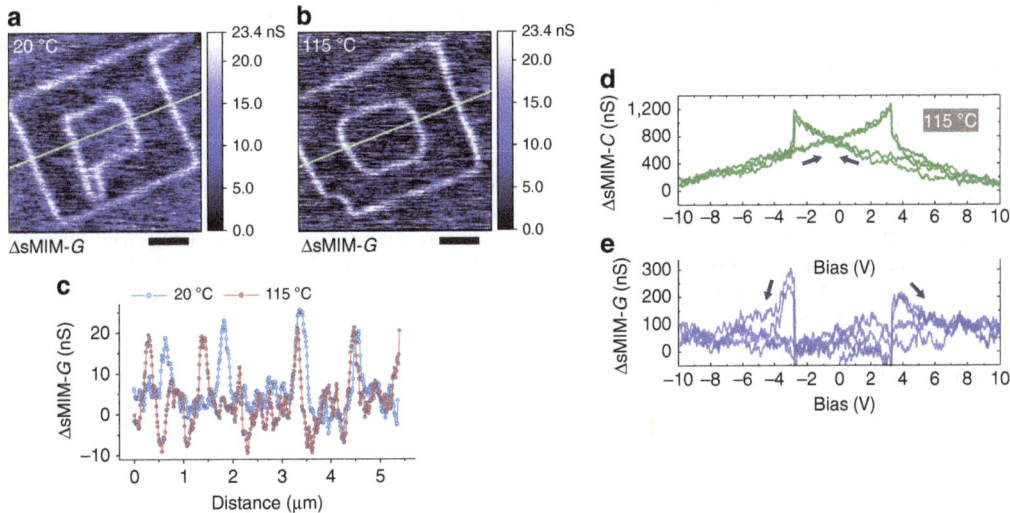

Figure 4 | Temperature dependence of the a.c. conduction of domain walls. (a,b) sMIM-*G* images of box-in-box domain structures written and imaged at different sample temperatures. The images were acquired with a zero d.c. bias of the probe. The image in **a** was obtained at room temperature, and the image in **b** was obtained at a sample temperature of 115 °C. As seen, the intensity of the domain wall response in respect to the background is the same for the images in **a** and **b**. This is further illustrated in **c**. **(c)** Signal profiles along the green lines in images **(a,b)**. Blue and red correspond to room temperature and 115 °C, respectively. The curves were slightly offset along the vertical axis for the ease of comparison. Note that the geometry of the box-in-box structure is distorted in **b** due to thermal drift. **(d)** Single-point signal from the sMIM capacitance channel and **(e)** simultaneous signal from the sMIM conductance channel at a sample temperature of 115 °C as functions of d.c. bias applied to the probe for the annealed film. Arrows indicate the direction of the hysteresis. The signal in **e** is shown in respect to a reference signal approximately corresponding to the domain bulk. Signals in **d** and **e** are stronger than in other experiments described in the paper because of an increased sample–probe force and, consequently, a larger sample–probe contact area. This was done to increase signal-to-noise ratio to compensate for the increase of the noise level associated with the elevated temperature of the probe. Scale bars, 1 μm **(a,b)**.

However, it should be taken into account that displacement of domain walls in response to the applied electric field contributes to material polarizability and permittivity[30]. When a domain wall is present under the sMIM probe, it is expected that the extrinsic contribution of the domain wall vibration to the intrinsic material permittivity is significant and, therefore, can be detected in the capacitance channel if the associated energy loss is detectable by the conduction channel. In our experiments, however, while changes of the film permittivity could be readily observed by tuning the permittivity with an applied bias, no contrast from conducting domain walls could be seen in the corresponding sMIM-C images (Fig. 2d; Supplementary Figs 2 and 4). This is

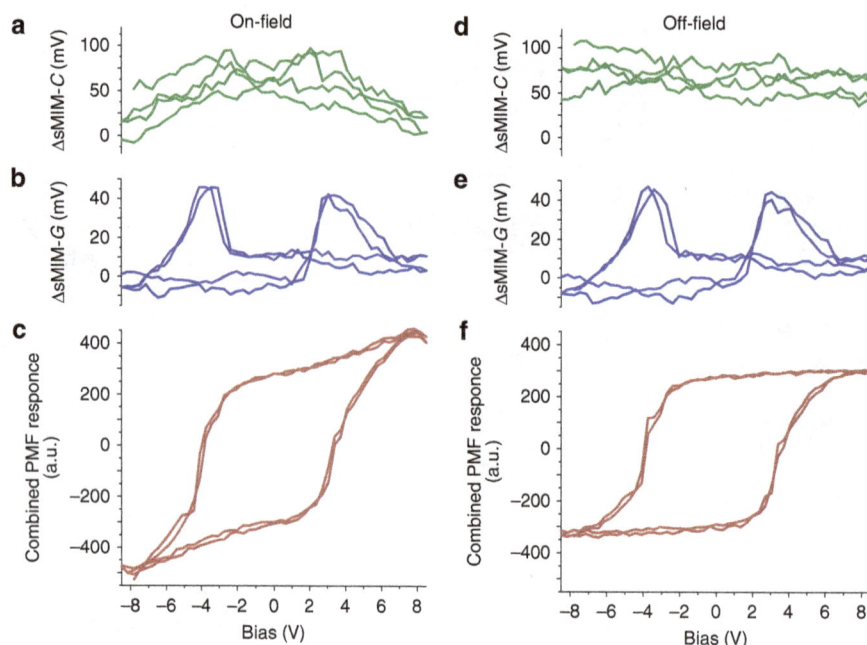

Figure 5 | Single-point sMIM measurements simultaneous with switching spectroscopy PFM. (**a**) On-field sMIM-C and (**b**) on-field sMIM-G signals corresponding to (**c**) on-field out-of-plane PFM hysteresis loop for a BiFeO₃ film. (**d**) Off-field sMIM-C and (**e**) off-field sMIM-G signals corresponding to (**f**) remnant out-of-plane PFM hysteresis loop of the film. The sMIM signals are uncalibrated.

further strengthened by single-point sMIM measurements combined with simultaneous switching spectroscopy PFM. In an switching spectroscopy PFM experiment, voltage bias is applied to the probe in a series of pulses of varying amplitude, and PFM as well as sMIM responses are measured during the application of a pulse and in between consecutive pulses (when the bias is set to zero)[31]. The responses during pulse application (on field) and in between corresponding pulses (off field or remnant) are plotted separately as functions of the pulse amplitude. As illustrated in Fig. 5 for a film of BiFeO₃ as an example, while the on-field sMIM-C signal (Fig. 5a) shows the trend expected for the dielectric tunability, the remnant sMIM-C response (Fig. 5d) is constant over probe bias, being the same at the polarization switching (when domain walls are present under the probe) and away from it (when the probe is surrounded by a uniformly polarized material). At the same time, the on- and off-field sMIM-G signals (Fig. 5b,e) are nearly identical, with conduction peaks around switching evens. A very similar behaviour was observed with PZT and BiFeO₃ films, evidencing that the effect is weakly dependent on the specific nature of the material, and hence surrounding and dynamic characteristics of the domain walls. It can be concluded that the domain wall vibrations, if present, do not contribute enough to be detectable and to explain the observed a.c. conduction of domain walls.

To explain the large a.c. conduction of domain walls, we note that domain wall pinning by lattice defects, and associated strain and field disorder will disrupt the idealized straight shape of the wall making it locally curved[32–34]. The curvature in respect to polarization will translate into bound charges distributed along the roughened domain wall and compensated by localized clouds of mobile carriers, which are responsible for the enhanced a.c. conductivity. This effect is reminiscent of a.c. conduction in metal–insulator composites with metal concentrations below the critical value for percolation threshold[24]. To model this effect, we implemented phase-field modelling[35] in the presence of random field disorder (see the Methods section for details). As seen in Fig. 6a–c, the disorder indeed significantly roughens the

otherwise smooth domain wall, creating local head-to-head and tail-to-tail polarization configurations along the 180° walls, which is revealed in the modelled electric potential distribution map in Fig. 6d. Above a certain disorder strength, substantial electron accumulation is observed along the domain wall (Fig. 6e,f).

In conclusion, we revealed a large a.c. conductance of nominally uncharged ferroelectric domain walls in common oxide ferroelectrics. We explain the a.c. conduction by the domain wall roughness associated with disorder in the ferro-electric films. Taking into account the universal applicability of the a.c. conduction, implications of our findings can be extended to other materials, especially where the a.c. conductivity can be controlled with external fields. This may lead to the emergence of novel device paradigms based on a large spectrum of function-alities potentially offered by complex oxide and nanoscale material systems.

Methods

Pb(Zr₀.₂Ti₀.₈)O₃ film sample fabrication. The ferroelectric Pb(Zr₀.₂Ti₀.₈)O₃ (100) thin film was grown using pulsed laser deposition on a single-crystal TiO₂-terminated SrTiO₃ (001) substrate with 50 nm SrRuO₃ as a bottom electrode. The growth temperatures were 700 and 630 °C for SrRuO₃ and PZT layers, respectively. The growth oxygen pressure was the same for both materials—0.13 mbar. After growth, the sample was cooled down at 1 bar oxygen with a cooling rate of 5 °C min⁻¹. The thickness of the PZT film was controlled by the growth rate and was confirmed with X-ray reflectometry.

BiFeO₃ film sample fabrication. The 100 nm BiFeO₃/50 nm SrRuO₃/DyScO₃ (110) heterostructure was grown using pulsed laser deposition from a Bi₁.₁FeO₃ target. The SrRuO₃ bottom electrode was grown at 645 °C in a dynamic oxygen pressure of 100 mtorr at a laser fluence 1.8 J cm⁻² and a frequency of 17 Hz. The BiFeO₃ film was grown at 700 °C in a dynamic oxygen pressure of 100 mtorr at a laser fluence of 1.0 J cm⁻² and a frequency of 20 Hz. Following growth, the heterostructure was cooled in a static oxygen pressure of 760 torr at a rate of 5 °C min⁻¹.

PFM and c-AFM imaging. PFM imaging in ambient was performed in the same set-up and with the same probes as the sMIM imaging (see below, the Microwave imaging and measurements section) using the Asylum Research Dual AC Reso-nance Tracking mode. A 1.5-V_pp a.c. voltage at a frequency close to the contact

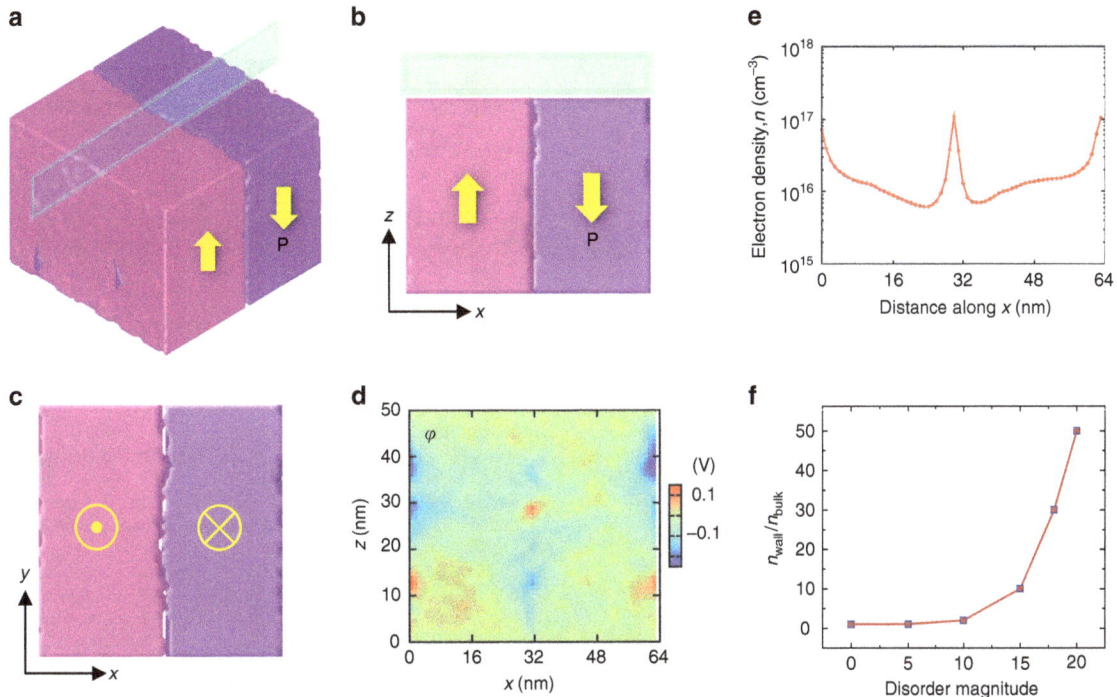

Figure 6 | Phase-field simulation of the ferroelectric domain structure. The simulations were performed for a Pb(Zr$_{0.2}$Ti$_{0.8}$)O$_3$ film with a uniform defect disorder. (**a**) Equilibrium three-dimensional domain structure with rough 180° domain walls. Parallel domain walls are in the centre (visible in the figure) as well as at the sides due to periodic boundary conditions. (**b,c**) Two-dimensional plots of domain structure in the model x–z plane at $y = n_y /2$ (indicated by a green rectangle in **a**) and on the top surface (x–y plane), respectively. (n_y is the model size along the y direction.) The effect of the random field on the wall roughness is less obvious in the film interior than in the surface vicinity. (**d**) Electric potential distribution in the x–z plane at $y = n_y/2$. The electric potential reaches 0.1 and -0.2 V in the wall region. (**e**) Electron density averaged along the normal to the film surface (z direction) throughout the film thickness as a function of the position perpendicular to the walls (x direction). The plot shows a higher electron density in the wall vicinity than in the domain bulk. (**f**) Averaged ratio of electron density in the wall and bulk regions versus disorder magnitude. The disorder magnitude was set to $M = 15$ for calculation of the data for the curve in **e** and for the map in **d**.

resonance frequency of the probe (~250 kHz) was applied for imaging. The c-AFM and corresponding PFM imaging were performed in an Omicron VT AFM in ultrahigh vacuum at a background pressure of 2×10^{-10} torr using Budget Sensors ElectriMulti75-G cantilevers.

sMIM imaging and signal calibration. Microwave imaging and measurements were performed in ambient with a ScanWave (Prime Nano, Inc.) sMIM add-on unit installed on an Asylum Research MFP-3D atomic force microscope. The sMIM microwave output power was set to 100 μW (-10 dBm), at which the upper estimate for the a.c. amplitude at the probe tip is ca. 300 mV$_{a.c.}$. Fully shielded sMIM cantilever probes (Prime Nano, Inc., see also ref. 36) had spring constants in a range 7–8 N m^{-1}. The probes allow both microwave and PFM imaging. The sensing pyramid of the sMIM probes is made of Ti/W alloy, and this was used to increase the sensitivity of sMIM imaging at the expense of spatial resolution. Namely, the set point of the microscope was increased for a short time by a factor of 3–5 in comparison with normal imaging regime to increase the tip–sample loading force. Effectively, this increased the tip–sample contact area, and increased the probe–sample capacitive coupling.

To correctly define the sMIM-C and -G channels, the ScanWave's detector phase was adjusted while imaging a set of 1-μm diameter capacitors of a SMM calibration kit from MC2 Technologies (France). The phase was adjusted so that the variation ΔsMIM-G of the reflectometer sMIM-G signal from the capacitors was zeroed. Supplementary Fig. 10 illustrates sensitivity of the image contrast to the detection phase setting.

The sMIM-C channel sensitivity to admittance changes was calibrated using an MC2 Technologies SMM calibration kit based on the approach of ref. 37. The calibrated sensitivity value is applicable to the sMIM-G channel as well and can be used to measure the real part of the admittance of the tip–sample system on contact with a sample. The microscope sensitivity was determined as $\Delta V_{out}/\Delta Re,Im(Y)$, where ΔV_{out} is the raw, uncalibrated, voltage outputs at sMIM channels (in mV), and $\Delta Re,Im(Y)$ are the corresponding changes of the tip–sample admittance real and imaginary parts (in nS).

sMIM measurements of permittivity and conductivity. The PZT relative permittivity ε_{PZT} was measured using the sMIM-C microscope response in respect

to air (probe is out of contact with a sample) as a reference signal and an MC2 calibrations kit. The measurements yield $\varepsilon_{PZT} \approx 70 \pm 5$, which is in a good agreement with the literature value 70–100 for the same composition[38]. The conductivity of the domain bulk was determined from the sMIM-G signal change in respect to air. Since the sMIM outputs depend on the probe–sample contact area, the contact diameter was measured using the apparent domain wall thickness in PFM images. The real part of the admittance over domain walls was measured from domain wall images based on the sMIM-G signal change in respect to the domain bulk. The data for measurements were taken from images of the domain walls obtained after writing. While creating a suitable domains structure, a special attention was paid to stability of tip–sample contact through all writing sequence to preserve the validity of the probe calibration. Quantifying measurements for the film bulk were carried out with larger tip–sample contact areas to ensure a sufficiently large signal-to-noise ratio. The calibration measurements and measurements of the tip–sample contact diameter were performed before and after each image intended for quantification of permittivity and/or conductivity.

Measurement of the probe-sample contact size. PFM images of the domain walls were used to determine the tip–sample contact diameter. Since the domain wall thickness is 1–2 nm, that is, much smaller than the contact diameter, the apparent width of the domain walls in the PFM images can be taken as the tip–sample contact diameter with a high accuracy. This method was used for quantification of the film conductivity from both sMIM and c-AMF data. In sMIM measurements, the contact diameter was 100–350 nm, depending on the probe wear. In turn, in c-AFM measurements, the contact diameter was 25–30 nm.

sMIM imaging at elevated temperatures. For imaging at elevated temperatures, the sample was glued to a heater with a silver paint. Sample temperature was measured with a miniature thermocouple attached to the sample. The sample temperature of 115 °C was maximal, when properties of the sMIM probe remained stable and imaging was possible. For better thermal stability, the experiments were started following a 10-min waiting period after the temperature of 115 °C was reached. Because heating of a sample warms up the probe as well, the phase setting of the sMIM electronics was adjusted at 115 °C after that. The probe was kept retracted from the sample surface by 3–5 μm during heating and waiting periods.

To avoid tip-apex ware and contamination (to ensure the tip–sample contact stability for a valid comparison of the results), we did not perform a search of the structure written at room temperature after a temperature point was reached, but prepared another similar structure. The domain structures were prepared and imaged at electrically identical conditions at room temperature and at 115 °C.

Numerical (finite elements) modelling. The calibrated measurement values were applied to quantify the conductivities of the PZT film bulk and domain walls via numerical modelling. Numerical modelling for a frequency $f = 3$ GHz was performed using an a.c./d.c. module of the COMSOL v.4.2a Multiphysics finite elements analysis package (COMSOL AB). Two-dimensional axisymmetric models were used in the simulations. The conductivity of the SrRuO$_3$ film was set to 10^5 S m^{-1} (ref. 39). The domain wall in the model (Supplementary Fig. 8) was represented by a cylinder placed coaxially with the tip. The cylinder radius was set so that the contact area between the tip and cylinder was equal to the area of the straight-wall cross-section along the tip–sample contact diameter. Taking the wall thickness equal to 3 nm and tip–sample contact radius equal to 160 nm, this area is 660 nm^2. The model does not account for the surface roughness, which may reduce the contact area in the experiments.

In the simulations, first, the PZT film permittivity was set to a certain value in a range $\varepsilon_{PZT} = 70 \pm 5$, and conductivity of the uniform PZT film was swept to find a conductivity matching the measured value of the sMIM-G response for the domain bulk. Then, the herewith determined conductivity for the film was fixed, while the domain wall conductivity was swept to match an experimental value of the tip–sample conductance over a domain wall. With $\varepsilon_{PZT} = 70$ and the other parameter values as listed above, the a.c. conductivities of the PZT domain bulk and domain wall were determined to be ~0.6 and 6 S m^{-1}, respectively, for the pristine film. The ranges 0.4–0.7 S m^{-1} for the pristine film bulk and 4–8 S m^{-1} for the domain wall, as listed in the main text, take into account uncertainty of the ε_{PZT} measurement as well as uncertainties of the sMIM-G signal due to the measurement noise and instabilities in the tip–sample contact.

Phase-field simulations. The spatial and temporal evolution of Pb(Zr$_{0.2}$Ti$_{0.8}$)O$_3$ domain structure was simulated using a phase-field model by numerically solving the time-dependent Landau–Ginzburg–Devonshire equations[35]:

$$\frac{\partial P_i}{\partial t} = -L\frac{\delta F}{\delta P_i} (i = 1, 2, 3), \tag{1}$$

where **P** is the ferroelectric polarization vector, L is a kinetic coefficient related to the domain motion, t is time and F is the total free energy, which is expanded as:

$$F = \int_V \left[f_{\text{bulk}}(P_i) + f_{\text{grad}}(\nabla P_i) + f_{\text{elas}}(P_i, \varepsilon_{kl}) + f_{\text{elec}}(P_i, E_i)\right] dV, \tag{2}$$

where $f_{\text{bulk}}(P_i)$, $f_{\text{grad}}(\nabla P_i)$, $f_{\text{elas}}(P_i, \varepsilon_{kl})$ and $f_{\text{elec}}(P_i, E_i)$ represent the Landau–Ginzburg–Devonshire free energy density of the uniform strain-free bulk, the gradient energy density, the elastic energy density and the electrostatic energy density, respectively. Details of the expressions for each of the energy density terms are summarized in refs 40,41. Without loss of generality, to investigate the role of the defect disorder on the 180° domain walls, we introduce disorder as a fixed random built-in electric field $\mathbf{E}_{\text{bi}}(x,y,z)$, which can be associated with charged defects randomly distributed in the film volume. The x, y and z components of $\mathbf{E}_{\text{bi}}$ at a point in the film volume are:

$$\begin{aligned} E_{\text{bi}-x} &= A\cos(\alpha)\sin(\beta), \\ E_{\text{bi}-y} &= A\cos(\alpha)\cos(\beta), \\ E_{\text{bi}-z} &= A\sin(\alpha), \end{aligned} \tag{3}$$

where α and β describe the random field orientation and are uniformly distributed random numbers between -2π and 2π. A is the amplitude of the random field $A = M \cdot f(\mu)$, in which M is the disorder magnitude and $f(\mu)$ is the Gaussian function with μ being a uniformly distributed random number between -1 and 1. To determine the local electric potential and the spatial distribution of free charge carrier in the presence of the charged defects, coupled equations:

$$n = N_c F_{1/2}\left(\frac{E_f - E_c + q\varphi}{k_B T}\right), \tag{4}$$

$$p = N_v F_{1/2}\left(\frac{E_v - E_f - q\varphi}{k_B T}\right), \tag{5}$$

$$-\nabla^2 \varphi = \frac{q \cdot z_i(N_d + p - n - N_a) - \nabla \cdot P_i}{\varepsilon_0 \varepsilon_r} \tag{6}$$

are solved, where N_d, N_a, n and p denote the local concentrations of donors, acceptors, electrons and holes, respectively. N_c and N_v are the effective density of states of electrons in the conduction and holes in the valance bands, respectively (~10^{21} cm^{-3}; ref. 42). E_c, E_v and E_f are the energies of the conduction band edge, valance band edge and Fermi level of PZT, respectively, q is the unit charge and φ is the electric potential. $F_{1/2}$ is the Fermi–Dirac integral, k_B is the Boltzmann constant, T is the absolute temperature, z_i is the charge number of each charged

species, and ε_0 and ε_r are the vacuum permittivity and the relative permittivity of PZT, respectively.

The PZT film is assumed to be heavily n-doped. Equation (1) is solved using a semi-implicit spectral method[43] with periodic boundary conditions along x, y and z directions. The simulation volume is discretized into a three-dimensional mesh $64\Delta x \times 64\Delta x \times 64\Delta x$, in which Δx is set to 1 nm. The thicknesses of the film and substrate are assumed to be $50\Delta x$ and $10\Delta x$, respectively. The in-plane epitaxial compressive strain is chosen to be 1.0%. The energies of the conduction band edge, valence band edge and the Fermi level are set to -4.0, -7.4 and -4.2 eV, respectively. The PZT relative permittivity is assumed to be 50. The Landau coefficients, electrostrictive coefficients and elastic compliance constants of PZT have been found in refs 44–46.

References

1. Balke, N. *et al.* Enhanced electric conductivity at ferroelectric vortex cores in BiFeO$_3$. *Nat. Phys.* **8**, 81–88 (2012).

2. McGilly, L. J., Yudin, P., Feigl, L., Tagantsev, A. K. & Setter, N. Controlling domain wall motion in ferroelectric thin films. *Nat. Nanotechnol.* **10**, 145–150 (2015).

3. Padilla, J., Zhong, W. & Vanderbilt, D. First-principles investigation of 180° domain walls in BaTiO$_3$. *Phys. Rev. B* **53**, R5969–R5973 (1996).

4. Jia, C.-L. *et al.* Atomic-scale study of electric dipoles near charged and uncharged domain walls in ferroelectric films. *Nat. Mater.* **7**, 57–61 (2008).

5. Hlinka, J., Ondrejkovic, P. & Marton, P. The piezoelectric response of nanotwinned BaTiO$_3$. *Nanotechnology* **20**, 105709 (2009).

6. Geng, Y., Lee, N., Choi, Y. J., Cheong, S. W. & Wu, W. Collective magnetism at multiferroic vortex domain walls. *Nano Lett.* **12**, 6055–6059 (2012).

7. Zubko, P., Gariglio, S., Gabay, M., Ghosez, P. & Triscone, J.-M. Interface physics in complex oxide heterostructures. *Annu. Rev. Condens. Matter Phys.* **2**, 141–165 (2011).

8. Catalan, G., Seidel, J., Ramesh, R. & Scott, J. F. Domain wall nanoelectronics. *Rev. Mod. Phys.* **84**, 119–156 (2012).

9. Seidel, J. *et al.* Conduction at domain walls in oxide multiferroics. *Nat. Mater.* **8**, 229–234 (2009).

10. Maksymovych, P. *et al.* Dynamic conductivity of ferroelectric domain walls in BiFeO$_3$. *Nano Lett.* **11**, 1906–1912 (2011).

11. Guyonnet, J., Gaponenko, I., Gariglio, S. & Paruch, P. Conduction at domain walls in insulating Pb(Zr$_{0.2}$Ti$_{0.8}$)O$_3$ thin films. *Adv. Mater.* **23**, 5377–5382 (2011).

12. Meier, D. *et al.* Anisotropic conductance at improper ferroelectric domain walls. *Nat. Mater.* **11**, 284–288 (2012).

13. Farokhipoor, S. & Noheda, B. Conduction through 71° domain walls in BiFeO$_3$ thin films. *Phys. Rev. Lett.* **107**, 127601 (2011).

14. Crassous, A., Sluka, T., Tagantsev, A. K. & Setter, N. Polarization charge as a reconfigurable quasi-dopant in ferroelectric thin films. *Nat. Nanotechnol.* **10**, 614–618 (2015).

15. Sluka, T., Tagantsev, A. K., Bednyakov, P. & Setter, N. Free-electron gas at charged domain walls in insulating BaTiO$_3$. *Nat. Commun.* **4**, 1808 (2013).

16. Oh, Y. S., Luo, X., Huang, F.-T., Wang, Y. & Cheong, S.-W. Experimental demonstration of hybrid improper ferroelectricity and the presence of abundant charged walls in (Ca,Sr)$_3$Ti$_2$O$_7$ crystals. *Nat. Mater.* **14**, 407–413 (2015).

17. Vul, B. M., Guro, G. M. & Ivanchik, I. I. Encountering domain walls in ferroelectrics. *Ferroelectrics* **6**, 29–31 (1973).

18. Maksymovych, P. *et al.* Tunable metallic conductance in ferroelectric nanodomains. *Nano Lett.* **12**, 209–213 (2012).

19. Strukov, D. B., Snider, G. S., Stewart, D. R. & Williams, R. S. The missing memristor found. *Nature* **453**, 80–83 (2008).

20. Ma, E. Y. *et al.* Charge-order domain walls with enhanced conductivity in a layered manganite. *Nat. Commun.* **6**, 7595 (2015).

21. Ma, E. Y. *et al.* Mobile metallic domain walls in an all-in-all-out magnetic insulator. *Science* **350**, 538–541 (2015).

22. Mott, S. N. Electrons in glass. *Rev. Mod. Phys.* **50**, 203–208 (1978).

23. Elliott, S. R. a.c. conduction in amorphous chalcogenide and pnictide semiconductors. *Adv. Phys.* **36**, 135–217 (1987).

24. van Staveren, M. P. J., Brom, H. B. & de Jongh, L. J. Metal-cluster compounds and universal features of the hopping conductivity of solids. *Phys. Rep.* **208**, 1–96 (1991).

25. Jonscher, A. K. Dielectric relaxation in solids. *J. Phys. D: Appl. Phys.* **32**, R57 (1999).

26. Seeger, A. *et al.* Charge carrier localization in La$_{1-x}$Sr$_x$MnO$_3$ investigated by a.c. conductivity measurements. *J. Phys. Condens. Matter* **11**, 3273 (1999).

27. Lunkenheimer, P. & Loidl, A. Response of disordered matter to electromagnetic fields. *Phys. Rev. Lett.* **91**, 207601 (2003).

28. Lunkenheimer, P. *et al.* Dielectric properties and dynamical conductivity of LaTiO$_3$ from d.c. to optical frequencies. *Phys. Rev. B* **68**, 245108 (2003).

29. Pollak, M. & Geballe, T. H. Low-frequency conductivity due to hopping processes in silicon. *Phys. Rev.* **122**, 1742–1753 (1961).

30. Jin, L., Porokhonskyy, V. & Damjanovic, D. Domain wall contributions in Pb(Zr,Ti)O$_3$ ceramics at morphotropic phase boundary: A study of dielectric dispersion. *Appl. Phys. Lett.* **96**, 242902 (2010).

31. Jesse, S., Baddorf, A. P. & Kalinin, S. V. Switching spectroscopy piezoresponse force microscopy of ferroelectric materials. *Appl. Phys. Lett.* **88**, 062908 (2006).

32. Kontsos, A. & Landis, C. M. Computational modeling of domain wall interactions with dislocations in ferroelectric crystals. *Int. J. Solids Struct.* **46**, 1491–1498 (2009).

33. Gao, P. *et al.* Atomic-scale mechanisms of ferroelastic domain-wall-mediated ferroelectric switching. *Nat. Commun.* **4**, 2791 (2013).

34. Ziegler, B., Martens, K., Giamarchi, T. & Paruch, P. Domain wall roughness in stripe phase BiFeO$_3$ thin films. *Phys. Rev. Lett.* **111**, 247604 (2013).

35. Chen, L. Q. Phase-field method of phase transitions/domain structures in ferroelectric thin films: a review. *J. Am. Ceram. Soc.* **91**, 1835–1844 (2008).

36. Yang, Y. *et al.* Batch-fabricated cantilever probes with electrical shielding for nanoscale dielectric and conductivity imaging. *J. Micromech. Microeng.* **22**, 115040 (2012).

37. Huber, H. P. *et al.* Calibrated nanoscale capacitance measurements using a scanning microwave microscope. *Rev. Sci. Instrum.* **81**, 113701 (2010).

38. Isarakorn, D. *et al.* The realization and performance of vibration energy harvesting MEMS devices based on an epitaxial piezoelectric thin film. *Smart Mater. Struct.* **20**, 025015 (2011).

39. Ito, A., Masumoto, H. & Goto, T. Microstructure and electrical conductivity of SrRuO$_3$ thin films prepared by laser ablation. *Mater. Trans.* **47**, 2808–2814 (2006).

40. Li, Y. L., Hu, S. Y., Liu, Z. K. & Chen, L. Q. Phase-field model of domain structures in ferroelectric thin films. *Appl. Phys. Lett.* **78**, 3878–3880 (2001).

41. Li, Y. L., Hu, S. Y., Liu, Z. K. & Chen, L. Q. Effect of substrate constraint on the stability and evolution of ferroelectric domain structures in thin films. *Acta Mater.* **50**, 395–411 (2002).

42. Baiatu, T., Waser, R. & Härdtl, K.-H. d.c. electrical degradation of perovskite-type titanates: III, a model of the mechanism. *J. Am. Ceram. Soc.* **73**, 1663–1673 (1990).

43. Chen, L. Q. & Shen, J. Applications of semi-implicit Fourier-spectral method to phase field equations. *Comput. Phys. Commun.* **108**, 147–158 (1998).

44. Haun, M. J., Zhuang, Z. Q., Furman, E., Jang, S. J. & Cross, L. E. Thermodynamic theory of the lead zirconate-titanate solid-solution system, part III: Curie constant and 6th-order polarization interaction dielectric stiffness coefficients. *Ferroelectrics* **99**, 45–54 (1989).

45. Haun, M. J., Furman, E., Jang, S. J., Mckinstry, H. A. & Cross, L. E. Thermodynamic theory of PbTiO$_3$. *J. Appl. Phys.* **62**, 3331–3338 (1987).

46. Haun, M. J., Furman, E., Jang, S. J. & Cross, L. E. Thermodynamic theory of the lead zirconate-titanate solid-solution system, part I: phenomenology. *Ferroelectrics* **99**, 13–25 (1989).

Acknowledgements

We thank Stuart Friedman for technical support and Ramamoorthy Ramesh for assistance with film growth, and Long-Qing Chen for input in the phase-field model. This research was sponsored by the Division of Materials Sciences and Engineering, Office of Science, Basic Energy Sciences, US Department of Energy (A.T., S.V.K. and P.M.). Scanning probe measurements were conducted at the Center for Nanophase Materials Sciences, which is sponsored at Oak Ridge National Laboratory by the Scientific User Facilities Division, Office of Basic Energy Sciences, US Department of Energy. P.Y. was financially supported by the National Basic Research Program of China (grant 2015CB921700) and National Natural Science Foundation of China (grant 11274194). L.R.D. and L.W.M. acknowledge support from the Office of Basic Energy Sciences, US Department of Energy under grant no. DE-SC0012375.

Author contributions

A.T. and P.M. conceived the experiments and the model, performed measurements and wrote the paper. P.Y., L.R.D. and L.W.M. fabricated thin-film samples. Y.C. performed phase-field modelling. All authors discussed the results and commented on the manuscript.

Additional information

25

The formation mechanism for printed silver-contacts for silicon solar cells

Jeremy D. Fields[1,*,†], Md. Imteyaz Ahmad[2,*,†], Vanessa L. Pool[2], Jiafan Yu[3], Douglas G. Van Campen[2], Philip A. Parilla[1], Michael F. Toney[2] & Maikel F.A.M. van Hest[1]

Screen-printing provides an economically attractive means for making Ag electrical contacts to Si solar cells, but the use of Ag substantiates a significant manufacturing cost, and the glass frit used in the paste to enable contact formation contains Pb. To achieve optimal electrical performance and to develop pastes with alternative, abundant and non-toxic materials, a better understanding the contact formation process during firing is required. Here, we use *in situ* X-ray diffraction during firing to reveal the reaction sequence. The findings suggest that between 500 and 650 °C PbO in the frit etches the SiN_x antireflective-coating on the solar cell, exposing the Si surface. Then, above 650 °C, Ag^+ dissolves into the molten glass frit – key for enabling deposition of metallic Ag on the emitter surface and precipitation of Ag nanocrystals within the glass. Ultimately, this work clarifies contact formation mechanisms and suggests approaches for development of inexpensive, nontoxic solar cell contacting pastes.

[1] National Renewable Energy Laboratory, 15013 Denver West Pkwy, Golden, Colorado 80401, USA. [2] SLAC National Accelerator Laboratory, 2575 Sand Hill Road, Menlo Park, California 94025, USA. [3] Department of Electrical Engineering, Stanford University, 350 Serra Mall, Stanford, California 94305, USA. * These authors contributed equally to this work. † Present address: SolarWorld Americas, 25300 NW Evergreen Road, Hillsboro, Oregon 97124, USA (J.D.F.); Department of Ceramic Engineering Indian Institute of Technology (BHU), Varanasi 221005, India (M. I. A.). Correspondence and requests for materials should be addressed to M.F.T. (email: mftoney@slac.stanford.edu) or to F.A.M.v.H. (email: Maikel.van.Hest@nrel.gov).

By far the most commercially viable option for photovoltaic energy generation, crystalline silicon (c-Si) continues to dominate the industry with over 90% market share. Optimally designed silver (Ag) front-contacts in the majority of c-Si solar cells utilize narrow grid lines (approximate width of 50 µm) to minimize shading loss and achieve high current, high fill factor, and hence, high photo-conversion efficiency. While screen-printing provides an economically attractive means for making these contacts, owing to suitability for high-volume manufacturing, the use of Ag adds significantly to the solar cell cost. Furthermore, despite more than three decades of use in photovoltaic manufacturing, the mechanism of contact formation during the firing of screen-printed contacts remains a subject of debate. Achieving optimal firing conditions to minimize contact resistance, and to develop new pastes with alternative materials (that is, earth-abundant and non-toxic), requires a detailed understanding of the contact formation process. *In situ* X-ray diffraction (XRD) data obtained during rapid-thermal processing (RTP) of Ag-paste materials reveal that there are multiple competing anti-reflective layer burn-through and Si oxidation reactions occurring with different temperature thresholds. The primary reactive species responsible for burn-through is lead-oxide (PbO), and these PbO driven redox reactions lead to liquid Pb formation and reversible Ag–Pb alloying during firing. Our *in situ* experiments also elucidate factors affecting dissolution of Ag into the frit and subsequent nanocrystal precipitation – key for achieving low contact resistance.

Typically, screen-printing contacts onto the n-type emitter of a p-type c-Si solar cell employs a paste consisting of Ag particles, organic binders and metal-oxide glass frit. The frit, usually a PbO-based borosilicate glass, promotes a series of reactions allowing electrical contact to the emitter after deposition of the silicon-nitride (SiN_x) anti-reflective and passivation layer during contact firing. Although the frit has long been known to enable this process[1], conflicting arguments bring into question the specific role of the frit to enable SiN_x burn-through and the mechanism of Ag transport to the c-Si surface. Hence, the actual electrical contact formation pathway remains scarcely understood – largely because previous investigations lacked *in situ* measurement capabilities with the temporal-resolution and operating temperature range necessary to observe these processes in real-time.

Reactions responsible for contact formation occur during rapid heating, with ramp-rates of 50–100 °C s^{-1}, at temperatures of 500–800 °C. In general, the Ag-paste is well known to penetrate the SiN_x layer during firing by oxidation, and the resulting silicon-oxide (SiO_2) reaction product becomes incorporated into the molten frit. However, the associated oxidation reaction remains in question, since this can occur in one of several ways. The two most frequently cited arguments suggest either that SiN_x is oxidized by PbO in the frit by[2–4]:

$$2PbO + SiN_x \rightarrow 2Pb + SiO_2 + \frac{x}{2}N_2 \qquad (1)$$

or, that Ag, which can dissolve in frit as silver oxide (Ag_2O) up to approximately 5 wt.%(refs 5,6), etches the SiN_x by[7,8]:

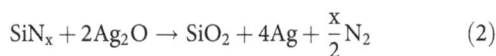

$$SiN_x + 2Ag_2O \rightarrow SiO_2 + 4Ag + \frac{x}{2}N_2 \qquad (2)$$

The second case is perhaps less obvious, since Ag_2O itself readily decomposes under thermal activation, and so may not seem an obvious reactant at contact firing temperatures. However, Ag_2O becomes more stable when bound in certain glasses, and therefore, technically both reactions (1) and (2) are thermodynamically feasible. *In situ* results presented in this work reveal the dominant SiN_x burn-through reaction.

After SiN_x opening, Ag contacts the Si emitter. Electrical conduction in fired contacts relies largely on tunneling through a layer of re-solidified glass frit, which forms between the c-Si surface and Ag bulk during firing. Distributed precipitation of Ag and Pb nanocrystals within the glass enables the tunneling[9–11]. Minimizing the glass intermediate layer thickness, inducing Ag sintering and promoting a controlled amount of Ag deposition on the c-Si surface (without spike-shunting the junction), are said to impart low contact resistance[12]. Here, certain additional details remain in question. For instance, the means by which Ag deposits onto the emitter is not obvious. According to some authors the deposition of Ag on the emitter requires the Ag to first oxidize by:

$$2Ag + \frac{1}{2}O_2 \rightarrow Ag_2O \qquad (3)$$

and then dissolve into the glass frit before a subsequent redox reaction between Ag_2O and c-Si (refs 13,14):

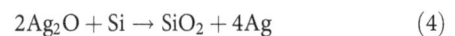

$$2Ag_2O + Si \rightarrow SiO_2 + 4Ag \qquad (4)$$

Not all agree with the Ag_2O-based deposition model. Schubert (ref. 15) and Horteis[16] argue that Pb formed in reaction 1 lowers the Ag–Si interaction temperature such that when the molten system phase separates upon cooling Ag deposits on the Si surface and within the glass layer. Sopori *et al.*[17] agree with an Ag–Si alloy-assisted contact formation mechanism, and suggest involvement of other metal solvent species to further depress the melting temperature. In addition, the extent of Pb formation, argued to assist in Ag sintering, has been questioned. Thus, ambiguity persists regarding details of metal transport and precipitation, the extent of Ag–Pb alloying, the extent of Ag–Pb–Si alloying if any occurs at all, and therefore the keys for achieving low contact resistance.

Here, direct observation of SiN etching by PbO and Ag_2O during different stages of firing using *in situ* XRD reveals the burn-through reaction and contact formation sequence. Studies on Ag, frit, c-Si and SiN model systems show a strong dependence of Pb formation, and hence Ag–Pb alloying, on oxygen (O_2) partial pressure during firing. In addition, Ag dissolution into the glass frit, and hence the resulting abundance of Ag precipitate formation, shows a strong O_2-dependence. This investigation finds the Ag redox reactions only occur with sufficient O_2 at temperatures higher than 650 °C – negating a role of Ag in opening of the antireflective-coating at lower temperatures. Our findings also explain increased Pb formation observed when firing in a N_2 atmosphere compared with air. Understanding the temperature thresholds for each of these competing mechanisms accurately clarifies the nature of screen-printed, fast-fired contacts.

Results

Powder model system. Model systems containing Ag nanocrystals, PbO-based frit, SiN_x and c-Si, in varying combinations, have been fired in an air or nitrogen (N_2) atmosphere and characterized using an *in situ* XRD/RTP apparatus[18]. Use of powder model systems enabled characteristic XRD data acquisition with short collection intervals (<1 s), with high signal-to-noise achievable due to high interface area between reactants of interest within the sampled volume. This allowed detection of varying reaction species abundance with high sensitivity during rapid heating and cooling with temperature profiles analogous to solar cell contact firing conditions. This provides a unique opportunity to observe the reaction sequence in real time; such detailed analysis is impossible to perform on an actual Ag contact on a solar cell during firing in a belt-furnace. Results at the end of this section show that the powder mixture model systems fired by RTP in the present study behave analogous to fast-fired contacts made by screen-printing with conventional Ag pastes. Sample preparation and measurement procedures are explained further in the Methods section.

Pb formation. XRD patterns measured during firing of a frit/SiN$_x$ (1:1 molar ratio) mixture in air to 800 °C (Fig. 1a) demonstrate the *in situ* measurement capability. The frit and amorphous-SiN$_x$ generate only weak, broad diffraction, in the initial patterns. However, sharp crystalline-Pb peaks emerge after firing, as the system cools below 327 °C (the melting temperature of Pb), which substantiates evidence of reaction 1. Results from analogous measurements on Ag/frit/SiN$_x$ mixtures (Fig. 1b), fired to 500, 600, 700 and 800 °C, in air and N$_2$, provide insights concerning the temperature threshold to initiate SiN$_x$ etching, the accompanying Pb formation, and hence the extent of Ag–Pb alloying during contact firing.

As shown in Fig. 1b, which plots the ratio of Pb(111) integrated intensity to that of Ag(111), varying amounts of Pb form depending on both the maximum temperature and the firing atmosphere. Pb is not detected upon heating to 500 °C, whereas heating to 600 °C induces measurable Pb in both air and N$_2$. Trend-lines show the amount of Pb increasing progressively with maximum firing temperature. Heating in N$_2$ promotes Pb formation significantly compared with firing in air. Analogous

behavior is observed in experiments with Ag/frit/c-Si, wherein PbO oxidizes c-Si by:

$$2PbO + Si \rightarrow 2Pb + SiO_2 \qquad (5)$$

Enhanced Pb formation in N$_2$ is explained by irreversible PbO reduction. In air, when Pb forms by reaction 1, it tends to react with O$_2$, or Ag$_2$O in the frit if present, and oxidize back to PbO by:

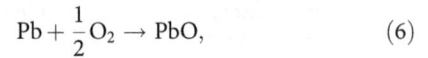

$$Pb + \frac{1}{2}O_2 \rightarrow PbO, \qquad (6)$$

or:

$$Ag_2O + Pb \rightarrow 2Ag + PbO \qquad (7)$$

respectively. Thus, it seems regeneration of PbO by reactions 6 and 7 allows complete SiN$_x$ etching despite a very small loading of PbO in the Ag paste (about 2 wt.%), as these reactions perpetuate the burn-through process.

Burn-through reaction sequence revealed. Figure 2 shows diffraction patterns obtained from an Ag/frit/SiN$_x$/Si mixture at different stages of firing. The Si diffraction decreases during heating between 550 and 750 °C, and does not recover upon cooling. This suggests oxidation of c-Si by Ag$_2$O (reaction 4) and/or PbO (reaction 5). The Ag peaks initially narrow during

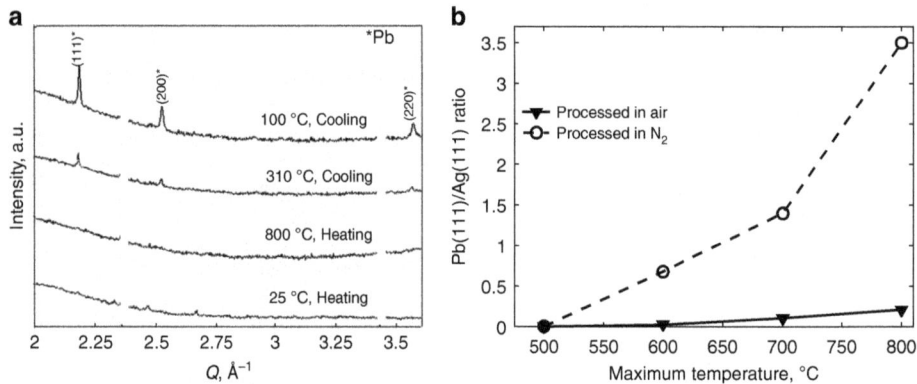

Figure 1 | Lead formation dependence on firing temperature and atmosphere. (**a**) XRD patterns measured from frit/SiN$_x$ during firing and cooling in air and (**b**) integrated Pb(111) diffraction intensity normalized by Ag(111) intensity observed upon firing Ag/frit/SiN$_x$ mixtures to varying temperatures in air and N$_2$.

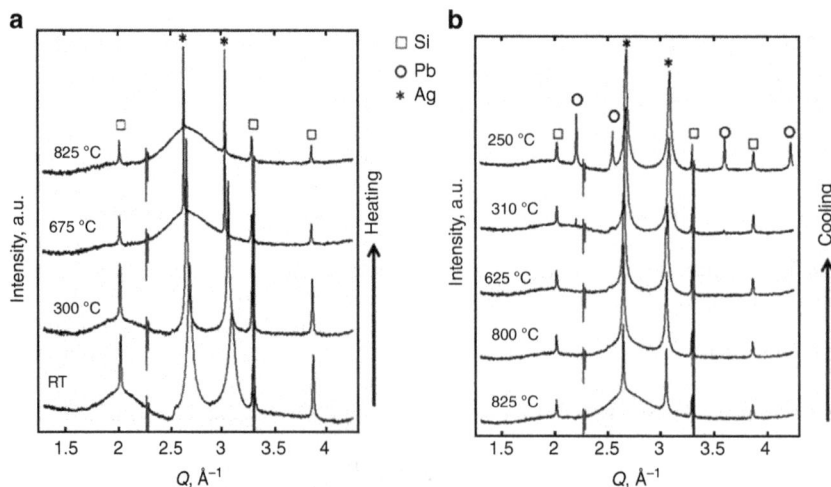

Figure 2 | *In situ* diffraction measured from Ag/frit/SiN$_x$/Si during firing. (**a**) XRD profiles obtained during heating to a max temperature of 825 °C at 50 °C s^{-1} and (**b**) cooling in air. Squares, circles, and stars indicate Si, Pb and Ag peaks, respectively. Note the changes in diffuse scattering from amorphous phases, near $Q = 2$ Å^{-1} (attributed to glass frit) at low temperatures, and near $Q = 2.7$ Å^{-1} (attributed to molten metal) at ~675–825 °C.

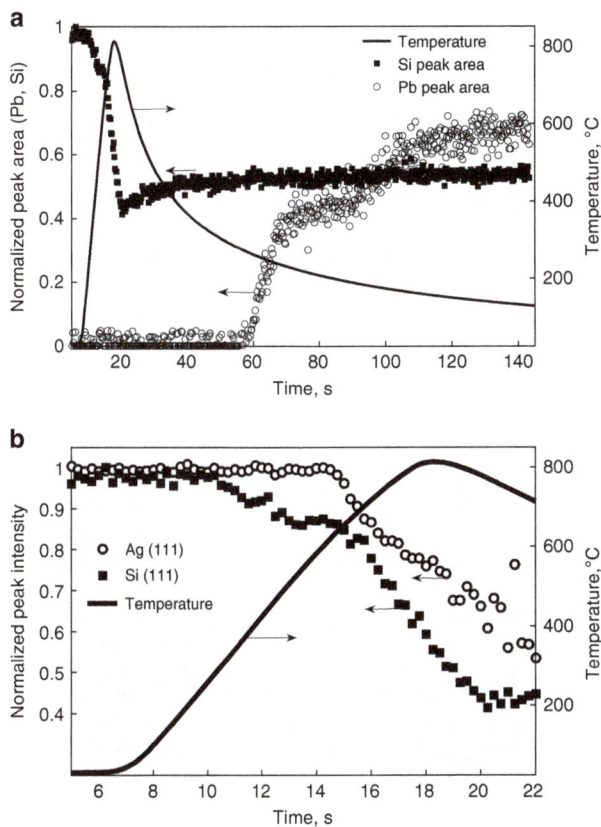

Figure 3 | *In situ* diffraction from a model frit system during firing in air.
(**a**) Temperature profile versus time (solid line) and corresponding variation in Si and Pb diffraction intensity during temperature ramping and cooling. (**b**) Si and Ag peak-area observed during firing, along with the temperature ramp profile (solid line).

Figure 4 | *In situ* measurement of Ag diffraction during isothermal heating. Ag(111) diffraction intensity as function of time during heating of Ag/frit (5 wt.% Ag) in air suggests Ag dissolution into the frit at 700 °C, but not at 600 °C.

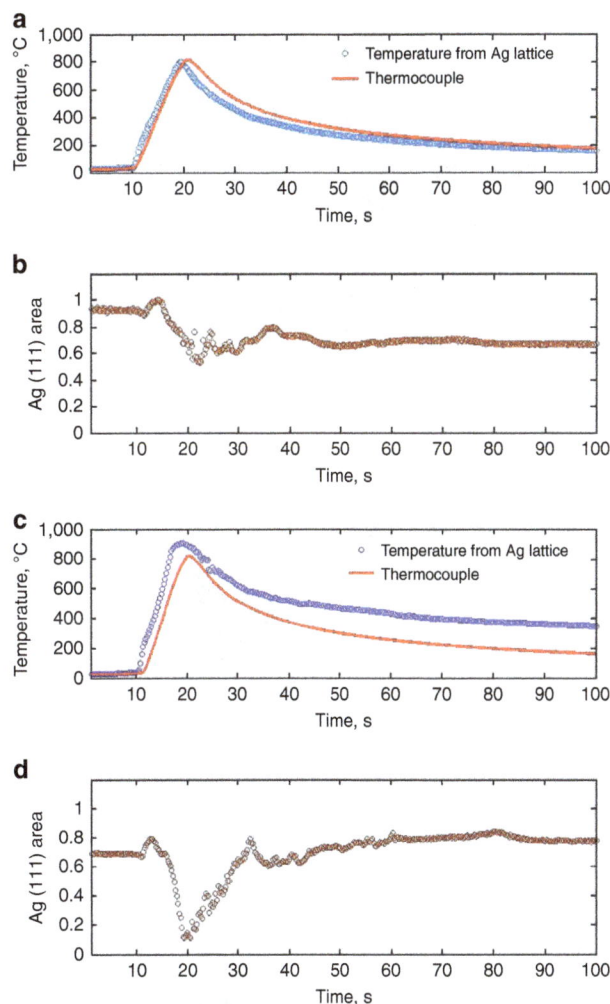

Figure 5 | Temperature profiles and Ag(111) diffraction measured from Ag/frit/SiN$_x$/Si during firing (100 °C s^{-1}) and cooling in air and N$_2$. (**a,c**) shows the temperature measured by thermocouple (red lines) versus calculated based on the Ag peak position (blue lines), and (**b,d**) shows the Ag diffraction intensity dependence on time. (**a,b**) corresponds to firing in air, while (**c,d**) is in N$_2$.

temperature ramping as a consequence of Ag grain growth by sintering and ripening. Then on further heating to temperatures above 650 °C the Ag peak intensity diminishes, and a diffuse peak emerges near the Ag (111) peak (that is, at $Q \cong 2.7\,\text{Å}^{-1}$). The emerging diffuse peak is attributed to scattering from an amorphous phase, such as liquid metal (possibly Pb or Ag–Pb alloy), which is not present at room temperature and becomes most intense at the max temperature (825 °C) where the Ag peak intensity is a minimum. On cooling, the diffuse peak disappears while the Ag peaks gain intensity and broaden. Most likely, Ag peak broadening reflects contributions from newly formed Ag nanocrystals that precipitate within the glass during cooling. Pb peaks appear below 327 °C.

Loss of Ag diffraction intensity during temperature ramping most likely follows from one of two possible mechanisms: Pb formed by redox reaction with SiN$_x$ alloys and melts Ag, or Ag oxidizes and dissolves into the molten frit by reaction 3. In fact, the present study finds that both mechanisms occur, though with different onset temperatures. Examining the intensity trends versus time and temperature allows decoupling of the competing mechanisms. Figure 3 shows the corresponding temperature profile versus time along with Ag, Si and Pb integrated intensities. The Si diffraction intensity begins to decrease at about 10 s (500 °C). Then, the Ag peak intensity begins to diminish 5 s later, at about 14 s (650 °C). As the Ag signal begins to drop, the rate of Si signal loss accelerates markedly. Loss of Si signal results from Si oxidation by either PbO or Ag$_2$O in the frit. Ag dissolution into the molten frit by reaction 3 must occur before etching by Ag$_2$O (reaction 4). Since the Si signal decreases before the onset of Ag

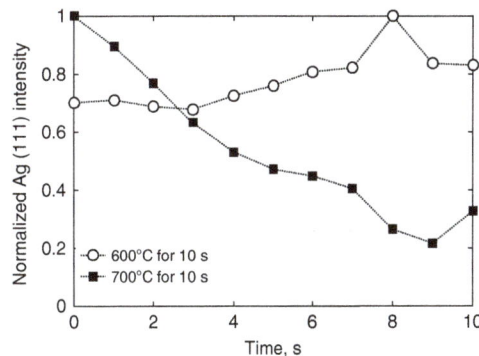

intensity loss, the low temperature oxidation (<650 °C) must be dominated by reaction with PbO in the frit (that is, by reaction 1). The observation of Pb diffraction upon cooling below 320 °C supports this interpretation, as this confirms formation of Pb.

Excellent agreement between the onset of Ag intensity decline and increased rate of Si oxidation at 15 s (650 °C) suggests a change in the dominant oxidation reaction during temperature ramping – most likely the onset of Si etching by Ag_2O (reaction 4). Indeed, with Ag being the most noble metal in the system, Si oxidation by Ag_2O, if present, should dominate given its thermodynamic favorability compared to reaction 1 (ref. 4).

Involvement of Ag in SiN_x burn-through unlikely. Examining the nature of Ag dissolution into PbO-based frit under varying temperature conditions negates the possibility of Ag participation in SiN_x removal during the early stages of contact firing. To test this, Ag/frit mixtures (5 wt.% Ag) were isothermally annealed at 600 and 700 °C for 10 s. As shown in Fig. 4, annealing at 600 °C

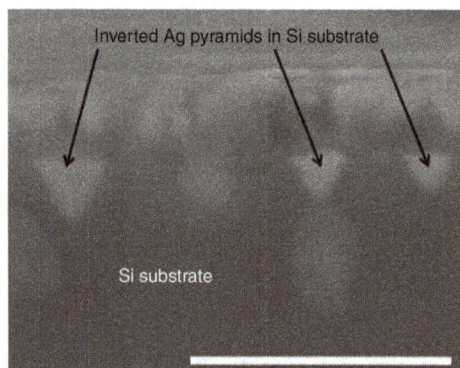

Figure 6 | Cross section SEM showing Ag pyramids in a fired test sample. SEM image at ×100,000 magnification where the scale bar is 1 μm, shows a cross-section of an Ag/frit/SiN_x sample fired to 800 °C at 100 °C s^{-1} and reveals formation of inverted Ag pyramids penetrating the surface of the c-Si substrate.

induces only a modest increase in Ag(111) intensity, probably due to grain growth. On the other hand, at 700 °C, the Ag intensity decreases and only 30% of the initial intensity remains after 10 s. The loss of Ag diffraction intensity implies about 70% Ag dissolution into the frit by reaction 3. Therefore, Ag dissolution into the frit must occur with a temperature threshold between 600 °C and 700 °C, and hence, oxidation of SiN_x or Si (observed onset temperature is approximately 500 °C) below this temperature range cannot involve Ag^+ ions. The majority of SiN_x oxidation during temperature ramping must therefore be attributed to redox reactions with PbO.

Silver transport. Monitoring the Ag peak intensity during firing reveals the extent of Ag dissolution into the glass under varying conditions. Notably, the response of an Ag/frit/SiN_x/Si mixture strongly depends on whether firing occurs in air or N_2 (Fig. 5). During heating, the Ag peaks initially become sharper as a consequence of grain growth, and the intensity increases slightly. Then, above 650 °C (about 15 s), the Ag intensity diminishes continuously with increasing temperature up to the maximum temperature of 810 °C (about 20 s). In both cases, air and N_2 atmosphere, the Ag peak intensity recovers during cooling, albeit to a different extent. In air, as applies in solar cell manufacturing, the Ag signal intensity decreases to approximately 50% at maximum temperature, and then recovers to about 75% of the initial value on cooling. On the other hand, when firing in N_2, the Ag signal intensity drops to just 10% of the initial value at the maximum temperature, and then recovers completely upon cooling.

The different behavior in air versus N_2 reflects the different forms of Ag during firing, respectively. According to the equilibrium Ag–Pb phase diagram, metallic Pb formed by redox reactions between PbO and SiN_x readily forms a liquid alloy with Ag in this temperature range. As was previously shown, more Pb forms in N_2 due to irreversible reduction of PbO (Fig. 1b). Hence more Ag–Pb alloying occurs in N_2, which is almost entirely

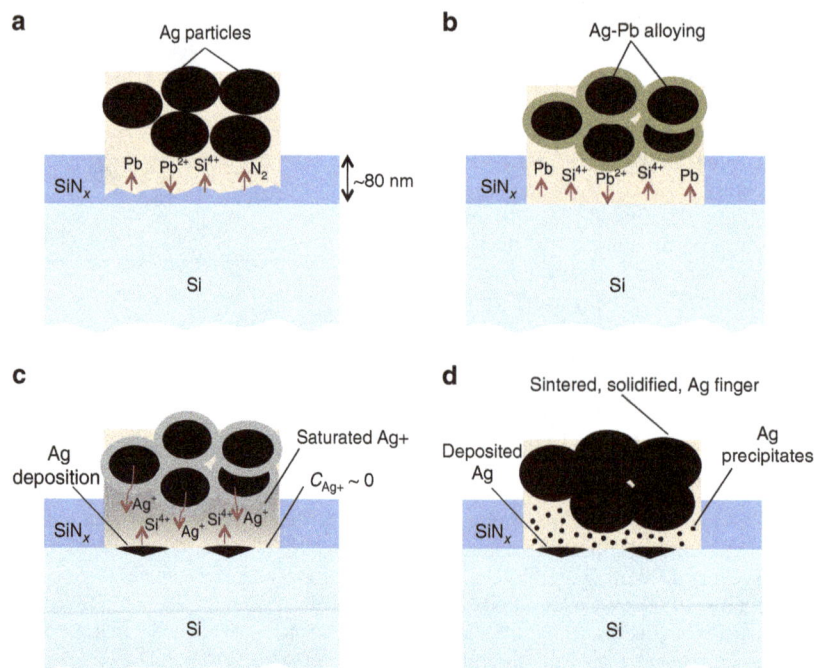

Figure 7 | Illustrations showing stages of contact formation during firing. (**a**) SiN_x etching by PbO in the frit, (**b**) Ag-Pb alloying, (**c**) Ag^+ transport through molten frit and deposition at the Si surface, and (**d**) resulting fired-contact morphology, with inclusion of small Ag precipitates within the glass intermediate layer.

reversible due to low miscibility of Ag and Pb solid phases at lower temperatures. On the other hand, Fig. 4 showed that Ag dissolves readily into the glass in the presence of O_2 above 650 °C. Thus, in the case of firing in air, the dip in Ag signal reflects both Ag–Pb alloying and Ag dissolution, whereas only the former occurs in N_2. In air, some of the Ag signal recovers as Ag nanocrystals precipitate within the glass upon cooling as a consequence of the temperature-dependent solubility. However, weak scattering from nanocrystals having a diameter below 2 nm, as are commonly seen in cross-sections of fired contacts[11–14], and Ag remaining bound as Ag^+ within the glass, are both consistent with lower Ag signal detected after firing in air.

Before considering implications of these results on screen-printed solar cell contacts, it is important to show that the model systems used in the present work behave representatively. Figure 6 shows a cross-section SEM of our model frit system with inverted Ag pyramids penetrating the surface of the c-Si substrate beneath the Ag/frit/SiN$_x$ mixture after firing. The Ag pyramid formation observed here is analogous to that observed in numerous prior studies on fired Ag contacts[3,7,13,15,16]. Therefore, our frit, powder mixtures and firing conditions clearly mimic the screen-printed contact firing process. Interestingly, in this case, the mixture contained a larger proportion of frit (1:1 frit/Ag) compared with common Ag pastes used in solar cell contacts (typically lower than 0.1:1), which seems to have induced a larger characteristic Ag pyramid size compared with most other studies.

Discussion

A complete picture of the contact formation process, consistent with all previous findings and informed by the present *in situ* results, emerges as follows: during firing, above 500 °C, the frit melts and wets the Ag/SiN$_x$ interface. Between 500 and 650 °C, PbO in the frit reacts with and penetrates the SiN$_x$ anti-reflective layer by reaction 1, as shown in Fig. 7a. Pb formed during the burn-through process alloys with and assists in liquid-phase assisted sintering of Ag (Fig. 7b). Above 650 °C, Ag dissolves into the frit and diffuses towards the emitter surface. In the vicinity of the Si emitter surface, Ag ions are consumed by redox reaction 4, which oxidizes Si to SiO_2 and deposits metallic Ag on the emitter surface (Fig. 7c). SiO_2 formed by reaction 4 is incorporated into the molten frit. On cooling, the solubility of Ag in the melt decreases and nanocrystals precipitate from within the glass matrix. Kinetic constraints limit grain growth during fast cooling, resulting in a high density of distributed nanocrystals within the glass intermediate layer (Fig. 7d).

In summary, Ohmic contacts with low resistivity result when Ag crystals form within a thin glass intermediate layer and on the emitter surface; thus, in general, it would seem optimal contacts result with adequate heating to achieve complete SiN$_x$ opening and abundant Ag precipitation within the glass. At the same time, over-firing must be avoided both to prevent excessive Ag deposition onto the emitter – Ag spikes – which can shunt the *pn*-junction in extreme cases, and to minimize SiO_2 generation, which increases the thickness of the glass layer and diminishes carrier tunneling.

In the larger effort to replace costly Ag and toxic Pb in the solar cell manufacturing sector, the present study offers several insights. To replace Ag with an alternative, low-resistivity metal, assuming suitable work-function alignment for contacting Si, analogous use will require the metal have variable solubility in a frit system within a comparable temperature range (below 800 °C). The PbO-based frit may be replaced by a frit system containing another metal-oxide that alloys with the primary contact metal upon metal-oxide reduction and provides a low glass transition temperature. This is a necessary condition to achieve wetting and adhesion at the contact interface. The frit must contain a reactive species with a negative redox potential in relation to SiN$_x$, and preferably to both SiN$_x$ and Si. Ultimately, metal solubility and transport kinetics enabling nanocrystal precipitation within the glass upon cooling, which allows carrier tunneling (that is, ohmic contact formation), depends strongly on the frit composition. Thus, the metal and accompanying frit used in the paste must be developed in tandem.

Methods

Model system. Ag, frit, SiN$_x$ and c-Si model systems in this study have been characterized as powder mixtures. Our model, controlled PbO-based frit was synthesized following ref. 15 by quenching a molten mixture of PbO (60 mol%), SiO_2 (30 mol%) and B_2O_3 (10 mol%), and then milling to form a powder. Ag nanocrystals and SiN$_x$ powder was purchased from Sigma-Aldrich, and c-Si nanocrystals were from Nanostructured and Amorphous Materials, Inc. Samples of varying composition were mixed to desired proportions in dry form, and then deposited onto 20 × 20 mm Si wafer substrates using isopropyl alcohol as a dispersing agent. Multiple samples were characterized for each test condition to ensure the reliability of the data.

Measurement and heating. Our model systems were fired at heating rates around $100 °C s^{-1}$ and characterized by XRD in real-time using a synchrotron-mounted, RTP apparatus, equipped with a Pilatus, Dectris 300 K area detector, with 250 ms integration time (4 Hz sampling frequency). Specifics of the *in situ* XRD/RTP apparatus can be found elsewhere[18]. Briefly, the *in situ* RTP chamber was gold-coated aluminium enclosure with four high-power quartz-tungsten-halide lamps (750 W each). The chamber is sealed and plumbed to allow atmosphere control. Al coated Kapton windows were used to allow high X-ray transmission while containing heat during thermal processing. The RTP chamber body was water cooled by means of a chiller. Firing under N_2 atmosphere was accomplished by purging the RTP chamber for 5 min prior to heating. *In situ* temperature measurements were obtained using a thermocouple, and cross-referenced based on calculations using the Ag diffraction signal (that is, the temperature dependence of the Ag lattice). This method provides a high level of accuracy in temperature measurement.

Data analysis. Analysis of *in situ* XRD datasets was accomplished by fitting peaks with Pearson VII functions. The integrated area of the diffraction peak for a given phase corresponds directly with its abundance, which allowed extraction of phase-evolution trends versus time for the individual species of interest.

Cross-section SEM was performed using an FEI Nova NanoSEM, with a working-distance of 4.4 mm, accelerating potential of 10 kV and 1.3 mA operating current.

References

1. Mertens, R. *et al.* Critical processing parameter optimization for screen printed semicrystalline silicon solar cells. *Proc. 17th IEEE PVSC* 1347–1351 (1984).
2. Schubert, G., Fischer, B. & Fath, P. Formation and nature of Ag thick film front contacts on crystalline silicon solar cells. *Proc. PV Eur. Rome* 343, 343–346 (2002).
3. Schubert, G., Huster, F. & Fath, P. Physical understanding of printed thick-film front contacts of crystalline Si solar cells–review of existing models and recnet developments. *Sol. Energy Mater. Sol. Cells* 90, 3399–3406 (2006).
4. Horteis, M., Gutberlet, T., Reller, A. & Glunz, S. W. High-temperature contact formation on n-type silicon: basic reactions and contact model for seed-layer contacts. *Adv. Funct. Mater.* 20, 476–484 (2010).
5. Ueda, S., Kumagai, T. & Yamaguchi, K. Thermodynamic study on the Ag–Pb–O system at 1273 K. *Mater. Trans.* 46, 1861–1864 (2005).
6. Ueda, S., Kumagai, T. & Yamaguchi, K. Activity coefficient of $AgO_{0.5}$ in the PbO–SiO$_2$ melt at 1273 K. *Mater. Trans.* 48, 1458–1462 (2007).
7. Hong, K. K. *et al.* Mechanism for the formation of Ag crystallites in the Ag thick-film contacts of crystalline Si solar cells. *Sol. Energy Mater. Sol. Cells* 93, 898–904 (2009).
8. Chung, B. M. *et al.* Influence of oxygen on Ag ionization in molten lead borosilicate glass during screen-printed Ag contact formation for Si solar cells. *Electrochim. Acta* 106, 333–341 (2013).
9. Prudenziati, M., Moro, L., Morten, B., Sirotti, F. & Sardi, L. Ag-based thick-film front metallization of silicon solar cells. *Act. Pass. Elect. Comp.* 13, 133–150 (1989).
10. Ballif, C., Huljic, D. M., Willeke, G. & Hessler-Wyser, A. Silver thick-film contacts on highly doped n-type silicon emitters: structural and electrical properties of the interface. *Appl. Phys. Letts.* 82, 1878–1880 (2003).
11. Hilali, M. M. *et al.* Understanding the formation and temperature dependence of thick-film Ag contacts on high-sheet-resistance Si emitters for solar cells. *J. ECS* 152, 742–749 (2005).

12. Psych, D., Mette, A., Filipovic, A. & Glunz, S. W. Comprehensive analysis of advanced solar cell contacts consisting of printed fine-line seed layers thickened by silver plating. *Prog. Photovoltaics* **17,** 101–114 (2009).

13. Hong, K. K., Cho, S. B., Huh, J. Y., Park, H. J. & Jeong, J. W. Role of PbO-based glass frit in Ag thick-film contact formation for crystalline Si solar cells. *Met. Mater. Int.* **15,** 307–312 (2009).

14. Cho, S. B., Hong, K. K., Huh, J. Y., Park, H. J. & Jeong, J. W. Role of the ambient oxygen on the silver thick-film contact formation for crystalline silicon solar cells. *Curr. Appl. Phys.* **10,** 222–225 (2010).

15. Schubert, G. *Thick Film Metallization of Crystalline Silicon Solar Cells* (PhD thesis, University of Konstanz, 2006).

16. Horteis, M. *Fine-Line Printed Contacts on Crystalline Silicon Solar Cells* (PhD thesis, University of Konstanz, 2009).

17. Sopori, B. *et al.* Fundamental mechanisms in the fire-through contact metallization of Si solar cells: a review. in *17th Workshop on Crystalline Silicon Solar Cells and Modules: Materials and Processes; Workshop Proceedings.* No. NREL/BK-520-42056 (National Renewable Energy Laboratory (NREL), Golden, CO, 2007).

18. Ahmad, M. I. *et al.* Rapid thermal processing chamber for *in-situ* x-ray diffraction. *Rev. Sci. Inst.* **86** 130902-1–130902-7 (2015).

Acknowledgements

This project is funded through the Bridging Research Interactions through collaborating the Development Grants in Energy (BRIDGE) program under the SunShot initiative of the Department of Energy (DE-EE0005951). Sample preparation and SEM analysis were performed at the National Renewable Energy Laboratory, which is operated under the prime contract no. DE-AC36-08GO28308. *In situ* characterization was performed at the Stanford Synchrotron Radiation Laboratory, a national user facility operated by Stanford University on behalf of the US Department of Energy, Office of Basic Energy Sciences, under contract no. DE-AC02-76SF00515. We thank Bobby To at NREL for performing the SEM analysis, and Ron Marks and Bart Johnson at SSRL for assistance with beam line 7-2.

Author contributions

J.D.F., M.I.A., V.L.P., M.F.A.M.v.H. and M.F.T. conceived and designed the experiments. J.D.F. synthesized and prepared the samples. J.D.F., M.I.A., V.L.P., J.Y. and M.F.A.M.v.H. conducted X-ray experiments. M.I.A., D.V.C. and J.Y. designed and built the RTP chamber with input from M.F.A.M.v.H. and P.A.P. M.I.A. and M.F.T. analyzed the diffraction data. All authors discussed the results and contributed to the manuscript.

Additional information

Competing financial interests: The authors declare no competing financial interests.

26

Polarized three-photon-pumped laser in a single MOF microcrystal

Huajun He[1,*], En Ma[2,*], Yuanjing Cui[1,*], Jiancan Yu[1], Yu Yang[1], Tao Song[1], Chuan-De Wu[3], Xueyuan Chen[2], Banglin Chen[1,4] & Guodong Qian[1]

Higher order multiphoton-pumped polarized lasers have fundamental technological importance. Although they can be used to *in vivo* imaging, their application has yet to be realized. Here we show the first polarized three-photon-pumped (3PP) microcavity laser in a single host–guest composite metal–organic framework (MOF) crystal, via a controllable *in situ* self-assembly strategy. The highly oriented assembly of dye molecules within the MOF provides an opportunity to achieve 3PP lasing with a low lasing threshold and a very high-quality factor on excitation. Furthermore, the 3PP lasing generated from composite MOF is perfectly polarized. These findings may eventually open up a new route to the exploitation of multi-photon-pumped solid-state laser in single MOF microcrystal (or nanocrystal) for future optoelectronic and biomedical applications.

[1] State Key Laboratory of Silicon Materials, Cyrus Tang Center for Sensor Materials and Applications, School of Materials Science and Engineering, Zhejiang University, Hangzhou 310027, China. [2] Key Laboratory of Optoelectronic Materials Chemistry and Physics, Fujian Institute of Research on the Structure of Matter, Chinese Academy of Sciences, Fuzhou, Fujian 350002, China. [3] Department of Chemistry, Zhejiang University, Hangzhou 310027, China. [4] Department of Chemistry, University of Texas at San Antonio, San Antonio, Texas 78249-0698, USA. * These authors contributed equally to this work. Correspondence and requests for materials should be addressed to X.C. (email: xchen@fjirsm.ac.cn) or to B.C. (email: banglin.chen@utsa.edu) or to G.Q. (email: gdqian@zju.edu.cn).

Polarization has been used in various fields, particularly in the field of biophotonics due to its ability to reduce multiple scattering, while to enhance the contrast and to improve tissue imaging resolution[1-3]. Through the measurement of the polarization state of the scattered light, a wealth of structural information of scatters (for example, lesions information in the tissue) can be collected given the fact that the microscopic structure of a scattering media is closely related to changes in the polarization state of the photon during a scattering process[1,3]. On the other hand, high-order multiphoton excitation can offer stronger spatial confinement, deeper tissue penetration and less Rayleigh scattering, which are significantly beneficial to the biological imaging[4-8]. To make use of the uniqueness of both polarization and high-order multiphoton excitation, the polarized three-photon and/or higher order pumped laser in single solid-state microcrystal is potentially useful for a new kind of biological imaging, so called multiphoton pumped (MPP) polarized emission biological imaging (Supplementary Fig. 1), but has never been realized. In order to produce such a unique laser, the gain medium not only needs to have a high multiphoton absorption (MPA) cross-section and lasing efficiency[4,5], but more importantly needs to be assembled into a suitable microcavity of high concentration and orientation (especially in the case that the absorption transition moment of gain medium is anisotropic) without significant luminescent quenching to enforce the high optical gain and to generate controllable and directional laser. This is really a daunting challenge. In fact, although extensive research endeavours have been pursued to target such a goal, progress has been very slow. The initial effort to diminish the significant quenching effects on the solid state was to homogeneously disperse the gain medium such as the dye molecules with high multiphoton absorption cross-section into its solution[6,9]. By employing such a strategy, it still remains extremely difficult to provide with a sufficiently high quenching concentration, which limits the realization of necessary optical gain for compensating the losses. The quenching concentration means that the aggregation-caused quenching (ACQ) gradually becomes dominant when the concentration of gain medium is higher than the quenching concentration. Furthermore, the molecules in the solution are randomly oriented, which would limit their capacities to maximize the optical gain. So far, this dispersed solution methodology can only lead to the amplified spontaneous emission instead of generating three-photon or higher order pumped laser[4,5]. Although the luminescent properties of quantum dots are intriguing, they only have generated three-photon-pumped (3PP) random lasing in which the emission direction, position and numbers of mode frequency, and the uniformity of light-emitting region of such lasers are very difficult to control[7,10]. Recently, the 3PP lasing from colloidal nanoplatelets in solution has been demonstrated by Li et al.[11]; however, no polarization property of the 3PP lasing has been realized. Furthermore, its liquid nature has limited practical applications. To take advantage of the pore confinement of porous materials, zeolites and nanoporous silica have been explored to incorporate dye molecules and semiconducting polymers into the corresponding crystals and thin films to develop solid-state lasing[12,13]. However, zeolite/dye composites can only generate single-photon pumped lasing, mainly due to the incompatibility between the inorganic framework and organic guest, leading to the low loading concentration (0.005 ~ 0.0005 M), uneven distribution of dye molecules and poor crystal morphology; while nanoporous silica/semiconducting polymer matrix basically leads to the single-photon pumped polarized amplified spontaneous emission.

Previously, we have used a porous metal–organic framework (MOF) for its pore confinement of a dye molecule bearing

moderately high two-photon absorption cross-section, and realized the two-photon pumped lasing from a composite crystal bio-MOF-1 ⊃ DMASM (DMASM = 4-[p-(dimethylamino)styryl] -1-methylpyridinium) at room temperature[14]. However, the pores (two types of channels along the c-axis of about 7.0 and 10.0 Å, respectively) within bio-MOF-1 are still too large to exactly match the dye molecules of DMASM, thus the orientation of the dye molecules inside the pore cavities is not of a high order, particularly when the high concentration of the dye molecules are applied. Such a moderate pore confinement of bio-MOF-1 apparently has limited us to further realize the higher order multiphoton pumped laser in this solid-state crystal. In order to enhance the pore confinement efficiency of a porous MOF crystal, the pore sizes within a porous MOF need to be tuned to match the size of the dye molecule better. But the dilemma is that when the pore sizes of a porous MOF can exactly match the size of the dye molecule, the dye molecules cannot diffuse into the pore channels through the simple post-synthetic exchange process. To overcome this problem, we have developed an in situ self-assembly strategy[15,16]: the components for building a MOF crystal (metal ion and organic linker) and the organic dye molecule are simultaneously assembled together to form the MOF/dye single crystals. Such a methodology has enabled us to tightly incorporate the dye molecules into the porous MOF crystals, and thus the dye molecules are highly ordered and oriented. We have also managed to immobilize high concentration of the dye molecules into the MOF crystal ZJU-68 ⊃ DMASM ($(DMASM)_{0.33}H_{1.67}[Zn_3O(CPQC)_3]$, CPQC, 7-(4-carboxyphenyl)quinoline-3-carboxylate) (the average pore size of the one-dimensional channel along the c-axis is 6.0 Å) with the dye content over 0.4 M. Furthermore, the suitable refraction index and well-faceted MOF composite crystals of certain morphology symmetries can be naturally and efficiently utilized as the laser resonant cavities without any other fabrications. The powerful in situ self-assembly strategy, highly efficient pore confinement of ZJU-68 for DMASM dye molecule, and suitable refraction index as well as perfect crystal morphology have enabled us to target the first example of polarized three-photon-pumped laser in single solid-state microcrystal.

Results

Synthesis and characterization. Reaction of a new organic linker 7-(4-carboxyphenyl)quinoline-3-carboxylic acid (H_2CPQC) containing quinolone group and $Zn(NO_3)_2 \cdot 6H_2O$ in N,N-dimethylformamide/acetonitrile/H_2O/HBF_3 at 100 °C affords colourless hexagonal prism crystals of $H_2[Zn_3O(C_{17}H_9NO_4)_3]$ $\cdot 2.5H_2O \cdot 0.5DMF \cdot MeCN$ (ZJU-68, Fig. 1a). Single crystal X-ray diffraction studies reveal that ZJU-68 crystallizes in the $P\bar{3}$ space group (see Supplementary Table 1 for detailed crystallographic data). As shown in Fig. 2a, trinuclear secondary building units (SBUs) of $[Zn_3O]^{4+}$ are linked by the ligands $CPQC^{2-}$ to form an anionic framework of $[Zn_3O(C_{17}H_9NO_4)_3]^{2-}$. In this structure, nine coordination sites of $[Zn_3O]^{4+}$ are completely occupied by six carboxylates and three of nitrogen atoms from the quinoline moieties, which are different from most of metal–organic frameworks with $[M_3O]^{3n-2}$ ($n=3$ for $M=Cr^{3+}$, Fe^{3+} and so on or $n=2$ for $M=Zn^{2+}$, Cu^{2+}) SBUs in which three sites are occupied by small capping ligands such as water and hydroxide[17,18].

The crystal has one-dimensional (1D) sub-nano channels along the c-axis with a hexagonal cross-section (Fig. 2b; Supplementary Fig. 2). The edge of the hexagon is about 3.0 Å. For the synthesis of laser dye functionalized crystals, we tried to introduce linear-shaped laser dye cations DMASM via an ion-exchange process, as described in our previous work[14], but failed. This is because the DMASM molecule (about 6.3 Å in the width, Supplementary

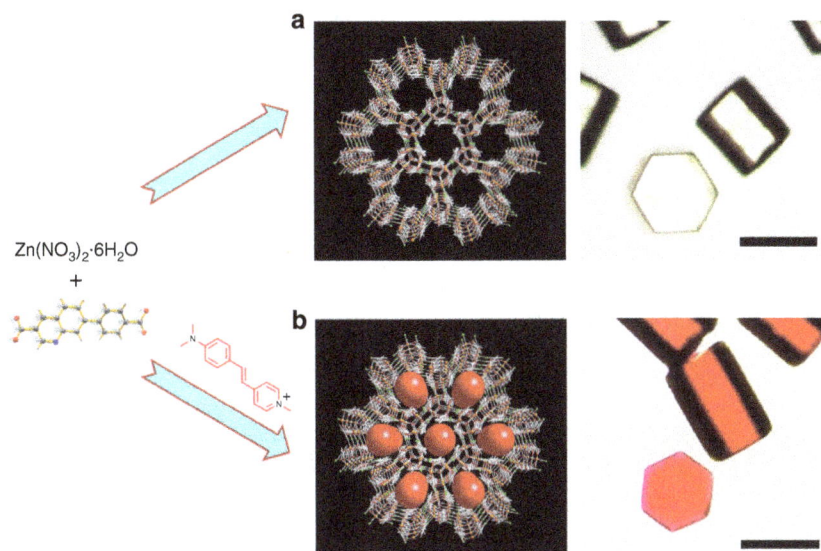

Figure 1 | Schematic synthesis of ZJU-68 and ZJU-68 ⊃ DMASM. (a) The synthesis and micrograph of a novel metal–organic framework ZJU-68. **(b)** *In situ* synthesis of laser dye incorporated metal–organic framework crystals ZJU-68 ⊃ DMASM. The inclusion of the red dye DMASM molecules leads to the color change from the original colourless ZJU-68 to red ZJU-68 ⊃ DMASM. Scale bar, 50 μm.

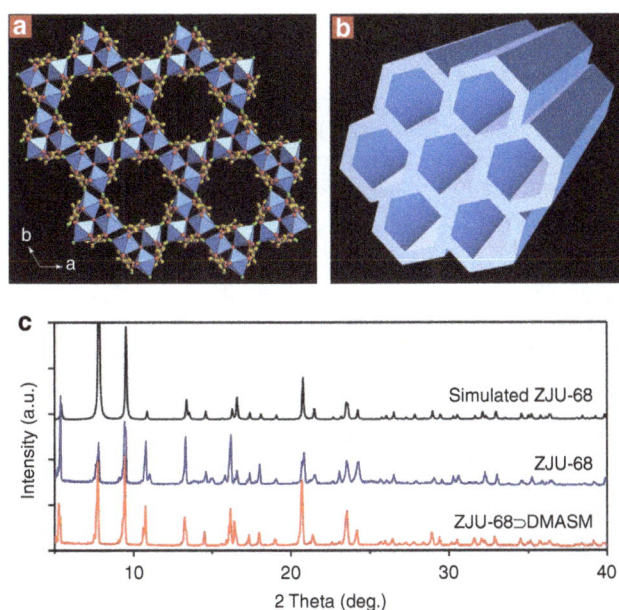

Figure 2 | The structure of a novel metal–organic framework crystal ZJU-68. (a) Crystal structure of ZJU-68 viewed along the crystallographic c direction (C, orange; N, green; O, red; Zn, blue polyhedra). H atoms and solvent molecules are omitted for clarity. In this structure, nine coordination sites of a trinuclear SBU $[Zn_3O]^{4+}$ are completely occupied by six carboxylates and three of nitrogen atoms from the quinoline moieties, which may play a crucial role in the stabilization of the resulting MOF, ZJU-68. **(b)** The simplified network structure of ZJU-68, displaying 1D channels along the c-axis. Different objects are not drawn to scale. **(c)** PXRD patterns of ZJU-68 and ZJU-68 ⊃ DMASM, which indicate that the ZJU-68 ⊃ DMASM has the identical framework structure with ZJU-68.

solution (Fig. 1b). The resulting dye DMASM included ZJU-68 ⊃ DMASM has the same hexagonal prism crystal morphology. The inclusion of the red dye DMASM molecules leads to the colour change from the original colourless ZJU-68 to red ZJU-68 ⊃ DMASM. Both single crystal and the powder X-ray diffraction studies (Fig. 2c) confirmed that the ZJU-68 ⊃ DMASM has the identical framework structure with ZJU-68. Furthermore, both ZJU-68 and ZJU-68 ⊃ DMASM demonstrate excellent stability in the air and in the common solvents such as water, ethanol and dimethylformamide (Supplementary Fig. 3). Of course, most of the 1D hexagonal channel spaces have been occupied by DMASM molecules in ZJU-68 ⊃ DMASM. Supplementary Fig. 4 shows the fluorescence micrographs of ZJU-68 ⊃ DMASM, taken by confocal laser scanning microscope. The flat and uniform intensity profiles suggest that the DMASM dyes are homogeneously distributed inside the ZJU-68 ⊃ DMASM composite crystals. The dye contents in this composite can be finely tuned by the addition of different amount of DMASM dyes during the *in situ* self-assembly solvothermal synthesis. Generally speaking, the relatively weak MPA responses require high dye content for MPA lasing measurements[5,6], it is thus necessary to encapsulate as much dye molecules as possible into the pore space of ZJU-68. However, the high dye contents (ingredient mole ratio of $n_{DMASM}/n_{H_2CPQC} \geq 70\%$) in the reaction mixtures not only affect the *in situ* self-assembly process (formation of other MOF phases) but also lead to the formation of poor crystalline ZJU-68 ⊃ DMASM. As such, we adjusted the dye concentration in the reaction solution, which produced the optimized ZJU-68 ⊃ DMASM crystals when the ingredient mole ratio of n_{DMASM}/n_{H_2CPQC} is 35%. Accordingly, per gram of resulting ZJU-68 ⊃ DMASM crystals contain 67.7 mg dye molecules corresponding to the concentration of 0.46 M (the molar amount of dye in per unit volume of solid composite; Supplementary Fig. 5). The optimized ingredient mole ratio (35%) is determined by the measurement of fluorescence quantum yield. Among the ZJU-68 ⊃ DMASM samples with different dye loading concentrations, the ZJU-68 ⊃ DMASM composite crystals with ingredient mole ratio of 35% exhibit the strongest emission at around 635 nm with the highest quantum

Fig. 2e) is too large to diffuse into the channels of ZJU-68 (ref. 19). We thus developed the *in situ* self-assembly synthetic approach in which the dye DMASM molecules were simultaneously incorporated into framework during the solvothermal synthesis of ZJU-68 by simply adding the dye molecules into the reaction

yield φ of $24.28 \pm 5\%$ on excitation at 450 nm (Supplementary Fig. 6). This is much higher than the quantum yield of 0.45% in dye solutions and of 1.48% solid powder[14]. These results demonstrate that the good confinement of the DMASM molecules within the size-matched channels of ZJU-68 can effectively restrain the intramolecular torsional motion and increase the conformational rigidity of the dye, thus diminishing the ACQ and populating its radiative decay pathway[20].

Multiphoton-excited fluorescence in ZJU-68 ⊃ DMASM. Figure 3a compares the single-photon-, two-photon- and three-photon-excited fluorescence spectra of a single ZJU-68 ⊃ DMASM crystal with the dye concentration of 0.46 M under the excitation of a femtosecond laser at different wavelengths. The ZJU-68 ⊃ DMASM shows a strong emission peaked at 627 nm on excitation at 532 nm, whereas the emission peak is red-shifted by 11 to 638 nm when excited at 1,064 nm. The full-width at half-maximum (FWHM) is 53.6 and 42.5 nm, respectively, in the single-photon-, two-photon-excited fluorescence spectra of ZJU-68 ⊃ DMASM. The emission spectrum on excitation at 1,380 nm is basically similar to that excited at 1,064 nm except one additional small peak at around 690 nm attributed to the second harmonic generation response. The red shift of 11 nm under multiphoton excitation can be ascribed to the reabsorption effect[14]. The diffuse reflectance ultraviolet-visible (vis) spectrum of ZJU-68 ⊃ DMASM was shown in Supplementary Fig. 7. There exists overlap between the long wavelength side of the absorption band and the short wavelength side of the fluorescence band (Fig. 3a). Furthermore, all MPP fluorescence bands in Fig. 3a are asymmetric with their left part seeming to be cut off[21]. In addition, the emission peaked at 627 nm from a single ZJU-68 ⊃ DMASM crystal on excitation at 532 nm is blue shifted relative to the spontaneous emission (maximum at 635 nm, see Supplementary Fig. 14) from multiple ZJU-68 ⊃ DMASM crystals, which also suggests the presence of reabsorption effect in ZJU-68 ⊃ DMASM (Supplementary Fig. 8). Fig. 3b shows the fluorescence intensity of the crystal with respect to the pump polarization direction when excited at 1,380 nm. The ZJU-68 ⊃ DMASM exhibits a strong emission when the pump polarization direction is parallel to the crystal channels (along the c-axis, denoted as 0°), but hardly emits any light when the pump polarization direction is perpendicular to the excitation direction (90°). Such significant directional fluorescence (dicroic ratio ~ 365 (ref. 15)) behaviours indicate that the absorption transition moments (approximately along the dye molecule axis[22]) are highly oriented along the crystal channels.

3PP lasing in ZJU-68 ⊃ DMASM. 3PP lasing properties were investigated on an isolated single crystal of ZJU-68 ⊃ DMASM with the dye concentration of 0.46 M under a microscope. A femtosecond laser at 1,380 nm was used to pump the crystal at room temperature. This laser beam was directed from a femtosecond optical parametric amplifier (OPA), and then was coupled to the microscope. The emission beam from the crystal was focused and collected with a fibre optic spectrometer. Representative emission spectra near the lasing threshold are shown in Fig. 4a. Under the low-pump energy (E) of 113 nJ, the emission spectrum shows a broad peak centred at ~ 649 nm with a FWHM of 57.6 nm, which corresponds to the spontaneous emission. The pump energy is defined as the laser energy directly received by the MOF crystal (after going through the objective lens and before being incident on the MOF crystal). As the pump energy increases to ≥ 230 nJ, a highly progressional emission pattern centred at 642.7 nm appears and grows rapidly with increasing pump energy, while the intensity of the broad spontaneous emission remains almost constant. The visible stimulated emission spectrum centred at 642.7 nm is between one half and one-third of the pumped wavelength of 1,380 nm, which means that the sum energy of two photons at 1,380 nm is not large enough to overcome the bandgap between the ground state (S_0) and excited state (S_1) of ZJU-68 ⊃ DMASM. The stimulated emission of ZJU-68 ⊃ DMASM is therefore induced by the simultaneous absorption of more than two near-infrared photons. To unravel how many photons involved in such a simultaneous absorption process, we further measured the dependence of the stimulated emission intensity on the pump energy. The right inset in Fig. 4a illustrates the pump energy dependence of the fluorescence intensity and FWHM plot as a function of pump energy, giving rise to a linear relationship with cubic pump energy and a low lasing threshold of $E_{th} \sim 224$ nJ as compared with other 3PP stimulated emission[4,6,7]. The FWHM plot shows a constant value below E_{th} and a sudden drop by more than two orders of magnitude when above E_{th}. The presence of a significant spectral narrowing and a threshold energy coupled with the linearly rapid increase in intensity with cubic pump energy suggest that the 3PP lasing has occurred in the ZJU-68 ⊃ DMASM crystal. The quality factor (Q) is given by $Q = \lambda/\delta\lambda$, where λ and $\delta\lambda$ are the peak wavelength and its FWHM, respectively. At pump energy of 369 nJ, the FWHM of lasing peaked at 642.7 nm is ~ 0.38 nm. This records a high-quality factor $Q \sim 1,691$ for 3PP lasing, which indicates the high crystal quality supported by our simple chemical approach without etching and coating.

For hexagonal ZJU-68 ⊃ DMASM crystal, the opposing facets can act as the mirrors of a Fabry–Pérot (F–P) cavity, or the six

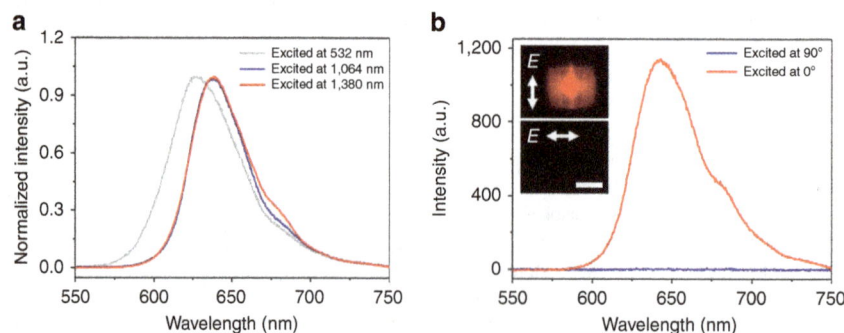

Figure 3 | Multiphoton-pumped fluorescence performance of a ZJU-68 ⊃ DMASM single crystal. (**a**) Single-photon-(532 nm), two-photon-(1,064 nm) and three-photon-(1,380 nm) excited emission spectra of ZJU-68 ⊃ DMASM. (**b**) The emission intensity versus pump polarization at two angles $\theta = 0°$ (parallel to the crystal channels) and $\theta = 90°$ (perpendicular to the crystal channels), excited at 1,380 nm. Insets: micrographs of a ZJU-68 ⊃ DMASM single crystal ($R = 36.5\,\mu m$) with different pump polarizations excited at 1,380 nm, Scale bar, 50 μm. The high intensity ratio between the two angles indicates the high orientation of dye molecules within the channels of ZJU-68.

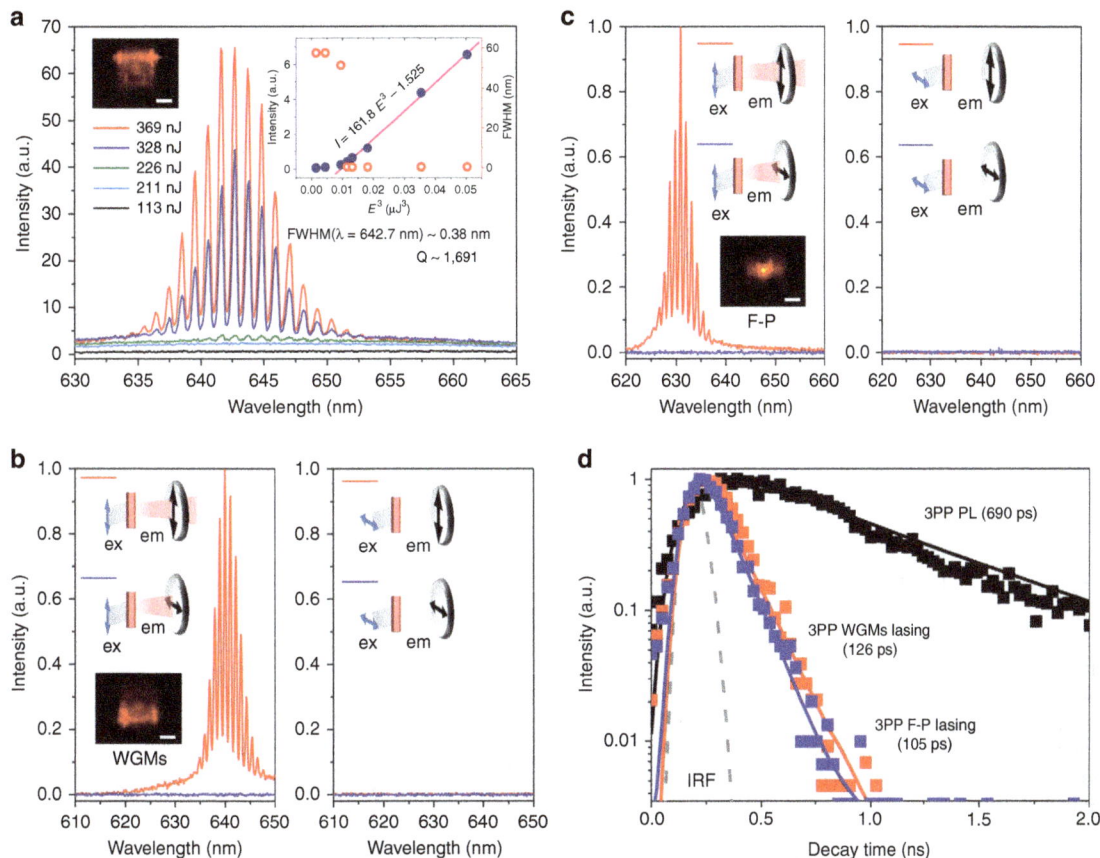

Figure 4 | Three-photon-pumped lasing performance of ZJU-68 ⊃ DMASM. (**a**) 1,380 nm pumped emission spectra of ZJU-68 ⊃ DMASM around the lasing threshold. Insets: the micrograph of a ZJU-68 ⊃ DMASM single crystal ($R = 26.2$ μm) excited at 1,380 nm (left) and emission intensity and FWHM as a function of pump energy showing the lasing threshold at ~ 224 nJ (right). At pump energy of 369 nJ, the FWHM of lasing peaked at 642.7 nm is ~ 0.38 nm, corresponding to a Q factor ~ 1,700. (**b,c**) Intensity-dependent emission spectra of 3PP WGMs (**b**, pump energy at 433 nJ) and F–P lasing (**c**, pump energy at 837 nJ) from two isolated crystals ($R = 26.1$ and 27.6 μm, respectively) with pump/emission-detected polarization combinations at two angles $\theta = 0°$ (parallel to the crystal channels) and $\theta = 90°$ (perpendicular to the crystal channels), excited at 1,380 nm. Insets: schematic diagrams of the measurement geometry for an individual crystal and micrographs of two kinds of lasing spot patterns due to the F–P (one-spot) and WGMs (two-spot) mechanisms, Scale bar, 20 μm. Both 3PP WGMs and F–P lasing exhibit perfectly polarized emission with DOP > 99.9%, which are attributed to the highly oriented assembly of dye molecules within the host–guest composite ZJU-68 ⊃ DMASM microcrystal. (**d**) TRPL decay kinetics measurements of ZJU-68 ⊃ DMASM under photoluminescence (PL), F–P and WGMs lasing excited at 1,380 nm. TRPL, time-resolved photoluminescence

facets can form a whispering gallery modes (WGMs) or other quasi-WGMs cavities[23] (Supplementary Fig. 9). We observed two kinds of lasing spot pattern on isolated ZJU-68 ⊃ DMASM crystals when excited at 1,380 nm: (1) the strong emission with spatial interference from two-side facets of a hexagonal prism crystal (two-spot pattern)[24], as shown in the insets of microscopy image in Figs. 4a and 4b; the strong emission from a round bright spot at the central facet of the crystal (one-spot pattern)[14], as shown in the inset of Fig. 4c. Such two lasing patterns can be attributed to the WGMs and F–P feedback mechanisms, respectively, as confirmed later. Fig. 4b,c show the anisotropic study of 3PP WGMs and F–P lasing from two crystals with side lengths of 26.1 and 27.9 μm, respectively. The red emission light from the crystal passed through a polarizer first and then was focused and collected with a fibre optic spectrometer. The schematic diagrams of the measurement geometry for an individual crystal are shown in Fig. 4b,c, where the polarization directions of the pump light and the polarizer are parallel (0°)/perpendicular (90°) to the crystal channels (along c-axis). We can see that both 3PP WGMs and F–P lasing with highly structured spectra occur when excited at 0° and emission polarization detected at 0°, while hardly any emission intensity

can be detected in all other configurations. The corresponding pump energy is 433 nJ for WGMs lasing and 837 nJ for F–P lasing. It should be noted that the pump energy at almost 3.2 E_{th} of 3PP WGMs lasing can realize the 3PP F–P lasing in our experiments, indicating that the WGMs mechanism is more conducive to the realization of 3PP lasing due to the total internal reflection for less loss of light in such size of the crystal. The degree of polarization can be defined as $\mathrm{DOP} = (I_{max} - I_{min})/(I_{max} + I_{min})$ in our experiments[25], and we calculated that both 3PP WGMs and F–P lasing exhibited DOP > 99.9% (limited by the spectral intensity sensitivity of our measurement system) when the excitation polarization is fixed parallel to the crystal channels, indicating a perfectly polarized 3PP lasing operation. Compared with the WGMs lasing, the parallel mirrors in a F–P cavity cannot be utilized as Brewster windows for the polarization selectivity. Therefore, the perfectly polarized 3PP F–P lasing is attributed to the highly oriented assembly of dye molecules within the host–guest composite ZJU-68 ⊃ DMASM microcrystal given the fact that the angle between the absorption transition moment and emission transition moment is close to zero in the dye molecule DMASM[26]. These anisotropic results indicate that ZJU-68 ⊃ DMASM can only be excited at the polarization direction parallel to the crystal channels, and can produce lasing perfectly polarized along the crystal channels, which exhibit a great potential for bioimaging, optical sensing and future optoelectronic integration.

To further confirm the optical-feedback mechanisms for 3PP lasing in these hexagonal ZJU-68 ⊃ DMASM crystals, several single crystals with different side lengths R were chosen for the further measurements (Supplementary Fig. 10a and b). For both feedback mechanisms, the spectra of lasing exhibit an increased mode spacing with the decrease of side length of the MOF crystals. For possible resonant modes, the mode spacing $\Delta\lambda_s$ is defined as[27]

$$\Delta\lambda_s = \frac{\lambda^2}{Ln_g} \quad (1)$$

where λ is the resonant wavelength, L is the cavity path length ($3\sqrt{3}R$ for WGMs and $2\sqrt{3}R$ for F–P cavity), and n_g is the group index of refraction. The measured mode spacing, $\Delta\lambda_s$, around 635 nm, demonstrates a linear relationship with $1/R$ for each feedback mechanism, which agrees well with Equation (1) (see Supplementary Fig. 10c,d). This result indicates that the lasing with the similar output spot pattern (one-spot or two-spot) can be attributed to the same feedback mechanism. According to the fitting formula, we calculated the ratio of slopes ($S_{one\text{-}spot}/S_{two\text{-}spot}$) to be 1.47, which is very close to the ratio of the cavity path lengths ($L_{WGMs}/L_{F\text{-}P} = 1.5$), verifying that the lasing with these two kinds of output pattern (one-spot and two-spot) should be attributed to the F–P cavity and WGMs, respectively. On the basis of Equation (1), we also derived $n_g \sim 3.27$ for F–P cavity, and $n_g \sim 3.21$ for WGMs at the wavelength of 635 nm. The relatively high group index n_g value may result from the unusual dispersion relation near the absorption band or the strong exciton–photon coupling in organic materials[28]. Further insight into the 3PP emission performance of the single crystal ZJU-68 ⊃ DMASM arises from time-resolved photoluminescence measurements (Fig. 4d). The pulse durations of the 3PP F–P lasing and WGMs lasing were determined to be 105 and 126 ps, respectively, which are much shorter than the corresponding 3PP fluorescence (below the E_{th}) decay time of 690 ps. Such temporal narrowing can be ascribed to the depletion in the population inversion of the gain medium with photon-stimulated amplification[6]. The lasing pulse durations from WGMs and F–P are almost in the same order of magnitude, indicating that the optical-feedback mechanism may have little effect on the pulse duration. Subtle differences in our measured decay times of lasing may be ascribed to the proportion of stimulated emission and spontaneous emission in the emitted light, which depends on multiple factors, for example, pump energy, crystal size and crystalline quality[29,30].

Discussion

In summary, we have achieved an unprecedented solid-state polarized frequency-upconversion lasing in a novel composite single microcrystal ZJU-68 ⊃ DMASM by simultaneous three-photon absorption in the near-infrared region. The tightly confined and highly oriented cationic DMASM dye molecules in anionic ZJU-68 nano-channels through an *in situ* assembly process efficiently increase the loaded concentration, minimize the aggregation and optimize the orientation of dye molecules within the framework, which fulfilled the high-gain lasing with highly polarized excitation response and perfectly polarized emission in a micro-sized laser cavity. Particularly, the 3PP lasing, with a low lasing threshold of ∼ 224 nJ centred at 642.7 nm on excitation at 1,380 nm, has been successfully achieved with a record high-quality factor of ∼ 1,700. Both F–P

and WGMs optical-feedback mechanisms have been confirmed to be responsible for 3PP lasing in ZJU-68 ⊃ DMASM microcrystals. Owing to the highly oriented assembly of dye molecules within ZJU-68 ⊃ DMASM, the 3PP WGMs and especially F–P lasing show a perfect emission polarization with DOP > 99.9%. The observed solid-state frequency-upconversion polarized lasing induced by 3PP may find great potentials in practical applications such as photonics, information storage and biomedicine, to name a few. For instance, the wavelength of 1,380 nm belongs to the near-infrared-IIa window (1,300-1,400 nm), which is very promising in biological applications (especially for *in vivo* imaging) because such wavelength region not only can reach deeper penetration depths and minimize the scattering/auto-fluorescence of biological tissues, but also avoid an increased light absorption from water above 1,400 nm (ref. 31). Because the MOF strategy and design can provide us with rich structures of the systematically tuned pore/channel sizes to encapsulate various chromophores with controlled concentration and orientation[32–34], we anticipate that higher order multiphoton-pumped lasing in solid state can also be realized given that the chromophores (or other nano-sized materials) with great multiphoton absorption properties are well incorporated into the structurally matched MOFs. These findings may eventually open up a new route to the exploitation of multiphoton-pumped solid-state laser in single MOF microcrystal (or nanocrystal) for future optoelectronic and biomedical applications.

Methods

Synthesis of ZJU-68 ⊃ DMASM. A mixture of $Zn(NO_3)_2 \cdot 6H_2O$ (0.34 mmol, 149 mg), H_2CPQC (0.17 mmol, 50 mg), DMF (10 ml), MeCN (2 ml), H_2O (0.05 ml), HBF_3 (0.05 ml) and DMASM iodide (0.03 mmol, 11 mg) were sealed in a 15 ml Teflon-lined stainless-steel bomb at 100 °C for 24 h, which was then slowly cooled to room temperature. After decanting the mother liquor, the fine red hexagonal crystalline product was rinsed three times with fresh DMF (5 ml × 3) and dried in air. The synthesis of the new organic linker H_2CPQC can be found in Supplementary Fig. 11 and Supplementary Methods.

Measurements. For MPP, an optical parametric amplifier (TOPAS-F-UV2, Spectra-Physics) pumped by a regeneratively amplified femtosecond Ti:sapphire laser system (800 nm, 1 kHz, pulse energy of 4 mJ, pulse width < 120 fs, Spitfire Pro-FIKXP, Spectra-Physics), which was seeded by a femtosecond Ti-sapphire oscillator (80 MHz, pulse width < 70 fs, 710-920 nm, Mai Tai XF-1, Spectra-Physics) was used for generating the excitation pulse (1 kHz, 240–2,600 nm, pulse width < 120 fs). The incident laser was coupled to the microscope (Ti-U, Nikon), focusing on crystals through an objective lens (CFI TU Plan Epi ELWD 50 ×, numerical aperture = 0.60, work distance = 11.0 mm) with an exposure region of diameter around 30 μm (supplementary Fig. 12). The excited red light was then focused and collected by the fibre optic spectrometer (QE65Pro, Ocean Optics).

The decay curves of multiphoton-pumped emissions were measured by a picosecond lifetime spectrometer (Lifespec-ps, Edinburgh Instruments). For the lifetime measurement of upconverted fluorescence, the pump energy was under the lasing threshold to ensure that no stimulated emission was generated. To measure the decay of the multiphoton-pumped lasing, the pump energy was enhanced over the threshold so that the ultra-strong lasing could be achieved.

Contents of well-dried dye-included ZJU-68 ⊃ DMASM crystals were determined by ${}^1H$ NMR. As shown in Supplementary Figs 5 and 13c, we calibrated and obtained peak area values of peaks that belong to H_2CPQC and DMASM, respectively. The ratio (R_a) of their peak area values represents the ratio of their contents in the crystal. The dye concentration of the ZJU-68 ⊃ DMASM composite is calculated from $c = 2R_a/N_A V$, where $V = 2403.91$ Å^3 and $N_A = 6.02 \times 10^{23}$ mol^{-1} is Avogadro's constant.

References

1. Jameson, D. M. & Ross, J. A. Fluorescence polarization/anisotropy in diagnostics and imaging. *Chem. Rev.* **110,** 2685–2708 (2010).
2. Ghosh, N. & Vitkin, I. A. Tissue polarimetry: concepts, challenges, applications, and outlook. *J. Biomed. Opt.* **16,** 110801 (2011).
3. Gurjar, R. S. *et al.* Imaging human epithelial properties with polarized light-scattering spectroscopy. *Nat. Med.* **7,** 1245–1248 (2001).
4. He, G. S., Tan, L. S., Zheng, Q. & Prasad, P. N. Multiphoton absorbing materials: molecular designs, characterizations, and applications. *Chem. Rev.* **108,** 1245–1330 (2008).

5. Guo, L. & Wong, M. S. Multiphoton excited fluorescent materials for frequency upconversion emission and fluorescent probes. *Adv. Mater.* **26,** 5400–5428 (2014).

6. Zheng, Q. D. *et al.* Frequency-upconverted stimulated emission by simultaneous five-photon absorption. *Nat. Photon.* **7,** 234–239 (2013).

7. Wang, Y. *et al.* Stimulated emission and lasing from CdSe/CdS/ZnS core-multi-shell quantum dots by simultaneous three-photon absorption. *Adv. Mater.* **26,** 2954–2961 (2014).

8. Hoover, E. E. & Squier, J. A. Advances in multiphoton microscopy technology. *Nat. Photon.* **7,** 93–101 (2013).

9. He, G. S., Markowicz, P. P., Lin, T. C. & Prasad, P. N. Observation of stimulated emission by direct three-photon excitation. *Nature* **415,** 767–770 (2002).

10. Gomes, A. S., Carvalho, M. T., Dominguez, C. T., de Araujo, C. B. & Prasad, P. N. Direct three-photon excitation of upconversion random laser emission in a weakly scattering organic colloidal system. *Opt. Express* **22,** 14305–14310 (2014).

11. Li, M. *et al.* Ultralow-threshold multiphoton-pumped lasing from colloidal nanoplatelets in solution. *Nat. Commun.* **6,** 8513 (2015).

12. Vietze, U. *et al.* Zeolite-dye microlasers. *Phys. Rev. Lett.* **81,** 4628–4631 (1998).

13. Martini, I. B. *et al.* Controlling optical gain in semiconducting polymers with nanoscale chain positioning and alignment. *Nat. Nanotechnol.* **2,** 647–652 (2007).

14. Yu, J. *et al.* Confinement of pyridinium hemicyanine dye within an anionic metal-organic framework for two-photon-pumped lasing. *Nat. Commun.* **4,** 2719 (2013).

15. Martinez-Martinez, V., Garcia, R., Gomez-Hortiguela, L., Perez-Pariente, J. & Lopez-Arbeloa, I. Modulating dye aggregation by incorporation into 1D-MgAPO nanochannels. *Chemistry* **19,** 9859–9865 (2013).

16. Martínez-Martínez, V. *et al.* Highly luminescent and optically switchable hybrid material by one-pot encapsulation of dyes into MgAPO-11 unidirectional nanopores. *ACS Photon.* **1,** 205–211 (2014).

17. Mao, C. *et al.* Anion stripping as a general method to create cationic porous framework with mobile anions. *J. Am. Chem. Soc.* **136,** 7579–7582 (2014).

18. Ferey, G. *et al.* A chromium terephthalate-based solid with unusually large pore volumes and surface area. *Science* **309,** 2040–2042 (2005).

19. Zhao, C. F., He, G. S., Bhawalkar, J. D., Park, C. K. & Prasad, P. N. Newly synthesized dyes and their polymer/glass composites for one-photon and 2-photon pumped solid-state cavity lasing. *Chem. Mater.* **7,** 1979–1983 (1995).

20. Cui, Y. J. *et al.* Dye encapsulated metal-organic framework for warm-white LED with high color-rendering index. *Adv. Funct. Mater.* **25,** 4796–4802 (2015).

21. Ren, Y. *et al.* Synthesis, structures and two-photon pumped up-conversion lasing properties of two new organic salts. *J. Mater. Chem.* **10,** 2025–2030 (2000).

22. Weiß, Ö. *et al.* in *Host-Guest-Systems Based on Nanoporous Crystals* 544–557 (Wiley-VCH Verlag GmbH & Co., 2005).

23. Wang, X. *et al.* Whispering-gallery-mode microlaser based on self-assembled organic single-crystalline hexagonal microdisks. *Angew. Chem. Int. Ed.* **53,** 5863–5867 (2014).

24. Braun, I. *et al.* Hexagonal microlasers based on organic dyes in nanoporous crystals. *Appl. Phys. B* **70,** 335–343 (2000).

25. Zhu, H. *et al.* Lead halide perovskite nanowire lasers with low lasing thresholds and high quality factors. *Nat. Mater.* **14,** 636–642 (2015).

26. Gozhyk, I. *et al.* Polarization properties of solid-state organic lasers. *Phys. Rev. A* **86,** 043817 (2012).

27. Choi, S., Ton-That, C., Phillips, M. R. & Aharonovich, I. Observation of whispering gallery modes from hexagonal ZnO microdisks using cathodoluminescence spectroscopy. *Appl. Phys. Lett.* **103,** 171102 (2013).

28. Takazawa, K., Inoue, J., Mitsuishi, K. & Takamasu, T. Fraction of a millimeter propagation of exciton polaritons in photoexcited nanofibers of organic dye. *Phys. Rev. Lett.* **105,** 067401 (2010).

29. Zhang, C. *et al.* Two-photon pumped lasing in single-crystal organic nanowire exciton polariton resonators. *J. Am. Chem. Soc.* **133,** 7276–7279 (2011).

30. Liu, X. *et al.* Whispering gallery mode lasing from hexagonal shaped layered lead iodide crystals. *ACS Nano* **9,** 687–695 (2015).

31. Hong, G. S. *et al.* Through-skull fluorescence imaging of the brain in a new near-infrared window. *Nat. Photon.* **8,** 723–730 (2014).

32. Furukawa, H., Cordova, K. E., O'Keeffe, M. & Yaghi, O. M. The chemistry and applications of metal-organic frameworks. *Science* **341,** 1230444 (2013).

33. Kitagawa, S., Kitaura, R. & Noro, S. Functional porous coordination polymers. *Angew. Chem. Int. Ed.* **43,** 2334–2375 (2004).

34. Chen, B., Xiang, S. & Qian, G. Metal-organic frameworks with functional pores for recognition of small molecules. *Acc. Chem. Res.* **43,** 1115–1124 (2010).

Acknowledgements

We acknowledge the financial support from the National Natural Science Foundation of China (Nos. 51229201, 51272229, 51272231, 51402259, 51472217, 51432001, U1305244 and 21325104) and Zhejiang Provincial Natural Science Foundation of China (Nos. LR13E020001 and LZ15E020001). This work is also partially supported by Welch Foundation (AX-1730) and National Science Foundation of United States (ECCS-1407443). X.C. and E.M. acknowledge the support from Special Project of National Major Scientific Equipment Development of China (No. 2012YQ120060) and the CAS/SAFEA International Partnership Program for Creative Research Teams. We also thank Dr Ghezai Musie for proof-reading the manuscript.

Author contributions

H.H., E.M., Y.C., J.Y. and G.Q. conceived and designed the experiments. H.H. synthesized the materials. J.Y. and C.W. Analysed the crystal structure. H.H. and E.M. performed the multiphoton experiments. Y.C., J.Y., Y.Y. and T.S. assisted with the linear optical property measurements and characterization of the material. H.H., E.M., Y.C., X.C., B.C. and G.Q. analysed the data and co-wrote the manuscript. All authors discussed the results and commented on the manuscript.

Additional information

Accession codes: The X-ray crystallographic coordinates for structure reported in this study has been deposited at the Cambridge Crystallographic Data Centre (CCDC), under deposition number 1046524. These data can be obtained free of charge from the Cambridge Crystallographic Data Centre via www.ccdc.cam.ac.uk/data_request/cif.

Competing financial interests: The authors declare no competing financial interests.

27

Efficient entanglement distillation without quantum memory

Daniela Abdelkhalek[1,2], Mareike Syllwasschy[2], Nicolas J. Cerf[3], Jaromír Fiurášek[4] & Roman Schnabel[1,2]

Entanglement distribution between distant parties is an essential component to most quantum communication protocols. Unfortunately, decoherence effects such as phase noise in optical fibres are known to demolish entanglement. Iterative (multistep) entanglement distillation protocols have long been proposed to overcome decoherence, but their probabilistic nature makes them inefficient since the success probability decays exponentially with the number of steps. Quantum memories have been contemplated to make entanglement distillation practical, but suitable quantum memories are not realised to date. Here, we present the theory for an efficient iterative entanglement distillation protocol without quantum memories and provide a proof-of-principle experimental demonstration. The scheme is applied to phase-diffused two-mode-squeezed states and proven to distil entanglement for up to three iteration steps. The data are indistinguishable from those that an efficient scheme using quantum memories would produce. Since our protocol includes the final measurement it is particularly promising for enhancing continuous-variable quantum key distribution.

[1] Institut für Laserphysik, Universität Hamburg, Hamburg 22761, Germany. [2] Institut für Gravitationsphysik, Leibniz Universität Hannover and Max-Planck-Institut für Gravitationsphysik (Albert-Einstein-Institut), Hannover 30167, Germany. [3] Quantum Information and Communication, Ecole Polytechnique de Bruxelles, CP 165, Université libre de Bruxelles, Brussels 1050, Belgium. [4] Department of Optics, Palacký University, Olomouc 77146, Czech Republic. Correspondence and requests for materials should be addressed to R.S. (email: roman.schnabel@physnet.uni-hamburg.de).

Light is the most suitable carrier of quantum information over long distances. Distribution of entangled states of light among distant nodes of a quantum communication network[1,2] may be used for various purposes, such as quantum cryptography[3,4] or quantum teleportation[5–7]. However, the transmission of quantum states of light is, in practice, unavoidably affected by losses and other decoherence effects that usually grow with distance. To fight these detrimental effects, entanglement distillation can be used, which extracts from a large number of noisy and weakly entangled states a smaller number of copies with increased entanglement and purity[8–16]. Crucially, entanglement distillation only requires local quantum operations and classical communication between the spatially separated parties holding parts of the entangled state. A canonical iterative entanglement distillation protocol[8,9] is illustrated in Fig. 1a. At each step, two copies of a decohered entangled state are consumed and, with some probability, one copy of a distilled state with improved properties is produced, which in turn serves as the input for the next round of the protocol. An efficient implementation of this iterative protocol requires a quantum memory[17], which motivates the extensive current effort for harnessing the quantum light-matter interface, ultimately leading to quantum repeaters[18]. Without quantum memory, all elementary two-copy distillation steps need to succeed simultaneously, which imposes an exponential overhead in terms of required resources, and hence drastically reduces the success rate of the protocol. Let us consider N iterations of the protocol and suppose, for simplicity, that the success probability of each elementary distillation step is the same and equal to P. A single attempt to perform N iterations requires 2^N input states, and all $2^N - 1$ elementary distillation steps must succeed simultaneously. To obtain 1 distilled copy, one thus needs to consume on average $2^N/P^{2^N-1}$ input states. With quantum memories, the distillation becomes far more efficient because the successfully distilled states after each step of the protocol can be stored and used in the subsequent step as required. Ideally, this reduces the required number of input states to $2^N/P^N$. Note that

the hardware-efficient pumping Gaussifier proposed in ref. 19 requires only a single quantum memory unit on each side, and enables sequential processing of the individual copies of distilled quantum state but achieves similarly inefficient distillation rate as a purely optical scheme without quantum memory.

To date, single-copy entanglement concentration[10,13,14,16,20], and elementary two-copy entanglement distillation[11,12] have been demonstrated for both discrete- and continuous-variable quantum states of light. A collective three-copy distillation of continuous-variable entangled states could even be implemented[15], but it remained inefficient in the absence of a quantum memory. Aside from this, an efficient realization of the full iterative multicopy entanglement distillation protocol could never be done because it is pending on operating an efficient quantum memory.

Here, we address this challenge from an opposite viewpoint and cancel out the need for quantum memories by exploiting the recently proposed concept of emulation of a quantum protocol[21,22]. The emulation replaces the actual physical implementation of a certain quantum operation by suitable postprocessing of measurement data and postselection, which offers a high potential to circumvent hardware implementation problems. In particular, the emulation of noiseless quantum amplification and attenuation by processing the data resulting from (eight-port) homodyne detection was proposed[21,22], and a proof-of-principle experimental emulation of single-mode noiseless quantum amplification was reported[23].

In this article, we apply such a strategy to multicopy continuous-variable entanglement distillation. Our scheme is solely based on the measurement data taken on a single beam of light that carries a continuous stream of copies of a decohered entangled state. Remarkably, due to the specific nature of the measurement and postprocessing, neither quantum memories nor the simultaneous physical realization of many copies are required, in contrast to the common knowledge on entanglement distillation. In principle, an arbitrary number of elementary (two-copy) distillation steps can be emulated by postprocessing the experimental data, solely limited by the length of the measured data stream. Specifically, we demonstrate the distillation of phase-diffused two-mode-squeezed states of light by iterative Gaussification[24–27] based on the interference of two copies of the state on balanced beam splitters, followed by projecting one output port on each side onto a vacuum state, see Fig. 1b. Here, we show that this scheme can be emulated by suitable processing of data obtained by eight-port homodyne detection on each mode of each phase-diffused state. Crucially, such emulation is completely indistinguishable from a full physical implementation with even ideal quantum memories to anyone outside Alice and Bob's labs. The only difference is that the distilled states are already detected and not physically available for further processing.

Nevertheless, the data may be used to fully characterize the distillation protocol and, for example, to extract a secret key from the distilled states. Our procedure is therefore particularly suitable for quantum cryptography, where it can convert seemingly useless highly noisy states into states that allow for the extraction of a secret key. It is similarly applicable to all related quantum communication protocols provided they terminate with suitable measurements.

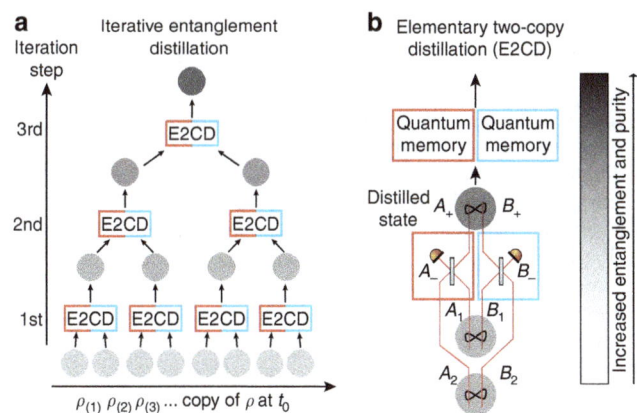

Figure 1 | Iterative entanglement distillation. (a) At each elementary step of the protocol, two copies of the input state, described by the density matrices ρ, are combined to produce, with some probability, one output copy with better properties. (b) Elementary step of iterative Gaussification of two-mode continuous-variable quantum states. Alice's and Bob's modes locally interfere on balanced beam splitters and one output mode on each side is measured with a photodetector. The procedure succeeds if both modes are projected onto vacuum. A quantum memory was previously deemed necessary for an efficient implementation of such an iterative protocol as one must store the successfully distilled states in a quantum memory to make them available for subsequent distillation steps.

Results

Emulation of iterative entanglement distillation. The starting point of our work is the iterative Gaussification scheme[24,25] illustrated in Fig. 1, which can serve for entanglement distillation of non-Gaussian quantum states of light. Each elementary step of

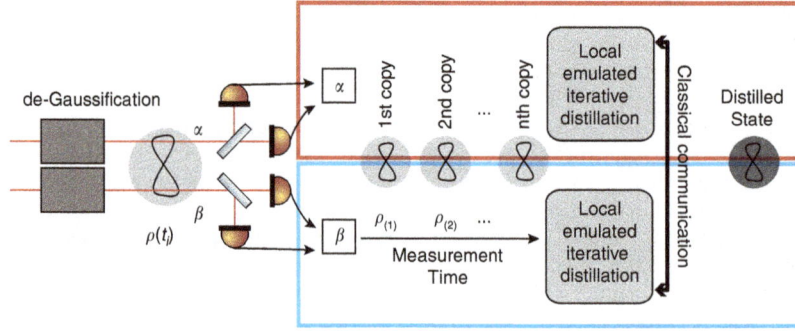

Figure 2 | Emulation of iterative Gaussification protocol. A single, continuously operated source distributes copies of the entangled and decohered states ρ to Alice and Bob. The states' coherent amplitudes are locally measured with two out-of-phase BHDs each, the so-called eight-port homodyne detectors. By prolonging the measurement time any quantity of copies may be generated. An emulated iterative distillation protocol based on the local operations and classical communication is realized as described in the main text. In the end, Alice and Bob share data that are equivalent to that from measurements on distilled entangled states.

the Gaussification protocol requires two copies of a two-mode entangled state, whose modes are labelled as A_1, B_1, and A_2, B_2, respectively. Modes A_1 and A_2 belong to Alice, and modes B_1 and B_2 are held by Bob. Both Alice and Bob combine their pairs of modes on a balanced beam splitter and obtain output modes A_+, A_- and B_+, B_-, where $+$ and $-$ refer to constructive and destructive interference, respectively. Subsequently, Alice and Bob perform photon number measurement on the modes A_- and B_- and exchange the measurement outcomes. This elementary distillation step is successful if both modes A_- and B_- are projected onto a vacuum state, and the distilled state in modes A_+ and B_+ provides the input for another round of the iterative protocol. Suppose now that Alice and Bob decide to measure the final distilled state with eight-port homodyne detectors, which perform projections onto coherent states. The generally impracticable implementation of the full iterative distillation protocol can be replaced by the following emulation procedure as depicted in Fig. 2. Alice and Bob perform the eight-port homodyne detection directly on the distributed decohered entangled state, and they repeat this measurement many times to collect a sufficiently large data set. Let α_j and β_j denote the measurement outcomes (complex amplitudes of coherent states) for the jth copy of the distributed state. Alice and Bob then combine the measurement results into pairs, say $(\alpha_{2n}, \alpha_{2n+1})$ and $(\beta_{2n}, \beta_{2n+1})$, and emulate the interference on beam splitters making the first layer of the protocol by performing simple additions and subtractions of the measurement results, which faithfully mimics interference of coherent states,

$$\alpha_{n,+} = \tfrac{1}{\sqrt{2}}(\alpha_{2n}+\alpha_{2n+1}), \quad \alpha_{n,-} = \tfrac{1}{\sqrt{2}}(\alpha_{2n}-\alpha_{2n+1}),$$
$$\beta_{n,+} = \tfrac{1}{\sqrt{2}}(\beta_{2n}+\beta_{2n+1}), \quad \beta_{n,-} = \tfrac{1}{\sqrt{2}}(\beta_{2n}-\beta_{2n+1}). \quad (1)$$

Projection of the output modes A_- and B_- onto vacuum is equivalent to conditioning on $\alpha_{n,-}=\beta_{n,-}=0$. After the conditioning we obtain a reduced set of pairs $\alpha_j^{(1)} = \alpha_{n_j,+}$ and $\beta_j^{(1)}=\beta_{n_j,+}$ where n_j represent the values of n for which the conditioning was successful. These pairs of data represent effective measurement outcomes of eight-port homodyne detections on modes A and B of the distilled two-mode state after the first iteration of the distillation protocol. This procedure can be repeated, and outputs of the kth round of the protocol, $\alpha_j^{(k)}$ and $\beta_j^{(k)}$, can be used as inputs of the next round of the protocol, resulting in $\alpha_j^{(k+1)}$ and $\beta_j^{(k+1)}$.

Since the probability to obtain the specific outcomes $\alpha_{n,-}=\beta_{n,-}=0$ in eight-port homodyning vanishes, we need to modify

the conditioning to make the protocol practicable. A theoretically appealing modification is to impose an acceptance probability that is a Gaussian function of the complex amplitudes,

$$P_{\mathrm{acc}}(\alpha_-,\beta_-) = \exp\left(-\frac{|\alpha_-|^2}{\bar{n}}\right)\exp\left(-\frac{|\beta_-|^2}{\bar{n}}\right). \quad (2)$$

Physically, this corresponds to a modified conditioning in Fig. 1b, where modes A_- and B_- are projected onto thermal states with a mean number of thermal photons equal to $\bar{n}$. The choice of Gaussian P_{acc} guarantees that the iterative Gaussification protocol indeed converges to a Gaussian state[25,27], which greatly simplifies its theoretical treatment and allows us to derive analytically the asymptotic state to which the protocol converges.

To be more specific, let us consider as an example the distillation of phase-diffused two-mode-squeezed states, as used in our experiment. A covariance matrix of the initial Gaussian symmetric two-mode-squeezed state before phase diffusion reads[1]

$$\gamma_{\mathrm{AB}} = \begin{pmatrix} a & 0 & b & 0 \\ 0 & a & 0 & -b \\ b & 0 & a & 0 \\ 0 & -b & 0 & a \end{pmatrix}. \quad (3)$$

Here, a denotes the (symmetric) variance of phase space projections of the complex amplitude (quadratures) of the individual modes, and b represents the correlations between quadratures of modes A and B. Suppose now that modes A and B are sent through noisy channels where random phase shifts ϕ_A and ϕ_B are imposed. For the sake of simplicity, we shall assume that the phase diffusions are independent and have the same statistics for both modes (although our results can be easily extended to more general scenarios). Note that the phase-noise does not modify the value of parameter a, since each mode is locally in a thermal state that is invariant under phase-shift. So the covariance matrix of the phase-diffused non-Gaussian state at the output of the noisy channels preserves the form (3), but the phase diffusion reduces the intermodal correlations,

$$a_{\mathrm{PD}} = a, \quad b_{\mathrm{PD}} = qb, \quad (4)$$

where $q = \langle\cos\phi_A \cos\phi_B\rangle$ quantifies the phase diffusion for uncorrelated noise. The stronger the phase diffusion, the smaller is $|q|$. We assume here that the phase noise is symmetric, $\langle\sin\phi_A\rangle = \langle\sin\phi_B\rangle = 0$. If no phase diffusion is present (ϕ_A and ϕ_B always zero) q equals unity. If the phases are fully random q equals zero. For given fixed values of the random phase shifts ϕ_A

and ϕ_B, the covariance matrix γ_{AB} of an input two-mode state is transformed according to $R\gamma_{AB}R^T$ with

$$R(\phi) = \begin{pmatrix} \cos\phi_A & \sin\phi_A & 0 & 0 \\ -\sin\phi_A & \cos\phi_A & 0 & 0 \\ 0 & 0 & \cos\phi_B & \sin\phi_B \\ 0 & 0 & -\sin\phi_B & \cos\phi_B \end{pmatrix} \quad (5)$$

describing the action of the particular phase-shift values ϕ_A and ϕ_B at the level of the covariance matrix. This expression has to be averaged over the random phase shifts to obtain the resulting covariance matrix of the de-phased state. Note that in this last averaging step we assume that the mean values of quadratures of the input state vanish, which is satisfied in our experiment where we utilize squeezed vacuum state. If non-zero, the mean values can be eliminated and set to zero by suitable local coherent displacements, which in our emulation-based scheme can be implemented simply as displacements of the measured data. With the help of the general theory of iterative Gaussification protocols presented in ref. 27, one can derive the following analytical expression for the covariance matrix of the asymptotic Gaussian state,

$$\gamma_{AB,\infty} = \langle R(\phi)[\gamma_{AB} + (2\bar{n}+1)I]^{-1}R^T(\phi)\rangle_\phi^{-1} - (2\bar{n}+1)I. \quad (6)$$

Where, I denotes the identity matrix and $\langle.\rangle_\phi$ represents statistical averaging over random phase diffusions.

The performance of the distillation protocol and its usefulness for quantum key distribution is illustrated in Fig. 3. The squeezing properties of the two-mode state can be characterized by the squeezed variance V_{sq} defined as the minimum eigenvalue of the covariance matrix, $V_{sq} = \min(eig(\gamma_{AB}))$, while entanglement can be detected using the Duan–Simon criterion[28,29]. Technically, one has to determine the minimum symplectic eigenvalue μ of a covariance matrix corresponding to the partially transposed state, and entanglement is witnessed if $\mu < 1$. For a symmetric two-mode-squeezed state as in equation (3), we find that $V_{sq} = \mu = a - |b|$, hence the presence of squeezing, $V_{sq} < 1$, also indicates that the state is entangled. In Fig. 3a, the dependence of μ on the dephasing parameter q is plotted for the phase-diffused state (blue solid line) and the asymptotic distilled state for different $\bar{n}$ (red lines). The enhancement of entanglement by distillation is clearly visible, and the closer we choose $\bar{n}$ to zero (which corresponds to projection on vacuum) the stronger is the effect. Importantly, the distillation protocol can even recover seemingly lost quadrature entanglement as the distilled state can exhibit $\mu < 1$ even if the initial phase-diffused state exhibited $\mu > 1$. Distillation of entanglement necessarily implies that the input state was (weakly) entangled too. However, its initial entanglement is not visible via the covariance matrix (which is consistent with a separable Gaussian state) but hidden in higher-order correlations of the quadratures.

We can determine the maximum tolerable phase noise (minimum $|q|$) for which we can still distil Gaussian entanglement. For the *pure* symmetric two-mode Gaussian state with covariance matrix given by equation (3), and $a = \cosh(2r)$ and $b = \sinh(2r)$ we find that the Gaussian entanglement is asymptotically distilled provided that

$$\frac{\bar{n}}{\bar{n}+1}\tanh r < |q|. \quad (7)$$

Here, r denotes the squeezing parameter of the squeezed input states. If we project on vacuum ($\bar{n} = 0$) then entanglement is distilled for any $r > 0$ and $|q| > 0$. For $\bar{n} > 0$, however, entanglement is only distilled in the asymptotic limit if the dephasing is not too large ($|q|$ not too small). For mixed states, the entanglement might not be distillable even if $\bar{n} = 0$, c.f. the red solid line in Fig. 3a which crosses the horizontal line $\mu = 1$ at non-zero q.

We can go further and calculate the distillable secret key in continuous-variable quantum key distribution $K = I_{AB} - \chi_{AE}$ (Fig. 3b), where I_{AB} is the mutual information between Alice's and Bob's data and χ_{AE} is Eve's classically accessible information on Alice's quantum state. The key rate can be calculated analytically using the optimality of Gaussian attacks[30–32] and the well-known formulas for entropies of Gaussian probability distributions and Gaussian quantum states (see refs 1,21,33 for more details). As shown in Fig. 3b, the distillation protocol can significantly enlarge the range of q values for which a secret key can be distilled. Crucially, the key rate for the distilled state can be positive even if the initial phase-diffused state does not exhibit squeezing, $V_{sq} > 1$, that is, if its covariance matrix is compatible with that of a Gaussian separable state. In such a case, no secure key could be obtained from the decohered state by any protocol involving Gaussian measurements and a security analysis based on the covariance matrix of that state. The emulation of iterative entanglement distillation can thus efficiently convert seemingly useless noisy data into data from which a non-zero secret key could be extracted.

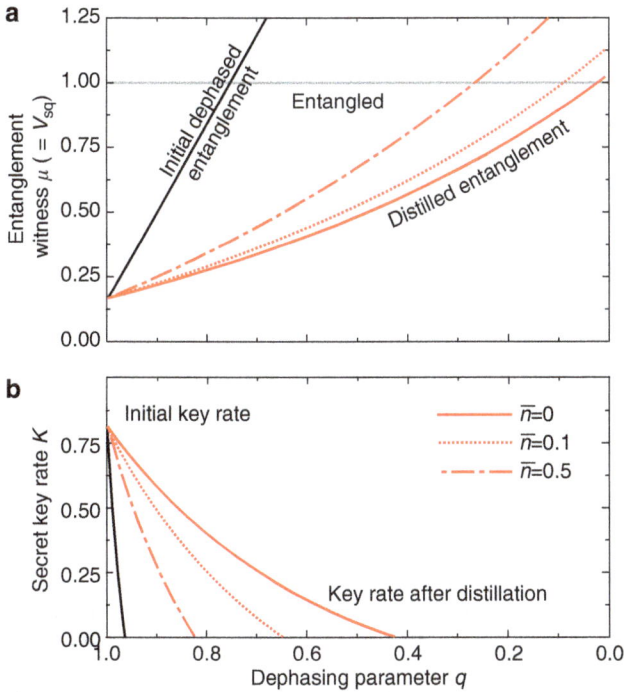

Figure 3 | Entanglement distillation of phase-diffused two-mode-squeezed states. Squeezing variance V_{sq} (**a**) and distillable secret key K (**b**) are plotted for the phase-diffused state (blue) and for the asymptotic distilled state for different $\bar{n}$ (red lines). $\bar{n}$ gives here the mean number of thermal photons of the state we project onto. The parameter q quantifies the strength of the phase diffusion: $q = 1$ indicates that no phase noise is present, whereas $q = 0$ corresponds to a completely randomized phase. The parameters of the input state before phase-diffusion read $a = 3.583$ and $b = 3.417$, which corresponds to variances of squeezed and anti-squeezed quadratures $V_{sq} = 1/6$ and $V_{antisq} = 7$.

Experimental setup. A schematic picture of the experimental setup is shown in Fig. 4. In a first step, we generated two Gaussian entangled light fields, which was achieved by superimposing two squeezed vacuum states on a balanced beam splitter with a 90° phase-shift (refs 6,34). Previous realizations of multicopy continuous-variable entanglement distillation protocols used the

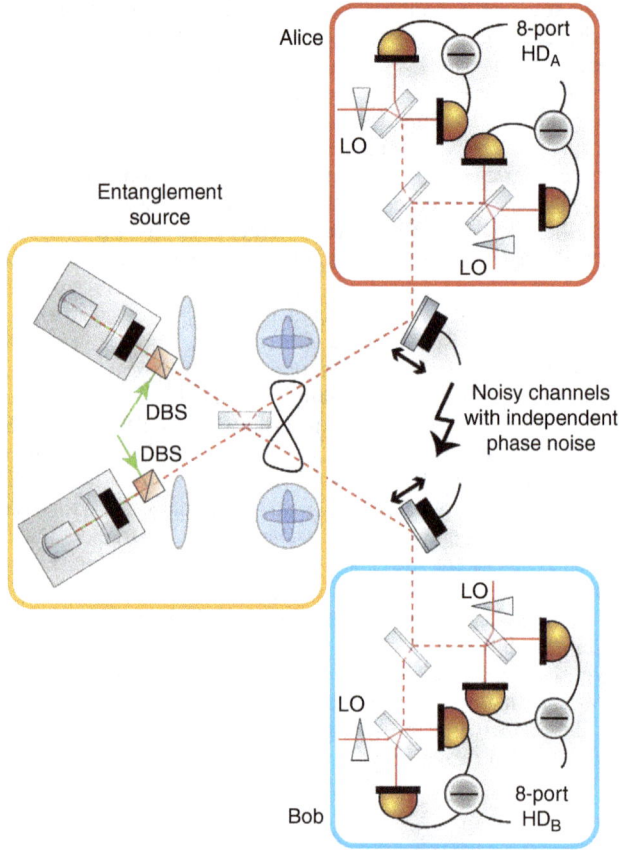

Figure 4 | Experimental setup. Two propagating squeezed-light modes are superimposed on a balanced beam splitter. The entangled outputs are sent to two separate sites. During transmission the modes are exposed to phase noise, which produces de-Gaussified mixed quantum states. For strong phase noise the two-mode squeezing, that is, the entanglement in the second moments of the quadratures, is completely destroyed. The decohered modes are continuously and unconditionally detected by eight-port homodyne detectors, which have negligible intrinsic phase noise and detection efficiencies of about $87 \pm 6\%$. DBS, dichroic beam splitter.

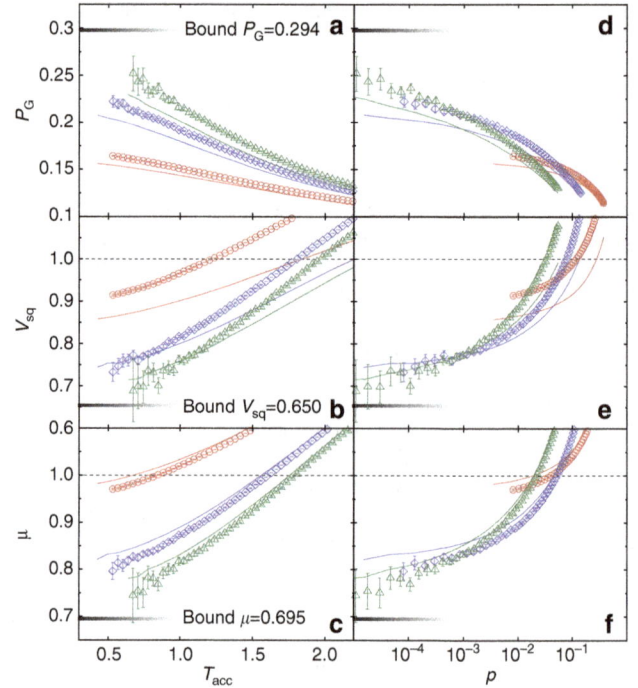

Figure 5 | Experimental distillation results. The panels **a**–**f** show the dependences of V_{sq}, μ and P_G on the threshold T_{acc} and success probability p for one (red, circle), two (blue, square) and three (green, triangle) iterations of the distillation protocol as discussed in the main text. The error bars give a 2σ confidence interval and are obtained by bootstrapping the samples 100 times where the length of each bootstrapping sample equals the length of the analysed data set. The solid lines give theoretical predictions obtained by Monte Carlo simulations for 10^{10} samples. These simulations use the measured undiffused covariance matrix (equation 8) and the experimentally determined phase-noise factor $q = 0.78$. The model takes into account additional losses induced by the implementation of our de-phasing channel, and, for simplicity, we assume an uncorrelated Gaussian distribution of the random phase shifts with equal variances. The general good agreement between simulation and experiment suggests that our model represents a good approximation. The discrepancy for V_{sq} can be explained by the fact that the actual phase noise does not precisely meet our assumptions. By a slight change of q we could get a good fit for V_{sq} at the expense of a worse fit for μ. The short fading horizontal lines represent the asymptotic limit for projection onto vacuum ($\bar{n} = 0$).

so-called v-class entangled states[12,15], which are simply achieved by superimposing a squeezed-vacuum mode and an ordinary vacuum mode on a balanced beam splitter. Replacing the vacuum mode by a second squeezed light field increases the experimental effort, but also allows, in principle, stronger entanglement, which is needed for QKD. Our scheme has the advantage that the (efficient) multistep iterative distillation of entanglement only requires at-most two squeezed light sources, whereas an actual hardware implementation of a (still inefficient) iterative distillation without quantum memories requires doubling the number of squeezed light sources with each iteration step. In our proof-of-principle experiment the squeezed-vacuum sources produced slightly different squeezing values, around 3 dB. Due to this asymmetry the covariance matrix of the entangled state did not hold the form given by equation (3) but was given by

$$\gamma_{AB,exp} = \begin{pmatrix} 3.20 & -0.13 & -2.90 & -0.04 \\ -0.13 & 6.24 & -0.03 & 6.08 \\ -2.90 & -0.03 & 3.70 & -0.06 \\ -0.04 & 6.08 & -0.06 & 6.83 \end{pmatrix}. \quad (8)$$

We reconstructed this covariance matrix by measuring the prepared state without additional phase diffusion with the

eight-port homodyne detectors and using the formula $\gamma_{AB,exp} = 2\gamma_{EHD} - I$, where γ_{EHD} is a covariance matrix calculated directly from the eight-port homodyne data, and I denotes the identity matrix. The small values at the off-diagonal elements are not exactly zero due to imprecisions in the quadrature phases. The different values of the diagonal elements are mainly caused by the different squeezing values. The matrix $\gamma_{AB,exp}$ was reconstructed without correcting for detection inefficiencies and detector dark noise. For this reason, the covariance matrix is a good description of the actual measurement data (but a less good description of the quantum state before detection). We note that two-mode squeezing of > 10 dB has already been demonstrated[34] and our scheme would work equally well with these higher squeezing strengths.

Our entangled modes were distributed between Alice and Bob's sites. During transmission, the modes were exposed to independent phase noise, which models a noisy transmission through, for example, optical fibres. The noise was applied by varying the position of steering mirrors in the path, driven by piezo electric transducers. We varied the strength of the noise by amplifying or attenuating the voltage given to the actuators to examine the behaviour of the protocol for different noise strengths. For all measurements shown in the next section we used the same phase diffusion which corresponded to $q = 0.78$. At the sites, the received light beams were absorbed in eight-port homodyne detectors, consisting of a balanced beam splitter and two conventional balanced homodyne detectors (BHDs), which measure the amplitude and phase quadrature on the first and the second beam, respectively. This measurement corresponds to a projection onto coherent states. The recorded data provided the basis to execute the emulated distillation protocol.

Experimental results. In the experiment, we collected up to 5×10^8 data points for a fixed strength q of phase noise, and we emulated up to three rounds of the iterative entanglement distillation protocol. To simplify data processing, we used a deterministic conditioning rule and the elementary entanglement distillation step was taken successful if $|\alpha_-| < T_{acc}$ and $|\beta_-| < T_{acc}$, where T_{acc} is a tunable threshold. While such conditioning may lead to some residual non-Gaussianity of the asymptotic state, it does not modify the qualitative properties of the distillation protocol. We have used the resulting data to reconstruct the covariance matrix γ of the two-mode state after each iteration of the distillation protocol, and we have used the covariance matrix to determine various properties of the distilled state. The results are plotted in Fig. 5, which shows the two-mode squeezing variance V_{sq}, the symplectic eigenvalue μ witnessing entanglement, and the quantity $P_G = 1/\sqrt{\det \gamma}$, which for Gaussian states coincides with the state purity.

Figure 5a–c illustrate that each iteration of the protocol increases squeezing and (Gaussian) entanglement of the distilled state for a fixed threshold T_{acc}, as indicated by the reduction of V_{sq} and μ. Furthermore, P_G increases with the number of iterations, which is a strong indication that the purity of the state is increased by the protocol. The absolute bounds presented in all panels are the asymptotic bounds for projection onto vacuum ($\bar{n} = 0$) calculated with equation (6). Statistical properties of the random phase shifts required for evaluation of the phase average in equation (6) were determined by comparing the covariance matrices of the initial and the de-phased states. It is remarkable how close the data come to the absolute bound with only three iteration steps and a finite threshold T_{acc}. Figure 5d–f show the practically more relevant dependence of V_{sq}, μ and P_G on the total success probability of the protocol p, which was determined as the ratio of the number of distilled copies of the state versus the total number of input copies of the state. These plots fully demonstrate the usefulness of an iterative distillation scheme. For a fixed total success probability, more iterations may be advantageous as they lead to better squeezing, entanglement and purity. The crossover between one and two iterations is clearly and unambiguously demonstrated by the data, while the crossover between two and three iterations can be seen in the region of $10^{-4} < P < 10^{-3}$, albeit the results in this region are already affected by statistical uncertainties. We have repeated the experiment for several different strengths of phase noise, and the results were in all cases qualitatively similar to those shown in Fig. 5.

We emphasize that the emulated iterative entanglement distillation presented here is inherently as efficient as if the protocol had been implemented with quantum memories, which would store successfully distilled states from previous rounds of the protocol to be utilized in the next rounds of the protocol. In the emulation, we naturally use in the subsequent rounds of the protocol only the successfully distilled states from previous rounds, which ensures that the average number of input copies

that are consumed to produce a single distilled copy scales exactly as if quantum memories had been used.

Discussion

Our proof-of-principle experiment paves the way to efficient quantum communication protocols, where a specific data processing mimicks quantum memories (hence, provides the same advantage in terms of resources) without actually requiring them. This strategy can be applied at the end points of any entanglement distribution scheme where the end users perform eight-port homodyne detection. Importantly, our protocol is not limited to schemes where entanglement is physically distributed, but is also applicable to a prepare-and-measure quantum key distribution schemes where Alice prepares Gaussian-modulated coherent states and sends them to Bob who performs eight-port homodyne detection[35]. Indeed, since the preparation of coherent states by Alice is indistinguishable from the preparation of two-mode-squeezed vacuum followed by eight-port homodyne detection on Alice's mode, the preparation can be equivalently interpreted as a measurement on a (virtual) shared entangled state. Similarly, our procedure is also applicable to the recently proposed and demonstrated measurement-device independent continuous-variable QKD protocol[36,37], where Alice and Bob both prepare randomly modulated coherent states and send them to an untrusted relay which performs a continuous-variable Bell measurement on the two modes and publicly announces the results of the measurement. Given its wide range of potential applications, our proposal represents a promising tool for improving the performance of various quantum communication systems.

Methods

Squeezed-light preparation. The main light source was a Nd:YAG laser that produced a continuous-wave field at a wavelength of 1064 nm, as well as frequency-doubled light at 532 nm. The infrared light provided the control fields for the active-length stabilization of two squeezed-light resonators containing 7% magnesium-oxid-doped lithium-niobate crystals (MgO:LiNbO3), as well as the optical local oscillators for homodyne detection. The green light field was mode-matched into the squeezed-light resonators to pump degenerate type I parametric-downconversion processes, which provided resonator output modes at 1064 nm in squeezed-vacuum states. The squeezed-light resonators were singly resonant for 1064 nm, had a standing-wave and half-monolithic design, and a length of about 40 mm. The resonators' coupling mirrors were attached to piezo electric transducers to stabilize the cavity lengths on resonance using the Pound–Drever–Hall locking scheme[38,39]. An active temperature control stabilized the crystal temperatures at phase matching of the fundamental and the harmonic fields at about 60 °C.

Light detection and data acquisition. The two-mode (squeezed) state was detected with two eight-port homodyne detectors incorporating altogether four conventional BHDs. Each BHD consisted of a balanced beam splitter, two high-quantum efficiency PIN photodiodes and used a homodyne local oscillator power of about 3 mW at 1064 nm. The difference photo-electric current of the two photodiodes was amplified and transferred to a voltage by a trans-impedance amplifier. The voltage signal was then mixed with a 6.4 MHz electronic local oscillator, anti-alias filtered with a corner frequency of 400 kHz and finally synchronously sampled at a frequency of 1 MHz.

References

1. Weedbrook, C. et al. Gaussian quantum information. Rev. Mod. Phys. **84**, 621–669 (2012).
2. Kimble, H. J. The quantum internet. Nature **453**, 1023–1030 (2008).
3. Ekert, A. K. Quantum cryptography based on Bell's theorem. Phys. Rev. Lett. **67**, 661–663 (1991).
4. Scarani, V. et al. The security of practical quantum key distribution. Rev. Mod. Phys. **81**, 1301–1350 (2009).
5. Bouwmeester, D. et al. Experimental quantum teleportation. Nature **390**, 575–579 (1997).
6. Furusawa, A. et al. Unconditional quantum teleportation. Science **282**, 706–709 (1998).

7. Bowen, W. P *et al.* Experimental investigation of continuous variable quantum teleportation. *Phys. Rev. A* **67**, 032302 (2003).

8. Bennett, C. H. *et al.* Purification of noisy entanglement and faithful teleportation via noisy channels. *Phys. Rev. Lett.* **76**, 722–725 (1996).

9. Deutsch, D. *et al.* Quantum privacy amplification and the security of quantum cryptography over noisy channels. *Phys. Rev. Lett.* **77**, 2818–2821 (1996).

10. Kwiat, P. G., Barraza-Lopez, S., Stefanov, A. & Gisin, N. Experimental entanglement distillation and 'hidden' non-locality. *Nature* **409**, 1014–1017 (2001).

11. Pan, J.-W., Gasparoni, S., Ursin, R., Weihs, G. & Zeilinger, A. Experimental entanglement purification of arbitrary unknown states. *Nature* **423**, 417–422 (2003).

12. Hage, B. *et al.* Preparation of distilled and purified continuous-variable entangled states. *Nat. Phys.* **4**, 915–918 (2008).

13. Dong, R. *et al.* Experimental entanglement distillation of mesoscopic quantum states. *Nat. Phys.* **4**, 919–923 (2008).

14. Takahashi, H. *et al.* Entanglement distillation from Gaussian input states. *Nat. Photon.* **4**, 178–181 (2010).

15. Hage, B., Samblowski, A., Diguglielmo, J., Fiurášek, J. & Schnabel, R. Iterative entanglement distillation: approaching the elimination of decoherence. *Phys. Rev. Lett.* **105**, 230502 (2010).

16. Kurochkin, Y., Prasad, A. S. & Lvovsky, A. I. Distillation of the two-mode squeezed state. *Phys. Rev. Lett.* **112**, 070402 (2014).

17. Simon, C. *et al.* Quantum memories. *Eur. Phys. J. D* **58**, 1–22 (2010).

18. Duan, L.-M., Lukin, M. D., Cirac, J. I. & Zoller, P. Long-distance quantum communication with atomic ensembles and linear optics. *Nature* **414**, 413–418 (2001).

19. Campbell, E. T., Genoni, M. G. & Eisert, J. Continuous-variable entanglement distillation and noncommutative central limit theorems. *Phys. Rev. A* **87**, 42330 (2013).

20. Ulanov, A. E. *et al.* Undoing the effect of loss on quantum entanglement. *Nat. Photon.* **9**, 764–768 (2015).

21. Fiurášek, J. & Cerf, N. J. Gaussian postselection and virtual noiseless amplification in continuous-variable quantum key distribution. *Phys. Rev. A* **86**, 060302 (2012).

22. Walk, N., Ralph, T. C., Symul, T. & Lam, P. K. Security of continuous-variable quantum cryptography with Gaussian postselection. *Phys. Rev. A* **87**, 020303 (2013).

23. Chrzanowski, H. M. *et al.* Measurement-based noiseless linear amplification for quantum communication. *Nat. Photon.* **8**, 333–338 (2014).

24. Browne, D., Eisert, J., Scheel, S. & Plenio, M. Driving non-Gaussian to Gaussian states with linear optics. *Phys. Rev. A* **67**, 062320 (2003).

25. Eisert, J., Browne, D. E., Scheel, S. & Plenio, M. B. Distillation of continuous-variable entanglement with optical means. *Ann. Phys.* **311**, 431–458 (2004).

26. Datta, A. *et al.* Compact continuous-variable entanglement distillation. *Phys. Rev. Lett.* **108**, 060502 (2012).

27. Campbell, E. T. & Eisert, J. Gaussification and entanglement distillation of continuous-variable systems: a unifying picture. *Phys. Rev. Lett.* **108**, 020501 (2012).

28. Duan, L., Giedke, G., Cirac, J. I. & Zoller, P. Inseparability criterion for continuous variable systems. *Phys. Rev. Lett.* **84**, 2722–2725 (2000).

29. Simon, R. Peres-Horodecki separability criterion for continuous variable systems. *Phys. Rev. Lett.* **84**, 2726–2729 (2000).

30. Grosshans, F. & Cerf, N. J. Continuous-variable quantum cryptography is secure against non-gaussian attacks. *Phys. Rev. Lett.* **92**, 047905 (2004).

31. Navascues, M., Grosshans, F. & Acin, A. Optimality of Gaussian attacks in continuous-variable quantum cryptography. *Phys. Rev. Lett.* **97**, 190502 (2006).

32. Garcia-Patron, R. & Cerf, N. J. Unconditional optimality of Gaussian attacks against continuous-variable quantum key distribution. *Phys. Rev. Lett.* **97**, 190503 (2006).

33. Lodewyck, J. *et al.* Quantum key distribution over 25km with an all-fiber continuous-variable system. *Phys. Rev. A* **76**, 042305 (2007).

34. Eberle, T., Händchen, V. & Schnabel, R. Stable control of 10 dB two-mode squeezed vacuum states of light. *Opt. Express* **21**, 11546–11553 (2013).

35. Weedbrook, C. *et al.* Quantum cryptography without switching. *Phys. Rev. Lett.* **93**, 170504 (2004).

36. Zhang, Y.-C. *et al.* Continuous-variable measurement-device-independent quantum key distribution using squeezed states. *Phys. Rev. A* **90**, 052325 (2014).

37. Pirandola, S. *et al.* High-rate measurement-device-independent quantum cryptography. *Nat. Photon.* **9**, 397–402 (2015).

38. Drever, R. *et al.* Laser phase and frequency stabilization using an optical resonator. *Appl. Phys. B* **31**, 97–105 (1983).

39. Black, E. D. An introduction to Pound-Drever-Hall laser frequency stabilization. *Am. J. Phys.* **69**, 79 (2001).

Acknowledgements

This research was partially supported by the Deutsche Forschungsgemeinschaft (project SCHN 757/5-1). J.F. acknowledges financial support from the EU FP7 under Grant Agreement No. 308803 (project BRISQ2) cofinanced by MSMT CR (7E13032). N.J.C. acknowledges financial support from the Fonds de la Recherche Scientifique (F.R.S.-FNRS) under grant T.0199.13 and from the Belgian Science Policy Office (BELSPO) under grant IAP P7-35 Photonics@be. D.A. acknowledges financial support from the DFG Graduate School 'Quantum mechanical noise in complex systems' (RTG1991). We thank Tobias Gehring and Aiko Samblowski for helpful discussions.

Author contributions

J.F. and N.J.C. developed the theory, D.A. and M.S. performed the experiment and all measurements under the supervision of R.S. The data were analysed by J.F. and D.A., and J.F., D.A., R.S. and N.J.C. wrote the paper.

Additional information

Competing financial interests: The authors declare no competing financial interests.

28

A multiplexed light-matter interface for fibre-based quantum networks

Erhan Saglamyurek[1,2], Marcelli Grimau Puigibert[1,2], Qiang Zhou[1,2], Lambert Giner[1,2,†], Francesco Marsili[3], Varun B. Verma[4], Sae Woo Nam[4], Lee Oesterling[5], David Nippa[5], Daniel Oblak[1,2] & Wolfgang Tittel[1,2]

Processing and distributing quantum information using photons through fibre-optic or free-space links are essential for building future quantum networks. The scalability needed for such networks can be achieved by employing photonic quantum states that are multiplexed into time and/or frequency, and light-matter interfaces that are able to store and process such states with large time-bandwidth product and multimode capacities. Despite important progress in developing such devices, the demonstration of these capabilities using non-classical light remains challenging. Here, employing the atomic frequency comb quantum memory protocol in a cryogenically cooled erbium-doped optical fibre, we report the quantum storage of heralded single photons at a telecom-wavelength ($1.53\,\mu m$) with a time-bandwidth product approaching 800. Furthermore, we demonstrate frequency-multimode storage and memory-based spectral-temporal photon manipulation. Notably, our demonstrations rely on fully integrated quantum technologies operating at telecommunication wavelengths. With improved storage efficiency, our light-matter interface may become a useful tool in future quantum networks.

[1] Institute for Quantum Science and Technology, University of Calgary, 2500 University Drive NW, Calgary, Alberta, Canada T2N 1N4. [2] Department of Physics and Astronomy, University of Calgary, 2500 University Drive NW, Calgary, Alberta, Canada T2N 1N4. [3] Applied Physics Division, Jet Propulsion Laboratory, California Institute of Technology, 4800 Oak Grove Drive, Pasadena, California 91109, USA. [4] National Institute of Standards and Technology, Boulder, Colorado 80305, USA. [5] Battelle, 505 King Avenue, Columbus, Ohio 43201, USA. † Present address: Department of Physics, University of Ottawa, 150 Louis Pasteur, Ottawa, Ontario, Canada K1N 6N5. Correspondence and requests for materials should be addressed to W.T. (email: wtittel@ucalgary.ca).

Multiplexing, particularly in the form of wavelength division multiplexing, is key for achieving high data rates in modern fibre-optic communication networks. The realization of scalable quantum information processing demands adapting this concept, if possible using components that are compatible with the existing telecom infrastructure. One of the challenges to achieve this goal is to develop integrated light-matter interfaces that allow storing and processing multiplexed photonic quantum information. In addition to efficient operation, ease of integration and feed-forward controlled recall, the suitability of such interfaces depends on the storage time for a given acceptance bandwidth, that is, the interface's time-bandwidth product, as well as its multimode storage and processing capacities. It is important to note that although the multimode capacity of a light-matter interface determines the maximum number of simultaneously storable and processable photonic modes, the time-bandwidth product only sets an upper bound to the multimode capacity for temporally and/or spectrally multiplexed photons.

Over the past decade there has been significant progress towards the creation of such light-matter interfaces. One promising approach is based on a far off-resonant Raman transfer in warm atomic vapour, for which a time-bandwidth product of 5,000 has been reported[1]. However, owing to noise arising from (undesired) four-wave mixing, the storage and recall of quantum states of light have proven elusive[2]. This problem can be alleviated by implementing the Raman protocol in other media, for example, diamonds (with storage in optical phonon modes) and laser-cooled atomic ensembles, for which time-bandwidth products around 22 have been obtained with non-classical light[3,4]. However, the multimode operation of any Raman-type memory and hence the utilization of the potentially achievable large time-bandwidth products will remain challenging because of unfavourable scaling of this scheme's multimode capacity with respect to optical depth[5].

Another promising avenue for a multiplexed light-matter interface is the atomic frequency comb (AFC)-based quantum memory scheme in cryogenically cooled rare-earth ion-doped materials[6,7]. An attractive feature of this approach is that, unlike in other protocols, the multimode storage capacity is independent of optical depth; it is solely given by the time-bandwidth product of the storage medium, which can easily go up to several thousands because of the generally large inhomogeneous broadening (enabling large storage bandwidth) and narrow homogeneous linewidth (allowing long storage times) of optical transitions in rare-earth ion-doped materials. This aspect has already allowed several important demonstrations, including the simultaneous storage of 64 and 1,060 temporal modes by AFCs featuring pre-programmed delays[8,9], 5 temporal modes by an AFC with recall on-demand[10] and 26 spectral modes supplemented with frequency-selective recall[11], respectively. Despite the importance of these demonstrations, they were restricted to the use of strong or attenuated laser pulses rather than non-classical light, as required in future quantum networks. The only exception is the very recent demonstration of the storage of photons, emitted by a quantum dot, in up to 100 temporal modes[12].

In this paper, we present a spectrally multiplexed light-matter quantum interface for non-classical light. More precisely, we demonstrate large time-bandwidth product and multimode storage of heralded single photons at telecom wavelength by implementing the AFC protocol in an ensemble of erbium ions. As an important feature for future quantum networks, our demonstrations rely on fully integrated quantum technologies, that is, a fibre-pigtailed LiNbO$_3$ waveguide for the generation of heralded single photons by means of parametric down-

conversion, a commercially available, cryogenically-cooled erbium-doped single-mode fibre for photon storage and manipulation, and superconducting nanowire devices for high-efficiency single-photon detection.

Results

Experimental setup. Our experimental setup, illustrated in Fig. 1, is composed of an integrated, heralded single-photon source, an AFC-based erbium-doped fibre memory and a measurement unit including two superconducting nanowire single-photon detectors (SNSPDs).

We generate pairs of energy-time quantum correlated telecom-wavelength photons—commonly referred-to as 'signal' ('s') and 'idler' ('i')—by sending pump light from a continuous-wave laser operating at 766 nm wavelength to a periodically poled lithium niobate (PPLN) waveguide, as shown in Fig. 1a. Spontaneous parametric down conversion (SPDC) based on type-0 phase matching results in the creation of frequency-degenerate photon pairs centred at 1,532 nm wavelength and having a bandwidth of ~40 nm. We note that the PPLN waveguide is fibre pigtailed at both input and output faces (that is, for the pump light as well as the down-converted photons), which makes the source alignment free. After filtering away the remaining pump light, the spectra of the generated photons are filtered down to 50 GHz resulting in a photon-pair generation rate of 0.35 MHz. The ensuing photons are probabilistically separated into two standard telecommunication fibres using a 50/50 fibre-optic beam splitter (BS2). One member of each split pair is sent into an SNSPD featuring a system detection efficiency of around 70% (see the Methods for details of the SNSPDs). Its electronic output heralds the other member, which travels through standard telecommunication fibre to our light-matter interface (for more details about the heralded single-photon source see the Supplementary Note 1).

To store or manipulate the heralded single photons, we prepare an AFC-based memory in a 20-m-long erbium-doped silica fibre maintained at a temperature of 0.6–0.8 K and exposed to a magnetic field of 600 G (see Fig. 1b and the Methods for details). The memory relies on spectral tailoring of the inhomogenously broadened, $^4I_{15/2} \leftrightarrow {}^4I_{13/2}$ transition in erbium into a comb-shaped absorption feature characterized by the teeth spacing, Δ, as shown in Fig. 2a. The spectral tailoring is performed by frequency-selective optical pumping of ions into long-lived auxiliary (spin) levels. When an input photon is absorbed by the ions constituting the comb, a collective atomic excitation is created. It is described by:

$$|\Psi\rangle = \frac{1}{\sqrt{N}} \sum_{j=1}^{N} c_j e^{i2\pi m_j \Delta t} e^{-ikz_j} |g_1, \cdots e_j, \cdots g_N\rangle, \qquad (1)$$

where N is the total number of addressed atoms, k is the optical wave number and $|g_j\rangle$ and $|e_j\rangle$ are the j'th atom's ground and excited states, respectively. The detuning of the atom's transition frequency from the photon carrier frequency is given by $m_j\Delta$, whereas z_j is the position of the atom measured along the propagation direction of the light, and the factor c_j depends on both the resonance frequency and position of the atom. Owing to the periodic nature of the AFC, the atomic excitation is converted back to photonic form and the input photon is re-emitted in the originally encoded state after a storage time given by the inverse of the peak spacing, $t_{storage} = 1/\Delta$, as shown in Fig. 2b. The use of erbium-doped fibre is particularly attractive for AFC-based quantum memory because of its polarization insensitive operation at wavelengths within the telecom C-band[13], its ability to store photonic entanglement in combination of ease of integration with standard fibre infrastructure[14], and its large

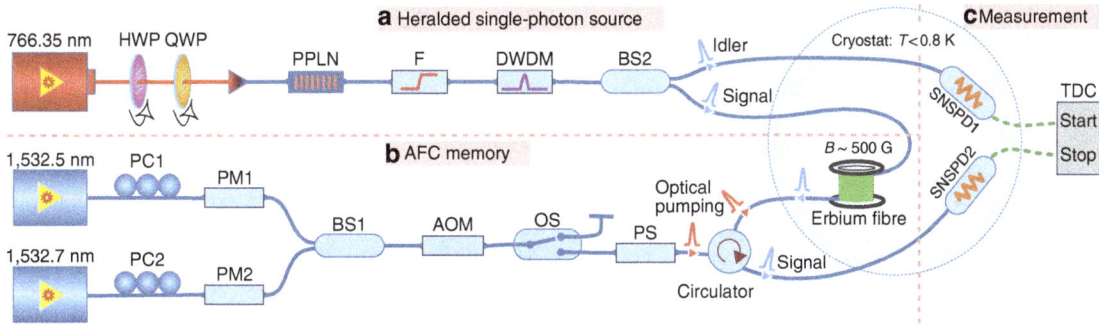

Figure 1 | Experimental setup. (**a**) Heralded single-photon source. Narrow linewidth continuous-wave (CW) light at 766.35 nm wavelength and with 100 μW power is sent to a fibre-pigtailed, periodically poled lithium niobate (PPLN) waveguide that is heated to 52.8 °C. A quarter-wave-plate (QWP) and a half-wave-plate (HWP) match the polarization of the light to the crystal's C axis to maximize the nonlinear interaction. Spontaneous parametric down conversion (SPDC) in the PPLN crystal results in frequency degenerate photon pairs with 40 nm bandwidth, centred at 1,532 nm wavelength. The residual pump light at 766.35 nm is suppressed by 50 dB by a filter (F) and the bandwidth of the created photons is filtered down to 50 GHz using a dense-wavelength-division-multiplexer (DWDM). The filtered photon pairs are probabilistically split using a beam-splitter (BS2). The detection of one member (the 'idler' photon) heralds the presence of the other (the 'signal' photon), which is directed to the input of the AFC memory. (**b**) Quantum memory. The quantum memory is based on an erbium-doped fibre that is exposed to a magnetic field of 600 G and cooled to a temperature below 1 K. Light from two independent CW lasers with wavelengths of 1,532.5 and 1,532.7 nm, respectively, is used to spectrally tailor the inhomogeneously broadened 1,532 nm absorption line of erbium through frequency-selective optical pumping into one or several atomic frequency combs. Towards this end, phase-modulators (PM1 and PM2) followed by an acousto-optic modulator are used to generate chirped pulses with the required frequency spectrum. The optical pumping light from the two lasers is merged on a beam-splitter (BS1) and enters the erbium fibre from the back via an optical circulator. Polarization controllers (PM1 and PM2) match the polarization to the phase-modulators' active axes, and the polarization scrambler ensures uniform optical pumping of all erbium ions in the fibre[13]. (**c**) Measurement unit. The detection of the heralding photon ('idler') and subsequently the 'signal' photon is performed by two superconducting nano-wire single-photon detectors (SNSPD1 and SNSPD2) maintained at the same temperature as the memory. The coincidence analysis of the detection events is performed by a time-to-digital converter (TDC).

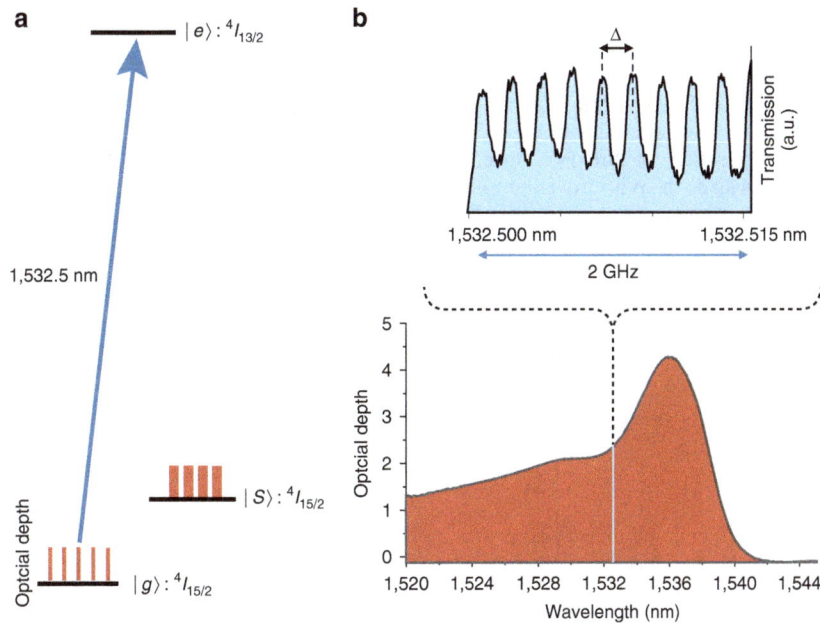

Figure 2 | Quantum memory. (**a**) Simplified level scheme of Er^{3+} in silica glass. Frequency-selective optical pumping from the $^4I_{15/2}$ electronic ground state ($|g\rangle$) via the $^4I_{13/2}$ excited state ($|e\rangle$) into an auxiliary (spin) state ($|s\rangle$) allows spectral tailoring. (**b**) Inhomogeneous broadening and AFC structure. The inhomogeneously broadened optical absorption line of erbium ions in silica fibre at 1 K extends roughly from 1,500 to 1,540 nm wavelength. A 2-GHz-wide section of a 16-GHz-wide comb at 1,532.5 nm with teeth spacing $\Delta = 200$ MHz is shown.

usable inhomogeneously broadened absorption line[15], which allows for multimode storage and manipulation of photons with large bandwidth, as detailed below.

Finally, we detect the heralded photons after storage/manipulation using a second SNSPD, featuring similar performance as that used to detect the heralding photon. All detection signals are sent to a time-to-digital converter (TDC) to perform time-resolved coincidence measurements. This allows us to calculate the cross-correlation function

$$g_{si}^{(2)} = \frac{R_{si}}{R_i R_s},\qquad(2)$$

where R_{si} is the rate of coincidence detections, and R_s and R_i are the single detection count-rate for 'signal' and 'idler' photons, respectively. A classical field satisfies the Cauchy–Schwarz inequality $[g_{si}^{(2)}]^2 \leq g_s^{(2)} g_i^{(2)}$ where $g_s^{(2)}$ and $g_i^{(2)}$ are second-order autocorrelation functions for 'signal' and 'idler' modes. For photons derived from an SPDC process, the second-order autocorrelation is bounded by $1 \leq g_{s,i}^{(2)} \leq 2$ (ref. 16). Consequently, measuring a cross-correlation $g_{si}^{(2)} > 2$ violates the inequality and thus verifies the presence of quantum correlations between the members of the photon pairs. We characterize our source for different SPDC pump powers (shown in Supplementary Fig. 1), finding that the cross-correlation function exceeds 1,000 for all powers. As $g_{si}^{(2)} \gg 2$ implies that the heralded autocorrelation function of the stored signal photons is $\ll 1$ (ref. 17), we denote these as single photons. The details of how $g_{si}^{(2)}$ is obtained from the coincidence counts are given in the Methods.

Measurements. First, we investigate the storage of broadband heralded single photons in an AFC memory—prepared using a single optical pumping laser at 1,532.5 nm—with a total bandwidth of 8 GHz and 200 MHz tooth spacing (corresponding to 5 ns storage time), as shown in Fig. 2b. The recalled photons are shown as a light-blue trace in the histogram of coincidence detections in Fig. 3. Analysing the correlations between signal and recalled idler photons, we find $g_{si}^{(2)} = 8.33 \pm 0.47$, which shows that the heralded photons after—and hence also before—storage, are indeed non-classical, and thus confirms the quantum nature of our light-matter interface. To improve the storage efficiency and thus the signal-to-noise ratio in the coincidence counts, we increase the storage bandwidth of the memory to 16 GHz—recall that the bandwidth of the input photons is 50 GHz. Owing to restrictions imposed by the level structure of erbium (see Supplementary Fig. 2 for details), the bandwidth increase is achieved by generating an additional 8-GHz-wide section separated from the first by around 20 GHz from edge to edge. This entails using two independent lasers operating at 1,532.5 and 1,532.7 nm wavelength for preparation of the two spectrally separated AFCs. As expected, this leads to an improvement of the overall memory efficiency and thus coincidence count rate from

0.59 ± 0.06 to 1.29 ± 0.13 Hz (see Fig. 3 for details), and an increase of $g_{si}^{(2)}$ to 18.2 ± 0.9. The number of AFCs, and thus the total storage bandwidth, can be further increased by employing additional pump lasers.

Second, to demonstrate quantum storage with large time-bandwidth product, we extend the memory storage time from 5 to 50 ns. To this end, we programme a 16-GHz-wide spectral region with two AFCs having tooth spacing ranging from 200 to 20 MHz. For each case, we map heralded photons onto the double AFC and collect coincidence statistics for the recalled photons as shown in Fig. 4. As further detailed in the Supplementary Information of ref. 14, decoherence effects and imperfect preparation of the AFCs decreases the memory efficiency as the storage time increases. Nevertheless, as shown in the inset of Fig. 4, $g_{si}^{(2)}$ still remains above 2 up to the maximal storage time of 50 ns, and thus we demonstrate the storage of non-classical light with a time-bandwidth product up to 800 (that is, 16 GHz × 50 ns). This is an improvement by close to a factor of 40 over that obtained in Raman-based memories[4] and a factor of 3 over the recent AFC-based memory demonstration[12], respectively. We emphasize that this comparison of time-bandwidth products does not reflect the difference in multimode storage capacities of the Raman and AFC-based implementations.

Third, to establish the multimode operation of our quantum light-matter interface, we divide the total currently accessible bandwidth—18 GHz for this experiment—into different numbers of AFC sections, and programme each section with a different storage time ranging from 3 to 13 ns. This allows identifying photons stored in different frequency modes (that is, different AFC sections) through their recall time. In succession, we perform measurements with: two AFCs of 9 GHz bandwidth; then four AFCs of 4.5 GHz bandwidth (this case is shown in Fig. 5a); and finally six AFCs of 3 GHz bandwidth. The corresponding histograms of coincidence detections, depicted in Fig. 5b, confirm that two, four and six modes, respectively, have been stored simultaneously. For each recalled spectro-temporal mode, we find that $g_{si}^{(2)}$, listed in Table 1, exceed the classical limit of two, thereby confirming multimode storage of non-classical states of light in matter.

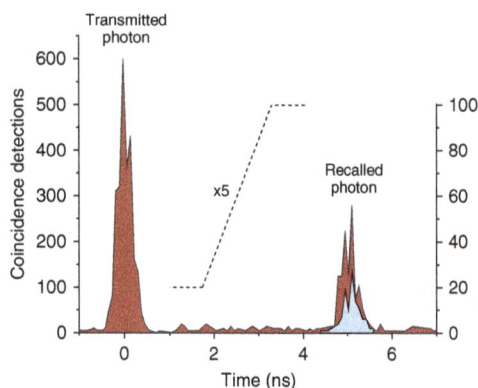

Figure 3 | Reversible mapping of broadband and heralded single photons. Heralded telecom-wavelength photons, centred at 1,532.7 nm wavelength and having a bandwidth of 50 GHz, are mapped onto the AFC memory and recalled after time $t_{storage} = \frac{1}{\Delta} = 5$ ns. The histogram shows time-resolved coincidence detections collected over 3 min. Owing to non-unit absorption probability by the AFC as well as bandwidth mismatch between the spectra of the photons and the AFC, a significant portion of the input photons is directly transmitted and detected at zero delay. The light blue highlighted section bounded by a dotted line in the trace of the recalled photon (counts scaled by a factor of five) corresponds to the measurement in which the total AFC bandwidth was 8 GHz; the red section corresponds to a measurement using a 16-GHz-wide AFC.

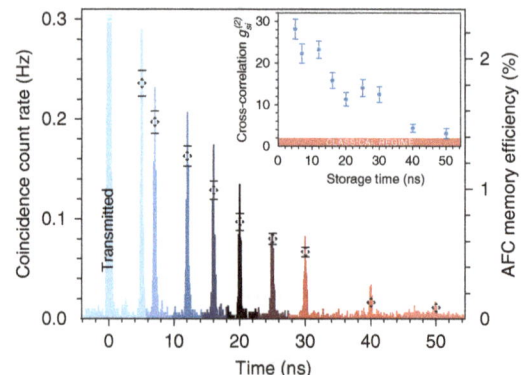

Figure 4 | Quantum storage with large time-bandwidth product. 16-GHz-wide AFC regions with teeth spacing ranging from 200 to 20 MHz are subsequently programmed to store heralded single photons for 5–50 ns. The histograms show measured coincidence rates for recalled photons for each storage time, and the dashed diamonds depict the corresponding memory efficiencies (see Supplementary Note 3 for details). Note that in this figure and henceforth the transmitted pulse at $t = 0$ exceeds the vertical scale and thus is capped at the top. Experimentally obtained $g_{si}^{(2)}$ for each storage time are shown in the inset. The measurement time varied between 5 and 15 min. All error bars are standard deviations derived from Poissonian counting statistics.

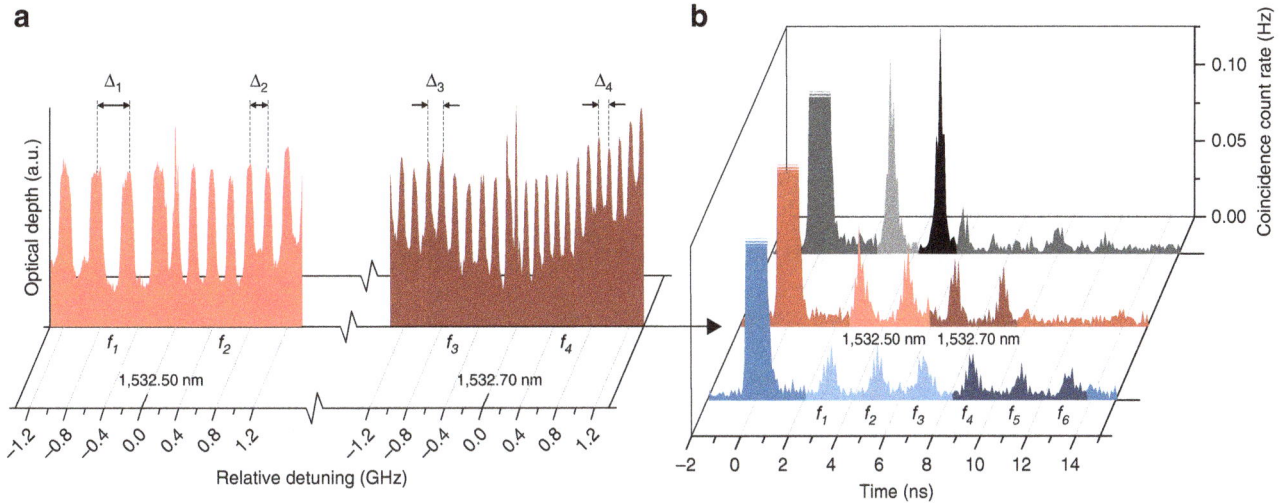

Figure 5 | Multimode quantum storage. (a) Creation of AFCs. The total, currently addressable bandwidth of 18 GHz is divided into different spectral sections, each featuring an AFC with distinct peak spacing. Depicted is the case of four, 4.5-GHz-wide AFCs, created using two lasers operating at 1,532.50 and 1,532.70 nm. The AFCs feature peak spacings of 333, 200, 143 and 111 MHz, corresponding to storage times of 3, 5, 7 and 9 ns, respectively. For each AFC, only a 1.3-GHz-wide section is shown. **(b)** Storage and recall. When broadband heralded photons with 50 GHz bandwidth are mapped onto a two-section AFC, where each section extends over 9 GHz bandwidth and has a peak spacing of 333 and 200 MHz, respectively, they are stored in two spectral modes and retrieved in two spectro-temporal modes, as shown in the back trace. Decreasing the bandwidth per AFC allows increasing the number of AFCs (spectral modes), as demonstrated with the storage of four and six spectral modes in the middle and the front trace, respectively. The modes are labelled f_i. For each mode, $g_{si}^{(2)}$ is measured to be larger than 2 (see Table 1).

Table 1 | Experimental cross-correlation values.

Mode	Storage	$g_{si}^{(2)}$ for spectro-temporal modes		
#	Time (ns)	9 GHz	4.5 GHz	3 GHz
f_1	3	16.8 ± 0.8	8.0 ± 0.6	5.7 ± 0.6
f_2	5	16.4 ± 0.8	7.2 ± 0.6	5.3 ± 0.6
f_3	7	—	6.3 ± 0.5	5.4 ± 0.6
f_4	9	—	5.2 ± 0.5	5.0 ± 0.6
f_5	11	—	—	3.4 ± 0.5
f_6	13	—	—	3.4 ± 0.5

Measured values of the cross-correlation function of each recalled spectral mode f_i for different AFC bandwidths and total numbers of modes. The storage times are given in column 2.

The frequency-to-time mapping described above was used to unambiguously distinguish between different frequency modes. This is convenient in our proof-of-principle demonstration, but may not be required in all applications. However, the ability to flexibly reconfigure our light-matter interface in terms of the number, bandwidth, storage time and frequency separation of the AFCs lends itself to versatile manipulation at the quantum level. In particular, it allows modifying the temporal shape of single photons by recalling different spectral modes at the same time and with adjustable phases. As an illustration, Fig. 6 shows two different mappings of six spectral modes onto three distinct temporal modes. We note that additional processing, for example, pulse sequencing and temporal compressing, relying on the frequency-to-time mapping is possible by combining AFCs with frequency shifters, as demonstrated in ref. 18.

Discussion

A central feature of our demonstration is the large multiplexed storage and manipulation capacity of our atomic interface for non-classical light. Yet, as noted above, there are clear avenues to increase this even further. Our additional investigations depicted in Supplementary Fig. 3 show that at least a 14-nm span of the

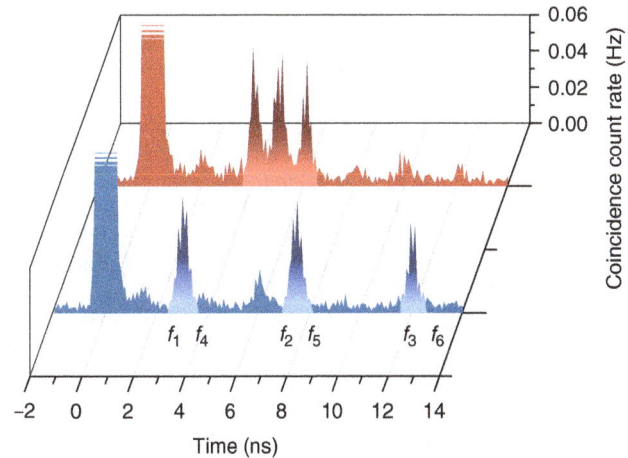

Figure 6 | Pulse manipulation. The AFC memory with total bandwidth of 2×8 GHz = 16 GHz is tailored such that simultaneously absorbed photons in six spectral modes (from f_1 to f_6), each of 2.65 GHz bandwidth, are recalled in three temporal modes spaced either by 4.5 ns (front trace), or by 1 ns (back trace).

erbium absorption line is suitable for quantum state storage. If we generate 8 GHz broad AFCs separated by 20 GHz, this yields a total of 64 AFCs covering a combined bandwidth of 64×8 GHz = 512 GHz. Assuming 50 ns storage time, one could thus reach a time-bandwidth product of more than 512 GHz × 50 ns ≈ 25,000, which is equal to the number of temporal—spectral modes that could be stored simultaneously.

On the other hand, a number of improvements are required for our multiplexed light-matter interface to be truly suitable for use in future quantum networks. First, imperfect optical pumping during the preparation of the AFCs currently results in a storage efficiency at 0.6 K of 1–2% (see the Supplementary Information of ref. 14). However, as we described in ref. 15 and in

Supplementary Note 2, we expect that it is possible to substantially improve the optical pumping, for example, with lower temperatures and smaller erbium concentrations, and hence to achieve a significantly higher efficiency under certain conditions.

Second, the $\sim$10-ms radiative lifetime of the $^4I_{15/2} \leftrightarrow {}^4I_{13/2}$ transition in erbium fundamentally limits the coherence time to 20 ms. However, coupling to so-called two-level systems[19], which are intrinsic to disordered materials such as glass fibres, reduces the coherence time, and the attainable storage time is hence currently restricted to a few tens of nanoseconds. Although we anticipate that this value increases with smaller doping concentration and lower temperature[20,21] (see Supplementary Information of ref. 14 for further discussions), it is an open question whether or not it is possible to extend storage times to hundreds of microseconds, which would, for example, allow building a quantum repeater based on spectral multiplexing[11]. Yet, we note that not all applications of quantum memory require long storage times, and that the figure of merit for multiplexing schemes is the time-bandwidth product. Hence, our light-matter interface can be useful even without long storage times.

Finally, we point out that our storage device provides pre-programmed delays, which is sufficient for increasing the efficiency of multi-photon applications if one incorporates external elements that allow feed-forward-based mode-mapping, for example, frequency shifters[11]. However, for some applications, it may be desirable to perform the mode mapping during storage—a familiar example being storage combined with read-out on demand, that is, the possibility to select the recall time after the photon has been stored. One option to enable this feature is to utilize the Stark effect, which allows 'smearing' out the AFC peaks—whose spacing determines the recall time—after photon absorption[22]. This would inhibit the rephasing of the collective excitation and hence reemission of light after $t_{\text{storage}} = 1/\Delta$, and lead to the possibility to recover the original photon a time $t_{\text{storage}} = n/\Delta$ after absorption, where n is an integer. Another possibility is to map the optical coherence—created by the photon absorption—onto a long-lived spin level (spin-wave storage)[23,24], for example, a hyperfine level (caused by the non-zero nuclear spin of $^{167}\text{Er}^{3+}$)[25,26] or a superhyperfine level (caused by the interaction of Er with co-dopants in the host)[27]. However, this possibility remains challenging because of the complex and not fully characterized structure of these levels in erbium-doped fibres. Moreover, this approach would limit the memory bandwidth appreciably because of the intrinsically small splitting of nuclear spin levels. But provided these levels feature long coherence times, and thus storage duration, the large time-bandwidth product needs not be affected significantly.

Provided that the performance of our light-matter interface is improved as discussed, it can advance several quantum photonic applications. For instance, by programming AFCs with different storage times into different spectral regions, it can serve as a time-of-flight spectrometer, which would, for example, allow fine-grained, spectrally resolved photon measurements, including two-photon Bell-state measurements in a quantum repeater architecture based on frequency multiplexing[11]. As a second example, our interface can be used to spectrally and temporally tailor single photon wave packets, as exemplified in our demonstrations, and thereby adapt their properties for subsequent interfacing with other quantum devices. More generally, by combining the interface with feed-forward-controlled frequency shifters, for example, broadband electro-optic phase modulators, it can be turned into a programmable atomic processor for arbitrary manipulation of photonic quantum states encoded into time and/or frequency[18]. This would find application in optical quantum computing in a single-spatial mode[28] or in atomic-interface-based photonic quantum state

processing[29–32]. In addition, we note that the large multiplexing capacity of our light-matter can be increased further by adapting multiplexing in spatial degree[33], and/or angular orbital momentum degree[34] with use of multimode erbium-doped fibres, as envisioned for future fibre-based classical and quantum networks[35].

In conclusion, we have presented a quantum light-matter interface that is suitable for multiplexed quantum-information-processing applications. More precisely, we have demonstrated a large time-bandwidth-product memory capable of multimode storage of non-classical light created via spontaneous parametric down-conversion, and showed that our light-matter interface can be employed for quantum manipulation of broadband heralded single photons. The fully pigtailed photon source and the fibre-based memory, along with telecommunication-wavelength operation and commercial availability, make our demonstration important in view of building future fibre-based quantum networks.

Methods

Erbium doped silica fibre. We use a 20-m long, commercially available single-mode, 190 p.p.m.-wt erbium-doped silica fibre specified to have 0.6 dB m^{-1} absorption at 1,532 nm wavelength at room temperature. At 1 K, we measure 0.1 dB m^{-1} absorption at 1,532 nm wavelength (see Fig. 4 for details). In addition to Er, the fibre is co-doped with P, Al and Ge. The fibre is spooled in layers around a copper cylinder with $\sim$4 cm diameter that is thermally contacted with the base plate of an adiabatic demagnetization refrigerator maintained at about 0.6–0.8 K and exposed to a field of $\sim$600 G inside a superconducting solenoid magnet. This setup induces $\sim$70% bending loss from input to output, mainly because of the insufficiently large diameter of the erbium-fibre spool. The fibre is fusion-spliced to standard single-mode fibres (SMF-28) at each end with less than 5% loss per splice.

Preparation of AFC. The memory is prepared using the setup described in Fig. 1. Frequency chirped laser pulses, applied 500 times per pumping cycle of 500 ms duration, allow frequency-selective optical pumping of erbium ions into a long-lived Zeeman level, and hence the formation of peaks and troughs of the AFC. A polarization scrambler randomly changes the polarization of the pump light every 500 μs to ensure that heralded photons, which propagate counter to the optical pumping light, are absorbed regardless their polarization state[13]. Upon completion of the optical pumping, a significant portion of the erbium ions accumulate in the auxiliary ground-state Zeeman-level $|s\rangle$ (see Fig. 2 for details), whose decay at 0.7 K and under a magnetic field of 600 G is characterized by two lifetimes of 1.3 and 26 s (ref. 15). However, a considerable amount of atoms remains in excited level $|e\rangle$, and subsequent spontaneous emission from these would mask all recalled photons. Thus, to eliminate this spontaneous emission noise, we set a wait time of 300 ms, which is significantly larger than the 11-ms exited level lifetime. Following the wait time, we store and retrieve many heralded single photons during a 700-ms measurement time. The retrieval efficiency of our AFC is about 1–2%, which is primarily limited by residual absorption background as a result of incomplete population transfer during the optical pumping, (see Supplementary Information of ref. 14 for further details)

Superconducting nanowire single-photon detectors. The detection of the telecom-wavelength photons is carried out by a set of SNSPDs[36] attached to the base plate of the adiabatic demagnetization refrigerator and maintained at the same temperature as the memory. In our setup, the tungsten silicide (WSi) SNSPDs have efficiencies of $\sim$70%, which includes the loss due to fibre-splices and bends. Their detection efficiencies exhibit a small polarization dependence of $\sim$5%. The time jitter of the detectors is around 250 ps, which allows us to resolve detection events separated by 1 ns (see Fig. 6 for details). We measure the dark count rate of the detectors to be $\sim$10 Hz, which results in a negligibly small contribution to accidental coincidences and hence high signal-to-noise ratios for the coincidence detections of the recalled photons.

Coincidence and $g^{(2)}$ measurements. All detection signals are directed to a TDC, which allows recording detection times with a resolution of 80 ps. The detection of the heralding ('idler') photon on SNSPD1 is used to trigger (start) the TDC. The other member of the pair (the 'signal' photon) is detected by SNSPD2—either as recalled (delayed) photon after storage, or as a directly transmitted photon through the erbium fibre that features no extra delay—and sent to the TDC, which records the time interval from the trigger signal. This allows us to generate histograms of the time-resolved coincidence detections between 'signal' and 'idler' photons (see, for example, Fig. 3).

To calculate $g_{si}^{(2)}$ using equation 2, we extract the values for the rates of the coincidence detection $R_{si}(t=0)$ and the individual detections R_s and R_i from coincidence histograms. Here we have explicitly stated the time difference between the coincidence detections to be 0. We note that, as the photon-pair generation process is spontaneous, there is no statistical correlation between subsequent pair generation events. Hence we can re-write the product $R_s R_i$ as $R_{si}(t \neq 0)$, that is, as the coincidence detection rates for 'signal' and 'idler' photons that are not members of the same photon pair, which is often referred to as accidental coincidence count-rate[37]. We extract $R_{si}(t=0)$ from a coincidence histogram by counting all detections within the 'coincidence peak' that is centred at time t_0 and has width t_p, and normalizing this number by measurement time and coincidence window width t_p. Similarly, $R_{si}(t \neq 0)$ is evaluated by appropriate normalization of coincidence counts taken in a window of width t_{bg} that is adjacent to the 'coincidence peak'. Hence, to evaluate $g_{si}^{(2)}$ from experimental histograms of coincidence counts $R_{si}(t)$, we use

$$g_{si}^{(2)} = \frac{t_{bg} \int_{t_0 - t_p/2}^{t_0 + t_p/2} R_{si}(t) dt}{t_p \int_{t_0 + t_p/2}^{t_0 + t_p/2 + t_{bg}} R_{si}(t) dt}. \tag{3}$$

For all the cross-correlation values in this paper, we used coincidence windows $t_p = t_{bg} = 0.8$ ns.

References

1. England, D. G. *et al.* High-fidelity polarization storage in a gigahertz bandwidth quantum memory. *J. Phys. B At. Mol. Opt. Phys.* **45**, 124008 (2012).
2. Michelberger, P. S. *et al.* Interfacing ghz-bandwidth heralded single photons with a warm vapour raman memory. *N. J. Phys.* **17**, 043006 (2015).
3. England, D. G. *et al.* Storage and retrieval of thz-bandwidth single photons using a room-temperature diamond quantum memory. *Phys. Rev. Lett.* **114**, 053602 (2015).
4. Ding, D.-S. *et al.* Raman quantum memory of photonic polarized entanglement. *Nat. Photon* **9**, 332–338 (2015).
5. Nunn, J. *et al.* Multimode memories in atomic ensembles. *Phys. Rev. Lett.* **101**, 260502 (2008).
6. de Riedmatten, H., Afzelius, M., Staudt, M. U., Simon, C. & Gisin, N. A solid-state light -matter interface at the single-photon level. *Nature* **456**, 773–777 (2008).
7. Afzelius, M., Simon, C., de Riedmatten, H. & Gisin, N. Multimode quantum memory based on atomic frequency combs. *Phys. Rev. A* **79**, 052329 (2009).
8. Usmani, I., Afzelius, M., de Riedmatten, H. & Gisin, N. Mapping multiple photonic qubits into and out of one solid-state atomic ensemble. *Nat. Commun.* **1**, 12 (2010).
9. Bonarota, M., Gouët, J. L. L. & Chanelière, T. Highly multimode storage in a crystal. *N. J. Phys.* **13**, 013013 (2011).
10. Gündoğan, M., Mazzera, M., Ledingham, P. M., Cristiani, M. & de Riedmatten, H. Coherent storage of temporally multimode light using a spin-wave atomic frequency comb memory. *N. J. Phys.* **15**, 045012 (2013).
11. Sinclair, N. *et al.* Spectral multiplexing for scalable quantum photonics using an atomic frequency comb quantum memory and feed-forward control. *Phys. Rev. Lett.* **113**, 053603 (2014).
12. Tang, J.-S. *et al.* Storage of multiple single-photon pulses emitted from a quantum dot in a solid-state quantum memory. *Nat. Commun.* **6**, 8652 (2015).
13. Jin, J. *et al.* Telecom-wavelength atomic quantum memory in optical fiber for heralded polarization qubits. *Phys. Rev. Lett.* **6**, 140501 (2015).
14. Saglamyurek, E. *et al.* Quantum storage of entangled telecom-wavelength photons in an erbium-doped optical fibre. *Nat. Photon* **9**, 83–87 (2015).
15. Saglamyurek, E. *et al.* Efficient and long-lived zeeman-sublevel atomic population storage in an erbium-doped glass fiber. *Phys. Rev. B* **92**, 241111 (2015).
16. Tapster, P. R. & Rarity, J. G. Photon statistics of pulsed parametric light. *J. Mod. Opt.* **45**, 595–604 (1998).
17. Bashkansky, M., Vurgaftman, I., Pipino, A. C. R. & Reintjes, J. Significance of heralding in spontaneous parametric down-conversion. *Phys. Rev. A* **90**, 053825 (2014).
18. Saglamyurek, E. *et al.* An integrated processor for photonic quantum states using a broadband light-matter interface. *N. J. Phys.* **16**, 065019 (2014).
19. Macfarlane, R. M., Sun, Y., Sellin, P. B. & Cone, R. L. Optical decoherence in Er^{3+} doped silica fiber: Evidence for coupled spin-elastic tunneling systems. *Phys. Rev. Lett.* **96**, 033602 (2006).
20. Staudt, M. U. *et al.* Investigations of optical coherence properties in an erbium-doped silicate fiber for quantum state storage. *Opt. Commun.* **266**, 720–726 (2006).
21. Sun, Y., Cone, R. L., Bigot, L. & Jacquier, B. Exceptionally narrow homogeneous linewidth in erbium-doped glasses. *Opt. Lett.* **31**, 3453–3455 (2006).
22. Lauritzen, B., Minar, J., de Riedmatten, H., Afzelius, M. & Gisin, N. Approaches for a quantum memory at telecommunication wavelengths. *Phys. Rev. A* **83**, 012318 (2011).
23. Gündoğan, M., Ledingham, P. M., Kutluer, K., Mazzera, M. & de Riedmatten, H. Solid state spin-wave quantum memory for time-bin qubits. *Phys. Rev. Lett.* **114**, 230501 (2015).
24. Jobez, P. *et al.* Coherent spin control at the quantum level in an ensemble-based optical memory. *Phys. Rev. Lett.* **114**, 230502 (2015).
25. Hashimoto, D. & Shimizu, K. Population relaxation induced by the boson peak mode observed in optical hyperfine spectroscopy of $^{167}Er^{3+}$ ions doped in a silica glass fibre. *J. Opt. Soc. Am. B* **28**, 2227–2235 (2011).
26. Baldit, E. *et al.* Identification of Λ-like systems in $Er^{3+}:Y_2SiO_5$ and observation of electromagnetically induced transparency. *Phys. Rev. B* **81**, 144303 (2010).
27. Thiel, C. W. *et al.* Optical decoherence and persistent spectral hole burning in $Er^{3+}:LiNbO_3$. *J. Lumin* **130**, 16031609 (2010).
28. Humphrey, P. C. *et al.* Linear optical quantum computing in a single spatial mode. *Phys. Rev. Lett.* **111**, 150501 (2013).
29. Hosseini, M., Sparkes, B. M., Longdell, G. H. J. J., Lam, P. K. & Buchler, B. C. Coherent optical pulse sequencer for quantum application. *Nature* **461**, 241–245 (2009).
30. Buchler, B. C., Hosseini, M., Hetet, G., Sparkes, B. M. & Lam, P. K. Precision spectral manipulation of optical pulses using a coherent photon echo memory. *Opt. Lett.* **35**, 1091–1093 (2010).
31. Nunn, J. *et al.* Enhancing multiphoton rates with quantum memories. *Phys. Rev. Lett.* **110**, 133601 (2013).
32. Campbell, G. T. *et al.* Configurable unitary transformations and linear logic gates using quantum memories. *Phys. Rev. Lett.* **113**, 063601 (2014).
33. Lan, S.-Y. *et al.* A multiplexed quantum memory. *Opt. Exp.* **17**, 13639 (2009).
34. Zhou, Z.-Q. *et al.* Quantum storage of three-dimensional orbital-angular-momentum entanglement in a crystal. *Phys. Rev. Lett.* **115**, 070502 (2015).
35. Bozinovic, N. *et al.* Terabit-scale orbital angular momentum mode division multiplexing in fibers. *Science* **340**, 1545–1548 (2013).
36. Marsili, F. *et al.* Detecting single infrared photons with 93% system efficiency. *Nat. Photon* **7**, 210–214 (2013).
37. Kuzmich, A. *et al.* Generation of nonclassical photon pairs for scalable quantum communication with atomic ensembles. *Nature* **423**, 731–734 (2003).

Acknowledgements

E.S., M.G., Q.Z., L.G., D.O. and W.T. thank Jeongwan Jin, Neil Sinclair, Charles Thiel and Vladimir Kiselyov for discussions and technical support, and acknowledge funding through the Alberta Innovates Technology Futures and the National Science and Engineering Research Council of Canada. Furthermore, W.T. acknowledges support as a Senior Fellow of the Canadian Institute for Advanced Research, and V.B.V. and S.W.N. partial funding for detector development from the DARPA Information in a Photon (InPho) programme. Part of the research was carried out at the Jet Propulsion Laboratory, California Institute of Technology, under a contract with the National Aeronautics and Space Administration. L.O. and D.N. acknowledge Srico Inc. for their assistance with the fabrication of PPLN wafers.

Author contributions

The experiment was conceived by W.T., E.S. and D.O. and the fibre quantum memory setup was developed by E.S. The SPDC waveguide source was designed and fabricated by D.N. and L.O. and characterized by L.O., L.G., Q.Z. and M.G.P. The SNSPDs were designed, fabricated and characterized by F.M., V.B.V. and S.W.N. The measurements were performed by E.S., Q.Z., M.G.P. and D.O. The manuscript was prepared by E.S., Q.Z., D.O. and W.T.

Additional information

Permissions

List of Contributors

T.O. Tasci and D.W.M. Marr
Chemical and Biological Engineering Department, Colorado School of Mines, Golden, Colorado 80401, USA

K.B. Neeves
Chemical and Biological Engineering Department, Colorado School of Mines, Golden, Colorado 80401, USA
Department of Pediatrics, University of Colorado, Denver, Colorado 80045, USA

P.S. Herson
Department of Anesthesiology, University of Colorado, Denver, Colorado 80045, USA
Department of Pharmacology, University of Colorado, Denver, Colorado 80045, USA

Dietmar Korn, Matthias Lauermann, Patrick Appel, Luca Alloatti and Robert Palmer
Institute of Photonics and Quantum Electronics (IPQ), Karlsruhe Institute of Technology (KIT), 76131 Karlsruhe, Germany

Sebastian Koeber, Wolfgang Freude and Christian Koos
Institute of Photonics and Quantum Electronics (IPQ), Karlsruhe Institute of Technology (KIT), 76131 Karlsruhe, Germany
Institute of Microstructure Technology (IMT), Karlsruhe Institute of Technology (KIT), 76344 Eggenstein-Leopoldshafen, Germany

Juerg Leuthold
Institute of Photonics and Quantum Electronics (IPQ), Karlsruhe Institute of Technology (KIT), 76131 Karlsruhe, Germany
Institute of Microstructure Technology (IMT), Karlsruhe Institute of Technology (KIT), 76344 Eggenstein-Leopoldshafen, Germany
Laboratory for Electromagnetic Fields and Microwave Electronics (IFH), Swiss Federal Institute of Technology (ETH), Zu"rich 8092, Switzerland

Pieter Dumon
Department of Information Technology, IMEC, 9000 Gent, Belgium

Adam W. Bushmaker, Vanessa Oklejas, Don Walker and Alan R. Hopkins
Physical Sciences Laboratories, The Aerospace Corporation, 355 S. Douglas Street, El Segundo, California 90245, USA

Jihan Chen and Stephen B. Cronin
Department of Electrical Engineering, The University of Southern California, 3601 Watt Way, Los Angeles, California 90089, USA

Aline Fluri, Daniele Pergolesi and Alexander Wokaun
Department for Energy and Environment, Paul Scherrer Institut, 5232 Villigen-PSI, Switzerland

Thomas Lippert
Department for Energy and Environment, Paul Scherrer Institut, 5232 Villigen-PSI, Switzerland.
Department of Chemistry and Applied Biosciences, Laboratory of Inorganic Chemistry, Vladimir-Prelog-Weg 1-5/10, ETH Zu"rich, Zürich CH-8093, Switzerland

Vladimir Roddatis
Institut für Materialphysik, Universita"t Go"ttingen, Friedrich-Hund-Platz 1, Göttingen 37077, Germany

Andreas F. Shick and Timothy J. White
Materials and Manufacturing Directorate, Air Force Research Laboratory, Wright-Patterson Air Force Base, Ohio 45433, USA

Taylor H. Ware
Department of Bioengineering, The University of Texas at Dallas, Richardson, Texas 75080, USA

John S. Biggins and Mark Warner
Cavendish Laboratory, Cambridge University, Cambridge CH3 0HE, UK

Soonil Hong, Geunjin Kim, Seongyu Lee, Seok Kim, Jong-Hoon Lee and Hyungcheol Back
School of Materials Science and Engineering, Gwangju Institute of Science and Technology, Gwangju 61005, Republic of Korea
Heeger Center for Advanced Materials, Gwangju Institute of Science and Technology, Gwangju 61005, Republic of Korea

Hongkyu Kang and Junghwan Kim
Heeger Center for Advanced Materials, Gwangju Institute of Science and Technology, Gwangju 61005, Republic of Korea
Research Institute for Solar and Sustainable Energies, Gwangju Institute of Science and Technology, Gwangju 61005, Republic of Korea

Jinho Lee
Heeger Center for Advanced Materials, Gwangju Institute of Science and Technology, Gwangju 61005, Republic of Korea
Department of Nanobio Materials and Electronics, Gwangju Institute of Science and Technology, Gwangju 61005, Republic of Korea

Minjin Yi and Jae-Ryoung Kim
Research Institute for Solar and Sustainable Energies, Gwangju Institute of Science and Technology, Gwangju 61005, Republic of Korea

Kwanghee Lee
School of Materials Science and Engineering, Gwangju Institute of Science and Technology, Gwangju 61005, Republic of Korea
Heeger Center for Advanced Materials, Gwangju Institute of Science and Technology, Gwangju 61005, Republic of Korea
Research Institute for Solar and Sustainable Energies, Gwangju Institute of Science and Technology, Gwangju 61005, Republic of Korea
Department of Nanobio Materials and Electronics, Gwangju Institute of Science and Technology, Gwangju 61005, Republic of Korea

David J. Gundlach
National Institute of Standards and Technology, Engineering Physics Division, 100 Bureau Drive, MS 8120, Gaithersburg, Maryland 20899, USA

Emily G. Bittle
National Institute of Standards and Technology, Engineering Physics Division, 100 Bureau Drive, MS 8120, Gaithersburg, Maryland 20899, USA
Department of Physics, Wake Forest University, 1834 Wake Forest Road, Winston-Salem, North Carolina 27109, USA

James I. Basham
National Institute of Standards and Technology, Engineering Physics Division, 100 Bureau Drive, MS 8120, Gaithersburg, Maryland 20899, USA
Department of Electrical
Engineering, The Pennsylvania State University, 121 Electrical Engineering East, University Park, State College, Pennsylvania 16802, USA

Thomas N. Jackson
Department of Electrical Engineering, The Pennsylvania State University, 121 Electrical Engineering East, University Park, State College, Pennsylvania 16802, USA

Oana D. Jurchescu
Department of Physics, Wake Forest University, 1834 Wake Forest Road, Winston-Salem, North Carolina 27109, USA

Chong Sheng, Hui Liu and Shining Zhu
National Laboratory of Solid State Microstructures & School of Physics, Collaborative Innovation Center of Advanced Microstructures, Nanjing University, Nanjing, Jiangsu 210093, China

Mordechai Segev and Rivka Bekenstein
Department of Physics and Solid State Institute, Technion, Haifa 32000, Israel

Johannes T.B. Overvelde, Twan A. de Jong and Sergio A. Becerra
John A. Paulson School of Engineering and Applied Sciences, Harvard University, Cambridge, Massachusetts 02138, USA

Yanina Shevchenko
Department of Chemistry and Chemical Biology, Harvard University, Cambridge, Massachusetts 02138, USA

George M. Whitesides
Department of Chemistry and Chemical Biology, Harvard University, Cambridge, Massachusetts 02138, USA
Wyss Institute for Biologically Inspired Engineering, Harvard University, Cambridge, Massachusetts 02138, USA

James C. Weaver
Wyss Institute for Biologically Inspired Engineering, Harvard University, Cambridge, Massachusetts 02138, USA

Chuck Hoberman
Wyss Institute for Biologically Inspired Engineering, Harvard University, Cambridge, Massachusetts 02138, USA
Hoberman Associates, New York, New York 10001, USA. 5 Graduate School of Design, Harvard University, Cambridge, Massachusetts 02138, USA

Katia Bertoldi
John A. Paulson School of Engineering and Applied Sciences, Harvard University, Cambridge, Massachusetts 02138, USA.
Kavli Institute, Harvard University, Cambridge, Massachusetts 02138, USA

Pavan Nukala, Chia-Chun Lin, Russell Composto and Ritesh Agarwal
Department of Materials Science and Engineering, University of Pennsylvania, Philadelphia, Pennsylvania 19104, USA

Seung Hyuk Back
KU-KIST Graduate School of Converging Science and Technology, Korea University, Seoul 02841, Korea

Jin Hyuk Park and Chunzhi Cui
Department of Chemical and Biological Engineering, Korea University, Seoul 02841, Korea

Dong June Ahn
KU-KIST Graduate School of Converging Science and Technology, Korea University, Seoul 02841, Korea
Department of Chemical and Biological Engineering, Korea University, Seoul 02841, Korea
Center for Theragnosis, Biomedical Research Institute, Korea Institute of Science and Technology, Seoul 02792, Korea

Yunlong Zi, Jie Wang, Sihong Wang, Shengming Li, Zhen Wen and Hengyu Guo
School of Materials Science and Engineering, Georgia Institute of Technology, Atlanta, Georgia 30332, USA

Zhong Lin Wang
School of Materials Science and Engineering, Georgia Institute of Technology, Atlanta, Georgia 30332, USA
Beijing Institute of Nanoenergy and Nanosystems, Chinese Academy of Sciences, Beijing 100083, China

F.-R. Chen
Department of Engineering and System Science, National Tsing-Hua University, 101 Kuang-Fu Road, Hsin Chu 300, Taiwan

D. Van Dyck
EMAT, Department of Physics, University of Antwerp, 2020 Antwerpen, Belgium

C. Kisielowski
Lawrence Berkeley National Laboratory, The Molecular Foundry and Joint Center for Artificial Photosynthesis, One Cyclotron Road, Berkeley California 94720 USA

Siqi Lin, Wen Li, Zhiwei Chen, Jiawen Shen and Yanzhong Pei
Key Laboratory of Advanced Civil Engineering Materials of Ministry of Education, School of Materials Science and Engineering, Tongji University, 4800 Caoan Road, Shanghai 201804, China

Binghui Ge
Beijing national laboratory for condensed matter physics, Institute of physics, Chinese academy of science, Beijing 100190, China

Joseph R. Corea, Anita M. Flynn, Balthazar Lechêne, Michael Lustig and Ana C. Arias
Department of Electrical Engineering and Computer Sciences, University of California, Berkeley, California 94720, USA

Greig Scott
Department of Electrical Engineering, Stanford University, Stanford, California 94305, USA

Galen D. Reed and Peter J. Shin
Department of Electrical Engineering and Computer Sciences, University of California, Berkeley, California 94720, USA
Department of Bioengineering, University of California, San Francisco, California 94722, USA

Ivaylo Nikolov, Paolo Cinquegrana, Bruno Diviacco, David Gauthier, Giuseppe Penco, Primož Rebernik Ribič, Eleonore Roussel, Marco Trovò, Michele Manfredda, Emanuele Pedersoli, Flavio Capotondi, Nicola Mahne, Lorenzo Raimondi, Alexander Demidovich, Miltcho Boyanov Danailov and Enrico Allaria
ELETTRA — Sincrotrone Trieste, Area Science Park, 34149 Trieste, Italy

Eugenio Ferrari
ELETTRA — Sincrotrone Trieste, Area Science Park, 34149 Trieste, Italy
Dipartimento di Fisica, Università degli Studi di Trieste, 34127 Trieste, Italy

Carlo Spezzani
ELETTRA — Sincrotrone Trieste, Area Science Park, 34149 Trieste, Italy
Laboratoire de Physique des Solides, Université Paris-Sud, CNRS-UMR 8502, Bât. 510, 91405 Orsay, France

Franck Fortuna
Centre de Sciences Nucléaires et de Sciences de la Matière, Université Paris-Sud, CNRS UMR 8609, Bât. 104-108, 91405 Orsay, France

Renaud Delaunay
Laboratoire de Chimie Physique Matière et Rayonnement, Sorbonne Universités, UPMC Univ Paris 06, CNRS UMR 7614, 75005 Paris, France

Franck Vidal
Institut des NanoSciences de Paris, Sorbonne Universités, UPMC Univ Paris 06, CNRS UMR 7588, 75005 Paris, France

Jean-Baptiste Moussy
Service de Physique de l'Etat Condensé, DSM/IRAMIS/SPEC, CNRS UMR 3680, CEA Saclay, 91191 Gif-sur-Yvette, France

Tommaso Pincelli
Dipartimento di Fisica, Università degli Studi di Milano, 20133 Milano, Italy

Lounès Lounis
Institut des NanoSciences de Paris, Sorbonne Universités, UPMC Univ Paris 06, CNRS UMR 7588, 75005 Paris, France
Ecole Normale Supe´rieure, PSL Research University, 75231 Paris, France

Cristian Svetina
ELETTRA—Sincrotrone Trieste, Area Science Park, 34149 Trieste, Italy
Graduate School of Nanotechnology, Universita` degli Studi di Trieste, 34127 Trieste, Italy

Marco Zangrando
ELETTRA—Sincrotrone Trieste, Area Science Park, 34149 Trieste, Italy.
Istituto Officina dei Materiali, Consiglio Nazionale delle Ricerche, 34149 Trieste, Italy

Luca Giannessi
ELETTRA—Sincrotrone Trieste, Area Science Park, 34149 Trieste, Italy
ENEA, Centro Ricerche Frascati, Via E. Fermi 45, 00044 Frascati, Italy

Giovanni De Ninno
ELETTRA—Sincrotrone Trieste, Area Science Park, 34149 Trieste, Italy
Laboratory of Quantum Optics, University of Nova Gorica, 5001 Nova Gorica, Slovenia.

Maurizio Sacchi
Institut des NanoSciences de Paris, Sorbonne Universités, UPMC Univ Paris 06, CNRS UMR 7588, 75005 Paris, France
Synchrotron SOLEIL, L'Orme des Merisiers, Saint-Aubin, B.P. 48, 91192 Gif-sur-Yvette, France

Justin Olamit and Kai Liu
Physics Department, University of California, Davis, One Shields Avenue, Davis, California 95616, USA

Dustin A. Gilbert
Physics Department, University of California, Davis, One Shields Avenue, Davis, California 95616, USA
NIST Center for Neutron Research, Gaithersburg, Maryland 20899, USA

Randy K. Dumas
Physics Department, University of California, Davis, One Shields Avenue, Davis, California 95616, USA
Department of Physics, University of Gothenburg, Gothenburg 412 96, Sweden

B.J. Kirby, Alexander J. Grutter, Brian B. Maranville and Julie A. Borchers
NIST Center for Neutron Research, Gaithersburg, Maryland 20899, USA

Elke Arenholz
Advanced Light Source, Lawrence Berkeley National Laboratory, Berkeley, California 94720, USA

Ryan C. Rollings
Department of Physics, Harvard University, Cambridge, Massachusetts 02138, USA.

Aaron T. Kuan
School of Engineering and Applied Sciences, Harvard University, Cambridge, Massachusetts 02138, USA

Jene A. Golovchenko
Department of Physics, Harvard University, Cambridge, Massachusetts 02138, USA
School of Engineering and Applied Sciences, Harvard University, Cambridge, Massachusetts 02138, USA

P.-Y. Hou, Y.-Y. Huang, X.-X. Yuan, X.-Y. Chang, C. Zu and L. He
Center for Quantum Information, Institute for Interdisciplinary Information Sciences, Tsinghua University, Beijing 100084, China

L.-M. Duan
Center for Quantum Information, Institute for Interdisciplinary Information Sciences, Tsinghua University, Beijing 100084, China
Department of Physics, University of Michigan, Ann Arbor, Michigan 48109, USA

Yu Hui, Zhenyun Qian and Matteo Rinaldi
Department of Electrical & Computer Engineering at Northeastern University, 360 Huntington Avenue, Boston, Massachusetts 02115, USA

Juan Sebastian Gomez-Diaz and Andrea Alù
Department of Electrical & Computer Engineering at The University of Texas at Austin, 1616 Guadalupe St., UTA 7.215, Austin, Texas 78701, USA

Robert J. Chapman, Zixin Huang and Alberto Peruzzo
Quantum Photonics Laboratory, School of Engineering, RMIT University, Melbourne, Victoria 3000, Australia
School of Physics, The University of Sydney, Sydney, New South Wales 2006, Australia

Matteo Santandrea, Giacomo Corrielli, Andrea Crespi and Roberto Osellame
Istituto di Fotonica e Nanotecnologie, Consiglio Nazionale delle Ricerche, Piazza Leonardo da Vinci 32, Milano I-20133, Italy
Dipartimento di Fisica, Politecnico di Milano, Piazza Leonardo da Vinci 32, Milano I-20133, Italy

Man-Hong Yung
Department of Physics, South University of Science and Technology of China, Shenzhen 518055, China

M.A.A. Hafiz, L. Kosuru and M.I. Younis
Physical Sciences and Engineering Division, King Abdullah University of Science and Technology, Thuwal 23955-6900, Saudi Arabia

Pu Huang, Dong Hou, Shaochun Lin, Chenyong Ju, Xiao Zheng and Jiangfeng Du
National Laboratory for Physics Sciences at the Microscale, University of Science and Technology of China, Hefei 230026, China
Department of Modern Physics, University of Science and Technology of China, Hefei 230026, China
Synergetic Innovation Center of Quantum Information and Quantum Physics, University of Science and Technology of China, Hefei 230026, China

Wen Deng
National Laboratory for Physics Sciences at the Microscale, University of Science and Technology of China, Hefei 230026, China
Department of Modern Physics, University of Science and Technology of China, Hefei 230026, China
High Magnetic Field Laboratory, Chinese Academy of Science, Hefei 230026, China

Jingwei Zhou, Liang Zhang, Chao Meng and Changkui Duan
National Laboratory for Physics Sciences at the Microscale, University of Science and Technology of China, Hefei 230026, China
Department of Modern Physics, University of Science and Technology of China, Hefei 230026, China

Fei Xue
High Magnetic Field Laboratory, Chinese Academy of Science, Hefei 230026, China

Haoxue Han and Sebastian Volz
Laboratoire EM2C, CNRS, CentraleSupélec, Université Paris-Saclay, Grande Voie des Vignes, 92295 Châtenay-Malabry, France

Yong Zhang and Johan Liu
SMIT Center, School of Automation and Mechanical Engineering and Institute of NanomicroEnergy, Shanghai University, 20 Chengzhong Road, Shanghai 201800, China
Electronics Materials and Systems Laboratory, Department of Microtechnology and Nanoscience, Chalmers University of Technology, Kemivägen 9, SE-412 96 Gothenburg, Sweden

Nan Wang, Majid Kabiri Samani, Michael Edwards and Murali Murugesan
Electronics Materials and Systems Laboratory, Department of Microtechnology and Nanoscience, Chalmers University of Technology, Kemivägen 9, SE-412 96 Gothenburg, Sweden

Yuxiang Ni
Department of Mechanical Engineering, University of Minnesota, 111 Church Street SE, Minneapolis, Minnesota 55455, USA

Hatef Sadeghi, Steven Bailey and Colin J. Lambert
Quantum Technology Center, Physics Department, Lancaster University, Lancaster LA1 4YB, UK

Zainelabideen Y. Mijbil
Quantum Technology Center, Physics Department, Lancaster University, Lancaster LA1 4YB, UK
Science Department, Veterinary Medicine College, Al-Qasim Green University, Babylon, Iraq

Shiyun Xiong
Max Planck Institute for Polymer Research, Ackermannweg 10, D-55128 Mainz, Germany

Kimmo Sääskilahti
Department of Biomedical Engineering and Computational Science, Aalto University, FI-00076 Aalto, Finland

Yifeng Fu
Electronics Materials and Systems Laboratory, Department of Microtechnology and Nanoscience, Chalmers University of Technology, Kemivägen 9, SE-412 96 Gothenburg, Sweden
SHT Smart High Tech AB, Ascherbergsgatan 46, SE-411 33 Gothenburg, Sweden

Lilei Ye
SHT Smart High Tech AB, Ascherbergsgatan 46, SE-411 33 Gothenburg, Sweden

Yuriy A. Kosevich
Laboratoire EM2C, CNRS, CentraleSupélec, Université Paris-Saclay, Grande Voie des Vignes, 92295 Châtenay-Malabry, France
Department of Polymers and Composite Materials, Semenov Institute of Chemical Physics, Russian Academy of Sciences, Kosygin Street 4, 119991 Moscow, Russia

Alexander Tselev, Ye Cao, Sergei V. Kalinin and Petro Maksymovych
Center for Nanophase Materials Sciences, Oak Ridge National Laboratory, Oak Ridge, Tennessee 37831, USA

Pu Yu
State Key Laboratory for Low-Dimensional Quantum Physics, Department of Physics and Collaborative Innovation Center for Quantum Matter, Tsinghua University, Beijing 100084, China
RIKEN Center for Emergent Matter Science (CEMS), Wako, Saitama 351-0198, Japan

Liv R. Dedon
Department of Materials Science and Engineering and Department of Physics, University of California, Berkeley, Berkeley, California 94720, USA

Lane W. Martin
Department of Materials Science and Engineering and Department of Physics, University of California, Berkeley, Berkeley, California 94720, USA
Materials Science Division, Lawrence Berkeley National Laboratory, Berkeley, California 94720, USA

Philip A. Parilla and Maikel F.A.M. van Hest
National Renewable Energy Laboratory, 15013 Denver West Pkwy, Golden, Colorado 80401, USA

Jeremy D. Fields
National Renewable Energy Laboratory, 15013 Denver West Pkwy, Golden, Colorado 80401, USA
SolarWorld Americas, 25300 NW Evergreen Road, Hillsboro, Oregon 97124, USA (J.D.F.)
Department of Ceramic Engineering Indian Institute of Technology (BHU), Varanasi 221005, India (M. I. A.)

Md. Imteyaz Ahmad
SLAC National Accelerator Laboratory, 2575 Sand Hill Road, Menlo Park, California 94025, USA
SolarWorld Americas, 25300 NW Evergreen Road, Hillsboro, Oregon 97124, USA (J.D.F.)
Department of Ceramic Engineering Indian Institute of Technology (BHU), Varanasi 221005, India (M. I. A.)

Vanessa L. Pool, Douglas G. Van Campen and Michael F. Toney
SLAC National Accelerator Laboratory, 2575 Sand Hill Road, Menlo Park, California 94025, USA

Jiafan Yu
Department of Electrical Engineering, Stanford University, 350 Serra Mall, Stanford, California 94305, USA

Huajun He, Yuanjing Cui, Jiancan Yu, Yu Yang, Tao Song and Guodong Qian
State Key Laboratory of Silicon Materials, Cyrus Tang Center for Sensor Materials and Applications, School of Materials Science and Engineering, Zhejiang University, Hangzhou 310027, China

En Ma and Xueyuan Chen
Key Laboratory of Optoelectronic Materials Chemistry and Physics, Fujian Institute of Research on the Structure of Matter, Chinese Academy of Sciences, Fuzhou, Fujian 350002, China

Chuan-De Wu
Department of Chemistry, Zhejiang University, Hangzhou 310027, China

Banglin Chen
State Key Laboratory of Silicon Materials, Cyrus Tang Center for Sensor Materials and Applications, School of Materials Science and Engineering, Zhejiang University, Hangzhou 310027, China
Department of Chemistry, University of Texas at San Antonio, San Antonio, Texas 78249-0698, USA

Daniela Abdelkhalek and Roman Schnabel
Institut für Laserphysik, Universität Hamburg, Hamburg 22761, Germany
Institut für Gravitationsphysik, Leibniz Universität Hannover and Max-Planck-Institut für Gravitationsphysik (Albert-Einstein-Institut), Hannover 30167, Germany

Mareike Syllwasschy
Institut für Gravitationsphysik, Leibniz Universität Hannover and Max-Planck-Institut für Gravitationsphysik (Albert-Einstein-Institut), Hannover 30167, Germany

Nicolas J. Cerf
Quantum Information and Communication, Ecole Polytechnique de Bruxelles, CP 165, Université libre de Bruxelles, Brussels 1050, Belgium

Jaromír Fiurášek
Department of Optics, Palacký University, Olomouc 77146, Czech Republic

Erhan Saglamyurek, Marcelli Grimau Puigibert, Qiang Zhou, Daniel Oblak and Wolfgang Tittel
Institute for Quantum Science and Technology, University of Calgary, 2500 University Drive NW, Calgary, Alberta, Canada T2N 1N4
Department of Physics and Astronomy, University of Calgary, 2500 University Drive NW, Calgary, Alberta, Canada T2N 1N4

Lambert Giner
Institute for Quantum Science and Technology, University of Calgary, 2500 University Drive NW, Calgary, Alberta, Canada T2N 1N4
Department of Physics and Astronomy, University of Calgary, 2500 University Drive NW, Calgary, Alberta, Canada T2N 1N4

Present address: Department of Physics, University of Ottawa, 150 Louis Pasteur, Ottawa, Ontario, Canada K1N 6N5

Francesco Marsili
Applied Physics Division, Jet Propulsion Laboratory, California Institute of Technology, 4800 Oak Grove Drive, Pasadena, California 91109, USA

Varun B. Verma and Sae Woo Nam
National Institute of Standards and Technology, Boulder, Colorado 80305, USA

Lee Oesterling and David Nippa
Battelle, 505 King Avenue, Columbus, Ohio 43201, USA

Index

www.ingramcontent.com/pod-product-compliance
Lightning Source LLC
Chambersburg PA
CBHW080534200326
41458CB00012B/4436